PLATE I (*Frontispiece*)

Geology of Pontypridd and Maesteg (*Mem. Geol. Surv.*)

A 9768

RHONDDA FACH VALLEY LOOKING SOUTH-SOUTH-EAST FROM BLAEN-LLECHAU

BRITISH GEOLOGICAL SURVEY

MEMOIRS OF THE GEOLOGICAL SURVEY OF GREAT BRITAIN

THE GEOLOGY OF THE SOUTH WALES COALFIELD

Part IV: The country around Pontypridd and Maesteg

Explanation of one-inch geological sheet 248
(England and Wales)

THIRD EDITION

A W Woodland and W B Evans

London: The Stationery Office

Stationery Office publications are available from:

The Stationery Office Publications Centre
(Mail, fax and telephone orders only)
PO Box 276, London SW8 5DT
Telephone orders 0171-873 9090
General enquiries 0171-873 0011
Queuing system in operation for both numbers
Fax orders 0171-873 8200

The Stationery Office Bookshops
49 High Holborn, London WC1V 6HB
(counter service only)
0171-873 0011   Fax 0171-831 1326
68–69 Bull Street, Birmingham B4 6AD
0121-236 9696   Fax 0121-236 9699
33 Wine Street, Bristol BS1 2BQ
0117-9264306   Fax 0117-9294515
9 Princess Street, Manchester M60 8AS
0161-834 7201   Fax 0161-833 0634
16 Arthur Street, Belfast BT1 4GD
01232-238451   Fax 01232-235401
71 Lothian Road, Edinburgh EH3 9AZ
0131-228 4181   Fax 0131-229 2734
The Stationery Office Oriel Bookshop,
The Friary, Cardiff CF1 4AA
01222-395548   Fax 01222-384347

The Stationery Office's Accredited Agents
(see Yellow Pages)

And through good booksellers

Keyworth, Nottingham NG12 5GG
0115-936 3100

Murchison House, West Mains Road, Edinburgh, EH9 3LA   0131-667 1000

London Information Office, Natural History Museum, Earth Galleries, Exhibition Road, London SW7 2DE
0171-589 4090

The full range of Survey publications is available through the Sales Desks at Keyworth and at Murchison House, Edinburgh, and in the BGS London Information Office in the Natural History Museum (Earth Galleries). The adjacent bookshop stocks the more popular books for sale over the counter. Most BGS books and reports can be bought from the Stationery Office and through Stationery Office agents and retailers. Maps are listed in the BGS Map Catalogue, and can be bought together with books and reports through BGS-approved stockists and agents as well as direct from the BGS.

The British Geological Survey carries out the geological survey of Great Britain and Northern Ireland (the latter as an agency service for the government of Northern Ireland), and of the surrounding continental shelf, as well as its basic research projects. It also undertakes programmes of British technical aid in geology in developing countries as arranged by the Overseas Development Administration.

The British Geological Survey is a component body of the Natural Environment Research Council.

Printed in the UK for the SO
Dd 301953   C3   10/96
ISBN 0 11 884542 X

Bibliographic reference

WOODLAND, A W, and EVANS, W B. 1964. The geology of the South Wales Coalfield, Part IV, the country around Pontypridd and Maesteg (4th edition). *Memoir of the British Geological Survey,* Sheet 248 (England and Wales).

PREFACE TO THE FIRST EDITION

THIS, THE fourth part of the Memoir descriptive of the South Wales Coal Field, is an explanation of the New Series Map Sheet 248, which includes the important part of the Coal Field extending from Pontypridd to Cwmavan, and the greater part of the range of the best steam-coal.

The ground was originally surveyed on the Old Series One-inch Map, Sheet 36, by Sir H. T. De la Beche, Sir W. E. Logan and Mr. D. H. Williams. Most of it was probably the work of De la Beche himself, but it is likely that the western part was included in the survey made by Logan before he became connected with the Geological Survey. The maps which he had prepared were exhibited in 1837 at a meeting of the British Association at Liverpool, and attracted the attention of De la Beche, with the result that they were handed over to the Geological Survey and incorporated in the official maps, while Logan himself became a member of the staff. Sheet 36 was published in or before the year 1845.

The re-survey was made on the six-inch scale under the superintendence of Mr. Strahan, and was published in 1899. The south-eastern part of Sheet 248 was surveyed by Mr. Strahan, the south-western part by Mr. Tiddeman, and the northern part by Mr. Gibson. Each geologist contributes an account of the area surveyed by himself to the present volume, and the whole has been edited by Mr. Strahan.

Two sub-divisions of the Coal Measures, namely, the Pennant Series and the Lower Coal-series, occupy nearly the whole area. In the latter occurs a group of seams which yield the well-known smokeless steam-coal. The measures are illustrated in the three sheets of Vertical Sections, Nos. 83, 84, 85, to which frequent reference is made in these pages, and in the Plate bound up in this volume. It is noticeable that though the seams can be recognised for long distances in an east and west direction, yet both they and the measures associated with them change so rapidly southwards that only a general correlation by groups is possible. In this part of the Coal Field some massive sandstones, much resembling Pennant, develop in the upper part of the Lower Coal-series.

The Pennant Series shows a no less remarkable expansion both in a southward and westward direction. At the same time its upper part becomes split up by shales and coal-seams, which have yielded the bulk of the coal near Neath, Swansea and Llanelly. Thus both the top and the bottom of the Pennant is less distinct than in Monmouth-shire, where the sub-division consisted almost exclusively of sandstone.

The Upper or Llantwit Measures occur in two small outliers only in the area comprised in Sheet 248.

In the South Crop, where the high dip usual in the southern margin of the Coal Field prevails, the Millstone Grit and what may be the top of the Carboniferous Limestone come into view, but they are partly overspread by Triassic strata, which have been laid down unconformably upon their truncated edges. The Coal Measures themselves are partly thus covered, and it is worthy of note that the Triassic conglomerates consist not of Coal Measure rocks, but of fragments of Carboniferous Limestone.

The abnormal condition of the Rhaetic Group in this part of South Wales is well exhibited near Pyle, but a full account of these interesting rocks is reserved for the description of the region round Bridgend, where they are more widely distributed.

The structural geology has been treated in great detail in view of its importance in connection with mining. The course of the Pontypridd anticline and the development

iii

of the Moel Gilau Fault along the same line of disturbance are among the most important features. Mr. H. K. Jordan's conclusions with respect to this great fracture receive
confirmation from the recent work.

By an examination of the glacial deposits it has been possible to trace the route
taken by an ice-flow which had its source in Brecknock. It followed the main valleys
across the Coal Field, such as those of the Taff, Neath and Tawe, but, though it filled
them to overflowing, it failed to surmount the great Pennant scarp of Carn Mosyn.
Here, on the other hand, was generated a subordinate ice-flow which followed the
Rhondda, Ogmore, Garw and Llynfi valleys, but which, both to the east and west,
amalgamated with the Brecknock ice and joined in the general press southwards.

Chapter X summarises the observations on the principal economic products. The
character of the coals, being a matter which affects the Coal Field as a whole, will be
treated in a general memoir more suitably than in the explanation of a single sheet.

The Map is published in two Editions, on one of which (the Solid Edition) glacial
deposits are omitted, while on the other (the Superficial Edition) such deposits are
indicated by colour, as well as those portions of the Solid Geology which are not
concealed by Drift. MS. six-inch Maps, geologically coloured, are deposited in this
office, where they can be consulted. Copies can be obtained at cost price.

We have again to express our thanks for valuable assistance rendered by Colliery
Managers, Mining Engineers and Surveyors.

J. J. H. TEALL
Director.

Geological Survey Office,
28, Jermyn Street, London,
28th February, 1903

PREFACE TO THE SECOND EDITION

THE preparation of a Second Edition of this Memoir gives the opportunity of adding
recent information, more especially with relation to the mining developments which
have taken place since the publication of the First Edition in 1903. Thus, while the
workings on the steam-coals along the south crop west of the Ogwr have come to a
standstill, with the exception of the establishment of the Cribbwr Fawr Colliery near
Pyle, the Rock-Fawr Seam is now extensively mined by slopes along the southern
margin of the Pennant. East of the Ogwr several coals at about the same horizon are
being worked, but it is still uncertain which of them corresponds to the No. 2 Rhondda
Seam. North and north-west of Llantrisant, deep shafts have recently been sunk
through the Pennant to reach the steam-coals at the Cwm Colliery near Llantwit
Fardre, and at the Coed-Ely Colliery in the Ely Valley. In the Ogwr Valley, and
especially in the Avan Valley and its tributaries, developments have followed the
establishment of docks at Port Talbot; but in the Rhondda Valleys there has been
but little change in mining exploration.

The revisions of the Memoir, apart from the descriptions of the works referred to,
have been mainly editorial. Mr. R. H. Tiddeman having left the staff in February, 1902,
the task of collecting fresh information was entrusted to Dr. Gibson and Mr. Cantrill.
Their duties have been greatly facilitated by the readiness with which mining engineers

and colliery officials have supplied details. Valuable assistance has also been given by
Mr. H. J. Randall and Mr. W. A. Williams, of Bridgend. Special reference also must
be made to the completion by Mr. H. K. Jordan of his researches on the South Wales
Coalfield.

Some new collieries referred to in this volume will not be found on the edition of
the one-inch map, issued in 1902; but sufficient details are given in the text to enable
the reader to locate them.

The six-inch geological maps covering the area included in the one-inch Sheet 248,
and the sections of the principal shafts in the area, are quoted in lists on pp. vii and viii.

A. STRAHAN,
Director.

Geological Survey Office,
    28, Jermyn Street, London,
        19th June, 1916

PREFACE TO THE THIRD EDITION

THE YEARS since the publication of the Second Edition of this memoir have seen a
considerable expansion in mining in the Pontypridd district, as well as a greatly increased
understanding of Coal Measures stratigraphy, palaeontology and sedimentation
generally. It was therefore decided that a resurvey of the district should be undertaken
instead of a revision similar to that of the Abergavenny (Sheet 232) and Merthyr Tydfil
(Sheet 231) memoirs, published in 1927 and 1933 respectively. During the course of
this resurvey, particular attention has been given to the correlation of the Pennant
strata, and this has resulted in a revised classification of these rocks. Detailed examination of the Lower and Middle Coal Measures, with their productive coals, has
involved much time in the examination of underground exposures at all the working
collieries. In addition, many hundreds of new records of shafts, boreholes and mine-
roadways have been acquired.

The present account together with the published six-inch and one-inch maps represents
a great advance in the understanding of the geology of the district. The structure of
the Coal Measures has received much attention. Major thrusts, and gravity slides of
a type not previously recognized in Britain, are described for the first time. These,
together with the cross-faults and innumerable incompetence-structures which affect
the productive measures nearly everywhere, are interpreted as reflecting the continuous
evolution of Armorican movements in South Wales.

The resurvey was started in 1945 by Dr. A. W. Woodland, who in 1947 was joined
by Mr. W. B. Evans. The work was carried out under the superintendence first of the
late R. W. Pocock, and then of Dr. V. A. Eyles, and was completed in 1954. The areas
surveyed by the two officers are shown in the list of six-inch maps on p. xiii. Thirty-six
county maps have been published between the years 1957-60. The new one-inch map
was published in a Drift Edition in 1960, and a Solid Edition in 1963.

The memoir has been written by Dr. Woodland and Mr. Evans, and edited by
Dr. F. B. A. Welch. Animal fossils have been identified, and much help given throughout the work by Mr. M. A. Calver, Dr. W. H. C. Ramsbottom, and Dr. C. J. Stubblefield: Dr. J. Weir has also identified some of the non-marine shells. The plants were
named for the most part by Dr. R. Crookall, but Dr. W. G. Chaloner has determined
those more recently collected. The fossils were mainly collected by the two authors.

Throughout the resurvey great assistance has been forthcoming from numerous officials of the coal industry. Before nationalization, the surveyors, in particular those of the Powell Duffryn and Ocean Coal Companies, gave unstinted help, and in later years co-operation by the National Coal Board has been on a no less generous scale; and we gratefully acknowledge our thanks for this assistance. We also acknowledge the assistance given by the officers of the Coal Survey Laboratory at Cardiff, and in particular to Mr. H. F. Adams, the present Officer-in-Charge, who has supplied a short section (pp. 293–6) on the quality and utilization of the coals of the district.

C. J. STUBBLEFIELD,

Director.

Geological Survey Office,
  Exhibition Road,
    South Kensington,
      London, S.W.7
        7th November, 1963

CONTENTS

ILLUSTRATIONS

TEXT FIGURES

viii

PLATES

EXPLANATION OF PLATES

PLATE I.   Rhondda Fach Valley looking south-south-east from Blaen-llechau.
              A typical mining valley with Tylorstown and Pontygwaith occupying
              the lower slopes. The spur on left of valley is Twyn-llechau with a
              strong 'slack' feature associated with the Brithdir Seam. The central
              sky-line is Mynydd Troed-y-Rhiw, capped by basal Brithdir Beds.
              Ferndale Nos. 6 and 7 pits are in foreground (A 9768).

PLATE II.   A. Cwmparc looking north-east from Bwlch-y-Clawdd. On the hillsides
              lines of old coal-diggings can be seen, and several abandoned Pennant
              quarries occur on the higher slopes. The middle sky-line marks the
              surface of the Pennant plateau; the Brecknock Beacons form the
              distant sky-line (A 9766).

              B. Rolling Pennant plateau between Rhondda Fach and Cwmaman
              valleys. The peaty plateau surface, formed by the dip-slope of the
              sandstone overlying the No. 2 Rhondda Seam, is deeply incised by
              the head-streams of the Aman. The feature beyond the wall on the
              left marks the general course of a fault (A 9812).

PLATE III.   Details of correlation of seams from the Gellideg to Two-Feet-Nine.

PLATE IV.   Details of correlation of measures from the Two-Feet-Nine to the
              Upper Cwmgorse Marine Band.

PLATE V.   Comparative sections in the Pennant Measures.

LIST OF SIX-INCH MAPS

The following list shows the six-inch 'county' maps included, wholly or partly, in the area described by this memoir, with the initials of the surveyors and the dates of re-survey. In those marginal maps without asterisk, only the areas lying within the One-inch Sheet boundary were re-surveyed. The officers concerned with the re-survey were W. B. Evans and A. W. Woodland. All maps marked with an asterisk were published uncoloured between 1957 and 1960.

9 S.W.	Mynydd March-Hywel	.. .. ..	W.B.E.	1950
9 S.E.	Resolven	..	W.B.E.	1950
10 S.W.	Cwm-gwrach	..	W.B.E.	1947–50
10 S.E.	Llyn Fawr	..	W.B.E.	1947–48
11 S.W.	Cwm Dare	..	W.B.E.; A.W.W.	1946–48
11 S.E.	Aberdare	..	A.W.W.	1946
12 S.W.	Troed-y-rhiw	..	A.W.W.	1945
16 N.W.	Aberdulais	..	W.B.E.	1950
16 N.E.*	Clyne	..	W.B.E.	1947–50
16 S.W.	Neath	..	W.B.E.	1950
16 S.E.*	Michaelston Higher	..	W.B.E.	1947–49
17 N.W.*	Glyncorrwg	..	W.B.E.	1947–50
17 N.E.*	Blaenrhondda—Tynewydd ..	..	W.B.E.	1947–48
17 S.W.*	Cymmer	..	W.B.E.	1947–49
17 S.E.*	Aber-gwynfi	..	W.B.E.	1947–53
18 N.W.*	Treherbert—Maerdy	..	W.B.E.; A.W.W.	1946–48
18 N.E.*	Cwmaman—Aberaman ..	..	A.W.W.	1946
18 S.W.*	Cwm-parc—Ystrad Rhondda ..	..	W.B.E.; A.W.W.	1946–53
18 S.E.*	Ferndale—Tylorstown ..	..	A.W.W.	1946–51
19 N.W.*	Mountain Ash—Merthyr Vale ..	..	A.W.W.	1945–46
19 S.W.*	Penrhiwceiber—Abercynon ..	..	A.W.W.	1946
25 N.W.	Cwmavon	..	W.B.E.	1952
25 N.E.*	Pont Rhyd-y-fen—Bryn ..	..	W.B.E.	1948–49
25 S.W.	Aberavon	..	W.B.E.	1952
25 S.E.*	Cwm Wernderi	..	W.B.E.	1949
26 N.W.*	Caerau—Nantyffyllon ..	..	W.B.E.	1948–49
26 N.E.*	Blaengarw—Nant-y-moel	..	A.W.W.	1949
26 S.W.*	Maesteg	.... ..	W.B.E.	1949
26 S.E.*	Pontycymmer—Ogmore Vale ..	..	A.W.W.	1949
27 N.W.*	Gelli—Cwm Clydach ..	..	A.W.W.	1952–53
27 N.E.*	Llwyn-y-pia—Ynyshir ..	..	A.W.W.	1947–52
27 S.W.*	Gilfach Goch	..	A.W.W.	1952–53
27 S.E.*	Pen-y-graig—Porth	..	A.W.W.	1947–53
28 N.W.*	Ynys-y-bwl	..	A.W.W.	1947
28 S.W.*	Pontypridd	..	A.W.W.	1947
33 N.E.*	Margam Park	..	A.W.W.	1949
33 S.E.*	Pyle	..	A.W.W.	1948
34 N.W.*	Llangynwyd Middle	..	A.W.W.	1948
34 N.E.*	Bettws—Llangeinor	..	A.W.W.	1949
34 S.W.*	Kenfig Hill	..	A.W.W.	1949
34 S.E.*	Aberkenfig—Bryncethin ..	..	A.W.W.	1950
35 N.W.*	Llandyfodwg	..	A.W.W.	1952
35 N.E.*	Tonyrefail	..	A.W.W.	1952

35 S.W.*	Heol-y-Cyw—Penprysg	..	..	..	A.W.W. ..	..	1951
35 S.E.*	Llanharan ..	..	..	..	A.W.W. ..	..	1951–52
36 N.W.*	Beddau	..	..	..	A.W.W. ..	..	1947
36 N.E.	Llantwit	..	..	..	A.W.W. ..	..	1947
36 S.W.*	Llantrisant ..	..	..	..	A.W.W. ..	..	1952–53
36 S.E.	Taff's Well ..	..	..	..	A.W.W. ..	..	1953

THE GEOLOGY OF THE COUNTRY AROUND PONTYPRIDD AND MAESTEG

CHAPTER I

INTRODUCTION

GEOGRAPHICAL SETTING AND INDUSTRIAL BACKGROUND

THE PONTYPRIDD (248) SHEET of the One-Inch Geological Map of England and Wales depicts most of the central part of the South Wales Coalfield lying between the Taff and Neath valleys, together with the northern margins of the Vale of Glamorgan (Fig. 1). The greater part of the district is occupied by Coal Measures, but in the extreme south small areas of Millstone Grit crop out in the rim of the coal basin. This southern area includes small patches of Mesozoic rocks which, farther south between Bridgend and Cardiff have extensive outcrops.

Except for a narrow southern strip the entire district is formed by a deeply-dissected plateau of Pennant Measures, the so-called Blaenau Morgannwg, the surface of which falls gradually southwards. In the north, heights of 1500 to 1800 ft are attained on Cefntyle-brych (1756 ft), Mynydd Tynewydd (1694 ft), and Moel yr Hyrddod (1817 ft), but in the south the hills that overlook the

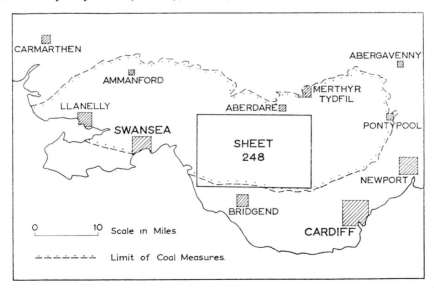

FIG. 1.   *Sketch-map showing the relation of the Pontypridd District (Sheet 248) to the South Wales Coalfield*

South Crop are only 800 to 1000 ft high. This southerly falling plateau is interrupted by a gentle ridge which runs southwards from Cefn Ffordd (1969 ft), in the Merthyr Tydfil district immediately to the north, towards the heads of the Ogwr Fawr and Garw valleys. Here are situated Werfa (1866 ft), the highest point within the district, the heights above Craig Fawr (1832 ft), Mynydd Caerau (1823 ft), and Mynydd William Meyrick (1769 ft). The plateau clearly post-dates the development of the major structures of the district, for its level is unaffected by both the folds and the faults. Lithological variations in the rocks that form its surface are reflected in minor steps which may have been accentuated by discontinuous uplift. It has been suggested that this high ground represents a series of unwarped platforms, but since the higher levels are those most perfectly preserved it appears simpler to regard it as a single warped surface, showing in this district a fall of rather less than 1° to the south.

The rivers draining the plateau are deeply incised to depths of a thousand feet or more. They are narrow and steep sided and, as yet, have destroyed little of the plateau surface. In the east, the Taff and its tributaries, the Cynon, the Clydach, the Rhondda Fach, and the Rhondda Fawr, drain south-eastwards. The Ogwr Fach and Fawr (or Ogmore), the Garw and the Llynfi all run more nearly southwards; while the Avan, Pelena, and Neath trend south-westwards. With the exception of the Taff, Cynon, and Neath, all rise within the Pennant plateau, and apart from the Neath, none shows any adjustment to the underlying structures. The river system is clearly superimposed, and it has been claimed by some to have originated upon a now-removed cover of Chalk; the apparent lack of erosion on the plateau surface, however, suggests that the streams are consequents which developed as the surface was uplifted, and reflect in their varying directions the changing slopes of that surface. In this connexion the Cefn Ffordd-Werfa ridge marks the divide between streams which flow south-eastwards, and those that drain southwards or south-westwards.

South of the Pennant plateau a continuous hollow extends from Margam to Rhiwsaeson along the strike of the mudstones of the Lower and Middle Coal Measures, and carries no significant strike streams. This is flanked still farther south by a series of low scarps and ridges formed by the sandstones of the lowest Coal Measures and Millstone Grit, which stretches from Kenfig Hill in the west, through Cefn Cribwr and Cefn Hirgoed to Penprysg, where it passes beneath glacial drift into the Bridgend district to the south.

Until the nineteenth century the district was settled only by scattered farming communities little visited by, or known to, the outside world. The small-scale digging of coal had for long supplemented wood as a source of fuel, certainly since Elizabethan times and possibly since as early as the thirteenth century, when the monks of Margam Abbey are reputed to have worked coal. These dayholes were as much a part of the rural scene as were the village forges or the small woollen mills.

The early years of the Industrial Revolution saw little change in this pattern of life. Despite the smelteries and collieries around Swansea and Neath, the ironworks along the North Crop eastwards from Hirwaun, and the growth of a major area of house-coal production in the Monmouthshire valleys, in only two parts of the present district did industries of any size exist. Near Maesteg and Cwmavon, and along the western part of the South Crop, where local iron-ores were easily accessible, small ironworks, presumably with their ancillary collieries, were established prior to 1830, and around Dinas there were sizeable

A 9766

A.　Cwm parc looking north-east from Bwlch-y-Clawdd

A 9812

B.　Rolling pennant plateau between Rhondda Fach and Cwmaman valleys

[*To face page 2*

mines exploiting the house-coals of the Pennant and sending them by canal, and later by railway, to Cardiff. Even in 1848 a traveller to the Rhondda could write 'The people of this solitudinous and happy valley are a pastoral race, almost entirely dependent on their flocks and herds for support . . . The air is aromatic with wild flowers and mountain plants—a sabbath stillness reigns.'

The real impetus to coal production came when it was realized that the Lower and Middle Coal Measures of much of the district contained seams of a quality ideal for steam-raising. The rapid increase in the number of steam-engines of all kinds, particularly coal-fired steam ships, led to an immense demand for these coals. In 1828 the first colliery producing steam-coal for sale outside the district was opened near Merthyr, and within two years workings began in the Cynon Valley. By 1843 Lletty Shenkin Colliery, apparently the first steam-coal pit in the district, had been sunk, followed in 1845 by Aberaman Colliery. In 1851 a trial pit at Cwmsaerbren (later to become Bute Colliery) proved the extension of these seams beneath the Rhondda, and within the next thirty years most of the collieries in this valley were established, though in the more isolated valleys of the Avan, the Garw and the Ogmore, new pits were still being sunk or old ones deepened in the early years of the twentieth century.

The industry reached its peak immediately before the First World War, at least 70 per cent of its output being for the overseas market. The dependence of the world's shipping on the coals of the district was reflected by the frequency with which the words 'Naval', 'Navigation' or 'Ocean' appeared in the names of collieries or colliery companies. Although production figures for the district are not available, the Rhondda valleys alone produced over 9 million tons in 1914.

After this date the gradual world-wide replacement of coal by oil as a marine-fuel led to an inevitable contraction in demand and production, which has continued ever since. In 1957 the total output from the Rhondda had fallen to 2,380,000 tons, and for the district as a whole it was only 8,247,000 tons in 1961, barely 4 per cent of the country's output. Even so, it still contributes over one third of the output of the South Wales Coalfield. Of late years the emphasis has shifted somewhat from the steam-coals to the coking-coals farther south.

GEOLOGICAL SEQUENCE

The formations represented on the map and sections, and described in this memoir are summarized below:

SUPERFICIAL FORMATIONS (DRIFT)

RECENT AND PLEISTOCENE
    Landslips
    Blown Sand
    Peat
    Alluvium and Alluvial Fans
    River Terrace Deposits
    Glacial Deposits, including sand and gravel, boulder clay and lake deposits

SOLID FORMATIONS

Thickness in feet
(generalized)

JURASSIC
    Lower Lias: clays and limestones   30

Thickness in feet
(generalized)

TRIASSIC
    Rhaetic: white sandstones and dark clays   75
    Keuper: Keuper Marl and Dolomitic Conglomerate ..   ..   up to 300

Major unconformity

CARBONIFEROUS
    Coal Measures
        Upper Coal Measures or Pennant Measures
            Grovesend Beds: measures above the Llantwit No. 3:
            mudstones and pennant sandstones with a few
            workable coals ..   ..   up to 600

Minor unconformity

            Swansea Beds (basal beds only present): mudstones,
            pennant sandstones and a few coals ..   ..   ..   ?
            Hughes Beds: measures between the Cefn Glas, Wenallt,
            etc. and the Graigola; pennant sandstones, mudstones
            and a few coals   300–800
            Brithdir Beds: measures between the Brithdir and Cefn
            Glas or Wenallt; pennant sandstones, mudstones and
            two or three thin coals   500–800
            Rhondda Beds: measures between No. 2 Rhondda and
            Brithdir; pennant sandstones, mudstones and a few
            coals, mostly thin   600–1100
            Llynfi Beds: measures between the Upper Cwmgorse
            Marine Band and the No. 2 Rhondda; mudstones,
            pennant sandstones and thin coals   250–700

        Middle Coal Measures
            Measures between Amman Marine Band and Upper
            Cwmgorse Marine Band; mudstones, subordinate
            sandstones and many workable coals, particularly in
            lower part   550–1600

        Lower Coal Measures
            Measures between the *Gastrioceras subcrenatum* Marine
            Band and the Amman Marine Band; mudstones,
            subordinate sandstones and many workable coals,
            particularly in upper part   400–1700

    Millstone Grit Series
        Mudstones and sandstones, including beds of *Homoceras*,
        Lower and Upper *Reticuloceras*, and Lower *Gastrioceras*
        ages ..   ..   up to 900

Unconformity

    Carboniferous Limestone Series
        Limestones with dark shale and chert at top in west of district
        (not exposed) ..   ..   —

GEOLOGICAL HISTORY

The oldest exposed rocks are of Millstone Grit age, though it is practically certain that at depth both the Carboniferous Limestone and the Old Red Sandstone are continuous beneath the district. The deposition of the Carboniferous Limestone was followed by minor uplift accompanied by progressive erosion of the uppermost beds towards the east. Renewed subsidence allowed the Millstone Grit sea to re-enter the area and to deposit its sediments above the plane of unconformity.

During the ensuing Coal Measures times, numerous episodes of limited subsidence and associated sedimentation were interspersed with periods of stability, during which luxurious vegetation spread across the silted area of deposition, producing accumulations of humic matter which on burial became coal seams. Unlike most other British coalfields the later Coal Measures were characterized by the entry of vast quantities of sand, now preserved as the Pennant Sandstones. During their accumulation, slight uplift resulted in erosion and unconformity in the eastern areas of the coalfield.

A major gap in the stratigraphical record separates the Coal Measures from the marls and conglomerates of the Keuper. Within this period the Armorican earth-movements produced in the earlier formations a large proportion of the structures observed at the present time. Uplift, accompanied by erosion, resulted in the removal of much of the strata already deposited, so that in the south of the district Keuper rocks now rest on the Millstone Grit, and in the area still farther south on beds as low as the Old Red Sandstone. It is also probable that considerable thicknesses of Upper Carboniferous rocks no longer preserved in the district were removed at this time.

The Triassic strata are believed to have formed under desert conditions. It is clear from their relationship with the Pennant near Llantrisant that they were laid down at the foot of a Pennant escarpment which occupied to a large extent its present position, though it does not follow that the present surface of the Pennant plateau is of Triassic origin.

The marine Rhaetic and basal Liassic rocks which follow were formed in a sea which lapped round this area of high ground, and in consequence they exhibit marginal facies.

Since the time interval between the Lias and the Pleistocene is not now represented by any deposits, any attempt to reconstruct the geological history must be largely speculative. It appears likely, however, that the district underwent at least one complete cycle of subaerial erosion, which removed any cover of later Mesozoic rocks and which cut the present surface of the coalfield plateau. At a relatively recent date this surface has been uplifted and tilted, and its slight degree of dissection makes it difficult to believe that this uplift could be earlier than the mid-Tertiary, to which period the post-Liassic faults may also belong. To certain platforms developed in the district to the south, a Pliocene age has been tentatively ascribed, which, if correct, sets an upward limit to the age of uplift.

Finally, during the Pleistocene the district was invaded by glaciers which produced minor modifications of the landscape; since the melting of the ice, few changes of significance have taken place.   A.W.W., W.B.E.

CHAPTER II

MILLSTONE GRIT SERIES

AND EARLIER CARBONIFEROUS BEDS

INTRODUCTION

THE MILLSTONE Grit Series crops out along a strip of country never more than three-quarters of a mile wide in the central southern margin of the district, extending from the neighbourhood of Pen-y-fai Common (half a mile south of Aberkenfig) in the west, to Penprysg in the east. In the former area the beds are faulted against Triassic rocks, and westwards to Swansea Bay they are known only in boreholes near Kenfig Pool, which lies in the north-western corner of the Bridgend district. With the exception of numerous lines of crags marking the outcrops of the harder sandstones, exposures are poor and no good continuous sections exist. Most of the knowledge of the shales and their accompanying faunas is derived from isolated exposures in the Ogmore Valley immediately south of Aberkenfig.

At the time of the original six-inch survey of the district little was known concerning the systematic palaeontology of the Upper Carboniferous. Few fossils were collected and no use was made of the faunal evidence for classifying and separating the various main divisions of strata. The upper limit of the 'Millstone Grit' was fixed arbitrarily at the top of a thick sandstone, the Cefn Cribbwr Rock (see Woodland and others 1957, p. 57), which forms the bold feature of Cefn Cribwr[1] and Cefn Hirgoed as far eastwards as Penprysg. This sandstone was believed to correspond with the Farewell Rock of the North Crop of the Coalfield which similarly marked the upper limit of the Millstone Grit in that area. In more recent years great advances have been made in the study of the fossils of the Upper Carboniferous and in accordance with international practice the Geological Survey now takes the dividing line between the Millstone Grit Series (Namurian) and the overlying Coal Measures (Westphalian) at the base of marine strata characterized by the presence of *Gastrioceras subcrenatum* C. Schmidt. During the course of the present survey the *G. subcrenatum* Marine Band has been identified at a level approximately 450 ft below the top of the Cefn Cribbwr Rock. This thickness of strata, formerly included as the uppermost beds of the Millstone Grit, is now accordingly assigned to the lowest Coal Measures.

Along this portion of the South Crop of the Coalfield the *G. subcrenatum* Marine Band is known only from the National Coal Board's Margam Park No. 1 Borehole, the site of which is some four miles west-north-west of the nearest outcrops at Pen-y-fai. This borehole serves to fix the position of the

---

[1] The ridge named 'Cefn Cribwr' on the present one-inch geological map (1960, 1963) was in previous editions called 'Cefn Cribbwr'. In this memoir the shorter spelling is used only in topographical references.

marine band with reference to the overlying beds, but unfortunately its relations to the beds below are obscure owing to the intervention of major structural complications. Using the evidence of this borehole it is nevertheless possible to infer the position of the top of the formation at outcrop, especially as the *Gastrioceras cumbriense* Marine Band, which should lie not far below, is known at several surface localities.

The basal beds of the Series, together with the uppermost beds of the Carboniferous Limestone, crop out in the area just south of the Pontypridd Sheet margin. This area, which lies between Litchard (1¼ miles north of Bridgend) and the main Bridgend–Maesteg road, on the margins of six-inch sheets Glamorgan 34 S.W. and 40 N.W., is included in the present account.

CLASSIFICATION

The systematic stratigraphy and palaeontology of the Millstone Grit Series is founded on the researches of W. S. Bisat and later workers in the Mid-Pennine Province of Northern England, and is based on the sequence of goniatite faunas. Comprehensive accounts of the classification at present in use in that area are described by Bisat (1924, 1928) and R. G. S. Hudson (1945) with later modifications. The main divisions are defined as follows (reading from above downwards):

G_1 Stage (Lower *Gastrioceras*) comprising the zones of *G. cumbriense* and *G. cancellatum*.

R_2 Stage (Upper *Reticuloceras*) comprising the zones of *R. superbilingue*, *R. bilingue* and *R. gracile*.

R_1 Stage (Lower *Reticuloceras*) comprising the zones of *R. reticulatum* s.s., *R. eoreticulatum* and *R. circumplicatile*.

H Stage (*Homoceras*) comprising the zones of *H. eostriolatum*, *H. beyrichianum* and *H. subglobosum*.

E_2 Stage (Upper *Eumorphoceras*).

E_1 Stage (Lower *Eumorphoceras*).

Following Bisat's work several workers (D. G. Evans and R. O. Jones 1929; Dix 1931; Robertson 1933 and Ware 1939) have described the Millstone Grit of the North Crop of the Coalfield and of North Gower to the west of the present area. These authors have shown that the goniatite sequences of these areas are closely comparable with those of the Pennines, and that the 'zonal' scheme outlined above is of general application to South Wales. In spite of the paucity and scattered nature of the exposures in the area at present under review, a fairly complete picture of the Millstone Grit Series can be pieced together. The generalized succession for the area south and south-east of Aberkenfig, together with the principal proved faunas, is given in Fig. 2. The compilation of the sequence from the mapping and location of fossil localities is not in doubt, but in two instances the exact super-position of the faunas is inferred by comparison with outside areas. Thus, the relationship of the *Gastrioceras cancellatum* horizon to that of *G. cumbriense* is not clear in the field and the *Homoceras eostriolatum* fauna is known only from spoil material.

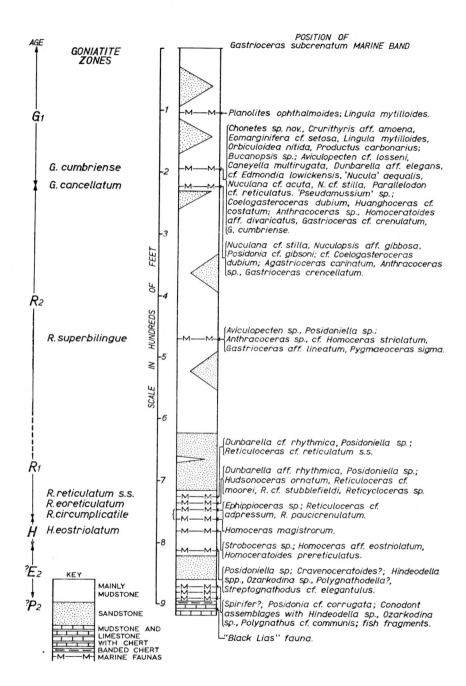

FIG. 2.   *Section of the Millstone Grit Series in the area south-east of Aberkenfig, showing generalized lithology and principal faunas*

RELATIONS OF THE MILLSTONE GRIT SERIES TO THE LOWER CARBONIFEROUS

It has long been established that the Millstone Grit in South Wales rests with a varying degree of unconformity on the Carboniferous Limestone. This was first demonstrated by the Geological Survey (see, for example, Strahan 1909, p. 20; Strahan and others 1914, p. 151) in the extreme eastern and north-western outcrops of the main coalfield where the effects of unconformity are now known to be at their greatest. Subsequently, numerous authors describing the zonal characters of both the Millstone Grit and the Carboniferous Limestone, have done much to elucidate the details of the unconformity; among these may be mentioned: Dixey and Sibly 1918, p. 123; Dixon 1921, p. 157; O. T. Jones 1925; Dixon and Pringle 1927, p. 124; George 1927, p. 51; Evans and R. O. Jones 1929, p. 164; Robertson 1927, p. 43, 1933, pp. 32–33; Robertson and George 1929, p. 4; Dix 1931; Ware 1939, p. 199.

The sequence appears most complete in the Tenby area and in North Gower, where there is little or no evidence of disconformity, and sedimentation has every appearance of having been continuous. Nevertheless shales (associated with cherts) of Upper *Eumorphoceras* (E_2) Age appear to rest directly on Upper Limestone shales ('Black Lias'), now known to be of Upper *Posidonia* (P_2) Age (see below), and thus non-sequence even here is inferred by the apparent absence of any strata representative of the Lower *Eumorphoceras* (E_1) Age. Northwards and eastwards from these areas the extent of the unconformity gradually increases by progressive erosion of Lower Carboniferous strata and by non-deposition of successive horizons of the Millstone Grit.

Difficulties in fixing a definite line of division between the two formations in the present area have been discussed by T. N. George (1933, pp. 254–7) and little can be added to the conclusions and opinions there expressed. The sequence of the area south of Aberkenfig shows strong similarities to that in North Gower. Limestones, associated with dark patchy chert, are interbedded with calcareous mudstones, and crop out between Litchard and the railway line north of Coity Junction. Some of the limestones are dark and argillaceous, some are fine-grained and compact, while a few are crinoidal; rotten-stone weathering is common. The lithology and fauna of these beds are typical of the 'Black Lias' of the Oystermouth area and it seems probable that they represent the same beds. In 1948 these strata were penetrated in a borehole near Kenfig Pool some $6\frac{1}{2}$ miles to the west, where they yielded goniatites of P_2 age (see Ramsbottom 1954, p. 54). This essentially calcareous facies is therefore confirmed as being of Lower Carboniferous age. Close above these beds in the area south-east of Aberkenfig occurs a thin layer of striped radiolarian cherts, followed by hard thinly bedded blue-grey siliceous mudstones about 30 ft thick. These beds are characterized by numerous conodonts and *Posidonia*, but so far have failed to yield any goniatites. There is an imperceptible passage from these beds into a massive quartzitic sandstone some 30 to 40 ft thick, immediately above which occurs a calcareous mudstone, weathering to rotten-stone and yielding abundant conodonts and *Cravenoceratoides?*, indicating a possible high E_2 age. A very short distance above this horizon goniatites of H age occur. Thus (excluding the sandstone) little more than 30 ft of strata at best can be referred to the E division.

In North Gower there are 400 to 500 ft of beds, mainly argillaceous, referable to E_2, the bottom part of which is characterized by black cherts, and beneath which occur striped cherts with radiolaria. At Kenfig Pool at least 110 ft of beds with numerous developments of black chert occur above similarly striped cherts. In the Litchard area no genuine black cherts or cherty mudstones have been noted, and it is inferred that nearly the whole of the E_2 strata are missing, the beds failing from the bottom upwards. Despite the failure of the black cherts the striped radiolarian cherts persist with the same relations to the 'Black Lias' as at Kenfig Pool and in North Gower; the striped cherts therefore appear to have a closer affinity to the Lower Carboniferous than to the E_2 division of the Millstone Grit. From this it would appear that the best place to draw a dividing line between the two formations is at the top of the striped cherts with radiolaria.

LITHOLOGY AND FAUNA

The Millstone Grit Series of the present area is composed mainly of mudstones and shales with intercalated beds of sandstone and gritstone up to rarely more than 50 ft in thickness. In the Aberkenfig area (see Fig. 2) the total thickness of strata is about 900 ft. East and north-east there must be considerable thinning caused by progressive cutting-out of the lower beds by unconformity. The sandstones in the top 600 ft (R_2 and G) are impersistent while those of the lower beds (E_2–R_1) seem to be more strongly developed. However, the division of the Series into two clearly defined lithological parts—the Shale Group above and the Basal Grit below—is not nearly so evident on the South Crop as it is to the north of the Coalfield (see Robertson, 1933, p. 49).

The sandstones are usually very massive and quartzitic, and in places coarse and even conglomeratic. White quartz grains and pebbles make up the bulk of the last-named types, but pink quartz and jasper have been observed. The quartzose beds are commonly coarsest nearest the top of the individual posts. The outcrops of these sandstones make parallel, bracken-covered bands with sporadic prominent crags, which stand out clearly from the wettish, bracken-free and rush-covered outcrops of the accompanying mudstones.

The mudstones are of typical Upper Carboniferous type. In the upper part of the Series (R_2 and G) marine mudstones and shales appear to be confined to definite horizons, probably separated by non-marine strata, the whole showing a marked degree of rhythmic sedimentation. The lower beds, however, appear to be almost wholly marine and the rhythmic character is less developed.

The lowest mudstones represent a conodont–*Posidonia* faunal phase in dark mudstones, but goniatites determined as *Cravenoceratoides?* were found at one locality. Abundant conodonts occur both as assemblages of the different form genera and as individuals, and it is possible that some of the assemblages are excretory (Rhodes, 1952, p. 888). The occurrence of *Posidonia* cf. *corrugata* (R. Etheridge jun.) in these beds tends to confirm the *Eumorphoceras* age indicated by the presence of *Cravenoceratoides?*.

The *Homoceras* and Lower *Reticuloceras* faunas from this area consist of typical goniatite-lamellibranch associations closely comparable with those of similar age in the north of England. They are confined to a single group of mudstones and shales and are probably not more than 80 ft thick. The only Upper *Reticuloceras* fauna collected—from the *Pygmaeoceras sigma* Bed—is of similar phase.

The principal marine bands of Lower *Gastrioceras* Age carry 'shelly' faunas, such as have been previously listed from the North Crop of the South Wales

Coalfield by Robertson (1927, pp. 48–50; 1933, pp. 63–5). They include, in addition to goniatites and lamellibranchs, many brachiopods, gastropods and nautiloids; faunas of such richness are uncommon in the Millstone Grit except in North and South Wales. Some of the species in these bands appear to belong to undescribed forms. The fauna collected from the *Gastrioceras cumbriense* Marine Band is considerably more varied than that from the *G. cancellatum* Band; it includes in particular the characteristic lamellibranch *Aviculopecten* cf. *losseni* (von Koenen) commonly found with *G. cumbriense* Bisat wherever the latter occurs.

Notable in this area is the occurrence of a horizon near the top of the Series which contains the worm burrows of *Planolites*, together with *Lingula mytilloides*. Such an association indicates that conditions may not have been favourable for the establishment of a fully marine fauna.   A.W.W.

DETAILS

As explained in the foregoing, the main outcrops of the Millstone Grit Series in the Pontypridd district occur in the area immediately south and south-east of Aberkenfig. Fig. 3 shows the solid geology of this area and the locations of the principal localities described below.

The basal beds of the Series crop out at two localities. The first is in a cutting [89848220] on the west side of the Bridgend–Maesteg road near the northern end of the Pen-y-fai Mental Hospital (locality 1 of Fig. 3). Here the following section, now much overgrown, can be made out:

	Ft
Buff sandstones, becoming quartzitic upwards .. .. seen to	10
Blocky buff silty and sandy mudstones	16
Hard black siliceous mudstones and shales, passing to ..	9
Buff silty mudstone	—

The following fauna has been obtained from the black mudstones: *Spirifer?*; *Posidonia* cf. *corrugata* (R. Etheridge jun.); fish fragments; and conodont assemblages including *Hindeodella sp.*, *Ozarkodina sp.* and cf. *Polygnathus communis* Branson and Mehl. The base of the formation is not exposed, and a short distance to the south the Quarella Sandstone (Rhaetic) is brought in by faulting. The same beds are poorly exposed on the east side of the Bridgend–Abergwynfi railway [90408220], about 600 yd N. of Coity Junction (locality 2). Here the same characteristic conodont assemblages are found in the lower beds close upon striped radiolarian cherts. The silty mudstones above are seen to pass gradually upwards into massive quartzitic sandstones about 20 ft of which are exposed in the cutting immediately south of the bridge over the road [90298244], 200 yd S.W. of Pen-y-cae Farm.

This sandstone is succeeded by about 100 ft of mudstones and shales, the top part of which is exposed in the road side [90388250] at the hairpin bend 150 yd W. by S. of Pen-y-cae (locality 3). Here 10 ft of blue and blue-black shales were noted, containing: *Sphenopteris sp.*; *Posidoniella sp.*; *Anthracoceras* or *Dimorphoceras*, *Reticuloceras* cf. *moorei* Bisat and Hudson, *R. stubblefieldi* Bisat and Hudson, and *Reticycloceras sp.*; this assemblage indicates the presence of the *R. nodosum* Subzone of the *R. eoreticulatum* Zone of R_1 Age.

Beds near the top of the same group of mudstones were cut in a sewer-trench [89458221], some 80 yd long and aligned almost due east-west, situated 420 yd N. by E. of All Saints' Church, Pen-y-fai (locality 4). The excavation throughout its length was in iron-stained, blue-black shales with a few nodules of ironstone, and three distinct horizons were recognizable. At the western end of the trench the shales yielded

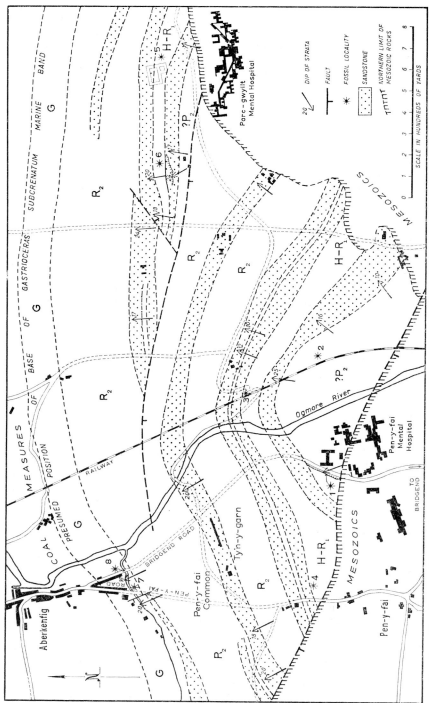

FIG. 3.   Sketch-map of the area between Aberkenfig and Parc-gwyllt, showing the outcrops of the principal divisions of the Millstone Grit
Series, and the fossil-localities described in the text

Ephippioceras sp., *Reticuloceras* cf. *adpressum* Bisat and Hudson and *R.* aff. *pauci-crenulatum* Bisat and Hudson; these forms are indicative of the *R. circumplicatile* Zone. In the middle portion of the trench *Dunbarella* aff. *rhythmica* (Jackson), *Hudsonoceras ornatum* (Foord and Crick), *Reticuloceras sp.* and *Reticycloceras sp.* were obtained; suggesting the development of the *R. eoreticulatum* Zone. In shallow foundations nearby, shales with goniatites of the *R. nodosum* Bisat and Hudson group, associated with crinoid columnals and *Posidoniella* cf. *minor* (Brown), confirmed the existence of this Zone, though the exact relationship of the two faunas was not clear. At the eastern end of the trench a fauna indicating the upper part of R_1 occurred: *Dunbarella* cf. *rhythmica*, *Posidoniella* cf. *minor*; *Homoceras?*, and *Reticuloceras* cf. *reticulatum* s.s.

At both this Pen-y-fai locality and near Pen-y-cae, exposures occurred in only the top portion of the shale-'slack', some 20 ft at most. However, about 200 yd N.N.W. of the main entrance to the Parc-gwyllt Mental Hospital a reservoir [91658275] has been excavated over a considerable portion of a shale-'slack' lying between two well-marked developments of quartzitic sandstone. The spoil from these excavations (locality 5), now grown over, yielded the following fauna: *Posidoniella sp.*; *Homoceras magistrorum* Hodson, *H.* aff. *eostriolatum* Bisat, *H. henkei* H. Schmidt, *Homoceratoides prereticulatus* Bisat, *Reticuloceras* cf. *adpressum*, *R. paucicrenulatum* and *R. sp.* (with delicate close non-crenulate transverse striae, lingua slight or absent). These collections show that several faunal horizons are present extending from the uppermost *Homoceras eostriolatum* Zone of H Age to the basal *R. circumplicatile* Zone of R_1.

The basal beds of this shale-'slack' are seen in a quarry [91308270], 400 yd to the W. by S. at a point 100 yd N. of Parc-gwyllt Fach Farm (locality 6). Here the following section can be seen:

	Ft	In
Buff-grey, very weathered shale	4	0
Blocky calcareous mudstone, weathering to rotten-stone.. ..	1	3
Very massive quartzitic sandstone	6	0
Hard bedded quartzitic sandstone	5	0
Thinly bedded silty sandstone	1	0
Blue-grey micaceous silty shale, passing to		5
Rusty weathered thinly bedded silty mudstone		9
Massive quartzitic sandstone	14	0

The lowest beds of weathered mudstone contain abundant conodonts including *Hindeodella sp.*, *Ozarkodina sp.*, *Polygnathodella?* and *Streptognathodus* cf. *elegantulus* Stauffer and Plummer, associated with *Posidoniella sp.* Close above, a bed with poorly preserved goniatites, identified as *Cravenoceratoides?*, is suggestive of E Age. This further supports the view that the whole of H should be present in the shales immediately above.

It appears from the contained fossils that the shales lying beneath the reservoir belong to the same horizon as those exposed in a parallel line of strike farther south in the Pen-y-fai sewer-trench and at the hairpin bend near Pen-y-cae. It therefore seems likely that strike-faulting of some considerable magnitude separates the two areas, though no positive evidence of this dislocation is to be seen in the field. The hypothetical position of such a fault necessary to explain the repetition of the faunas is shown on Fig. 3. This corresponds approximately with the position of a strike fault postulated, for different reasons, by R. H. Tiddeman during the original six-inch survey. In the area immediately north and north-west of the Parc-gwyllt Hospital Tiddeman mapped a small inlier of Carboniferous Limestone, based on the presence of chert fragments, which were correlated with the chert-beds of the 'Black Lias' in the area west of Litchard. During the present survey a number of shallow pits dug to bed-rock in the area of the supposed Carboniferous Limestone revealed only buff-weathered mudstone and silty mudstone of typical Millstone Grit type. Fragments of cherty limestone used as hard-core during the construction of the hospital were, however, observed, but were presumably imported. On the basis of this evidence, therefore, the inlier of Carboniferous Limestone is omitted from the present maps, though the strike fault of Tiddeman is retained.

The R_1 shales are succeeded by what is probably the thickest quartzitic sandstone of the Millstone Grit in this area. An impersistent band of mudstone or shale is developed near the middle of the sandstone, which crops out in the neighbourhood of Pen-y-cae (close to which a small quarry [90548248] is dug in about 12 ft of massive quartzite) and again along a roughly east-west line at the northern end of the reservoir referred to above. Small quarries [91108286] about 300 yd N.W. of Parc-gwyllt Fach show 6 to 8 ft of quartzitic sandstone, while sporadic crags continue the line of outcrop to the east and west.

Beds of definite R_2 Age are nowhere exposed in the Aberkenfig area. General stratigraphical considerations, however, suggest that they occupy most, if not all, of the broad featureless area of Pen-y-fai Common to the west of the Ogmore River, and much of the central portion of Cefn Hirgoed to the east. Features, associated with isolated crags, show the existence of several impersistent quartzitic sandstones in this portion of the sequence, which may have a total thickness of some 300 to 400 ft. Some three miles to the east of the Ogmore Valley, in a small stream [95508222] 100 yd E.N.E. of Gwastadwaun Farm (830 yd N.W. by W. of Pencoed Station), grey slightly silty shales yielded: *Aviculopecten sp.*, *Posidoniella sp.; Anthracoceras sp.*, cf. *Homoceras striolatum* (Phillips), *Gastrioceras* aff. *lineatum* Wright and *Pygmaeoceras sigma* (Wright). Grey and blue-grey shales and mudstones are exposed in the stream for about 200 yd to the north-west, but no other faunal horizons were found. The beds show evidence of minor folding and faulting, and their exact position in the sequence is not known, but they appear to be at a stratigraphically low level in the group here assigned to R_2

About 100 ft of beds, in which no fossils have been found, are to be seen in the west bank of the Ogmore River at the north-eastern corner of Pen-y-fai Common. These would appear to belong to the top part of R_2 or possibly the lowest part of G. The section may be summarized as follows:

	Ft
Grey mudstones, shales and siltstones with flaggy fine-grained sandstones in beds up to 18 in thick	about 50
Flaggy sandstones	about 10
Buff-grey silty mudstones with thin beds of fine-grained sandstone	about 40

Both the *Gastrioceras cancellatum* and *G. cumbriense* marine horizons are well exposed in the stream which marks the northern boundary of Pen-y-fai Common, at the southern end of Aberkenfig village. In the stream-bank immediately east of Pen-y-fai road [89438298], about 70 yd E. of Brookland House, the following section was measured (locality 7):

	Ft
Grey blocky mudstones	about 20
Grey silty mudstones with thin beds of fine-grained sandstone and ironstone nodules near the top	about 8
Thinly bedded grey mudstones	6

The basal grey mudstones yielded: *Nuculana* cf. *stilla* (McCoy), *Nuculopsis* aff. *gibbosa* (Fleming), *Posidonia* cf. *gibsoni* Salter; cf. *Coelogasteroceras dubium* (Bisat), *Agastrioceras carinatum* (Frech), *Anthracoceras sp.*, *Gastrioceras crencellatum* Bisat. The goniatites indicate the upper part of the *G. cancellatum* Marine Band or Zone.

On the west side of the Pen-y-fai Road [89408297], and only a short distance below the section just described, about 12 ft of grey mudstones and shales overlie some 10 ft of flaggy sandstones. Near the top the shales yielded: *Lingula mytilloides* J. Sowerby;

Caneyella multirugata (Jackson); cf. *Coelogasteroceras dubium*, poorly preserved *Gastrioceras sp.* Some 4 ft lower down a shelly horizon contained crinoid columnals, together with the following brachiopods: *Martinia?*, *Productus* (*P.*) *carbonarius* de Koninck, *P.* (*Dictyoclostus*) cf. *hindi* Muir-Wood, and *Schizophoria* aff. *hudsoni* T. N. George.

About 90 yd downstream, beneath the bridge carrying the main Bridgend–Maesteg road [89528302], the *Gastrioceras cumbriense* Marine Band crops out and the following section may be seen (locality 8):

	Ft	In
Grey micaceous shales	2	0
Hard dark blue-grey shales with ferruginous weathering; fossils abundant: crinoid columnals; *Chonetes sp. nov.* (smooth), *Crurithyris* aff. *amoena* T. N. George, *Lingula mytilloides*, *Martinia?*, *Orbiculoidea* cf. *nitida* (Phillips), *Productus* (*Eomarginifera*) cf. *setosus* Phillips; *Bucanopsis sp.*; *Aviculopecten* cf. *losseni*, *Caneyella multirugata* (Jackson), *Dunbarella* aff. *elegans* (Jackson), *Edmondia lowickensis* Hind, *Nuculana* cf. *stilla*, *Parallelodon* cf. *reticulatus* (McCoy), '*Pseudamussium*' *sp.*; *Huanghoceras* cf. *costatum* Hind, *Gastrioceras* cf. *crenulatum* Bisat, *G. cumbriense* Bisat, conodonts including *Ozarkodina*; fish remains including scales of *Rhabdoderma sp.* and Palaeoniscids	10	
Blocky dark blue mudstone weathering to soft clay, fossils numerous: *Crurithyris sp.*, *Lingula mytilloides*, *P.* (*Eomarginifera*) cf. *setosus*, *P.* (*P.*) *carbonarius*, *Schizophoria* cf. *hudsoni*; *Bucanopsis sp.*; *Aviculopecten* cf. *losseni*, '*Nucula*' *aequalis* J. de C. Sowerby, *Nuculana* cf. *acuta* (J. de C. Sowerby), *Parallelodon sp.*; cf. *Coelogasteroceras dubium*, *Gastrioceras* cf. *crenulatum*, and *G. cumbriense*	1	0
Grey shales with only a few goniatites	2	6
Hard dark blue ferruginous shale with: carbonized plant debris; *Lingula mytilloides*; *Posidoniella sp.*; *Homoceratoides* cf. *divaricatus* (Hind), *Gastrioceras* cf. *crenulatum*, and *G. cumbriense*	9	
Grey shales, virtually barren, weathering soft	4	0

Although the *G. cancellatum* and *G. cumbriense* marine bands presumably exist along the full extent of Cefn Hirgoed as far east as Penprysg (where the Millstone Grit Series plunges eastwards beneath Glacial Drift), no further exposures are known. Similarly there is no evidence at outcrop (apart from sandstone) of the 200 ft or so of beds presumed to lie between the *cumbriense* horizon and the base of the Coal Measures. The uppermost beds of the Series were proved in the Margam Park No. 1 Borehole [81948581], 2050 yd E. 14° S. of Margam Church, as follows:

	Ft	In
Base of *Gastrioceras subcrenatum* Marine Band	—	
Blue-grey mudstone with subordinate siltstone laminae; occasional ironstones up to 1 in	2	7
Dark blue-grey mudstone with scattered plant debris; ironstone layers up to ¾ in	22	3
Blocky medium grey silty mudstone	2	4
Blocky medium grey mudstone with roots	2	0
Strong pale grey silty mudstone with roots		11
Hard pale grey silty sandstone	1	9
Massive pale grey to white quartzite, generally fine-grained, but coarse with rolled clay-ironstone pellets near the base; thin muddy horizon about 19 ft from the top	53	3

	Ft	In
Blocky medium grey mudstone, silty near the base	2	4
Medium grey mudstone with a few ironstones; *Planolites ophthalmoides* Jessen and *Lingula mytilloides*	3	10
Tough blocky medium grey, somewhat silty mudstone with occasional thin layers and laminae of pale grey quartzose siltstone; *P. ophthalmoides* and *L. mytilloides* scattered throughout; recorded over a thickness of	56	4
Striped beds	2	1
Blocky pale grey silty mudstones, ending against fault		10

Down to the top of the mudstone 56 ft 4 in thick the dip was more or less normal—about 10° to 12°. Below this point, however, the cores exhibited rapid variation of dip up to vertical and the recorded thicknesses are obviously much too great. The borehole was continued a further 750 ft but the beds were highly disturbed and no lower horizons appear to have been proved. The thick quartzitic sandstone of the above section seems to have been repeated three times and the underlying mudstone with *Planolites* and *Lingula* at least twice; dips approached the vertical over much of this portion of the borehole which it is assumed was drilled into the lower limb of an underthrust.

A.W.W.

REFERENCES

BISAT, W. S. 1924. The Carboniferous goniatites of the north of England and their zones. *Proc. Yorks. Geol. Soc.*, **20**, 40–124.
——1928. The Carboniferous goniatite zones of England and their continental equivalents. *Cong. de Strat. Carb., Heerlen*, 117–33.

DIX, Emily. 1931. The Millstone Grit of Gower. *Geol. Mag.*, **68**, 529–43.

DIXEY, F. and SIBLY, T. F. 1918. The Carboniferous Limestone Series on the south-eastern margin of the South Wales Coalfield. *Quart. J. Geol. Soc.*, **73** [for 1917], 111–64.

DIXON, E. E. L. 1921. The unconformity between the Millstone Grit and the Carboniferous Limestone at Ifton, Monmouthshire. *Geol. Mag.*, **68**, 157–64.
—— and PRINGLE, J. 1927. The 'Penlan Quartzite'. *Sum. Prog. Geol. Surv.* for 1926, 123–6.

EVANS, D. G. and JONES, R. O. 1929. Notes on the Millstone Grit of the North Crop of the South Wales Coalfield. *Geol. Mag.*, **66**, 164–77.

GEORGE, T. N. 1927. The Carboniferous Limestone (Avonian) Succession of a Portion of the North Crop of the South Wales Coalfield. *Quart. J. Geol. Soc.*, **83**, 38–95.
——1933. The Carboniferous Limestone Series in the West of the Vale of Glamorgan. *Quart. J. Geol. Soc.*, **89**, 221–72.

HUDSON, R. G. S. 1945. The Goniatite Zones of the Namurian. *Geol. Mag.*, **82**, 1–9.

JONES, O. T. 1925. The base of the 'Millstone Grit' near Haverfordwest. *Geol. Mag.*, **62**, 558–9.

RAMSBOTTOM, W. H. C. 1954. In *Sum. Prog. Geol. Surv.* for 1953, 54.

RHODES, F. H. T. 1952. A Classification of Pennsylvanian Conodont Assemblages. *J. Paleont.*, **26**, 886–901.

ROBERTSON, T. 1927. The Geology of the South Wales Coalfield. Part II. Abergavenny. 2nd edit. *Mem. Geol. Surv.*

———1933. The Geology of the South Wales Coalfield. Part V. The Country around Merthyr Tydfil. 2nd edit. *Mem. Geol. Surv.*

——— and GEORGE, T. N. 1929. The Carboniferous Limestone of the North Crop of the South Wales Coalfield. *Proc. Geol. Assoc.*, **40**, 18–40.

STRAHAN, A. 1909. The Geology of the South Wales Coalfield. Part I. The Country around Newport. 2nd edit. *Mem. Geol. Surv.*

———, CANTRILL, T. C., DIXON, E. E. L., THOMAS, H. H. and JONES, O. T. 1914. The Geology of the South Wales Coalfield. Part IX. The Country around Haverfordwest. *Mem. Geol. Surv.*

WARE, W. D. 1939. The Millstone Grit of Carmarthenshire. *Proc. Geol. Assoc.*, **50**, 168–204.

WOODLAND, A. W., ARCHER, A. A. and EVANS, W. B. 1957. Recent Boreholes into the Lower Coal Measures below the Gellideg–Lower Pumpquart Coal horizon in South Wales. *Bull. Geol. Surv. Gt. Brit.*, No. 13, 39–60.

CHAPTER III

COAL MEASURES: GENERAL

EXCEPT FOR a narrow strip along the southern margin the entire district is underlain by Coal Measures rocks. These succeed the Millstone Grit Series conformably and upwards of 6500 ft of beds are preserved between Maesteg and Margam. The sequence is however incomplete, for the highest 1000 ft of measures, such as are seen west of Neath, have been removed by post-Carboniferous erosion, and a further 700 ft, somewhat lower in the succession, are apparently cut out by unconformity. A generalized section of the Coal Measures of the Pontypridd district is given in Fig. 4; it shows the distribution of the thicker sandstones, the principal coal seams, and the known horizons of marine strata, together with the classification used in this memoir.

The Coal Measures of South Wales, in common with those of most of Britain, consist of alternations of mudstones, siltstones and sandstones, together with numerous coals and associated seatearths. Sandstones, although occurring throughout the succession show a remarkable development in the top half, where in developments up to 500 ft or so in thickness, they make up by far the greater bulk of the strata. Persistent sandstones, formerly ascribed to the Millstone Grit (see p. 6), are also widespread in the lowest strata of the formation. Between these two groups the measures are largely argillaceous and it is in these that the major coals are concentrated.

CYCLIC SEDIMENTATION

The various sediments are arranged in a consistent order forming a succession of repetitive units termed rhythms, cycles, cyclic units or cyclothems. It is thought that sedimentation took place in shallow water on a shelf of considerable extent, and that the cyclothems resulted from intermittent subsidence of this shelf, possibly accompanied on occasions by eustatic changes. The clastic sediments represent the filling up of the depositional area to water-level, and the coals represent the decayed products of the vegetation which, during relatively lengthy stable periods, colonized the former area of deposition. The bulk of the sediments evidently accumulated under brackish or even fresh-water conditions, but the sporadic occurrence of comparatively thin but persistent layers of mudstone with typical marine fossils shows that occasional influxes of the sea took place, most of which spread throughout the area of sedimentation.

A typical Coal Measures cyclothem may be considered as starting with the first sediments accumulating after subsidence (the roof strata of a coal) and ending with the final products of silting up and emergence (the seatearth and its overlying coal). In general the clastic sediments coarsen upwards and the contained fossils suggest that the environment becomes increasingly non-marine. Truly marine forms, when present, lie at or near the base of the cyclothem; they tend to be succeeded upwards by such forms as *Lingula*, foraminifera and *Planolites ophthalmoides*, which may foreshadow oncoming brackish conditions; and these in turn by '*Estheria*' and the so-called non-marine lamellibranchs or

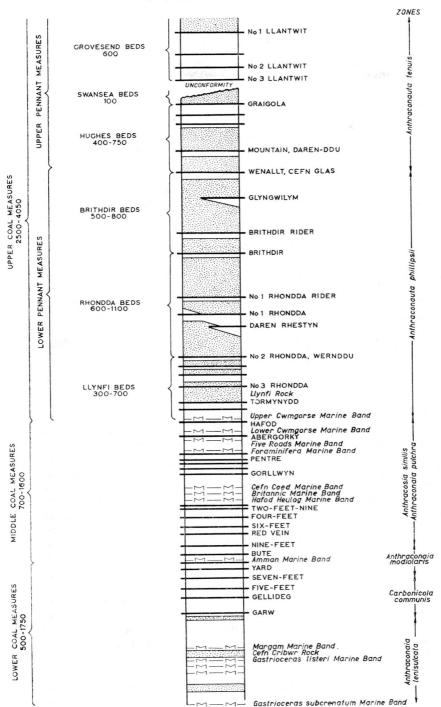

FIG. 4.   *Generalized vertical section of the Coal Measures of the Pontypridd District*

'mussels', suggestive of even less saline, estuarine facies. The overlying beds are characterized by plants and plant debris, and by the time vegetation became established, as witnessed by the presence of roots in seatearth, the environment was virtually one of fresh water.

Many attempts have been made to define a single 'standard' or 'ideal' cyclothem; but rarely do actual cases conform to this standard. Local variations in conditions of subsidence and deposition inevitably cause modifications to the

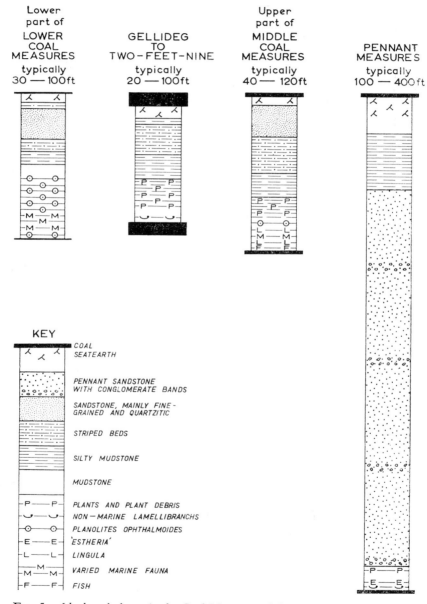

FIG. 5. *Ideal cyclothems in the Coal Measures of the Pontypridd District*

pattern, and produce a cycle of deposition differing widely in many cases from the hypothetical standard. Four main variations are recognizable in the Coal Measures of the present district (see Fig. 5).

Below the Gellideg Seam successive subsidences followed one another rapidly, and in consequence coals are insignificant. The south-western part of the district appears to have occupied a relatively 'off-shore' position, for the lowest portions of several cyclothems are occupied by thick marine strata, and these are overlain in turn by thick mudstones containing abundant traces of the burrowing worm, *Planolites ophthalmoides*, believed to indicate near-marine conditions. These two phases usually comprise the bulk of the individual cyclothems, which in general vary in thickness from 30 to 100 ft, though 200 ft are recorded at Margam Park.

From the Gellideg to the Two-Feet-Nine seams the periods of stability between successive subsidences were greatly prolonged, and consequently the typical cyclothem ends with a thick coal. Marine sediments are known from only one horizon; usually an 'estuarine' phase with 'mussels', up to a few feet thick, forms the base of the cyclothem; and these estuarine muds pass upwards into mudstones, silty mudstones and striped beds, generally rich in plant debris. Only rarely does this sequence end in a widespread sheet of sandstone. In thickness these cyclothems range from 20 to 100 ft and their lateral development is variable. Splitting of coals is common and mere partings in a seam at one locality may be represented by full cyclothems 100 ft thick a few miles away.

In the upper part of the Middle Coal Measures the cyclothems show a close approach to what is generally considered the 'ideal' Coal Measures form. Many of the units begin with marine strata carrying varied faunas; these are followed successively by *Lingula*, *Planolites*, 'mussel' (rare) and plant phases. The mudstones containing these fossils pass upwards into silty mudstones, striped beds, and, in several instances, into well-developed fine-grained sheet-quartzites, such as the Cockshot Rocks of the Maesteg area. Ganister-like seatearths at the top of the cyclothem are associated with coals of only moderate thickness, 1 to 3 ft being average. Commonly each unit has a total thickness of 40 to 120 ft; a generally more uniform subsidence is deduced from the decreasing importance of seam-splits.

In the Lower and Middle Coal Measures, and particularly among the main seams, sandstones are encountered which do not form part of the cyclothemic sequence as above described. They are characterized by sharp bases, and conglomeratic developments with rolled pellets of ironstone and mudstone are common in the lower layers. The strata on which they rest commonly show signs of local erosion, which was probably contemporaneous with the influx of the sand. They may interrupt any part of a cyclothem and in places the thicker sandstones may occupy the space of several units. They appear to be largely independent of the factors controlling true cyclic deposition and may reflect abnormal conditions in the adjoining land areas. These are the sandstones which are usually associated with 'washouts' in coal seams.

The typical Pennant cyclothems are markedly different, and it is debatable whether their mode of formation is the same as that of the sedimentary cycles previously described. They are seldom less than 100 ft thick, and they may reach over 500 ft, and at least 80 per cent is made up of a virtually continuous body of sandstone. This lies immediately or closely upon a coal, and the base is commonly conglomeratic. Mudstones, up to 20 ft or more thick, may intervene between the coal and sandstone, and may contain *Anthraconauta* and ostracods, associated at a few horizons with *Euestheria*, at their base; varied and abundant plants may also be present. The junction between the mudstones and the overlying sandstones is almost invariably sharp and there is commonly evidence of erosion of much or all of the mudstone. In contrast, the top of the sandstone always grades or interdigitates into silty mudstones and mudstones, which are unfossiliferous apart from plant debris. Several horizons of seatearth and associated coal streaks may occur before the cyclothem ends with a coal, commonly dirty, of 1 to 3 ft in thickness. These distinctive Pennant cyclothems are interpreted as reflecting a major change in the geography of the sedimentary basin and its surrounding landmass; reflecting possibly greater elevations of the land precursory to the onset of the Armorican orogeny.

LITHOLOGY

Argillaceous rocks.—These range from soapy clay to silty or sandy mudstones; they may be very fissile, especially in the roofs of coals, or hard and blocky. They vary from pale grey to black, the latter types usually occurring close to coals. The dark colour is usually due to the presence of finely divided carbonaceous material, but disseminated iron sulphide is responsible in some instances. Mica may be abundant, giving an apparent siltiness to the rock. Silty material is increasingly important higher in the cyclothem, and there is a gradual transition into siltstones. Marine mudstones and shales are usually dark bluish-grey and contain abundant pyrite in the form of cubes, granules, and 'fucoids', or replacing organic remains.

Nodules ('balls') and bands ('pins') of clay-ironstone are associated with the mudstones, especially in lower beds of the cyclothem; irregularly shaped nodules may also abound in the lower parts of seatearths. They are hard and compact and vary from light grey or brown to black; they consist mainly of ferrous carbonate with varying amounts of calcium and magnesium in solid solution, and may contain varying amounts of admixed clay material. Septaria are common, the cracks being filled with a variety of secondary minerals (see North and Howarth 1928). Hard compact carbonaceous sideritic mudstones, known as blackbands, occur widely at several horizons, usually in the roofs of coal seams. They may be blocky and massive, but are more generally thinly laminated; they are commonly rich in 'mussels' and ostracods.

Striped Beds.—These consist of alternations of pale grey quartzitic siltstone and medium grey mudstone or silty mudstone, the individual layers being commonly not more than about $\frac{1}{16}$ in. in thickness, though they may occur in bands up to several inches. The relative abundance of the two components is inconstant; some rocks are almost entirely muddy, with only a few silty stripes; others may consist largely of quartzitic silt with only subordinate wisps of mud. Lens-bedding is common and the rocks may also show evidence of turbulence and slumping.

Arenaceous rocks.—These fall into two main classes: quartzitic sandstone, often known as 'cockshot', and developed characteristically in the Lower and Middle Coal Measures; and 'pennant', largely restricted to the Upper Coal Measures. The former consist of angular quartz grains set in a crypto-crystalline groundmass of secondary silica; feldspars are rare or absent, and carbonaceous material is variable in amount and arranged usually on the bedding. Many sandstones are fine enough to merit the name siltstone, but coarse pebbly layers occur in some towards the base of the Lower Coal Measures. Individual beds are commonly about 5 ft thick; and false-bedding is not normally conspicuous.

Pennant sandstone when fresh is a green or bluish-grey rock, resistant to weathering, but which on prolonged exposure develops a yellowish or brownish surface. Large scale false-bedding is general, though many massive posts up to 20 ft or so thick show little trace of bedding. Locally thinly and regularly bedded tilestones and flagstones are developed, the fissility being commonly due to concentration of detrital mica on the bedding. The microscopic characters of the rock have been described by Flett (see, for example Strahan 1907, p. 58; Strahan and others 1907, p. 93; Strahan and others 1917, pp. 129–30) and Heard (1922). Typically it consists of subangular to sub-rounded grains of fairly even size set in a crypto-crystalline matrix of silica, white mica and clay minerals. The fragments consist mainly of quartz, some 'granitic' or sheared, but feldspars, largely kaolinized, are always present, and fragments of pre-existing rocks are not uncommon. Mica is abundant in some varieties. Carbonaceous material may be disseminated throughout the rock as specks of indeterminate character or it may be concentrated as layers of detrital coal. The casts of large prostrate tree-trunks and flattened pieces of timber are common, especially in some of the coarser bands.

Pebbly layers occur sporadically, most of the pebbles consisting of quartz, but shales, slates, cherts, quartzites, jaspers and sandstones are also present. Closely associated with such pebbly horizons, but also occurring at other levels, are beds containing rolled clay-ironstone pellets and 'rafts' and 'pebbles' of coal, pointing to the pene-contemporaneous erosion of only partly consolidated materials. Many of the coal 'pebbles' are rude cuboids, suggesting the break-up of a bed which already possessed a rectangular joint-pattern. The extensive tabular form of many of the 'rafts' makes it likely that the coal material was of a leathery consistency on burial, while the relationships of the raft and the enclosing sandstone show clearly that there has been no further reduction in volume upon subsequent coalification.

Seatearths.—The root-beds immediately beneath coals have been altered by the action of humic acids derived from the overlying vegetation. Lithologically they range from clays to coarse sandstones; good quality fireclays are rare and true ganisters virtually non-existent, and so the term seatearth is here used to cover all varieties of coal-floor strata.

Nature and Distribution of Fossil Bands

Marine Bands.—Within the district marine faunas are known from thirteen horizons (see Fig. 4). Apart from the Amman Marine Band, they occur in two main groups: five lie below the Garw and the remainder between the Two-Feet-Nine and the base of the Upper Coal Measures. Certain of the bands exhibit distinct faunal phases in sequence which are thought to correspond with variations in contemporary salinity. The band may start with a cannelly fish-phase, which, with increasing salinity, is followed first by horny brachiopods and foraminifera, and then by varied forms such as brachiopods, gastropods, lamellibranchs, and goniatites. Thereafter a decreasing salinity is indicated by a return of the *Lingula*- and foraminifera-phases followed by that of *Planolites ophthalmoides*.

The sub-Garw marine bands (see pp. 33–7) are known only along the western South Crop, where they commonly exceed 20 ft in individual thickness. A goniatite-phase is usual, and the bands are generally similar in their development to those in the upper part of the Millstone Grit Series. The *Planolites*-phase is particularly thick, 30 ft and 40½ ft being recorded in two of the bands.

The Amman Marine Band is for the most part little more than a *Lingula*-band, but towards the south-east the fauna becomes increasingly more varied and 'calcareous'.

Of the higher bands, the Hafod Heulog (p. 59), Britannic (p. 59) and Foraminifera (p. 62) horizons carry little other than horny brachiopods, foraminifera, and *Planolites*. The Five Roads Band (p. 63) is known from one locality, where it contains only *Myalina*. The others, the Cefn Coed (p. 60), the Lower Cwmgorse (p. 63) and the Upper Cwmgorse (p. 64), although locally *Lingula*-bands, more usually include varied faunas, especially in the west and south-west of the district.

Non-marine lamellibranchs.—These occur sporadically throughout the sequence, being especially common in the roof strata of coals, particularly those between the Gellideg and Two-Feet-Nine, where they are of particular value in correlation. Individual 'mussel'-bands are less continuous than marine bands and few persist throughout the district. Some, moreover, are associated with particular lithologies, and this association, while making them readily recognizable as markers over limited areas, also renders them liable to lateral failure. A zonal system based on variations in the 'mussel' fauna was erected by Davies and Trueman (1927) and extended by later workers (see Fig. 4). It has formed the basis of studies within the present area by Evans and Simpson (1934) and Moore and Cox (1943). While the zonal system is not sufficiently precise for detailed correlation, a number of restricted faunal assemblages, some indeed confined to single cyclothems, have been recognized within the present district and are described briefly below.

In the sub-Garw measures, 'mussel' faunas are known only from the Margam Park Boreholes (pp. 77–9). The lowest occurs about 250 ft above the base of the Coal Measures, where representatives of the *Carbonicola fallax* Wright group, characteristic of the lower part of the *lenisulcata* Zone, are found. Between the Margam Marine Band and the Garw, shells are known from several horizons. The commonest forms belong to *Curvirimula*, including variants of *C. trapeziforma* (Dewar), but *Anthraconaia sp. nov.* (cf. *modiolaris*), *Carbonicola* aff. *crispa* Eagar, *C.* aff. *pontifex* Eagar and *C.* aff. *proxima* Eagar have also been identified. These are generally considered to be at about the junction of the zones of *Anthraconaia lenisulcata* and *Carbonicola communis* (see also Eagar 1962, pp. 323, 330).

The '*pseudorobusta* fauna' is found between the Garw and Gellideg seams. Several more or less distinct layers of dark shaly mudstone carry a characteristic fauna of large *Carbonicola* with thick, often crushed, shells preserved in recrystallized calcite. These reach about 70 mm. in length and include variants of *C. communis* Davies and Trueman, *C. pseudorobusta* Trueman and *C. rhomboidalis* Hind; they are associated with *Curvirimula*, as well as with *Geisina arcuata* (Bean) and *Spirorbis sp.* Similar shell-bands are known from the same general position in most British coalfields.

The 'Yard fauna' is characterized by 'solids' dispersed through several feet of blocky grey mudstone. It consists of the last abundant *Carbonicola*, those grouped around *C. venusta* Davies and Trueman, and the earliest abundant

Anthracosia, variants of *A. regularis* (Trueman). These have a general similarity of shape and it is not always easy to differentiate between them.

The 'Amman fauna' is made up mainly of *Anthracosia* of the *aquilina* (J. de C. Sowerby)/*ovum* Trueman and Weir group, usually poorly preserved as 'ghosts' in smooth grey mudstone immediately overlying the Amman Marine Band. Many individuals show distinct Anthraconaioid characters, and this is presumed to result from the somewhat unusual environment consequent upon the marine incursion (Jenkins 1960, p. 116).

The 'Bute fauna' is widely developed over the western half of the district where 'solid' shells are usually abundant. It consists essentially of an association of *Anthraconaia* of the *modiolaris* (J. de C. Sowerby)/*williamsoni* (Brown) group and *Anthracosphaerium* of the *affine* (Davies and Trueman)/*exiguum* (Davies and Trueman)/*turgidum* (Brown) complex. The Bute roof is the lowest horizon at which these two genera are abundant.

The 'Nine-Feet fauna', best developed in the north-eastern areas, is essentially a facies fauna of large crushed *Anthracosia* (30 to 40 mm.) in dark carbonaceous and ferruginous mudstones. Variants and intermediates of *A. aquilina, beaniana* King, *disjuncta* Trueman and Weir, *ovum* and *phrygiana* (Wright) are common; shells of the *A. phrygiana* group showing a marked tilt of growth lines are especially characteristic.

The 'Red-Vein fauna' is less restricted; it occurs above the Nine-Feet and at several other horizons between this seam and the Caerau. Small 'solid' shells occur through blocky grey mudstone and characteristically possess a retiform wrinkling of the periostracum. The assemblage consists mainly of *Anthracosia* and *Anthraconaia*, but because of their small size it is difficult to separate the two genera with certainty: they may be referred to *Anthracosia* aff. *nitida* (Davies and Trueman) with variants approaching *A. angulata* (Chernyshev) and *Anthraconaia pulchella* Broadhurst.

The 'Six-Feet fauna' is best developed in a carbonaceous ironstone of blackband type, though similar but less abundant forms persist in grey mudstone. The shells, which reach 30 mm. in length, consist almost entirely of *Anthracosia* and are dominated by variants of *A. atra* (Trueman). Associated forms include *A. aquilinoides* (Chernyshev), *A. concinna* (Wright), *A. elliptica* (Chernyshev), *A. lateralis* (Brown) and *A. planitumida* (Trueman).

The 'Four-Feet fauna' is notable for the fine preservation of the 'solid' shells and the ease with which they can be removed from the enclosing blocky grey mudstone. *Anthracosia, Anthraconaia* and *Anthracosphaerium* are all represented in considerable variety. Particularly characteristic is a form identified as *Anthraconaia sp. nov.* (cf. *curtata*), and members of the *A. librata* (Wright)/*cymbula* (Wright) and *Anthracosphaerium exiguum/propinquum* (Melville) groups are almost equally typical. The *Anthracosia* element consists largely of variants of the species *A. atra, concinna* and *planitumida*.

The 'Two-Feet-Nine fauna' is composed of rather small (15 to 20 mm.) 'solids' dispersed through grey mudstone and usually associated with nodular ironstone bands carrying *Planolites montanus* Richter. The main elements are species of the genus *Anthracosia* including *acutella* (Wright), *atra, concinna, lateralis* (Brown), *planitumida* and *simulans* Trueman and Weir; *Anthracosphaerium propinquum* is well represented, as is also an elongate *Anthraconaia* of the *lanceolata* (Hind)/*wardi* (Hind *non* Salter) group. Similar faunas occur at several horizons close above the Two-Feet-Nine.

'Mussels' are not numerous in the measures between the Hafod Heulog and Upper Cwmgorse marine bands. Two horizons of note are recorded, both characterized by *Anthraconaia*. The Caedavid Seam of the Garw valley (p. 158) yields *A. adamsi* (Salter) preserved as squashed forms in blackband, and in the Rhondda Fawr and Llynfi valleys *A. hindi* (Wright) and *A. warei* (Dix and Trueman) have been obtained as 'solids' from grey mudstone above the Eighteen-Inch Seam.

Anthraconaia aff. *stobbsi* (Dix and Trueman) and *Naiadites* are found above the White Seam, but above this the 'mussel' faunas show a marked change. Apart from a few *Anthraconaia pruvosti* (Chernyshev) found near the middle of the Rhondda Beds, only *Anthraconauta* is recorded in the Upper Coal Measures. *A. phillipsii* (Williamson) is persistent, but *A. tenuis* (Davies and Trueman), which appears in the Rhondda Beds, becomes increasingly abundant upwards. Both species are confined to a relatively few horizons, at which, however, they are usually very abundant. Despite their individual lack of faunal distinctiveness, the bands tend to be confined to the roofs of particular seams and this gives them an added value as markers.

Other faunal bands.—At several horizons beneath the Gellideg there are thin bands of cannelly mudstone rich in fish debris. One of these, lying about 50 ft above the Margam Marine Band evidently passes into a marine band in the west of the coalfield (Woodland and others 1957b, p. 52). Another significant fish-band lies in the roof of the Garw; this persists throughout South Wales, and it may well extend much further afield for the roofs of the Kilburn and the Arley in the Pennine coalfields are almost identical with it.

'*Estheria*', although locally occurring at numerous levels, shows a remarkable concentration in the roof of the Abergorky. Here *Lioestheria vinti* (Kirkby) is virtually unaccompanied by any other fossils, and the band is traceable over most of the coalfield. A similar concentration is known from what may be the same horizon in other coalfields, notably the Main '*Estheria*' Band of the Nottingham–Derby area (Edwards and Stubblefield 1948, pp. 231–3). *L. striata* Pruvost (*non* Münster) is present widely on the roof of the Lower Six-Feet.

Floras.—Apart from the coal seams, which consist almost entirely of the products of decayed vegetation, plant remains are ubiquitous. Mostly they consist of indeterminate comminuted debris, but mudstones rich in well-preserved leaves and stems occur. Good roof floras have been obtained locally from the Yard, Bute, Lower Nine-Feet, Upper Nine-Feet, Caerau, No. 2 Rhondda, No. 1 Rhondda, Brithdir and Daren-ddu or Mountain seams. Protracted collecting can produce long lists of species even where the fossils are not abundant, and in this way D. Davies (1921) produced remarkable collections from all the worked seams in the collieries of Clydach Vale and Gilfach Goch.

Formerly plants were used as the main basis for correlation and classification (Kidston 1893–94, 1894). Working mainly in the western areas of the South Wales Coalfield, Dix (1934) proposed a nine-fold zonal subdivision of the Upper Carboniferous, and this scheme was followed closely by Moore and Cox (1943) in their work in the Taff and Rhondda valleys.

Most of the plant species are long-ranged, and the zones each cover considerable thicknesses of strata; the boundaries between the zones are hard to define, and individual seam floras are not readily distinguishable. Consequently no great attention has been paid to the plants during the present survey; they have been collected where readily available, and are listed in their appropriate places in the details.

CLASSIFICATION

Early classifications of the Carboniferous rocks in South Wales were wholly lithological. The traditionally arenaceous Millstone Grit was considered to extend upwards from the Limestone to the top of the 'Farewell Rock', the highest significant sandstone beneath workable coals and ironstones. In the succeeding Coal Measures the existence of a major sandstone formation, the Pennant, between two predominantly mudstone sequences led to a natural three-fold sub-division into Lower Coal Series, Pennant Series and Upper Coal Series, and, with minor variations, this classification was in general use until the present survey. Its drawbacks became apparent when detailed mapping in the present and adjoin-ing Swansea districts disproved the old idea that the Pennant lithology was bounded by two time planes, conveniently marked by the No. 2 Rhondda Seam below and the Mynyddislwyn or No. 3 Llantwit above. This old conception had led the earlier surveyors to serious mis-correlation. Thus in the earlier editions of this memoir the base of the Pennant Series was taken at the Brithdir Seam in the east, at the No. 2 Rhondda in the central areas, and at the Rock Fawr or No. 3 Rhondda along the western South Crop. These three distinct seams each mark the approximate base of massive pennant lithology locally, but they are now known to be spread over some 1800 ft of strata in the area of thickest development.

Over the last thirty or forty years detailed work in all the coalfield areas has demonstrated the continuity over wide areas of many of the bands containing marine fossils. These provide a secure basis for detailed correlation between the various coalfields and as a result the Geological Survey has recently adopted a single classification for the Coal Measures throughout England and Wales (Stubblefield and Trotter 1957). This recognizes three divisions, Lower, Middle and Upper: the division between the Lower and Middle is taken at the base of the marine band which occurs about the middle of the *Anthraconaia modiolaris* Zone, and which is characterized by *Anthracoceras vanderbeckei* (Ludwig); and that between Middle and Upper at the top of the highest marine band, that characterized by *Anthracoceras cambriense* Bisat. The application of this scheme to the South Wales sequence has already been discussed (see Woodland and others 1957a). Although palaeontological in its conception, it does, in South Wales, have the additional merit of restricting the Pennant sandstones, even where most fully developed, to the Upper Coal Measures. In order to retain such a deeply entrenched term this uppermost division has been given the local synonym of Pennant Measures, even though it includes the whole of the former Upper Coal Series and the uppermost part of the Lower Coal Series.

South Wales is unique among the major British coalfields in that the Upper Coal Measures occupy most of the surface and are generally a good deal thicker than the combined lower divisions. Further sub-division is therefore desirable, and this has been made possible by the tracing across the coalfield of several coal horizons. A separation into two major groups at the Hughes Seam and its equivalents is supported by both floral and faunal evidence. Crookall (1955, p. 3) suggests that this horizon marks the junction between Westphalian C and D (that is between Staffordian and Radstockian), and it may also be taken as the arbitrary boundary between the 'mussel' Zone of *Anthraconauta phillipsii* and that of *A. tenuis*. Both divisions have been further divided into three groups though these have been distinguished only on the six-inch maps. This classification and the reasons for its adoption have already been set out in detail

(Woodland and others 1957a) and the following table summarizes its application within the present district.

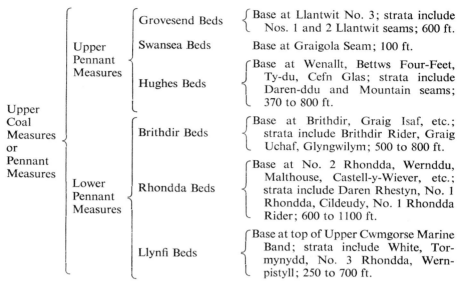

Upper Coal Measures or Pennant Measures	Upper Pennant Measures	Grovesend Beds	Base at Llantwit No. 3; strata include Nos. 1 and 2 Llantwit seams; 600 ft.
		Swansea Beds	Base at Graigola Seam; 100 ft.
		Hughes Beds	Base at Wenallt, Bettws Four-Feet, Ty-du, Cefn Glas; strata include Daren-ddu and Mountain seams; 370 to 800 ft.
	Lower Pennant Measures	Brithdir Beds	Base at Brithdir, Graig Isaf, etc.; strata include Brithdir Rider, Graig Uchaf, Glyngwilym; 500 to 800 ft.
		Rhondda Beds	Base at No. 2 Rhondda, Wernddu, Malthouse, Castell-y-Wiever, etc.; strata include Daren Rhestyn, No. 1 Rhondda, Cildeudy, No. 1 Rhondda Rider; 600 to 1100 ft.
		Llynfi Beds	Base at top of Upper Cwmgorse Marine Band; strata include White, Tormynydd, No. 3 Rhondda, Wernpistyll; 250 to 700 ft.

THICKNESS VARIATION

It has long been known that all divisions of the Carboniferous thicken markedly when traced across the coalfield from east to west. Much of this increase in the Coal Measures takes place within the Pontypridd district where, considering the measures up to the Daren-ddu and its equivalents, the highest level at which true comparison is possible, some 2900 ft of beds in the north-east expand to about 6250 ft near Margam. The thickening is not uniform throughout the sequence, being more marked in the lower divisions than the higher.

Detailed knowledge of the beds below the Garw is limited, but at Dowlais, some 5 miles north of Merthyr Vale, they are little more than 150 ft thick (Robertson 1933, pp. 69, 96); in the Margam Park No. 1 Borehole (p. 77) they were proved to be 1146 ft, a seven- or eight-fold increase in 22 miles. The higher beds of the Lower Coal Measures thicken in the same direction from about 200 ft in the Taff and Cynon valleys to about 600 ft at Kenfig Hill, a three-fold increase in 18 miles (Fig. 6). In the Middle Coal Measures both the beds between the Amman and Cefn Coed marine bands and those between the Cefn Coed and Upper Cwmgorse thicken westwards from about 300 ft around Pontypridd to 700 or 800 ft between Maesteg and Margam (Fig. 7). A similar thickening is seen in the Llynfi Beds, from 250 ft at Merthyr Vale to 700 ft at Margam, but above this the rate falls off progressively; the Rhondda Beds are rather less than 600 ft in the northern Rhondda and Taff valleys and about 1100 ft to the west of Aberkenfig; and the Brithdir Beds about 500 ft in the Rhondda and on the eastern South Crop and 800 ft south-west of Maesteg (Fig. 8). As far as can be estimated from their incomplete preservation the Upper Pennant Measures suffer little variation in thickness across the district.

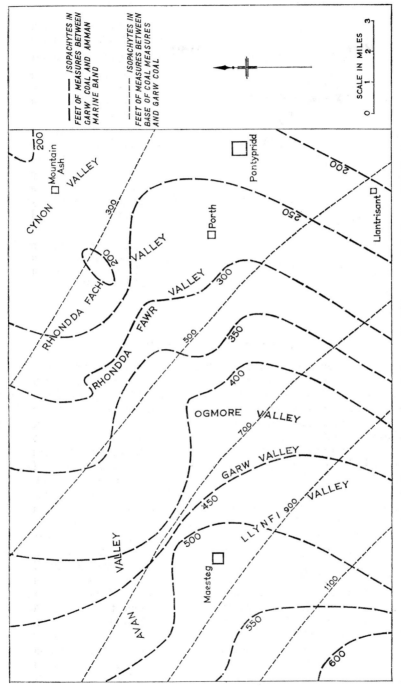

FIG. 6. *Thickness variations in the Lower Coal Measures of the Pontypridd District*

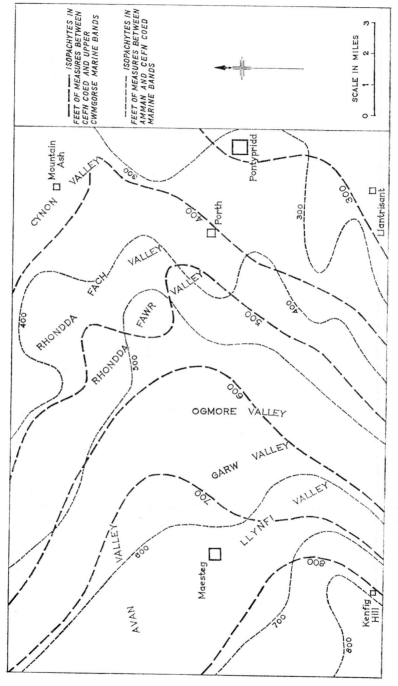

FIG. 7. Thickness variations in the Middle Coal Measures of the Pontypridd District

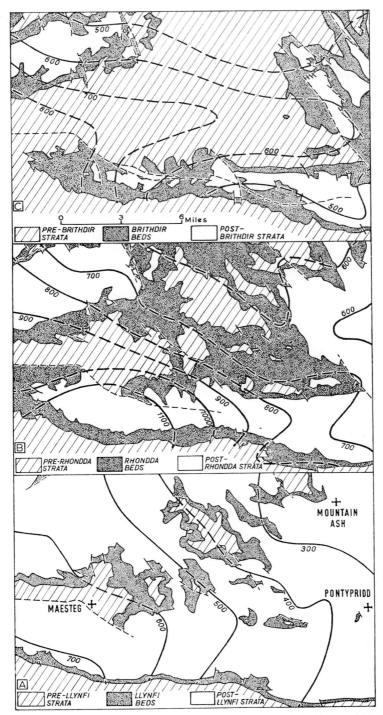

FIG. 8.   *Thickness variations in the Lower Pennant Measures of the Pontypridd District.*
A. Llynfi Beds; B. Rhondda Beds; C. Brithdir Beds

The variation in coal thickness is broadly analogous. Except below the Garw, coals are common throughout the sequence and upwards of 58, at least locally, reach 2 ft in thickness. Although they are fairly evenly distributed there is a concentration of the thicker ones between the Gellideg and Two-Feet-Nine, and almost all the present-day production in this central part of the coalfield comes from this group. In the Cynon Valley 12 coals occur in this portion of the sequence, spread through about 375 ft of measures; in all they total $46\frac{1}{2}$ ft or about one-eighth of the total thickness of strata; the individual seams average 3 ft $10\frac{1}{2}$ in thick and are about 34 ft apart. In the Margam–Kenfig area, on the other hand, splitting of the same coals has produced 18 significant seams totalling about $97\frac{1}{2}$ ft of coal through 1200 ft of measures; coal makes up about one-twelfth of the total thickness of strata and the individual seams, which average 5 ft 5 in, lie about 75 ft apart.

SEAM NOMENCLATURE

The naming of coal seams in South Wales has always been chaotic. Within much of the present district attempts were made to apply the names in use around Aberdare, at least to the main productive part of the sequence. Even so, local nomenclatures grew up; sometimes for a single seam, sometimes for an individual colliery, sometimes even for a region such as the South Crop. More confusing still were the many miscorrelations with the Aberdare sequence. Thus not only was the same seam given many names, but the same name was applied to many different seams. During the resurvey the seams between the Garw and Two-Feet-Nine were everywhere correlated with the Aberdare sequence and re-named accordingly, splitting being accommodated by use of the forenames Lower, Middle and Upper. These names appear on the six-inch and one-inch maps, are now in use by the National Coal Board, and form the basis of the following account of the stratigraphy. Plate III shows the detailed correlation of the coals from the Gellideg to Two-Feet-Nine, together with the new names and the local names in former use.

Above the Two-Feet-Nine the coals are less widely worked and the desirability for a single set of seam names is not so great. Simplification into two nomenclatures has, in general, proved sufficient within the Middle Coal Measures, and both sets of names are shown on the one-inch map and used in the following account, see Plate IV. Within the Pennant, only the No. 1 and No. 2 Rhondda were hitherto in use as regional names, and many coals were unnamed. Seams have been given the names in use where they are best developed, generally in the Taff–Rhondda district, and these names have, on the six-inch maps, been linked with any existing local names. The relationship of the various named sequences is shown in Plate V.

LOWER COAL MEASURES

The Lower Coal Measures outcrop within the district is limited to two areas: a narrow strip along the southern margin, and a small tract on the north side of the Moel Gilau Fault west of Bryn. Even in these areas the measures are poorly exposed and are covered by extensive drift deposits. The division falls naturally into two groups at the Garw, a seam which can be easily correlated throughout

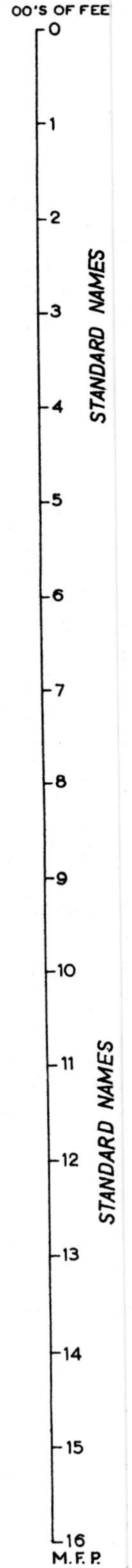

00'S OF FEE

0

STANDARD NAMES

1

2

3

4

5

6

7

8

9

10

STANDARD NAMES

11

12

STANDARD NAMES

13

14

15

16
M. F. P.

South Wales and which would seem to be readily correlated in other coalfields (e.g. the King of North Staffordshire, the Arley Mine of Lancashire, the Kilburn of the East Midlands and the Victoria of Durham).

Below the Garw, knowledge of the succession is virtually confined to the extreme south-west of the district, where the Margam Park No. 1 Borehole penetrated the full sequence, and where sporadic exposures are to be seen in the Cefn Cribwr area. Here these measures are about 1100 ft thick; eastwards and north-eastwards considerable thinning must take place, for in the Dowlais area, about 4 miles north of Merthyr Vale, the corresponding strata are leas than 200 ft thick (Robertson 1933, pp. 69, 96), while at Rudry in the Newport district, 8 miles east of Llantwit Fardre, the corresponding thickness is about 300 ft (Woodland and others 1957b, p. 50). The frequency of marine strata and the insignificant coal development illustrate the close links that these lowest Coal Measures have with the underlying Millstone Grit.

The measures above the Garw range in thickness from 200 ft in the north-east to 600 ft in the south-west. The sedimentation is of characteristic Coal Measures type: there are no marine strata and thick coals are developed at regular intervals throughout the sequence.

(a) *Measures below the Garw Seam*

Fig. 9 shows the lithological and faunal characters of these beds as proved in the Margam Park boreholes, and the following account is based on these sections except where clearly stated.

The **Gastrioceras subcrenatum Marine Band** is well developed through a thickness of 50 ft, while a few goniatite fragments are also recorded 16 ft above. The lowest 20 ft carries only *Lingula mytilloides* J. Sowerby, *Planolites ophthalmoides* Jessen, Orthotetid fragments and fish debris; the succeeding 30 ft contains a rich and varied fauna of goniatites (including *G. subcrenatum* C. Schmidt), mollusc spat and marine lamellibranchs. To the north of the district at Nant Melyn, one of the headstreams of the River Cynon, similar thick marine beds, nearly 50 ft, are recorded (Leitch and others 1958, p. 474), and the same is true to the west at Cynheidre, 80 ft, and to the east at Rudry, 49 ft (Woodland and others 1957b, pp. 43, 50). Near Glynneath the equivalent beds are only 4 ft thick (Leitch and others 1958, p. 466) and an important sandstone, the 'Farewell Rock', is developed immediately above. It would seem that, except possibly in the north-west, this marine band is present in considerable thickness throughout the Pontypridd district. Despite the great thickness of the band, which contrasts sharply with that in the Pennine province, it is noteworthy that it forms part of a normal cyclic sequence.

Nearly 80 ft of mudstones with sporadic ironstones separate the marine band from a flaggy sandstone 110 ft thick. This sandstone lies at approximately the same horizon as the 'Farewell Rock' of Carmarthenshire (Ware 1939, p. 173) and the head-waters of the River Neath (Jones and Owen 1956, p. 240), where it is about 130 ft thick; eastwards from Margam the sandstone must die out, for it is not represented in the Rudry Borehole (Woodland and others 1957b, p. 50).

The succeeding 120 ft or so of mudstones yield from numerous levels an abundant and rich flora, including such forms as *Alethopteris lonchitica* (Schlotheim), *Mariopteris acuta* (Brongniart), *Neuropteris schlehani* Stur and Sphenopterids, all of which are typical of the Floral Zone C of Dix (1934). Similar floras are recorded from the Nant Llech Plant Beds (Dix 1933, pp. 164–6). These plant beds are overlain by a further 42 ft of mudstone, the lowest 15 ft of which carry

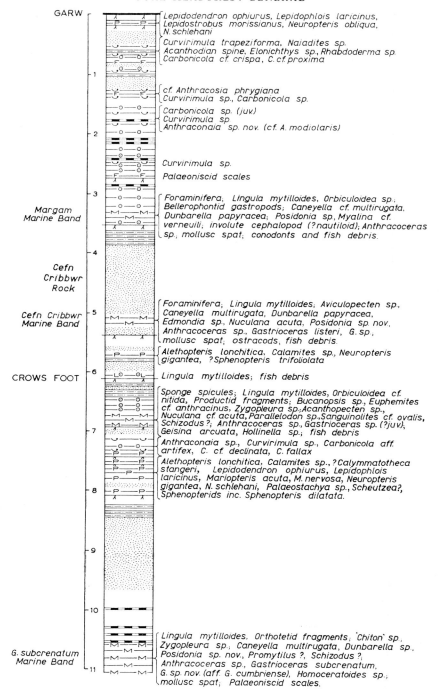

FIG. 9. *Generalized section of the measures between the* Gastrioceras subcrenatum *Marine Band and the Garw Seam; for key, see Fig. 5. Thick broken lines denote ironstones.*

'mussels', including *Curvirimula sp.*, *Carbonicola* aff. *artifex* Eagar, *C. declinata* Eagar and *C. fallax*, associated with sporadic *P. ophthalmoides* and fish debris. A similar fauna occurs in the same general position in the Pennine province, as for example, above the Bassy Mine in Lancashire, and the Two-Foot Coal in North Staffordshire.

The overlying cyclothem is 85 ft thick and culminates in a thin coal believed to be that worked in the Cefn Cribwr area under the name **Crows Foot** or Cors-y-fran. A 25-ft marine band occurs at the base and carries a rich fauna of burrowing and crawling molluscs accompanied by brachiopods, including horny forms and Productids; towards the top of the band a few non-diagnostic goniatites are associated with pyritized sponge spicules, *Lingula mytilloides*, *Hollinella sp.* and a few gastropods. The band is split towards the top by a 1 ft 7 in band in which the marine fauna is restricted and associated with *Geisina* cf. *arcuata* (an ostracod usually considered to be non-marine). It is noteworthy that at Cynheidre two distinct marine horizons separated by a near-seatearth occur in this position (Woodland and others 1957b, p. 42) and on the North Crop the M1 of Leitch and others (1958, pp. 466, 469) is similarly split by a thin sandstone and shale with plants.

At Margam the rest of the cyclothem above the marine band is represented largely by 33 ft of mudstones carrying *Planolites ophthalmoides* throughout. Along the crop east of Kenfig Hill sandstones up to 45 ft thick occur towards the top; these are commonly quartzitic with scattered pebbles. They have been quarried at Cefn Cribwr and Cefn Hirgoed.

The Crows Foot Coal is overlain by a 1 ft 7 in band containing *P. ophthalmoides*, *L. mytilloides* and fish debris. This marine band has been recorded at Cynheidre, where it is similarly impoverished, and along the North Crop, where it is believed to be the M2 horizon of Leitch and others (1958). The rest of the cyclothem consists of about 76 ft of mudstones with plants, interbedded with layers of quartzitic sandstone and siltstone.

The succeeding cyclothem is 185 ft thick. Near the base occurs 20 ft 9 in of dark marine mudstones carrying a varied fauna, and here called the **Cefn Cribbwr Marine Band**[1]. The band commences with a brief *Lingula*-phase followed by various lamellibranchs; towards the middle of the band abundant *Gastrioceras listeri* (J. Sowerby) and *G. sp.* (with faint spiral ornament) appear associated with *Anthracoceras sp.* and lamellibranchs, while the uppermost 8 ft yield only *Lingula* and *Planolites*. A similar fauna was obtained from a quarry and level spoil near Tycribwr Farm, west of Aberkenfig. This marine band is apparently that referred to as the Wernffrwd Marine Beds in West Gower (Jones, S. H. 1935, pp. 318–21), where *G. listeri* is also recorded. Away from the South Crop this goniatite is unknown in South Wales. At Cynheidre the horizon appears to be represented by 12 ft of mudstone with *Lingula* (Woodland and others 1957b, p. 42). On the North Crop its equivalent, the M3 horizon, is widely present though usually thin: at Cwm Gwrelech, for example, *Lingula*, Productids and Pectinids occur over a thickness of 2 ft 2 in (Leitch and others 1958, p. 465). Because of the occurrence of *G. listeri* this band has been correlated with the Bullion Mine and its equivalents in the Pennine province; certain significant differences, however, are seen when the full sequences of the two areas are compared and it is not impossible that the true correlative of the Bullion Mine lies in one or other of the marine horizons close below the Cefn Cribbwr Band.

---

[1] On the recently published one-inch maps this is termed the *Gastrioceras listeri* Marine Band.

The marine band is overlain by 121 ft of quartzitic sandstone, the **Cefn Cribbwr Rock** (Woodland and others, 1957b, p. 57), which range upwards through 33 ft of striped beds, sandy mudstone and seatearth, to a 5-in coal. The sandstone maintains its thickness from Kenfig Hill to Penprysg and has been quarried extensively.

The sequence is continued upwards by 83 ft of argillaceous measures to another thin seatearth. A *Planolites*/plant-debris phase nearly 8 ft thick occurs at the base and is succeeded by some 21 ft of strata containing marine fossils. This band, here termed the **Margam Marine Band**, begins with about 5 ft of mudstone containing foraminifera, horny brachiopods, *Myalina*, Bellerophontid gastropods and fish debris, associated throughout with *P. ophthalmoides*. This is overlain by about 12 ft of mudstones carrying much mollusc spat at top and bottom and abundant foraminifera with *Lingula* and *P. ophthalmoides* just above the middle. A 5-inch layer rich in lamellibranchs, such as *Dunbarella* and *Caneyella*, together with goniatites, chiefly *Anthracoceras sp.*, an involute cephalopod (cf. *Domatoceras*), and conodonts, is followed at the top of the band by 3 ft of dark pyritic mudstone containing only *Lingula*. The mudstones making up the rest of the cyclothem carry abundant *P. ophthalmoides* right up to the base of the overlying seatearth. The marine band crops out in part in a stream bed near Bryn-coch, where it is again characterized by abundant

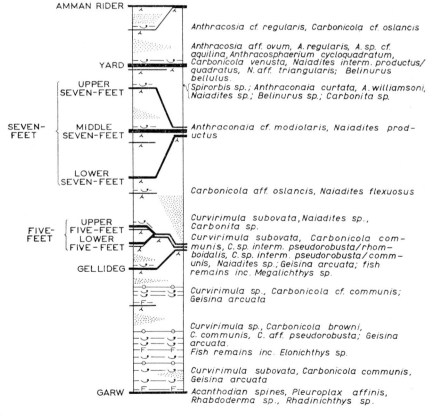

FIG. 10.   *Generalized section of the measures between the Garw and the Amman Rider seams; for key, see Fig. 5*

Anthracoceras and cf. *Domatoceras*. The horizon appears to be represented only by a *Lingula* band on the North Crop (the M4 horizon of Leitch and others) and in the west of the Coalfield, and its absence at Rudry suggests that it dies out eastwards.

The 300 ft of measures upwards to the Garw Seam are distinctive in containing no true seatearths. They are predominantly argillaceous in character, though thin impersistent sandstones characterize the upper portion and ironstones are numerous in the lower part. An abnormally thick cyclothem in this position is typical of the sequence in most British coalfields. *P. ophthalmoides* is common at intervals throughout these measures and several significant horizons of dark cannelly shale with fish debris also occur. One such fish-bed at the base also carries, in the west of the Coalfield, *Lingula* and foraminifera. 'Mussels' occur at numerous levels throughout the cyclothem. The lower bands yield *Curvirimula*. About midway in the group, shells referred to *Carbonicola* aff. *pontifex* occur, and similar forms are present at about the same horizon in the Dulais Valley (Woodland and others 1957b, p. 44). Not far below the Garw *Carbonicola crispa* and *C. proxima* occur and similar forms are known widely in South Wales, where they are generally taken as representing the highest fauna of the *Anthraconaia lenisulcata* Zone (Eagar 1962, p. 330).

(b) Garw to Amman Rider

A generalized vertical section of these measures showing seam-splitting and principal faunas is given in Fig. 10, while the lateral variation and detailed correlation of the coals is illustrated more fully in Fig. 27 and Plate III.

The **Garw** has been proved only at scattered points. About $1\frac{1}{2}$ ft thick at Aberaman, it thickens south-westwards to almost 3 ft near Kenfig Hill, where it has been worked on a small scale as the **Cribbwr Fach.** The roof appears to consist everywhere of dark fissile mudstones yielding much fish debris; sporadic *Planolites ophthalmoides* occur and suggest an approach to marine conditions.

Between the Garw and the Gellideg the measures are mainly argillaceous, though locally, as for example in the Rhondda valleys, thin sandstones are present in the lower part. In thickness they are less variable than most of the Lower Coal Measures sequence, being 80 to 100 ft in the Cynon Valley and reaching a known maximum of 190 ft west of Bryn. A thin coal is usually present not far below the Gellideg and its roof has also yielded *P. ophthalmoides*. As elsewhere (Robertson 1929, 1933) ironstone bands are common throughout and especially close above the Garw, and these were formerly worked at Bryndu on the South Crop and east of Cwmavon. Several bands of dark shale or mudstone with abundant thick calcite-shelled 'mussels' occur, the highest lying about 20 to 40 ft below the Gellideg and the lowest about 60 ft above the Garw. These are the '*pseudorobusta*' bands which yield variants of *Carbonicola communis*, *C. martini* Trueman and Weir and *C. pseudorobusta*, commonly associated with the ostracod *Geisina arcuata*. Between these conspicuous 'mussel'-bands the mudstones carry sporadic *Curvirimula* throughout.

The **Gellideg** (Fig. 11) is the lowest coal normally worked in the major collieries. Barely 2 ft thick in the east and north of the district, it thickens steadily south-westwards to reach about 9 ft at Newlands Colliery near Pyle. Typically it consists of a single clean coal, but in the north, as at Cwmaman and Abergorky, a thin parting lies near the base, while in the southern part of the Rhondda valleys up to 2 ft of bast are present at about the same position. In the Eastern and Western colliery-areas a further parting may be present near the top

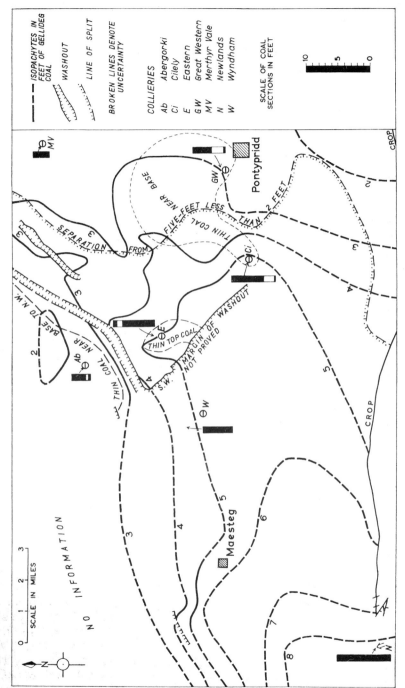

Fig. 11.   *Sketch-map showing the development of the Gellideg Seam*

of the seam. Over large areas the roof is sandy and under these conditions extensive washouts may occur. In general these trend from north-north-east to south-south-west, i.e. roughly at right-angles to the inferred land margin to the north-east. A major washout running from north-west to south-east lies astride the area between Eastern and Western collieries and extends southwards through Britannic Colliery; only its eastern margin has been proved with certainty but the area of barren ground is known to be more than a mile wide. Two distinct faunas are known from areas where mudstone forms the immediate roof: at Fernhill Colliery and Margam *Curvirimula subovata* (Dewar) is common, generally associated with plants carrying attached *Spirorbis*; at Maesteg an abundant fauna consisting of the internal moulds of variants of *Carbonicola communis* and related forms is preserved.

In the extreme north-east of the district, at Mountain Ash and Merthyr Vale, and in the south-east, at Llanharan and Llantrisant, the Gellideg and Five-Feet seams are only a foot or so apart and both are worked in the same roadways. Westwards they split rapidly, and parallel to the line of split a sandstone, often striped and up to 50 ft thick, is developed within the cyclothem. Still farther to the west this sandstone thins and is replaced by striped beds and mudstones with only thin beds of sandstone, the two seams here being usually 40 to 70 ft apart, a thickness rather less than where the sandstone is best developed.

The **Five-Feet** has a complex section and is subject to widespread splitting. The main elements of its seam-structure are shown in Fig. 12. At its simplest in the extreme east of the district it consists essentially of a single coal about 4 ft thick. To the west thin partings develop, and in the central Rhondda area it consists of two main leaves, each 2 to $2\frac{1}{2}$ ft thick, with a thin coal lying within the intervening dirt, and further thin coals present at the top and bottom of the complete section. In the north of the Rhondda Fawr the two coals split several feet apart. The Lower and Upper Five-Feet coals are clearly separate in the Ogmore Valley. At Maesteg and in the areas to the south and south-west further partings develop in both the main coals, while the thin middle coal of the Rhondda area thickens and forms a distinct seam. The roof of the Five-Feet rarely carries a fauna, but sporadic *Curvirimula* have been recorded from the collieries in the north-east of the district and on the South Crop at Llanharan.

Over much of the district the Upper Five-Feet and the lowest coal of the Seven-Feet group are separated by 20 to 30 ft of strata, usually argillaceous, but containing thin sandstones towards the base, especially in the east. At Abergorky and again west of Bryn the Upper Five-Feet and the Lower Seven-Feet are less than 2 ft apart and the two coals have been worked over small areas as a single seam.

The **Seven-Feet** group of seams behaves in a complicated fashion; representative sections of the group are given in Fig. 13 and these summarize the splitting and behaviour of the individual coals. Only in the Cynon Valley north of Mountain Ash are all the coals united so as to form a single seam. Three main leaves, totalling in all 5 to $5\frac{1}{2}$ ft, are clearly represented and this three-fold character remains after the seam splits. The top leaf, about 1 ft thick, and the bottom leaf, varying from 1 to 2 ft, split away southwards and westwards, and in the area south-west of Maesteg the three coals are spread over some 100 ft of measures. In much of the lower Rhondda valleys the Middle and Upper Seven-Feet coals are worked together as one seam; within a limited area around Cwmaman the same is true for the Lower and Middle coals. Over the rest of the

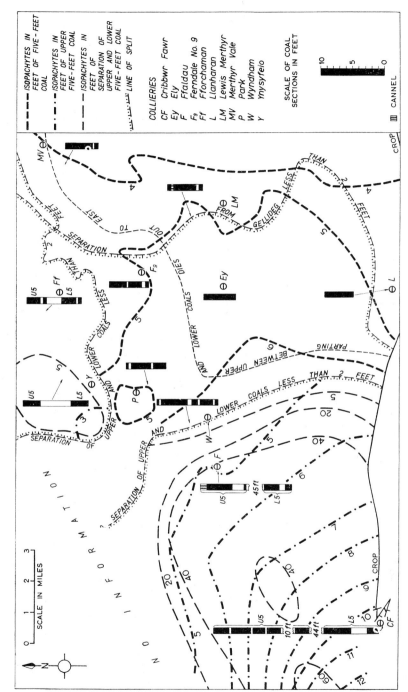

FIG. 12.   Sketch-map showing the development of the Five-Feet Seam; U5, Upper Five-Feet; L5, Lower Five-Feet

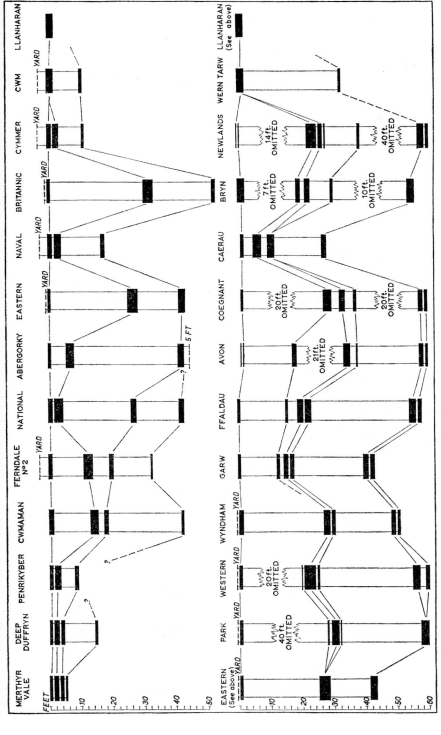

FIG. 13. *Vertical sections of the Seven-Feet group of seams. The Yard and Five-Feet seam-positions are shown only where these are closely associated*

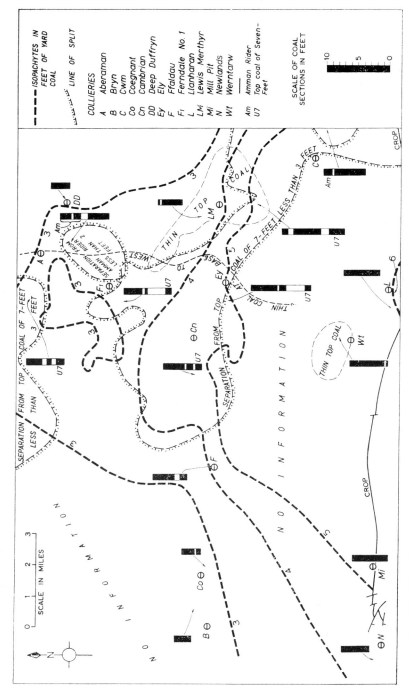

FIG. 14.   Sketch-map showing the development of the Yard Seam. In the coal sections the Amman Rider and the Upper
Seven-Feet coals are shown only where they are closely associated

district the three leaves form distinct seams of which only the Middle is in general workable; even this is of little value in the Maesteg area since it develops up to three thick seatearth partings. The south-westwards splitting is accompanied by a thickening of all the coals: around Maesteg and the western South Crop the Lower Seven-Feet is up to 3 ft thick and has been worked on a small scale; the Middle Seven-Feet contains up to 5 ft of coal; and the Upper Seven-Feet is 2 ft or so thick. Over considerable areas the Upper Seven-Feet lies in the floor of the overlying Yard Seam, and in the south-east of the district, around Cwm and Llanharan, the Lower Seven-Feet appears to die out, while the Middle and Upper seams combine to form a single coal only about 2 ft thick. Few fossils have ever been recorded from any of the Seven-Feet coals. In the Cynon Valley *Anthraconaia* cf. *curtata* (Brown) and *A. williamsoni* occur sporadically.

The Seven-Feet and the Yard are never far apart; indeed, over much of the southern Rhondda area the Upper Seven-Feet is worked as the bottom coal to the Yard, and at Fernhill Colliery the two seams are similarly associated. Elsewhere the separation is rarely more than 20 ft, except at Penrikyber Colliery where about 60 ft has been recorded. The strata are almost wholly argillaceous.

The **Yard** (Fig. 14) varies in thickness from about 2 to 6 ft, its thickest development being in the south-east Rhondda area and along the South Crop. Over much of the area the seam consists of two leaves, an upper one 2 to $3\frac{1}{2}$ ft, separated from a lower, 3 in to 1 ft 9 in thick, by up to 12 in of clod or rashings. Where the seam is thickest there is no separate bottom coal and it is presumed that the two coals have coalesced. A rich diagnostic fauna of *Anthracosia regularis* and *Carbonicola venusta* variants is present in much of the district.

Locally the **Amman Rider** lies close upon the Yard. At Deep Duffryn Colliery the separation is only a few inches and the two coals have been worked together as a single seam over a small area, where the multiple section was confused with that of the Seven-Feet; at Cwm Colliery the Amman Rider appears in the rippings of the Yard workings. Elsewhere varying amounts of mainly argillaceous strata separate the two seams, and in the area between Maesteg and the South Crop 80 to 100 ft of measures may be developed. The Amman Rider itself is a rather dirty sulphurous seam, and it has hardly ever been worked economically. In the south-east at Llanharan a single coal 10 to 24 in is present, but in the north-east the seam consists typically of two leaves, the upper one 12 to 22 in and the lower 8 to 14 in. Traced south-westwards these split and are, for example, 20 ft apart at Dare and 40 ft in the Garw Valley. At Maesteg and on the western South Crop the upper coal develops partings and presents typically a three-leaf section.

MIDDLE COAL MEASURES

For convenience of description the Middle Coal Measures have been divided into two groups at the Cefn Coed Marine Band.

The lower group, which comprises the upper part of the *Anthraconaia modiolaris* Zone and the Lower *Anthracosia similis–Anthraconaia pulchra* Zone, includes the bulk of the main group of worked seams. The generalized section of strata and the principal faunas are shown in Fig. 15, which also summarizes the splitting of the coals. The group thickens westwards from less than 300 ft in the Abercynon area and around Llantrisant to between 600 and 700 ft in the Avan Valley and 800 ft in the extreme south-west (see Fig. 7). Apart from the marked development of coal the measures are predominantly argillaceous; sandstones occur throughout the sequence but they are mostly impersistent. Non-marine

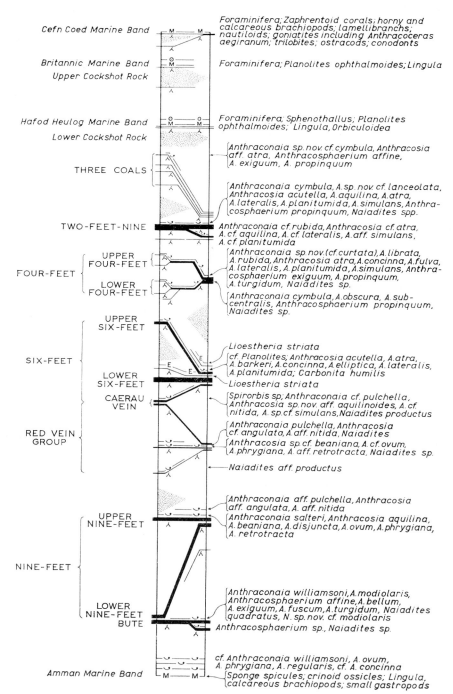

Cefn Coed Marine Band — Foraminifera; Zaphrentoid corals; horny and calcareous brachiopods; lamellibranchs; nautiloids; goniatites including Anthracoceras aegiranum; trilobites; ostracods; conodonts

Britannic Marine Band — Foraminifera; Planolites ophthalmoides; Lingula
Upper Cockshot Rock

Hafod Heulog Marine Band — Foraminifera; Sphenothallus; Planolites ophthalmoides; Lingula, Orbiculoidea
Lower Cockshot Rock

THREE COALS — Anthraconaia sp. nov. cf. cymbula, Anthracosia aff. atra, Anthracosphaerium affine, A. exiguum, A. propinquum

Anthraconaia cymbula, A. sp. nov. cf. lanceolata, Anthracosia acutella, A. aquilina, A. atra, A. lateralis, A. planitumida, A. simulans, Anthracosphaerium propinquum, Naiadites spp.

TWO-FEET-NINE — Anthraconaia cf. rubida, Anthracosia cf. atra, A. cf. aquilina, A. cf. lateralis, A. aff. simulans, A. cf. planitumida

FOUR-FEET
UPPER FOUR-FEET — Anthraconaia sp. nov. (cf. curtata), A. librata, A. rubida, Anthracosia atra, A. concinna, A. fulva, A. lateralis, A. planitumida, A. simulans, Anthracosphaerium exiguum, A. propinquum, A. turgidum, Naiadites sp.
LOWER FOUR-FEET — Anthraconaia cymbula, A. obscura, A. subcentralis, Anthracosphaerium propinquum, Naiadites sp.

SIX-FEET
UPPER SIX-FEET

Lioestheria striata
cf. Planolites; Anthracosia acutella, A. atra, A. barkeri, A. concinna, A. elliptica, A. lateralis, A. planitumida; Carbonita humilis
LOWER SIX-FEET — Lioestheria striata
CAERAU VEIN — Spirorbis sp; Anthraconaia cf. pulchella, Anthracosia sp. nov. aff. aquilinoides, A. cf. nitida, A. sp. cf. simulans, Naiadites productus

RED VEIN GROUP — Anthraconaia pulchella, Anthracosia cf. angulata, A. aff. nitida, Naiadites
Anthracosia sp. cf. beaniana, A. cf. ovum, A. phrygiana, A. aff. retrotracta, Naiadites sp.

Naiadites aff. productus

Anthraconaia aff. pulchella, Anthracosia aff. angulata, A. aff. nitida
NINE-FEET
UPPER NINE-FEET — Anthraconaia salteri, Anthracosia aquilina, A. beaniana, A. disjuncta, A. ovum, A. phrygiana, A. retrotracta

LOWER NINE-FEET — Anthraconaia williamsoni, A. modiolaris, Anthracosphaerium affine, A. bellum, A. exiguum, A. fuscum, A. turgidum, Naiadites quadratus, N. sp. nov. cf. modiolaris
BUTE — Anthracosphaerium sp., Naiadites sp.

Amman Marine Band — cf. Anthraconaia williamsoni, A. ovum, A. phrygiana, A. regularis, cf. A. concinna
Sponge spicules; crinoid ossicles; Lingula, calcareous brachiopods; small gastropods

FIG. 15. *Generalized section of the measures between the Amman and the Cefn Coed marine bands showing lithology and faunas*

lamellibranchs are found on most of the major seams and have been used extensively in correlation (see pp. 25–6): marine strata are confined to a single band at the base of the group, and to two, sometimes three, horizons in the uppermost 50 to 150 ft.

The upper group (Fig. 24), less than 300 ft thick in the Llantrisant area and more than 800 ft in the south-west, corresponds to the Upper *similis-pulchra* Zone. Although sandstones may be locally important, particularly in the higher part of the group, few are continuous throughout the district. Coals occur at intervals but they are not comparable in thickness with those of the lower group; most of them have been worked locally, but none maintains an economic section regionally. 'Mussel' bands are developed at only one or two levels, and more important as aids to correlation are the marine and *Euestheria* horizons which are found at fairly regular intervals.

(a) Amman Marine Band to Cefn Coed Marine Band

These measures underlie by far the greater part of the district; their outcrop is limited to comparatively narrow strips along the South Crop and on the north side of the Moel Gilau Fault around and to the west of Maesteg.

The **Amman Marine Band** (S. H. Jones 1934, p. 429–30) is virtually ubiquitous and is of particular significance since it comprises the only known marine strata associated with the main group of coals. Over most of the district it varies from 1 to 3 ft in thickness, but it may reach 6 ft or more in the eastern collieries and 8 ft in the west. Over large areas it contains little except *Lingula mytilloides* in fissile silty mudstones. In the Cynon Valley and in much of the southern Rhondda area these *Lingula* shales are underlain by blocky mudstones containing abundant pyritic burrows, 'fucoids' and granules, associated with sponge spicules, rare gastropods and a few lamellibranchs; still farther south-eastwards, at Cilely, Cwm and Llanharan collieries, calcareous brachiopods and crinoid columnals appear and presage the development of the rich calcareous faunas of the south-eastern parts of the Coalfield (Moore 1945, p. 175).

The marine strata are overlain by mudstones which are generally about 30 ft thick, though they are as little as 12 ft around Penrhiwceiber and Ynys-y-bwl and as much as 70 ft in the south-west. In the lowest few feet these mudstones contain abundant though stunted 'mussels', characterized nearly everywhere by a 'ghost'-like preservation; typical forms are cf. *Anthraconaia williamsoni*, and variants of *Anthracosia ovum*, *A. phrygiana* and *A. regularis*. Here and there solid shells resembling *A. concinna* occur. Locally to the south of Maesteg a thin coal occurs about 10 ft below the Bute and may be a split from that seam.

The **Bute** (Fig. 16) is everywhere of considerable economic importance. Over most of the Rhondda valleys and the areas to the east and south it consists of two distinct leaves separated by a parting more or less in the middle of the seam. The parting thickens steadily to the north-east; in the Rhondda Fach it has increased to about 17 in and in the Cynon Valley is thick enough to make the seam unworkable; genuine splitting takes place in the Taff Valley and at Merthyr Vale the two leaves are separated by as much as 15 ft of mudstone, which carry at their base a fauna of small *Anthracosphaerium* and *Naiadites*. To the west and south, into the Avan and Llynfi valleys and on the South Crop, the parting dies out and the seam consists of clean coal. The overall coal thickness varies from 3 ft in the north-west in the Glyncorrwg and Glyncastle colliery-areas, to rather more than 6 ft to the south of the Rhondda Valley; over most of the area the thickness varies from 4 to 5½ ft. Locally in the northern parts of the Rhondda Fach a thin

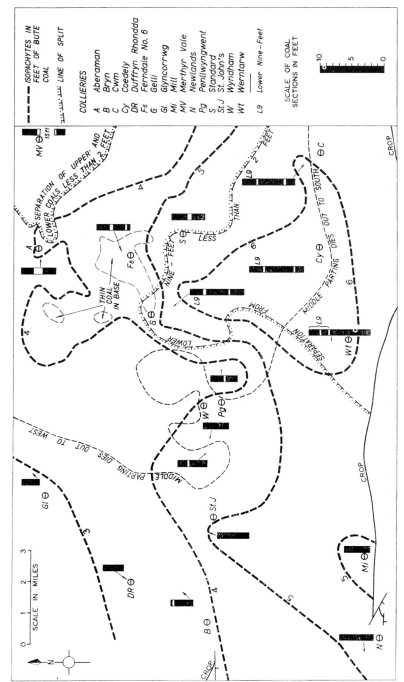

FIG. 16.   *Sketch-map showing the development of the Bute Seam. The Lower Nine-Feet is shown only where closely associated*

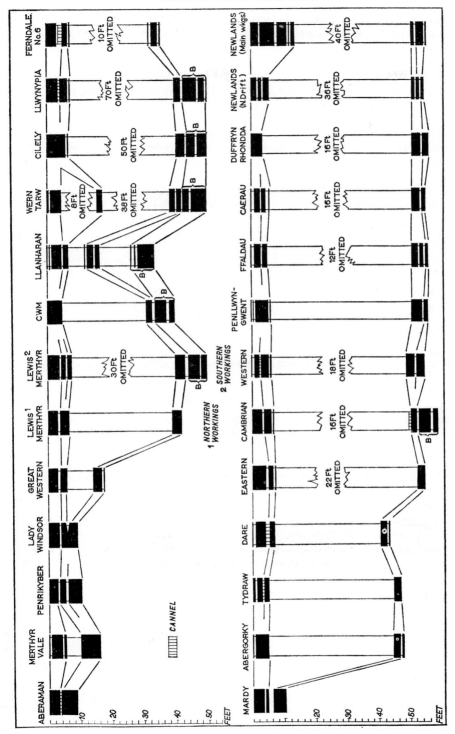

Fig. 17. *Vertical sections of the Nine-Feet group of seams. The Bute Seam (B) is shown only where closely associated*

coal is developed in the base of the seam, and a similar leaf of inferior coal appears at the base of the section in the Pontycymmer and Maesteg areas.

In much of the area south and south-east of Gelli the Lower Nine-Feet lies close above the Bute and the two coals were worked as one seam in the Cilely area. Elsewhere the two seams are distinct, with upwards of 50 ft of measures developed between them. The roof of the Bute is almost barren in the eastern areas, but in the west it carries an exceptionally rich and characteristic 'mussel' fauna. In the upper Rhondda Fawr and in the Corrwg the shells are all preserved as 'solids' and lie in about 12 in of dark mudstone in the immediate roof of the seam. South-westwards from Blaen-gwynfi to Kenfig Hill up to 5 ft of cannelly mudstones intervene between the 'mussel'-band and the coal; these, too, carry abundant shells, which although crushed, appear to consist essentially of the same forms as the solid fauna. Variants of *Anthraconaia williamsoni, A. modiolaris,* cf.*Anthracosia regularis, Anthracosphaerium affine, A bellum* (Davies and Trueman), *A. exiguum, A. fuscum* (Davies and Trueman), *A. turgidum, Naiadites quadratus* (J. de C. Sowerby) predominate together with *N. sp. nov.* cf. *modiolaris* Hind. The overlying beds consist mainly of mudstones, though thin impersistent sandstones may be present and locally cut out the 'mussel'-band. In the south-western part of the district and in the upper Rhondda Fawr a thin coal occurs some 10 to 20 ft below the Lower Nine-Feet.

A major split develops in the **Nine-Feet** Seam (Figs. 17, 18, 19), the line of the 2-ft parting running north-westwards from Pontypridd through the eastern workings of the Ferndale collieries to Maerdy, thence westwards to the area south of Fernhill and Glyncorrwg and into the north-western parts of the district. To the north of this line the single seam has an undisturbed thickness of 7 to 10 ft; but over wide areas it is so affected by intense structural deformation resulting from incompetence that no normal section can be recognized, and the seam may vary in thickness from virtually nothing up to 30 ft. Under these conditions the disturbed seam is associated with considerable developments of black rashings which have resulted from the complete breakdown of a highly carbonaceous roof-shale. These structural difficulties have hitherto precluded large scale development of the coal. A prominent cannel parting near the middle of the seam extends as far west as Fernhill, and continues southwards in the Upper Nine-Feet (see below). The split develops rapidly on another, and lower, dirt parting which thickens until, throughout much of the district, the Lower and Upper Nine-Feet coals are 40 to 60 ft apart, and locally the separation is more than 100 ft.

The **Lower Nine-Feet** (Fig. 18) has been worked extensively. In the collieries of the southern Rhondda Fawr and southwards towards Llanharan and Llantrisant it unites with the Bute Seam (see above). The coal is about 3 ft thick throughout most of the Rhondda area; it thickens south-westwards into the Garw and Llynfi valleys and it thins again to about 4 ft at Newlands. The thickening of the seam west of the Rhondda is accompanied by the development, 6 to 24 in above the base, of a parting, which increases southwards to produce genuine splitting of the seam. Thus at Bryn the lower leaf lies several feet below the upper; at Newlands what appears to be this bottom coal is 16 ft below the main Lower Nine-Feet. West of the Ogmore Valley further thin partings are present in the upper coal. Up to 18 in of cannel occur associated with the main parting in the Maesteg area, while at Western and Cambrian the roof of the coal carries up to 15 in of cannel.

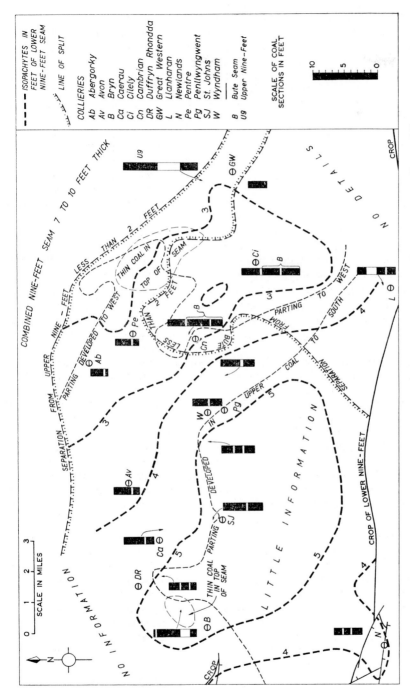

Fig. 18. *Sketch-map showing the development of the Lower Nine-Feet Seam. The Upper Nine-Feet and Bute seams are shown only where closely associated*

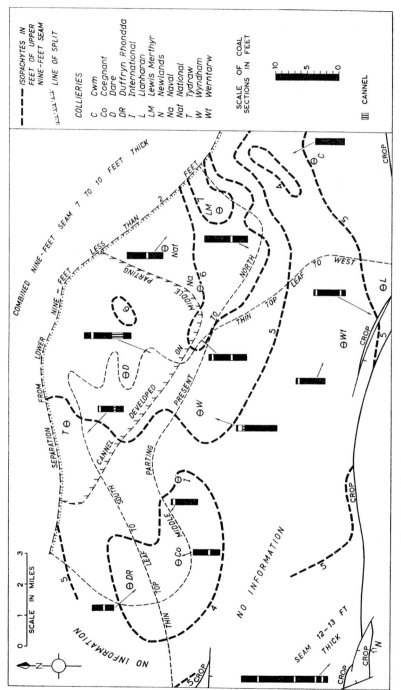

FIG. 19.　*Sketch-map showing the development of the Upper Nine-Feet Seam*

The measures between the Lower and Upper Nine-Feet coals consist mainly of sandy and silty mudstones, and sandstone bands are frequent. A thin coal lies close below the Upper Nine-Feet in the Upper Rhondda and at Newlands. These measures are almost devoid of animal fossils.

The **Upper Nine-Feet** (Fig. 19) is liable to internal disturbance and has not been so widely worked. It is thickest in the Porth area, where it exceeds 7 ft; over most of the district a thickness of around 5 ft is more normal, though in the Maesteg area it thins to less than 4 ft. At Newlands Colliery it has been worked extensively south of the Newlands Thrust, with an abnormally thick but consistent section of 12 to 13 ft of coal with several partings; about half a mile north of the thrust a more normal thickness of about 5 ft is recorded, and no explanation can be offered at present for the very sudden increase in thickness of the seam south of the thrust. Over much of the northern part of the district a parting, usually associated with cannel, appears in the middle of the seam. The cannel is 1 to 2 ft thick in the Rhondda valleys, and is clearly the same as that recorded in the combined Nine-Feet coal to the north-east. South-westwards into the Ogmore and Garw areas a parting, also locally associated with cannel, develops near the top of the seam. Near Maesteg the uppermost few inches of coal split away and a thin cannel is left in the roof of the main seam. In the area between the Rhondda Fawr and Ogmore valleys a thin bottom coal is recorded, and at Werntarw this may be represented by the 24 in of coal which lies 8 ft below the main Upper Nine-Feet.

In the north-east of the district the Nine-Feet roof, where it consists of dark carbonaceous shale, carries an abundant fauna of crushed and semi-crushed large 'mussels', among which have been recognized variants of *Anthraconaia salteri* (Leitch), *Anthracosia aquilina*, *A. beaniana*, *A. disjuncta*, *A. ovum*, *A. phrygiana* and *A. retrotracta* (Wright). These are overlain by paler blocky mudstone carrying the 'Red Vein fauna' of small solid 'mussels', such as *Anthraconaia* aff. *pulchella*, *Anthracosia* aff. *angulata* and *A.* aff. *nitida*.

The measures between the Nine-Feet and the Six-Feet are amongst the most variable in the sequence (Fig. 20), due in the main to the irregular behaviour of the Red Vein group of coals. Usually some 60 to 90 ft of strata are involved, but they may reach as much as 130 to 140 ft in the Clydach Vale–Cwm-parc–Llwyn-y-pia area, and as little as 20 ft at Coedely. However, this statement of variation is not as straightforward as it would seem, for at Cwmaman, where the separation between the Nine-Feet and Six-Feet is about 80 ft, the coals of the Red Vein have coalesced to form a single compound seam about 50 ft above the Nine-Feet. At Western Colliery, on the other hand, the separation is about the same, but the top leaves of the Red Vein lie within 3 ft of the Six-Feet while a thin coal which may represent the bottom leaf of the Cwmaman Red Vein is found only a few feet above the Nine-Feet.

The composite single **Red Vein** of the Maerdy–Cwmaman area totals about 3 to 5 ft of coal in three, rarely four, leaves. Southwards and south-eastwards the component coals split apart and thin markedly, and in the Mountain Ash area they are reduced to insignificance. After splitting in the central Rhondda the upper two leaves again unite on the western side of the Rhondda Fawr and make an important seam, here called the **Caerau**. This thickens steadily through the Ogmore, Garw and Llynfi valleys, where the double seam reaches 5 to 6 ft: the lower leaf of 33 to 41 in is separated from the upper leaf of 15 to 30 in by an inch or so of rashings. This well-developed seam always lies close below the

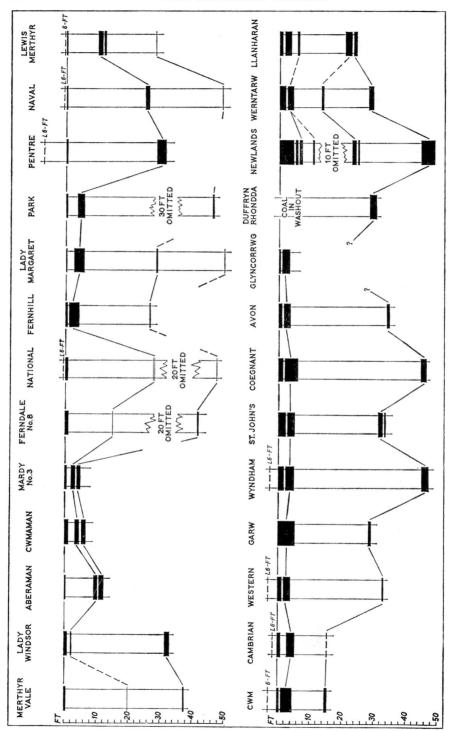

FIG. 20.   Sections showing variation of the Red Vein group of coals

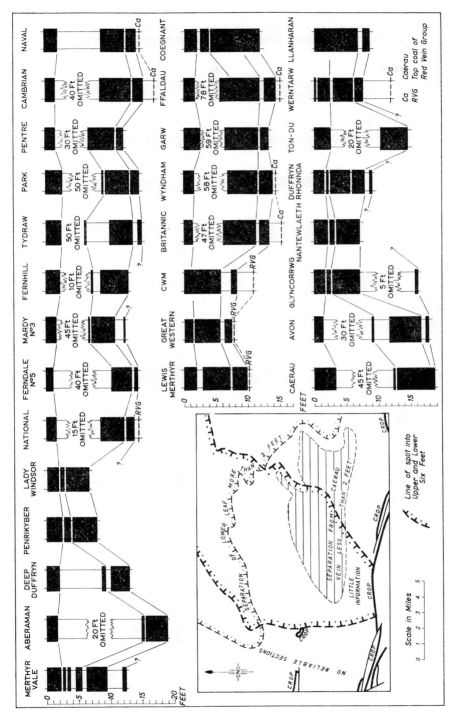

Fig. 21. Sections in the Six-Feet Seam

Six-Feet, and the separation is less than 3 ft over much of the area between the Garw Valley and Llantwit Fardre. One or two thin coals developed between the Nine-Feet and the Caerau throughout this central area represent the lower coals of the Red Vein group.

On the South Crop the Red Vein group comprises two distinct seams, which in the west show marked thickening. The **Caegarw** (Caerau) of Newlands totals $7\frac{1}{2}$ ft of coal with several dirt partings in the lower part: at Llanharan a double coal 4 to 5 ft thick includes 8 to 21 in of rashings in the middle. The lower seam, here, for the sake of convenience, retaining the name Red Vein, varies from $5\frac{1}{2}$ ft at Newlands to about 3 ft at Llanharan, and has two leaves at both localities.

The 'Red Vein fauna' of small Anthraconaioid *Anthracosia* occurs at various horizons within the group. At Western Colliery and certain of the northern Rhondda pits it is associated with an underlying development of the 'Nine-Feet' *A. phrygiana/aquilina* fauna on the second thin coal above the Nine-Feet. The first rider seam above the Nine-Feet in the area between the Ogmore and Rhondda Fach valleys carries a characteristic fauna of solid *Naiadites* of the *productus* (Brown)/*subtruncatus* (Brown) group.

The **Six-Feet** (Fig. 21) is probably the most important coal horizon in the district; almost certainly it has been worked on a greater scale than any other seam, and in many of the collieries has been completely exhausted. Major splitting is again an important feature and two distinct seams, the Lower and Upper Six-Feet, are present over large areas. A single compound seam exists to the south-east of a line drawn through Merthyr Vale, Mountain Ash, Wattstown, Dinas and Penrhiw-fer to Bryncethin. In most of this area a three-coal section is typical: top coal 22 to 28 in, parting up to 24 in, main coal 36 to 52 in, parting up to 7 in, bottom coal 12 to 24 in. In the Cynon Valley a further parting occurs about 9 to 15 in from the top of the main coal, while in this area also the bottom coal appears to have failed. The parting between the main and bottom coals is reduced to a mere plane in the Cilely–Porth–Pontypridd area; and in the extreme east and south-east, at Cwm and Llanharan collieries, that between the top and main coals dies out altogether, the seam consisting of $5\frac{1}{2}$ to 7 ft of clean coal overlying a bottom coal of 12 to 15 in.

The major split in the seam develops between the main and top coals, and the separating measures, which are mainly argillaceous, increase rapidly in thickness to 60 ft at Mardy and Dare collieries, and 50 to 70 ft in the Ogmore and Garw valleys. The sections of both split seams remain remarkably similar to the corresponding portions of the full seam just described. The **Lower Six-Feet** consists for the most part of a main coal, 48 to 66 in, and a bottom coal 16 to 30 in. In the northern parts of the Cynon and Rhondda valleys a parting is present 2 to 9 in from the top of the main coal, and is evidently the same as that already noted in the lower reaches of the Cynon Valley; this parting increases to about 2 to 5 ft in the Tydraw area. The **Upper Six-Feet** everywhere consists of clean coal usually about 20 to 26 in. in thickness, but locally in the Llynfi Valley it attains 40 in. The two seams reunite on the western side of the Llynfi and in the Corrwg and lower Avan valleys: the combined section is generally similar to that of the eastern parts of the district, though the main coal is rather thicker.

Because of structural complications the exact nature of the Six-Feet on the western reaches of the South Crop is obscure. It appears to consist mainly of a single seam about 6 ft thick, but local splitting has been observed during opencast operations in the Cefn area, west of Aberkenfig.

Over the northern part of the area where the Lower Six-Feet has a separate existence, its roof yields fragmentary 'mussels' and abundant *Lioestheria striata*. In most of the southern half of the district the Upper Six-Feet (both where it is separate and combined with the lower seam) carries in its immediate roof a striped black ferruginous mudstone, akin to blackband, which contains a characteristic fauna of crushed and semi-crushed *Anthracosia*: variants of *A. acutella*, *A. aquilina*, *A. atra*, *A. barkeri* Leitch, *A. concinna*, *A. elliptica*, *A. lateralis*, *A. planitumida* are associated with the ostracod *Carbonita humilis* (Jones and Kirkby). Farther north in the central Rhondda area the mudstone is only slightly ferruginous and well-preserved 'solids' of the same species abound. In the extreme north the ferruginous facies is absent; the rich fauna fails and is replaced by sporadic small shells locally associated with a small distinctive variety of *Planolites ophthalmoides*. At Glyncastle Colliery an inch or two of coal lie close above the composite seam and unite with it at Ynisarwed in the Neath Valley; this thin coal also carries abundant *L. striata*.

The measures up to the Four-Feet may be as little as 12 ft thick in the north-east and as much as 90 ft in the Garw Valley and at Llanharan. They are largely argillaceous, though impersistent sandstones are locally present in the northern and western areas.

The **Four-Feet** comprises yet another compound group of coals subject to complex splitting (Fig. 22). In the northern collieries of the Cynon Valley a single clean coal $5\frac{1}{2}$ to 6 ft thick has been worked long ago to virtual exhaustion; traced southwards and westwards this coal develops three distinct partings, and throughout the rest of the Cynon area, in the Rhondda Fach, and in the eastern parts of the Rhondda Fawr the combined four-leaved seam has been worked extensively. In the Rhondda Fach the four leaves vary as follows: (top) 12 to 21 in, 16 to 36 in, 27 to 34 in, and (bottom) 12 to 16 in; the partings vary up to 24 in. in thickness. In the north of the Rhondda Fawr, the Avan Valley and the northern collieries of the Garw and Llynfi valleys a single complex seam is again found, though throughout this area the section is more variable, and in places, as at Avon Colliery, comprises up to 12 ft of coal with numerous dirt partings. Over most of the southern Rhondda Fawr, the southern Llynfi and throughout the South Crop two distinct seams separated by as much as 40 ft of mudstone represent the Lower and Upper Four-Feet seams; the sections are variable but the four-fold make-up of the combined seam in the Rhondda Fach and Cynon valleys appears to be recognizable, the split having taken place between the two middle coals of the combined seam.

In the Duffryn Rhondda area thin streaks of coal appear in the seatearth of the combined seam and these thicken and split away when traced southwards through the Llynfi area to the South Crop: at Celtic Colliery the lowest is as much as 70 ft below the main Lower Four-Feet. In the Clydach Vale, Gilfach Goch and Ogmore Vale areas the Four-Feet coals split completely: four individual seams, thinner than their counterparts to the east, are spread over 50 to 80 ft of strata.

Along the South Crop and adjacent areas the Lower Four-Feet roof yields a rather poor fauna of *Anthraconaia cymbula*, *A. obscura* (Davies and Trueman), *A. subcentralis* (Salter), *Anthracosphaerium propinquum* and *Naiadites sp.* The top coal of the group, however, carries a highly diagnostic, abundant and varied fauna over almost the entire district; the well-preserved solid 'mussels' include variants and intermediates of the following species: *Anthraconaia sp. nov.* (cf. *curtata*), *A. cymbula*, *A. lanceolata*, *A. librata*, *A. rubida* (Davies and Trueman),

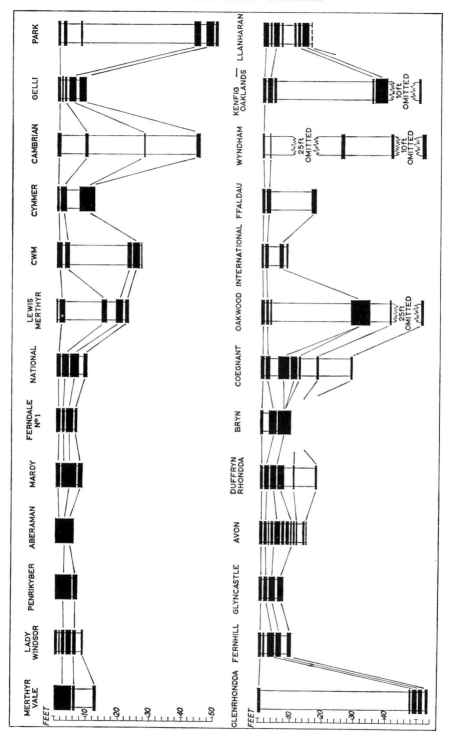

FIG. 22. *Vertical sections of the Four-Feet group of seams*

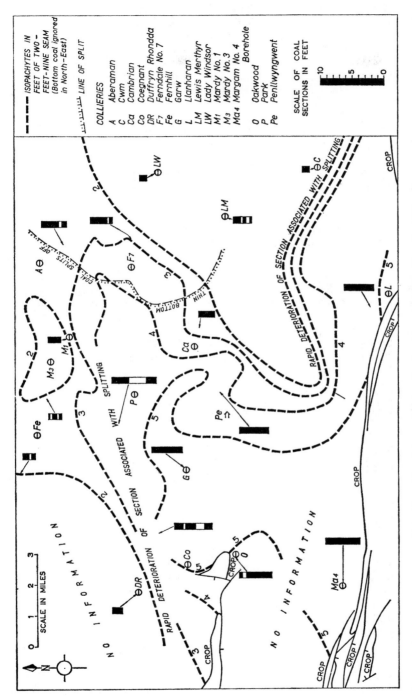

Fig. 23. *Sketch-map showing the development of the Two-Feet-Nine Seam*

Anthracosphaerium exiguum, *A. propinquum*, *A. turgidum*, *Anthracosia acutella*, *A. aquilina*, *A. atra*, *A. concinna*, *A. fulva* (Davies and Trueman), *A. lateralis*, *A. planitumida* and *A. simulans*. The blocky mudstone with these 'solids' is often underlain by a few inches of smooth dark shale containing crushed shells, mainly *Anthracosia* and *Naiadites* preserved as mere films.

The **Two-Feet-Nine** (Fig. 23) is separated from the top coal of the Four-Feet by some 20 to 50 ft. The seam is well developed in the Llynfi, Garw and Ogmore valleys and on the western side of Cambrian Colliery, and throughout this area it has been extensively worked; it consists of 4 to 5½ ft of clean coal with a further 1 to 1½ ft lying 6 in to 5 ft below, and up to 12 in of cannel may be developed in the roof. In the Llynfi Valley there is a plane-parting 9 to 18 in below the top of the seam and to the north-west of Coegnant and Caerau another parting develops near the middle of the main coal and thickens rapidly northwards. This line of split can be traced north-eastwards from Bryn to the Maerdy area: to the north-west the upper leaf is generally thin but it reaches a maximum of about 3 ft around Treherbert; the lower leaf remains associated with its bottom coal and in the north of the Rhondda valleys its roof has yielded many 'mussels' preserved in irony mudstone and reminiscent of the fauna of the Penny-pieces in the areas to the north-west (Ware 1930, p. 473). Characteristic forms are: *Anthraconaia* cf. *rubida*, *Anthracosia* cf. *atra*, *A*. cf. *aquilina*, *A*. cf. *lateralis*, *A*. aff. *simulans*, and *A*. cf. *planitumida*.

South of the line of split the seam has been widely mined in the Rhondda valleys north of Llwyn-y-pia and Wattstown; in this area a 3 to 4½ ft coal has one or two thin leaves in the floor. Southwards from Ferndale and Llwyn-y-pia and eastwards from Cambrian the seam suffers very rapid thinning and through most of the collieries of the southern Rhondda it is less than 2 ft thick. Southwards and westwards from Cwm Colliery it improves just as quickly and at Coedely 4½ ft of coal was worked, while at Llanharan and Llantrisant the section varies between 5½ and 6½ ft of clean coal. In this latter area cannel is again prominent at the top of the seam, and this passes upwards into several feet of black cannelly mudstone overlain by grey shelly mudstones.

Along the rest of the South Crop the seam is also thick. In the Heol-y-cyw area it has been proved by Opencast exploration to be 4 to 4½ ft, while, although intensely disturbed, it has also been extensively worked by opencast methods between Aberkenfig and Margam.

An abundant and varied fauna of small to medium-sized 'mussels' characterizes the grey mudstones for several feet above the seam along the South Crop and in the mid-Glamorgan valleys as far north as Avon and Park. Characteristic forms include: *Anthraconaia cymbula*, *A. sp. nov.* cf. *lanceolata*, *Anthracosia acutella*, *A. aquilina*, *A. atra*, *A. concinna*, *A. lateralis*, *A. planitumida*, *A. simulans* *Anthracosphaerium propinquum* and *Naiadites spp.* At Cambrian Colliery a rider coal (the exact correlation of which is not known), 7 to 8 in thick, is recorded 3 to 4 ft above the main coal. This carries a 'mussel' fauna in blackband similar to that from the Six-Feet: *Anthracosia atra*, *A. barkeri* and *A. fulva* are especially abundant.

The measures above the Two-Feet-Nine are poorly exposed along their outcrop and they can be seen in only a few of the collieries. They are only 60 ft thick at Cwm Colliery and rarely exceed 150 ft in the eastern parts of the district generally: westwards they expand to about 300 ft beyond Bryn and at Margam Park. Coals are rarely of great significance and they are seldom named. In the

Cynon Valley and the north of the Rhondda Fach an inferior composite seam of three to five leaves—the **Three Coals**—lies 2 to 50 ft above the Two-Feet-Nine. Westwards and south-westwards it splits and its behaviour is variable: upwards of four or five thin coals may be spread over 50 ft of strata; in places some of the coals fail altogether and at Llanharan only one coal is found between the Two-Feet-Nine and the Hafod Heulog Marine Band coal. In a few localities individual coals are named; as for example, the Fireclay and Coal Tar seams of Maesteg and the Lantern of the Cefn and Kenfig Opencast sites.

At the western end of the South Crop 'mussel' faunas closely comparable with that of the Two-Feet-Nine occur at several horizons including the roof of the Lantern; while at Cwm Cynon Colliery a varied fauna, including *Anthraconaia sp. nov.* cf. *cymbula, Anthracosia* aff. *atra* and abundant variants of *Anthracosphaerium affine, A. exiguum* and *A. propinquum,* occurs in the roof of the Three Coals.

Above the Three Coals and its equivalents a compact quartzitic sandstone, locally with a sharp base and showing signs of local erosion of the underlying beds, develops within and to the west of the Rhondda Fawr; in the Llynfi Valley it is 15 to 25 ft thick and is named the **Lower Cockshot Rock.** Immediately or close above lies a coal, which is usually thin but exceeds 2 ft in parts of the Llynfi and Ogmore valleys. The marine mudstone which forms the roof of this coal is known from several localities in the southern part of the district, and has been named the **Hafod Heulog Marine Band** (see revised six-inch maps of Pontypridd Sheet, 1957–60, e.g. Glam. 34 s.w.), after the farm 2 miles north-east of Kenfig Hill, near where it crops out. At Britannic Colliery the band is almost 11 ft thick, at Margam Park 7½ ft, at Penllwyngwent it has thinned to 3 ft and at Llanharan it is less than 2 ft. The fauna contains little other than horny brachiopods and foraminifera, associated with *Planolites ophthalmoides; Orbiculoidea* up to half an inch across is particularly characteristic. The occurrence of *Sphenothallus* at Britannic provides an interesting comparison with the Haughton Marine Band of the Pennine province. The band deteriorates in the north of the district: only *P. ophthalmoides* was found at Mardy Colliery and no evidence of marine conditions at this horizon was forthcoming at Fernhill and Duffryn Rhondda collieries.

At Britannic Colliery another marine horizon, called the **Britannic Marine Band,** (see Glam. 27 s.w.) lies about 25 ft above the Hafod Heulog Band. It consists of 2 ft of foraminifera-bearing mudstone in the roof of a thin coal; the only other fossils noted were *P. ophthalmoides* and doubtful *Lingula.* At Penllwyngwent and Margam Park the equivalent horizon yields only *Planolites,* but at the head of the Rhondda Fawr *L. mytilloides* was found. The band is not developed at Llanharan, nor has it been recognized at Mardy and Duffryn Rhondda collieries. Its equivalent in Yorkshire appears to be the Sutton Marine Band, and poorly developed marine strata are known from this horizon in most of the Pennine coalfields.

In the Llynfi and Avan valleys an impersistent thin coal often lies in the measures between these two marine bands; above it another compact quartzite—the **Upper Cockshot Rock**—reaches a thickness of almost 50 ft at Caerau.

In the north and east of the district the Cefn Coed Marine Band occurs in the roof of the first coal above the Britannic horizon. This coal splits to the south of a line running from Glyncorrwg through Cwm-parc towards Porth, and thence south-westwards towards the South Crop. The two leaves may be as much as

30 ft apart; *Naiadites sp.* has been recorded from the lower leaf at Duffryn Rhondda and *P. ophthalmoides* at Britannic. Locally quartzitic sandstones are present beneath each of the leaves, and the many anomalies in this part of the sequence may be explained by washouts associated with these sandstones. In the south-east of the district the top coal may die out, the marine band resting directly on seatearth.

As befits a horizon which correlates with the Dukinfield and Mansfield marine bands of the Pennines, Skipsey's Marine Band of Scotland and the Aegir of the Continent, the **Cefn Coed Marine Band** (Ware 1930, p. 478) is readily identified throughout the district. In the south-eastern areas it carries only *Lingula* in grey micaceous shaly mudstone. Elsewhere dark blocky pyritic mudstones occur, particularly towards the middle of the band, and are often associated with lenses of ankeritic siltstone. These mudstones commonly carry a rich and varied, though dwarfed, fauna of horny and calcareous brachiopods (including Athyrids, Chonetids, Productids, and Spiriferids), lamellibranchs, nautiloids, goniatites including *Anthracoceras aegiranum* H. Schmidt, trilobites, ostracods and conodonts. At Aberbaiden on the South Crop (see p. 141) the richest fauna of any Coal Measures marine band in Britain was found during the present survey. The band is about $1\frac{1}{2}$ ft thick, and at least 80 species were identified from a $5\frac{1}{2}$-in layer in the middle (Ramsbottom 1952).

(b) Cefn Coed Marine Band to Upper Cwmgorse Marine Band

Fig. 24 shows the generalized section of these strata as well as the principal faunas.

The cyclothem above the Cefn Coed Marine Band is consistently one of the thickest in the Middle Coal Measures. It is 60 ft in the eastern areas but thickens westwards to 80 to 100 ft over much of the district and reaches almost 180 ft locally in the extreme south-west. The beds are generally argillaceous, with an abundance of ironstone pins: at Fernhill these were formerly called the Black Pins Mine-ground. Quartzitic sandstone in the upper part was mis-called the Cockshot Rock in the southern Rhondda. A sandstone in a similar position is generally present throughout the Avan Valley, and around Bryn reaches 40 ft in thickness: it continues to the west and south-west, and at Margam is 33 ft thick.

The succeeding coal is called the **Caedavid** in the Llynfi, Garw, and Ogmore valleys, where it is best developed, and where it was formerly worked extensively. Here it is about 3 ft thick, with two distinct partings. South-westwards from Maesteg and northwards from Caerau the upper parting thickens, generally to about 20 ft, though it reaches 50 ft at Margam. From Caerau the split appears to follow an eastward course to Tonypandy and Porth. To the north the lower coal is usually called the **Gorllwyn** and the upper one the **Gorllwyn Rider.** The former has been worked around Treherbert and Maerdy and is $2\frac{1}{2}$ ft thick. In the south-eastern areas of the district only one thin coal is recorded in this general position. Shells are recorded from the seam at only one locality: at Glengarw Colliery abundant *Anthraconaia adamsi* is present in Caedavid spoil and is believed to come from one of the partings in the seam. Fish debris is recorded from cannelly mudstones in the roof of the Caedavid, and from the Gorllwyn Rider in the eastern parts of the district.

The cyclothem above is only 7 ft thick in the south-east, but elsewhere is generally 20 to 40 ft. Sandstones are common, especially towards the top: in the upper Rhondda they reach 20 ft, and in the Cynon Valley were referred to as the Big Rock.

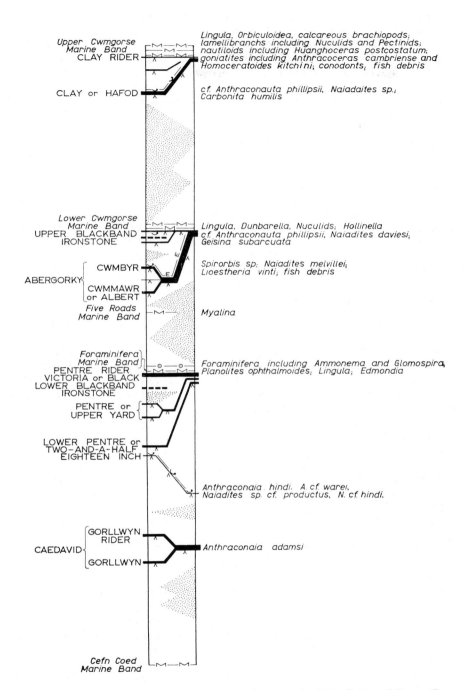

FIG. 24. *Generalized section of the Measures between the Cefn Coed and Upper Cwm-gorse marine bands, showing lithology and principal faunal elements. The thickness of strata involved varies from about 300 to 800 ft; for key see Fig. 5*

The coal next above was named only in the Llynfi Valley, where it was called the **Eighteen-Inch.** It has never been worked commercially. It averages 6 to 15 in over much of the district, but it thickens to 24 in around Tonypandy and westwards to Maesteg. In the upper Rhondda a thin blackband forms a parting in the seam. At Gelli and Naval collieries abundant 'mussels' occur in the roof, and in the Llynfi a similar fauna (Evans and Simpson 1934, pp. 460–2) formerly believed to lie above the Two-and-a-Half Seam is now thought to occur above the Eighteen-Inch (p. 000). Species include variants of *Anthraconaia hindi, A.* cf. *warei, Naiadites sp.* cf. *productus* and *N.* cf. *hindi* (Trueman and Weir).

The **Lower Pentre** or **Two-and-a-Half** Seam lies 20 to 40 ft above, the intervening measures being usually composed of mudstones with bands of sandstone which become more numerous upwards; locally in the Llynfi Valley and around Bryn these beds thin to 2 ft. The coal varies from 18 to 30 in. In the Cynon and upper Rhondda valleys a 'stone' parting 2 to 3 in thick lies about 9 in above the base: southwards and westwards this disappears and no persistent partings occur over the rest of the district. The coal has been worked near Gelli, in the Llynfi and Garw valleys, and along its crop west of Bryn. In the east the roof is planty, but at Duffryn Rhondda an associated thin rider a short distance above has yielded *A. hindi* from its roof.

Over most of the district the **Pentre** or **Upper Yard** lies 20 to 40 ft higher, the intervening strata being almost wholly mudstone. The seam is one of the most extensively worked in this group of measures. It is thickest in the area between Wern-tarw and Meiros, where it is over 4 ft, and it is over 3 ft as far west as Margam. In the Llynfi, Garw, Ogmore and Rhondda Fawr valleys it was also worked extensively at a thickness varying from 32 to 36 in, but the section is locally marred by two 'stone' partings. To the north of Ferndale and Treorky the upper parting thickens rapidly to 15 ft and a similar deterioration in the seam around Bryn may represent the western continuation of the same split. A plant roof is usual on the lower split, and indeterminate shell fragments have been recorded from the upper split.

The cyclothem above the Pentre is very variable. In the east it consists of only 3 to 6 ft of seatearth and mudstone. Westwards and north-westwards it thickens to about 120 ft in the Maesteg area. In the upper Rhondda a sandstone appears in the roof of the Pentre and is nearly 30 ft thick at Pentre Colliery; it is persistent throughout the Avan Valley and it thickens southwards to 60 ft at Duffryn Rhondda. In the Llynfi Valley it is usually 30 to 40 ft thick and it reaches a maximum of 90 ft at Garth. Massive, false-bedded, and resembling pennant, it has been quarried around Maesteg. West of the town and around Bryn 30 ft of mudstones separate the top of the sandstone from the overlying Victoria Seam, and a thin coal is present at the base of these mudstones. This coal is associated with the **Lower Blackband Ironstone,** up to 15 in thick, which in the last century was mined under Mynydd Bach.

The succeeding coal is known as the **Pentre Rider** in the Rhondda Fawr, the **Victoria** in the Llynfi and Garw valleys and the **Black** towards Cwmavon. It is normally a clean coal varying from 7 in. in parts of the Cynon Valley to 39 in east of Maesteg, but over much of the district it averages 18 in. Wherever seen the immediate roof consists of marine mudstone. This **Foraminifera Marine Band** (so-named during this resurvey) is characterized by abundant foraminifera in its ·lowest part, associated with only rare *Lingula* and *Edmondia; Planolites ophthalmoides* is profuse in the upper part of the band and ranges for several feet upwards.

The three coals, Lower Pentre, Pentre and Pentre Rider, all coalesce to form one seam, about 8 ft thick, at Llanharan Colliery. This has been worked extensively in recent years at that colliery under the name **Thick Seam,** and its crop is lined by numerous bell-pits in the area to the south and east.

The cyclothem above the Pentre Rider is everywhere a thick one; it increases steadily from 50 to 60 ft in the Cynon Valley and around Pontypridd to about 100 ft at Maesteg and Margam. Locally even greater thicknesses are recorded, as at Llanharan where almost 130 ft of beds are present. The lower strata are usually argillaceous, but sandstones become important upwards, especially in the north and east. At Coedely and Werntarw these are 50 to 70 ft thick and comprise most of the cyclothem; more generally they are 20 to 30 ft. In the Margam Park Nos. 2 and 3 boreholes *Myalina* was found in striped beds about 12 ft below the base of the Abergorky, and this represents the only record within the district of the **Five Roads Marine Band** (Archer, A. A. *in* Eyles 1956, p. 34).

The next coal is the **Abergorky**[1] of the Rhondda Fawr Valley, which equates with the **Graig** in the Cynon Valley. Around Treherbert and Gelli, where it has been worked on a small scale, it comprises two coals, the lower 12 in thick, separated from the upper, 21 to 24 in thick, by a foot or two of seatearth. Eastwards the parting dies out and in the Cynon Valley the Graig consists of 22 to 32 in of clean coal. Westwards from the Rhondda the two leaves split apart and are usually separated by 10 to 25 ft of sandy mudstone. In the Llynfi and Garw valleys the lower leaf appears to be the **Albert,** which is 18 in thick, and the upper leaf is thin and un-named. In the area east of Bryncethin and south of Porth, only one coal is present in this position, varying from $1\frac{1}{2}$ ft in the north to $2\frac{1}{4}$ ft in the south, and it is not certain whether this represents the full section or only the top coal. The roof of the top coal carries abundant *Lioestheria* cf. *vinti,* usually in dark shaly mudstone. This forms a useful marker band, for the fossils have been found wherever the seam-roof has been examined, except in the Cynon Valley where the dark mudstones are replaced by seatearth.

In the north-east at Mountain Ash a thin rider coal about 3 ft above the Graig carries the Lower Cwmgorse Marine Band. Westwards the separation increases to 30 to 50 ft in the Rhondda Fawr, and up to 60 ft in the south-west. Sandstones occur in the top of the cyclothem where it is thickest. The rider coal, which is nowhere named, is only a few inches thick in the north-east, but it thickens southwards to 21 in at Llanharan. To the west a second thin coal develops in the seatearth of the rider: it appears around Treherbert and carries a thin blackband in its roof. In the Llynfi and westwards to Cwmavon the lower coal is 6 to 9 in and the upper 12 to 15 in, and, between them, a 2-ft band of ironstone—the **Upper Blackband Ironstone**—was formerly extensively trenched and mined. Spoil from these workings has yielded abundant *Geisina? subarcuata* (Jones) associated with *Naiadites daviesi* Dix and Trueman and cf. *Anthraconauta phillipsii.*

The **Lower Cwmgorse Marine Band** (Trotter 1947, p. 92) cyclothem is 50 to 80 ft in the east and north-east of the district, 100 ft over much of the central area, and 125 ft in the south-west. In the eastern areas sandstones are prominent: in the Cynon Valley they are 30 to 60 ft thick in the upper part of the cyclothem; they figure largely at Fernhill, and at Llanharan 90 ft are present. Marine fossils

---

[1] This spelling is here preferred to that of 'Abergorki' used on the present one-inch geological map (1960, 1963), since it conforms with the usual rendering of the colliery of that name in the Rhondda Fawr Valley, after which the seam was named.

can usually be found at the base, and, where the sequence is argillaceous, they may range through a considerable thickness: at Mountain Ash they have been found through 35 ft, and at Ton Pentre in the Rhondda Fawr to within 30 ft of the overlying Hafod. The fossils are not numerous, and only a few species are represented: most typical are *Lingula*, *Dunbarella*, Nuculids and *Hollinella*.

The overlying **Hafod** Seam is a single coal only in the east—in the Cynon Valley and the area south of Porth as far west as Werntarw. It has been worked, together with its fireclay, in the Cynon Valley as the **Clay,** and in the Hafod– Cilely area, and at Werntarw, the thickness varying from 2 to 3 ft. The seam splits to the west into two, and locally, three coals, ranging over 20 to 40 ft, and to nearly 80 ft at Margam Park. The bottom coal is everywhere the thickest, and it has been worked as the Hafod at Aberbaiden on the South Crop. In the Maesteg–Bryn area it is again known as the Clay, reflecting the fact that it rests on a thick pale grey fireclay, and has been worked at several localities. In the northern Rhondda Fawr it carries in its roof a thin blackband yielding small crushed cf. *A. phillipsii* and *Naiadites sp.* together with ostracods including *Carbonita humilis*; a similar fauna was noted at Aberbaiden.

The **Upper Cwmgorse Marine Band** (Trotter 1947, p. 92) is carried in the roof of the Hafod or of its thin top split. In the eastern and south-eastern areas, where it varies from 2 to 8 ft thick, the fauna is poor: little other than *Lingula*, *Orbiculoidea* and a few Productids is found in a silty micaceous mudstone. In the Rhondda Fawr, where numerous surface exposures are known, it is about 5 ft thick and around Ton Pentre carries an abundant and varied fauna of lamellibranchs, nautiloids including *Huanghoceras postcostatum* (Bisat) and the goniatite *Anthracoceras cambriense* Bisat. In the Llynfi Valley, where only the basal part of the band is exposed, brachiopods are the commonest fossils. In the south-west of the area, and especially at Aberbaiden a marked thickening of the band takes place, and up to 42 ft of fossiliferous mudstones have been recorded. A large and varied fauna includes: Archaeocidarid spines; brachiopods including horny forms, *Crurithyris* and Productids; gastropods; lamellibranchs including *Curvirimula*, *Myalina*, Nuculids and Pectinids; nautiloids; goniatites including *Anthracoceras cambriense and Homoceratoides kitchini* Bisat; ostracods; conodonts; and fish debris.

UPPER COAL MEASURES OR PENNANT MEASURES

Up to about 4000 ft of Upper Coal Measures are preserved within the district, though the sequence is everywhere incomplete. A distinctive feature is the presence within them of numerous thick sandstones, the Pennant of tradition, separated by relatively thin mudstones, each of which usually contains one or more relatively thin coals. These sandstones occupy the greater part of the surface of the district, and in particular form the high ground between the valleys. Few shafts have penetrated other than the lowest beds; the coals, although worked locally in former days on a varying scale, are not much exploited today. Surface exposures are usually confined to the sandstones, and large continuous sections are almost non-existent; faunal horizons are never distinctive. Knowledge of the formation as a whole is therefore largely based on the mapping of features, and the degree of precision attained in the elucidation of the Lower and Middle Coal

Measures cannot be maintained. In particular the correlation of the individual leaves of coal within each mudstone 'slack' is rarely precise, though the identification of the 'slacks' themselves is generally reliable.

The characteristic massive pennant sedimentation was first established in the south-west of the district and gradually spread north-eastwards. On the western part of the South Crop it commenced with the Llynfi Rock, close above the Upper Cwmgorse Marine Band, and in this area about 700 ft of beds, consisting largely of pennant sandstone, had been deposited before similar conditions set in at the No. 2 Rhondda level in the Rhondda valleys, and a further 1100 ft had accumulated before the facies became firmly established in the extreme north-east, where massive and continuous pennant begins above the Brithdir. The lowest Pennant strata thus exhibit two contrasting lithologies: the one normally characteristic of the pennant litho-facies, the other scarcely different from that of the upper part of the Middle Coal Measures. Most of the problems of correlation in the past have been confined to the transitional belt between these two types of sedimentation.

A complication in Pennant stratigraphy arises from the theory that, over much of the district, the Swansea Beds of the coalfield west of Neath are largely or wholly cut out by unconformity beneath the Grovesend Beds (see Woodland and others 1957a, p. 9). This conclusion is based on the belief that the No. 3 Llantwit Seam of the south-eastern part of the district equates with the Mynyddislwyn of Monmouthshire, which is correlated on convincing faunal grounds with the Wernffraith or Swansea Four-Feet of the Swansea area. Acceptance of this view implies the absence in the Llantwit basin and in Monmouthshire of some 1500 ft of beds, including numerous cyclothems and coals, between the horizon of the Daren-ddu or Mountain and the Llantwit No. 3 or Mynyddislwyn: an unconformity appears to provide the most satisfactory explanation for this.

(a) Llynfi Beds

The Llynfi Beds have extensive outcrops along the lower slopes in the Cynon Valley around Aberaman, in the Rhondda Fach around Ferndale and Maerdy, in the Rhondda Fawr north of Pen-y-graig, in the upper reaches of the Ogwr Fawr, Garw and Llynfi, and north of the Moel Gilau Fault from Maesteg westwards to Cwmavon. Smaller inliers occur in the Ogwr Fach around Gilfach Goch and in the Ely Valley around Cilely. On the South Crop they can be traced continuously from east of Llantrisant to Margam Park. Traced from the north-east of the district to the south-west they show a thickening from little more than 250 ft around Mountain Ash and Merthyr Vale to over 700 ft in the area north and north-east of Margam (Fig. 8A). As already indicated, this thickening is to a large extent related to a marked south-westwards increase in the development of sandstones.

The measures from the Upper Cwmgorse Marine Band to the Tormynydd are about 120 ft thick at Cwmavon, a thickness which is maintained from here north-eastwards to the upper Rhondda Fawr. Eastwards and south-eastwards, they thin to some 80 ft in the central parts of the district, and to barely 60 ft around Merthyr Vale. A similar easterly attenuation to about 60 ft is shown when the measures are traced from Margam Park to Bryncethin, but the trend is reversed farther east along the South Crop, and from Wern-tarw to Llantrisant they expand to 110 or 120 ft. They are everywhere essentially argillaceous; any sandstones present are usually thin and impersistent. Two or three closely associated coals are distinguishable over most of the district. In the Cwmavon–Maesteg area, where these coals are best developed, the lowest, called the **White,** is succeeded in turn by the **Jonah** and the **Tormynydd.** All have been worked sporadically along their crops westwards from Maesteg, the White comprising 3 to 4½ ft of dirty coal, though the Jonah and the Tormynydd rarely exceed 2 ft. Away from this area the seams thin, and precise correlation of the individual leaves is not easy. Around Maesteg the White yields a characteristic fauna of *Anthraconaia* aff. *stobbsi, Naiadites daviesi* and ostracods preserved in ferruginous mudstone; shelly blackband is also present above the lowest coal of the group at Treherbert and Ferndale, and in the Taff Valley the Blackband Coal appears to be at the same horizon.

The sandstone directly overlying the Tormynydd throughout the central and south-western part of the district has been called the **Llynfi Rock** (see First Edition, p. 30); it is the lowest massive pennant-type sandstone in the succession. Locally between the Llynfi and Garw valleys, where it is immediately overlain by the No. 3 Rhondda Seam, its uppermost few feet are quartzitic and ganister-like. Around Margam Park, Bryn, and Maesteg, and in the Garw and Ogmore valleys, the Llynfi Rock is 150 to 180 ft thick, but it thins steadily away from these areas. In the extreme north-west at Glyncastle it is still 90 ft, but eastwards at the head of the Rhondda Fawr it is rarely more than 15 ft. Farther south a similar eastwards thinning is apparent. In the Rhondda Fawr it is 30 to 50 ft thick over much of the west side of the valley, but it is virtually absent in the **Ogwr Fach and Ely** valleys. In the Rhondda Fach it still maintains a thickness of 24 to 30 ft in the area south of Tylorstown, but it cannot be identified with certainty in the Cynon Valley (except possibly to the north-west of Aberaman) or in the Merthyr Vale area. Along the South Crop at least 75 ft remain at Bryncethin, but at Wern-tarw and to the east it is absent and presumably replaced by mudstones.

Lying close above the Llynfi Rock, the **No. 3 Rhondda** is the most valuable seam in this part of the sequence. It is well developed throughout the southern half of the district but it deteriorates northwards along a line from Pont-y-gwaith to Pentre, and thence through Ogmore Vale, Blaengarw and Garth to Margam Park with such abruptness as to suggest that it is affected by a regional washout. South of this line it has been worked extensively from many collieries, with a thickness varying from about 3 to 4½ ft. In the Ely Valley, at Gilfach Goch and Werntarw up to 1 ft of cannel lies in its immediate roof.

The measures between the No. 3 and No. 2 Rhondda are extremely variable and, in view of the paucity of detailed sections available for study, are difficult to correlate with certainty from one end of the district to the other.

The No. 3 Rhondda throughout much of the southern areas is overlain by sandstone up to 50 ft thick; this dies out northwards towards the heads of the Ogmore, Garw, and Llynfi valleys, and beneath the Avan–Neath watershed; and in the Ogwr Fach and Lower Rhondda areas the worked seam had a silty mudstone roof. In the mudstones overlying this sandstone there is a thin coal, unnamed except around Aberbaiden on the South Crop, where it is called the **Rock Fach.**

Above the Rock Fach, sandstones up to 150 ft or more in thickness are present at the western end of the South Crop and around Cwm Gwineu. These thin out northwards in much the same areas as the underlying sandstone, and to the east the lower portion is replaced by mudstones with several thin coals. A still higher sandstone is upwards of 100 ft thick in the Llynfi and parts of the Garw and

Ogmore valleys, and this apparently extends into the upper Rhondda where it is still up to 50 ft thick around Treherbert.

In the area from Mynydd Pen-hydd to Cwmavon, the Rock Fawr and Rock Fach together with their associated mudstones die out abruptly, and about 400 ft of sandstones lie above the Tormynydd, extending upwards to close below the Wernpistyll.

In the north and north-east the beds equivalent to these arenaceous measures above the No. 3 Rhondda are essentially argillaceous, but they include a number of sandstones, up to about 15 ft thick, which are probably impersistent, for they cannot be traced from one shaft to another, and their relationships with the major sandstones of the south-western areas are obscure. Coal development is also erratic, and little precise correlation can be achieved. Concurrently with the disappearance of the massive sandstones, the measures thin to less than 150 ft in the extreme eastern areas.

The highest Llynfi Beds are everywhere essentially argillaceous. They include in the lower Avan Valley two thin coals, the **Wernpistyll** and **Wernpistyll Rider,** both of which have been locally mined on a small scale; these lie respectively about 80 to 135 ft and 30 to 50 ft below the top of the group. Mudstones with thin sandstones and striped beds separate these coals. To the east thin coals occupy similar positions over much of the area but their exact correlation is not certain, and over much of the South Crop only one coal is developed.

The Llynfi Beds as a whole are almost devoid of marker horizons. Apart from the shelly blackband associated with the White and its probable equivalents, the only fauna recorded from an identifiable horizon is that from the roof of the Wernpistyll Rider. The species here identified are restricted to *Anthraconauta phillipsii* and *Carbonita sp.*

(b) Rhondda Beds

The Rhondda Beds have the most extensive outcrops of any of the Coal Measures groups. They form most of the high ground of the plateau in the central areas of the district, as well as a strip half a mile wide along the South Crop widening out in the west around the end of the Bettws syncline. The beds thicken south-westwards from less than 600 ft in the Taff Valley to more than 1100 ft on the South Crop between Coytrahen and Margam Park (Fig. 8B). This reflects a marked increase in the development of pennant sandstones: in the Taff and Cynon valleys the middle and upper parts of the group are largely argillaceous, but elsewhere sandstones are dominant and make up to about 90 per cent of the sequence.

At the base the **No. 2 Rhondda** is the most extensively worked of all the Pennant coals within the district. It is a single seam only in the central, north-eastern and south-eastern areas, where it varies from about 2 to 2½ ft in the Aberaman and Merthyr Vale areas, to over 6 ft at Llantrisant, though it more usually averages 3 to 4 ft. The seam has a variable section, and in places consists of three or more distinct leaves. One of the partings, generally 1 to 2 ft from the top, develops into a true split, and over large areas two distinct seams are present. Two separate areas of split coals are recognized. One extends through the lower Rhondda Fawr from Pontypridd towards Porth, and includes much of the Rhondda Fach and the Cynon Valley south of Mountain Ash. The bottom coal, known in the Rhondda valleys as the **No. 2 Rhondda,** is up to 3 ft thick, while the upper or **Fforest Fach,** 3 ft thick in the Pontypridd area, thins northwards and in the Rhondda Fach appears to die out as a result of pene-contemporaneous

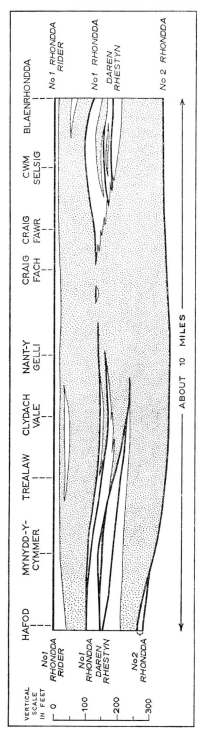

FIG. 25. *Section (somewhat diagrammatized) showing the lateral variations in the measures between the No. 2 Rhondda and No. 1 Rhondda Rider seams in the Rhondda Fawr; sandstones are shown stippled*

erosion. Two seams are again present west of a line traceable northwards from Heol-y-cyw, through Ogmore Vale and Mynydd Blaenafan, and thence westwards through Cymmer (Avan) towards the Neath Valley. Here, the lower coal, known variously as the **No. 2 Rhondda, Wernddu, Rock Fawr** or **Malthouse,** is generally about 3 to 4 ft thick. A local split develops in this seam in the Avan Valley west of Duffryn, the lower leaf being known as the **Wernddu Rider.** The upper coal, known generally as the **Field Vein,** is about 2 ft thick, and south of Cwmgwyneu Colliery is separated from the Wernddu by as much as 120 ft.

In most of the district massive pennant sandstones succeed the No. 2 Rhondda, though locally, as in the Ogmore Vale area, 40 ft of planty mudstones intervene. Over wide areas the lowest few feet of these sandstones consist of quartz conglomerate containing 'pebbles' and 'rafts' of coal and rolled pellets of clay-ironstone. Where fully developed, throughout most of the district to the south and west of the Rhondda Fawr, the sandstones continue the sequence up to the mudstones beneath the No. 1 Rhondda, broken only by a few thin and impersistent bands of silty mudstone. They are thinnest, about 150 ft, in the extreme south-east at Rhiwsaeson, but they thicken rapidly, when traced westwards, to about 500 ft along the western part of the South Crop.

In much of the lower Rhondda Fawr area, extending from Pontypridd to Clydach Vale and Ystrad Rhondda, the upper part of the sandstones is replaced, in part or wholly, by mudstones (see Fig. 25). Within these, the **Daren Rhestyn** coal, a dirty seam up to about 2 ft in thickness, is developed. North-east of Ystrad this coal dies out, and the associated mudstones are again replaced by sandstones in the Rhondda Fach. In the Cynon and Taff valleys a thin mudstone and coal horizon is traceable within the upper part of the sandstones overlying the No. 2 Rhondda, and this may be the equivalent of the Daren Rhestyn.

Similar lateral changes in lithology take place within the upper Rhondda Fawr, in the Corrwg Valley and in the Glyncastle area as far west as Melin Court Brook. Mudstones beneath the No. 1 Rhondda here reach 130 ft thick. A coal, once worked around Glyncorrwg and near Blaenrhondda, lies between 15 and 50 ft above the base of these mudstones, and is considered likely to equate with the Daren Rhestyn; both in the Rhondda Fawr and Corrwg it has locally a ferruginous mudstone roof carrying *Anthraconauta phillipsii* and ostracods. A thin smut or seatearth usually lies 15 ft above, and between Glyncorrwg and Cymmer two or three closely associated thin coals occur in this position. Distinctive, though thin, quartzitic sandstones are developed above, and in places below, this composite subsidiary coal horizon, separating it from the Daren Rhestyn and the No. 1 Rhondda.

Where best known in the lower Rhondda Fawr, the **No. 1 Rhondda** is 2 to 3 ft thick. In much of the district it is imperfectly known though its continuity is not in doubt, except locally west of Ton and Cwm-parc, to the north-east of Gilfach Goch, and again in the Avan Valley between Cymmer and Pont Rhyd-y-fen, where it appears to be affected by washouts. On the South Crop, north and north-east of Llanharan, the seam is thick and dirty, though over much of the west of the district it is thinner but cleaner. It has been worked on a moderate scale at numerous localities in the Rhondda Fawr, as well as on Moelcynhordy, and in Cwmcedfyw between the Garw and Llynfi valleys. To the north the seam varies from 12 to 18 in between Troedyrhiw and the head of the Rhondda Fawr, but westwards it improves to 2 to 2½ ft at Glyncorrwg and in the Neath Valley. At numerous localities throughout the district the seam is immediately overlain by as much as several feet of striped ferruginous mudstone or blackband containing abundant *A. phillipsii* and ostracods, which provide a useful marker.

In much of the lower Rhondda valleys the No. 1 Rhondda is closely overlain by a thin coal. The two seams are separated by 40 ft in Clydach Vale, 20 ft on Mynydd y Cymmer, and by only a few feet on the western slopes of Mynydd y Glyn between Cymmer and Trebanog. It may be that the upper coal is a split from the No. 1 Rhondda, but it is significant that it is the lower of the two seams which here carries the blackband fauna. In the south and west the No. 1 Rhondda is followed closely, and in some places immediately, by massive pennant sandstones 100 to 150 ft thick, which locally have a conglomeratic base. Eastwards and north-eastwards these become thin and are not recorded east of a line from Ferndale to Ynys-y-bwl. They also die out locally on the South Crop north-west of Llanharan and in the Corrwg Valley.

In most of the district a thin coal, the **No. 1 Rhondda Rider,** lies above about 20 to 30 ft of mudstone overlying the last mentioned sandstones. Near Pont Rhyd-y-fen it has been worked as the Graig Isaf, about 2 ft thick, but elsewhere few details are known concerning its section; both it and the underlying mudstones are thinner in the Corrwg and upper Rhondda valleys. Usually only one seam is developed within the mudstone, but locally, as in the Llantrisant shafts, a second thin coal lies near the base of the 'slack', about 30 ft below the 15 to 20-in Rider coal. 'Mussels' are recorded from the roof of the seam at a few places: in the Ferndale area *Anthraconauta phillipsii* is accompanied by *A.* cf. *tenuis*, and this represents the lowest occurrence of the latter in the district.

Throughout most of the district the No. 1 Rhondda Rider is succeeded by thick massive sandstones. At Llangynwyd these are 550 ft thick, continuing upwards to within 20 or 30 ft of the Brithdir horizon. They are 200 to 400 ft thick in the Avan Valley, and 400 ft in the Ely Valley.

In the Porth area of the Rhondda Fawr and generally throughout the Rhondda Fach a 50-ft mudstone is developed in the lower part of these sandstones, and is separated from the No. 1 Rhondda Rider by up to 30 ft of sandstone. A thin and impersistent dirty coal lies at the top of the mudstone and is immediately overlain by the main mass of sandstones. These thin from about 250 ft at Ynyshir to less than 70 ft on the slopes above Blaenllechau, and in the Cynon and Taff valleys the corresponding beds are essentially argillaceous; they contain two thin coals, the roof of the lower locally yielding *Anthraconaia pruvosti* and *Anthraconauta phillipsii* associated with abundant *Euestheria simoni* Pruvost.

In general, the highest beds of the group consist of 30 to 50 ft of argillaceous rocks locally carrying one or more insignificant coals. In the Taff Valley, however, a sandstone, up to 90 ft thick, appears to be developed in this part of the sequence, but this dies out westwards and cannot be mapped in the Cynon Valley.

Blue and red mudstones, recorded towards the top of the group at Penrikyber Colliery, represent the last appearance westwards of the coloured Deri Beds of Monmouthshire.

(c) Brithdir Beds

The Brithdir Beds occur in three main areas: in the north-east of the district they underlie much of the Taff, Cynon, and Clydach valleys, as well as capping the high ground of Cefn Gwyngul on the east side of the Rhondda Fach; in the north-west they are preserved in the Cefnmawr Syncline between the Avan and Neath valleys; and they have their most extensive development within the Llantwit–Tonyrefail–Bettws Syncline in the south. They vary in thickness from about 500 ft in the north and north-east to rather more than 800 ft to the west of Llangynwyd (Fig. 8c). In contrast to the Rhondda and Llynfi beds the measures are everywhere dominantly arenaceous, and the coals are thin and unimportant.

The **Brithdir** Seam at the base is widely worked in Monmouthshire, but it has largely deteriorated before entering the present district, where its precise equivalent is not known with certainty over much of the ground, for although the associated mudstone 'slack' is readily identifiable it contains several thin and dirty coals. The uppermost of these is usually the best defined and, for reasons of convenience, is equated with the Brithdir. The section is everywhere dirty, and the maximum known thickness of workable coal from the few trials on it is but 26 in.

Away from the Taff Valley, where locally a planty mudstone roof overlies the seam, the Brithdir is succeeded by massive pennant sandstones, which thicken from about 130 ft in the north-east of the district to at least 250 ft in the south-west. Usually these contain only insignificant lenses of silty mudstone, but in the north-western areas a more widespread development occurs about 60 to 120 ft from the top. Near Blaen-gwynfi and Cymmer this is associated with a thin coal. In the Upper Corrwg and Neath valleys the apparent equivalent of the mudstone is 35 ft thick, and a 2-ft coal, the **Graig,** lies towards the base. The roof of the coal carries *Anthraconauta phillipsii, A. tenuis* and ostracods.

The **Brithdir Rider** is also thin and, over most of the district, dirty. It has only been worked sporadically and never on a commercial scale. In the Cynon and Rhondda Fach valleys it consists of 15 to 18 in of clean coal, but elsewhere it is split by bands of rashings and seatearth. Generally its roof is sandstone, locally conglomeratic, but in the Rhondda Fach up to 20 ft of mudstones overlie the seam, and both here and in the Cynon Valley the immediate roof consists of dark shale with *A. phillipsii* and *A. tenuis*.

The succeeding sandstones are among the thickest in the sequence. They are upwards of 400 ft thick in the east of the district, and nearly 500 ft to the south of Maesteg. 'Tilestones' are widely developed towards the base in the north-east. Impersistent developments of thin mudstone, accompanied near the top by a thin coal, are known from widely separated parts of the district. Although these appear to vary in position, they probably all represent the same horizon, which may equate with the **Glyngwilym** seam of the north-west of the district. In the lower Rhondda the thin seam lies about 100 ft above the Brithdir Rider, in the Dimbath–Ogwr Fach area the figure is 200 ft, and south of Maesteg about 350 ft. The Glyngwilym Seam, west of the Avan, consists of up to 26 in of clean coal. Its roof carries fragmentary 'mussels' and ostracods. From Aber-cregan to Pont Rhyd-y-fen sandstones, up to 150 ft thick, closely overlie the seam, but these thin northwards and in the upper Cregan and Corrwg valleys and around Glyncastle only mudstones and siltstones separate it from the Wenallt some 30 to 70 ft above.

(d) Hughes Beds

The Hughes Beds are preserved in three broadly synclinal areas: around Abercynon, on the neighbouring high ground between the Taff, Cynon and Clydach valleys, and in the trough formed by the Daren-ddu and Llanwonno faults; in the Cefnmawr syncline; and along the axis of the Llantwit–Tonyrefail–Bettws Syncline from Llantwit Fardre to Llangynwyd. The group is completely preserved only on Cefnmawr, where it is about 800 ft thick. Elsewhere the original thickness cannot be determined since the highest beds are not present. Each area has its own seam nomenclature, the southern one possessing several variations.

At the base of the group a complex coal horizon is known variously in the east and south as the **Cefn Glas, Pen-y-coedcae, Ty-du, Lower Glynogwr** and **Bettws Four-Feet.** Throughout these areas it has been worked in a few places, with a section varying from about $2\frac{1}{2}$ ft near Abercynon to $4\frac{1}{2}$ to 5 ft at Tonyrefail, Llandyfodwg, Bettws and Llangynwyd. In the north-west several distinct seams are developed, presumably by splitting, the most important being the **Wenallt** at the base, and the **Wenallt Rider** some 15 ft above; both average 2 to $2\frac{1}{2}$ ft and are of considerable local importance. An abundance of large *Anthraconauta tenuis*, associated with *A. phillipsii* and ostracods, is known widely from the north-eastern outcrops and again around Bettws, while a similar fauna is known from the roof of the Wenallt Rider near Aber-cregan.

Thick and massive sandstones closely overlie this coal horizon, and locally infill small channels eroded in the underlying rocks. Around Bettws and in the Ogwr Fach up to 20 ft of mudstones may intervene between the coal and the sandstones, and west of Ton-mawr these expand to about 75 ft with a thin coal, the **Garlant,** at the top. The sandstones are thickest in the west of the district, 200 to 250 ft being measured in the Pelena Valley, but they thin eastwards and south-eastwards to 130 ft in the Cynon Valley and barely more than 100 ft in the Ogwr Fach. In the last-named area a thin mudstone and coal is locally present about 35 ft above the base.

The overlying mudstones are 20 to 40 ft thick. At or near their top a coal is widely present throughout the district, though it appears to fail north-eastwards in the Taff Valley. It is called the **Daren-ddu** near Pontypridd, the **Dirty** at Tonyrefail, the **Upper Glynogwr** near Llandyfodwg, and the **Mountain** north-west of the Avan. It is generally about 2 to 3 ft thick, but contains several dirt partings which are liable to thicken and render the seam unworkable. Mining has been most extensive in the trough between Pontypridd and Llanwonno, and around Ton-mawr. A massive sandstone roof is generally present, though planty mudstones intervene in a few places.

Yet another thick group of sandstones follow; these are 200 to 250 ft around Ton-mawr and this thickness is maintained in the north-east. Around Ynys-y-bwl a lens of thin mudstone with an associated coal streak occurs about 150 ft above the base; similar developments occur 90 ft above the base at Tonyrefail, and 50 to 100 ft above to the north-east of Llandyfodwg. In the Llantwit Syncline only about 150 ft of beds separate the Daren-ddu from the No. 3 Llantwit and this thinning may be due in part at least to unconformity.

On Cefnmawr the highest beds of the group consist of about 250 ft of mudstones. Near their base is a thin coal, the equivalent of the Westernmoor Seam of Neath, and 120 ft higher a second coal presumably represents the Shenkin of that area. No representatives of these beds have been recognized with certainty elsewhere in the district, though they may be present in the trough formed by the Moel Gilau and Felin-Arw faults between Pont Rhyd-y-cyff and Llangeinor.

(e) Swansea Beds

On Cefnmawr the **Graigola Vein,** a clean coal of about 6 ft, conformably overlies the Hughes Beds, and is succeeded by 80 ft of massive pennant. The Swansea Beds may also be in part preserved within the Felin-Arw trough, though the exact relationships of the strata here are obscure. A seam, possibly to be equated with the Graigola, is nearly 8 ft thick, and is also overlain by massive sandstones. Some distance above this sandstone a thick double coal, the **Bettws Nine-Feet,** may correlate with the Little and Hard veins of Gnoll Colliery, Neath.

(*f*) *Grovesend Beds*

These beds are restricted to the area of the Llantwit Syncline between Llantrisant and Llantwit Fardre. Here over 500 ft of measures are present including five coals. Of these the principal seams are: the **No. 3 Llantwit** at the base, about 4 ft thick; the **No. 2 Llantwit**, a dirty coal with an overall section of about 7 ft, some 110 ft above; and the **No. 1 Llantwit**, rather more than 250 ft above the latter, and consisting of two leaves 2 to 15 ft apart, together totalling about 8 ft of coal. Above the No. 1 Llantwit rather more than 100 ft of measures are preserved, including 50 ft or so of massive sandstone in the roof of the coal. Apart from this sandstone the measures are dominantly argillaceous. Red fireclays are recorded about 100 ft above the No. 2 Llantwit, and provide a point of similarity with the Grovesend Beds of the type area.

Only one faunal horizon is known, spoil from the No. 2 Llantwit yielding *Anthraconauta* cf. *tenuis*, *Spirorbis sp.* and ostracods.

A.W.W., W.B.E.

REFERENCES

CROOKALL, R. 1955. Fossil plants of the Carboniferous Rocks of Great Britain. *Mem. Geol. Surv. Palaeont.*, **4**, part 1.

DAVIES, D. 1921. The ecology of the Westphalian and the lower part of the Staffordian Series of Clydach Vale and Gilfach Goch. *Quart. J. Geol. Soc.*, **77**, 30–74.

DAVIES, J. H. and TRUEMAN, A. E. 1927. A revision of the non-marine lamellibranchs of the Coal Measures, and a discussion of their zonal significance. *Quart. J. Geol. Soc.*, **83**, 210–59.

DIX, Emily. 1933. The succession of fossil plants in the Millstone Grit and the lower portion of the Coal Measures of the South Wales Coalfield. *Palaeontographica*, **78**, 158–202.
———1934. The sequence of floras in the Upper Carboniferous, with special reference to South Wales. *Trans. Roy. Soc. Edin.*, **57**, 789–838.

EAGAR, R. M. C. 1962. New Upper Carboniferous non-marine lamellibranchs. *Palaeontology*, **5**, 307–39.

EDWARDS, W. and STUBBLEFIELD, C. J. 1948. Marine bands and other faunal marker-horizons in relation to the sedimentary cycles of the Middle Coal Measures of Nottinghamshire and Derbyshire. *Quart. J. Geol. Soc.*, **103**, 209–60.

EVANS, W. H. and SIMPSON, B. 1934. The Coal Measures of the Maesteg District, South Wales. *Proc. S. Wales Inst. Eng.*, **49**, 447–75.

EYLES, V. A. 1956. In *Sum. Prog. Geol. Surv.* for 1955.

HEARD, A. 1922. The Petrology of the Pennant Series. *Geol. Mag.*, **59**, 83–92.

JENKINS, T. B. H. 1960. Non-marine lamellibranch assemblages from the Coal Measures (Upper Carboniferous) of Pembrokeshire, West Wales. *Palaeontology*, **3**, 104–23.

JONES, D. G. and OWEN, T. R. 1956. The rock succession and geological structure of the Pyrddin, Sychryd and Upper Cynon valleys, South Wales. *Proc. Geol. Assoc.*, **67**, 232–50.

JONES, S. H. 1934. The correlation of the coal seams in the country around Ammanford. *Proc. S. Wales Inst. Eng.*, **49**, 409–38.

KIDSTON, R. 1893–4. On the fossil flora of the South Wales Coalfield and the relationship of its strata to the Somerset and Bristol Coalfield. *Trans. Roy. Soc. Edin.*, **37**, 565–614.
——1894. On the various divisions of the Carboniferous Rocks as determined by their fossil flora. *Proc. Roy. Phys. Soc. Edin.*, **12**, 183–257.

LEITCH, D., OWEN, T. R. and JONES, D. G. 1958. The basal Coal Measures of the South Wales Coalfield from Llandybie to Brynmawr. *Quart. J. Geol. Soc.*, **113**, 461–86.

MOORE, L. R. 1945. The geological sequence of the South Wales Coalfield: the 'South Crop' and Caerphilly basin, and its correlation with the Taff Valley sequence. *Proc. S. Wales Inst. Eng.*, **60**, 141–227.
—— and COX, A. H. 1943. The Coal Measure sequence in the Taff Valley, and its correlation with the Rhondda Valley sequence. *Proc. S. Wales Inst. Eng.*, **59**, 189–304.

NORTH, F. J. and HOWARTH, W. E. 1928. On the occurrence of millerite and associated minerals in the Coal Measures of South Wales. *Proc. S. Wales Inst. Eng.*, **44**, 325–48.

RAMSBOTTOM, W. H. C. 1952. The fauna of the Cefn Coed Marine Band in the Coal Measures at Aberbaiden, near Tondu, Glamorgan. *Bull. Geol. Surv. Gt. Brit.*, No. 4, 8–30.

ROBERTSON, T. 1927. The Geology of the South Wales Coalfield. Part II. Abergavenny. 2nd edit. *Mem. Geol. Surv.*
——1933. The geology of the South Wales Coalfield. Part V. The Country around Merthyr Tydfil. 2nd edit. *Mem. Geol. Surv.*

STRAHAN, A. 1907. The Geology of the South Wales Coalfield. Part VIII. The country around Swansea. *Mem. Geol. Surv.*
——, CANTRILL, T. C., DIXON, E. E. L. and THOMAS, H. H. 1907. The geology of the South Wales Coalfield. Part VII. The Country around Ammanford. *Mem. Geol. Surv.*
——, TIDDEMAN, R. H. and GIBSON, W. revised by GIBSON, W. and CANTRILL, T. C. 1917. The Geology of the South Wales Coalfield. Part IV. The Country around Pontypridd and Maesteg. 2nd edit. *Mem. Geol. Surv.*

STUBBLEFIELD, C. J. and TROTTER, F. M. 1957. Divisions of the Coal Measures on Geological Survey Maps of England and Wales. *Bull. Geol. Surv. Gt. Brit.*, No. 13, 1–5.

TROTTER, F. M. 1947. The Structure of the Coal Measures in the Pontardawe–Ammanford area, South Wales. *Quart. J. Geol. Soc.*, **103**, 89–133.

WARE, W. D. 1930. An account of the geology of the Cefn Coed sinkings. *Proc. S. Wales Inst. Eng.*, **46**, 453–501.
——1939. The Millstone Grit of Carmarthenshire. *Proc. Geol. Assoc.*, **50**, 168–204.

WOODLAND, A. W., EVANS, W. B. and STEPHENS, J. V. 1957a. Classification of the Coal Measures of South Wales with special reference to the Upper Coal Measures. *Bull. Geol. Surv. Gt. Brit.*, No. 13, 6–13.
——, ARCHER, A. A. and EVANS, W. B. 1957b. Recent boreholes into the Lower Coal Measures below the Gellideg–Lower Pumpquart horizon in South Wales. *Bull. Geol. Surv. Gt. Brit.*, No. 13, 39–60.

CHAPTER IV

LOWER COAL MEASURES: DETAILS

IN THE following account certain local terms, applied by the miners to particular lithologies, have been used for the sake of convenience, or have been quoted. The following list gives explanations of these terms:—

Balls: nodules of ironstone.

Bast: cannel or cannelly mudstone.

Bastard: applied usually to non-typical or intermediate lithologies; thus 'bastard clift', usually a strong sandy mudstone not sandy enough to be called 'rock'.

Blackband: ironstone or ferruginous mudstone with an appreciable amount of carbonaceous matter.

Brass: hard bands, lenses, or nodules rich in pyrite.

Clift (or cliff): blocky mudstone or silty mudstone.

Clod: soft mudstone forming a band in a coal seam; or that part of the roof which comes away with the coal during working.

Dirt: any band other than coal within a seam.

Engine coal: inferior coal.

Mine: ironstone; mine ground is clift with much ironstone in bands or nodules.

Pin: thin band of hard rock, usually ironstone.

Rashes or rashings: *either* soft carbonaceous shale with streaks of coal *or* highly disturbed, slickensided, comminuted shale or mudstone formed by movement parallel to the bedding and usually associated with the roof or dirt bands in coal seams. Normally 'rashes' should be retained for the former definition and 'rashings' for the latter.

Rider: a coal commonly thin and unworkable, and usually, but not always, overlying a coal referred to in the name; thus No. 1 Rhondda Rider—the seam above the No. 1 Rhondda.

Rock: hard massive sandstone.

Stone: a hard band usually within a coal seam.

To facilitate description of the details of stratigraphy in this and the following two chapters, the district has been divided into seven areas. These are depicted on Fig. 26, which also shows the positions of the major collieries; the grid-references of the collieries are given in the Appendix of colliery and borehole sections, p. 300 ff.

Much confusion exists over the rendering of Welsh place-names, particularly with regard to the use of hyphens in compound names. Throughout this memoir the representation used on the 6-in Geological Maps (published 1957–60) has, in general, been followed. In certain instances this has led to apparent anomalies, for the official style of a colliery carrying a locality name may differ from the name used on the map. Thus Avon Colliery lies in the Avan Valley; Llwynypia Colliery is located at Llwyn-y-pia; and Werntarw Colliery is in the vicinity of Wern-tarw Farm.

A. *Gastrioceras subcrenatum* MARINE BAND TO GARW SEAM

These measures crop out in a narrow strip, seldom more than half a mile wide, extending along the South Crop of the Coalfield from Kenfig Hill in the west, through Aberkenfig and along Cefn Hirgoed to Penprysg. Eastwards they plunge beneath glacial drift and pass into the Bridgend district to the south, where practically nothing is known concerning them.

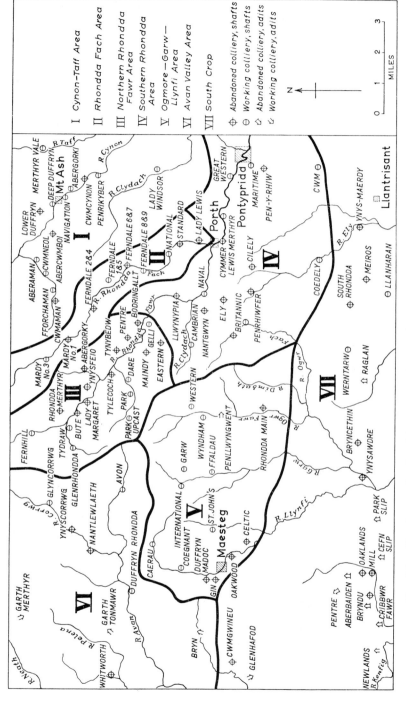

FIG. 26. Sketch-map showing delimitations of areas described in the text, and the sites of major collieries

In 1954 the Margam Park No. 1 Borehole [8194857̇8] penetrated these measures, and gave for the first time in this district a complete section of the strata. In addition, the No. 2 Borehole [81118632] reached a depth of 282ft below the Garw. The full logs of both these boreholes below the Gellideg Seam have already been published (Woodland and others 1957b, pp. 44–49), and the following record of the strata from the Garw to the base of the formation in the No. 1 Borehole is abridged from this source.

	Thickness (not corrected for dip)		Depth below the Garw Seam	
	Ft	In	Ft	In
GARW	—	—	—	—
Grey mudstone-seatearth, passing to silty and sandy mudstones with worm burrows	15	0	15	0
Massive pale grey sandstone	43	9	58	9
Striped beds	3	5	62	2
Grey mudstone with sporadic ironstones	17	9	79	11
Grey silty mudstone and striped beds with sandy layers	14	10	94	9
Grey to dark blue-grey mudstone; rare *Curvirimula sp.*; fish-debris; *Planolites ophthalmoides* at several horizons	44	4	139	1
Pale grey quartzitic sandstone	5	2	144	3
Blue-grey silty mudstone with *Carbonicola sp.*	14	9	159	0
Pale grey quartzitic sandstone	12	6	171	6
Blue-grey mudstone, darkening downwards; *P. ophthalmoides* at several horizons; sporadic *Anthraconaia sp.*, *Carbonicola sp.*, *Curvirimula sp.*; fish debris abundant near base	143	6	315	0
Pale grey seatearth passing into mudstones, darkening downwards; numerous ironstones up to 3 in; *P. ophthalmoides* throughout	54	0	369	0
Margam Marine Band: mudstones, mainly dark blue-grey with scattered pyrite granules; foraminifera including *Hyperammina sp.* and *Rectocornuspira?*; *P. ophthalmoides*; *Lingula mytilloides, Orbiculoidea sp.*; Bellerophontid gastropods and spat; *Caneyella* cf. *multirugata* (Jackson) (*? sp. nov.*), *Dunbarella* cf. *papyracea* (J. Sowerby), *Myalina* cf. *verneuili* Hind *pars* (*non* McCoy), *Posidonia* or *Caneyella sp.* (mainly juv.); involute nautiloid (cf. *Domatoceras*); *Anthracoceras sp.*; conodonts including *Hindeodella sp.* and *Ozarkodina sp.*; fish remains including *Elonichthys sp.* and *Rhabdoderma sp.*	21	2	390	2
Blue-grey mudstone; plant debris; *P. ophthalmoides* ..	7	10	398	0
COAL		5	398	5
Mudstone-seatearth with sphaerosiderite, passing into silty mudstone and striped beds	32	11	431	4
Cefn Cribbwr Rock: massive pale grey quartzitic sandstone	121	4	552	8
Cefn Cribbwr Marine Band: dark mudstones, silty in parts and sandy at base; *P. ophthalmoides* (at top); *L. mytilloides*; small gastropods; *Aviculopecten?*, *Caneyella* cf. *multirugata*, cf. *Edmondia sp.*, *Nuculana acuta*, *Posidonia sp. nov.* [right valve with prominent anterior ear]; *Anthracoceras sp.*, *Gastrioceras listeri*, *G. sp.* [with faint spiral ornament]; *Hollinella sp.*; *Hindeodella sp.*; fish remains	20	9	573	5

	Thickness (not corrected for dip)		Depth below the Garw Seam	
	Ft	In	Ft	In
Grey silty mudstone with plant debris	2	4	575	9
Pale grey quartzitic sandstone with striped partings ..	6	0	581	9
Core not recovered; said to include thin coal	1	4	583	1
Mudstone-seatearth passing into ganister-like mudstone with sphaerosiderite	21	3	604	4
Grey mudstone, sandy at top; *Alethopteris lonchitica, Calamites sp., Neuropteris gigantea* Sternberg, *?Sphenopteris trifoliolata* (Artis)	25	5	629	9
Massive pale grey quartzitic sandstone	16	3	646	0
Grey mudstone with thin layers of quartzitic siltstone; *P. ophthalmoides*; *L. mytilloides*; fish debris near base	15	1	661	1
CROWS FOOT: **Coal 1 in**		1	661	2
Mudstone-seatearth, carbonaceous at top, passing into grey sandy mudstone with layers of quartzitic sandstone; comminuted plant debris	26	0	687	2
Grey mudstone, silty above and smooth below, with frequent thin ironstones; *P. ophthalmoides* throughout	33	3	720	5
Dark blue-grey silty mudstone; pyritized sponge spicules; *P. ophthalmoides*; *L. mytilloides, Orbiculoidea* cf. *nitida,* Productid fragments; *Bucanopsis sp., Euphemites* cf. *anthracinus* (Weir), *Naticopsis?, Zygopleura sp.; Aviculopecten sp., Nuculana* cf. *acuta* (J. de C. Sowerby), *Parallelodon sp., Sanguinolites* cf. *ovalis* Hind, *Schizodus sp.; Anthracoceras sp., Gastrioceras?* [juv.]; *Geisina arcuata, Hollinella sp.;* fish scales	24	10	745	3
Dark grey silty mudstone; plant debris	1	3	746	6
Grey mudstone, silty in parts; roots at top; frequent thin ironstones in bottom 13 ft; *Anthraconaia sp.* [elongate form] and indeterminate 'mussels' 12 to 14 ft from base; *Curvirimula sp., Carbonicola* aff. *artifex* and *C.* cf. *declinata* 8 ft 11 in above base; *C.* cf. *fallax* near base; *P. ophthalmoides* 9 ft above base; fish remains 8 ft above base	41	11	788	5
Rashings		4	788	9
Grey mudstone, silty in parts, with several layers of massive quartzitic silstone; roots near top; well-preserved plants throughout, especially at 113 to 119 ft, 81 to 110 ft and 48 to 55 ft above base; *Alethopteris lonchitica, Calamites sp., ?Calymmatotheca stangeri* Stur, *Lepidodendron ophiurus* (Brongniart), *Lepidophloios laricinus* Sternberg, *Mariopteris acuta, M. nervosa* (Brongniart), *Neuropteris gigantea, N. schlehani* Stur, *Palaeostachya sp., Scheutzea?* and Sphenopterids including *Sphenopteris dilatata* Lindley and Hutton ..	118	10	907	7
Massive pale grey micaceous sandstone, quartzitic in part, with layers of grey silty mudstone with plant debris up to 3 ft thick in bottom 32 ft	110	3	1017	10
Blue-grey mudstone, slightly silty in parts; sporadic ironstones	76	8	1094	6

	Thickness (not corrected for dip) Ft In	Depth below the Garw Seam Ft In
Gastrioceras subcrenatum Marine Band: dark blue-grey and grey, shaly and silty mudstone with two distinct faunal assemblages. In top 32 ft: '*Chiton*' *sp.*; mollusc spat; *Zygopleura sp.*; *Caneyella* cf. *multirugata*, *Dunbarella sp.*, *Posidonia sp. nov.* [juv.]; *Anthracoceras sp.*, *Gastrioceras subcrenatum*, *G. sp. nov.* (aff. *cumbriense*), *Homoceratoides?* [juv.]; fish scales. In bottom 20 ft: *P. ophthalmoides*; *L. mytilloides*, Orthotetid fragments; *Promytilus?*, *Schizodus?* ..	52 0	1146 6

The evidence provided by the mapping of the western portion of the South Crop suggests that most of the significant horizons persist throughout this limited area. The *G. subcrenatum* Marine Band is not known at crop. The thick sandstone, about 80 ft above the band in the above borehole (equivalent to the 'Farewell Rock' in the western parts of the North Crop), persists eastwards to beyond Penprysg; it has been quarried [88658285] in the northern part of Ffwyl Wood, 1100 yd S. 38° W. of Aberkenfig Church, where thinly bedded fine-grained sandstones were formerly worked for flags and tiles, and on the central part of Cefn Hirgoed [940828], 500 yd W. 30° N. of Ffos-felen. The 30 ft of flaggy sandstones exposed in the cutting on the Llynfi Valley Railway [89858337], 150 yd E.N.E of Wern-dew, Aberkenfig, are probably part of the same sandstone.

The shales and mudstones immediately above are poorly exposed. The marine band, with base 745 ft 3 in below the Garw in the above borehole, appears to be represented in the stream just east of Bryn-coch, at a point [91148350] 400 yd W. 15° N. of Pistyllarian, where grey shales yielded *P. ophthalmoides*, *L. mytilloides* and *Euphemites?* through a thickness of 10 ft (see below). *Productus (P.) carbonarius* de Koninck from the spoil dug in the Mid-Glamorgan Water Board's reservoir [939829], 650 yd W. 30° N. of Ffos-felen, may also have come from this horizon.

A sandstone close below the Crows Foot Coal is very variable in its development. It is poorly represented in the Margam Park No. 1 Borehole, but on the west side of Cefn Cribwr and again on Cefn Hirgoed it is well developed, becoming quartzitic towards the east. A disused quarry [842827] near the Old Camp, 300 yd S.E. of Kenfig Hill Church shows the following section:

	Ft In
Flaggy sandstone	3 0
Blocky sandy mudstone	1 6
Strong sandstone..	1 0
Blocky sandy mudstone	2 6
Quartzitic sandstone .. .. 6 in to	11
Sandy mudstone .. .. 6 in to	9
Massive false-bedded sandstone	30 0
Thinly flaggy silty sandstone	9 0
Massive sandstone with bleached weathered surface	6 0

On Cefn Hirgoed the crop of what appears to be the same sandstone is marked by a line of crags of coarse whitish quartzitic grits containing scattered jasper pebbles, extending from the reservoir [939829], mentioned above, to the bare feature of Garn Fawr [956827], 700 yd E. by N. of Perth Celyn.

The Crows Foot Coal is reputed to be 24 to 32 in thick, but it has not been exposed for many years. A level [83358307] by the old quarry on the east side of the Porthcawl Railway, 750 yd W.N.W. of Kenfig Hill Church, presumably won the coal, and an old pit on Cefn Cribwr [85438262], 1130 yd W. 5° S. of Cefn Cross, is said to have struck it at about 60ft. A coal, 2 ft thick, exposed at the back of a quarry [915835], near Caehelyg, 800 yd S. 10° E. of Bryncethin Church, and worked from bell-pits nearby, is presumed to be the same seam. The 1-in coal, 661 ft 2 in below the Garw in the Margam Park No. 1 Borehole, is roofed by mudstone carrying *L. mytilloides* and fish debris; no evidence of this marine band has been found at crop.

The measures from the Crows Foot downwards to a horizon presumably not far above the base of the Coal Measures are exposed in the stream running S.S.E. from the small bridge [91128353] on the south-east of Bryncoch, 800 yd S. 20° W. of Bryncethin Church:

	Ft	In
Massive sandstone	4	0
Gap (presumably shales)	5	6
CROWS FOOT COAL, dirty and passing down into carbonaceous shale	1	6
Gap (presumably shales)	12	0
Coarse quartzitic grit passing down into flaggy sandstone	35	0
Grey shales (poorly exposed)	23	0
Grey shales: *P. ophthalmoides*; *L. mytilloides*; *Euphemites?*	10	0
Hard dark grey sandy mudstone	2	6
Grey blocky mudstone	17	0
Dark blue-grey sandy mudstone; *L. mytilloides*	5	0
Gap	14	0
Buff-grey silty shales, passing into	6	0
Buff flaggy sandstone	7	0
Buff-grey silty shales	9	0
Grey-buff flaggy sandstone	10	6
Carbonaceous mudstone		1
COAL		2
Grey seatearth passing down into grey shales	17	0
COAL		2
Grey and buff-grey shales with a few sandy layers	22	6
Buff-grey flaggy sandstone	18	0
Grey shales	6	0
Hard black shale		2
Grey shale		6
Buff-grey flaggy sandstones, thinly bedded in part	75	0
Buff-grey siltstone with spheroidal weathering	5	0
Gap	22	0
Grey silty mudstones	20	0

The beds associated with the Cefn Cribbwr Marine Band are seen in a much over-grown quarry [88128298], 1500 yd W. 30° S. of Aberkenfig Church. After excavation of the poorly exposed parts the following section was measured:

	Ft	In
Cefn Cribbwr Rock: impure quartzitic sandstone	10	0+
Cefn Cribbwr Marine Band: blue and blue-grey shales about	15	0
Buff blocky, slightly quartzitic sandstone with impersistent silty layers	9	0
COAL	thin	
Ganister-like seatearth	2	0
Blocky silty sandstone in beds up to 2 in	18	0
Ganister-like siltstone, weathering to soft tenaceous clay, passing down into	5	0
Purplish grey soft clay (? derived from dark blue mudstone)	2	0
Hard massive quartzitic sandstone, coarse-grained in part, in beds up to 6 ft in thickness	30	0+

Certain of these beds were also formerly worked for brick-making from an old level [87968317], 100 yd N.W. of Tycribwr, and much fossiliferous material from the Cefn Cribbwr Marine Band was found in the spoil. The following list combines the collections made from both spoil and quarry: *Crurithyris?*, *L. mytilloides*; *Caneyella* cf. *multirugata*, *Dunbarella* cf. *elegans* (Jackson), *D. papyracea*, *Posidonia* cf. *gibsoni* Salter, *P. sp. nov.* [right valve with prominent anterior lobe]; Orthocone nautiloid; *Anthracoceras sp.*, *Gastrioceras coronatum* Foord and Crick, *G. listeri*, *Homoceratoides divaricatus* (Hind); *Listracanthus sp.*, Palaeoniscid scales and *Rhadinichthys sp.*

This marine band is also poorly exposed in the south-east corner of the quarry near Caehelyg referred to above. A thin bed of blue-grey shale, very disturbed, lies between 40 ft of Cefn Cribbwr Rock, which was quarried, and massive quartzitic sandstone. A few fragmentary and poorly preserved goniatites and *L. mytilloides* were obtained.

The Cefn Cribbwr Rock, which varies in thickness from about 100 to 120 ft, is the first major sandstone below the main coals, and throughout its outcrop from Kenfig Hill to Penprysg has weathered into a marked ridge. The steep northern slopes of Cefn Cribwr and Cefn Hirgoed are formed by the dip slope of this sandstone, the clearly defined top of which is marked by a continuous spring-line. The sandstone varies from being flaggy and false-bedded, with a few silty partings towards the top, to being locally massive and quartzitic, and it has been much quarried. Several quarries near Waterhall Junction, on the north side of Kenfig Hill, show 30 ft of blocky and false-bedded sandstones, and similar beds are exposed in the side of the road from Fountain to Cefn Cribwr. Several adjacent quarries [891834] about 350 yd S.W. of Aberkenfig Church show 30 ft of well-bedded fine-grained sandstone, and similar sections are seen in quarries near Caehelyg [91528360], half a mile S. of Bryncethin. On the northern slope of Cefn Hirgoed shallow quarrying was almost continuous from near Derwengopa [93088333] eastwards for about $1\frac{1}{4}$ miles. The sandstone was extensively worked at Penprysg, where quarries [962825] on the west side of the Heol-y-Cyw road at the northern end of the village show:

	Ft	In
Thinly flaggy silty sandstone	12	0
Massive brown-weathering sandstone, with a few pellets of rolled clay-ironstone	7	6
Flaggy fine-grained sandstone	4	0
Soft flaggy sandstone, ferruginous in parts	5	0
Hard massive sandstone	20	0
Gap	about 4	0
Carbonaceous shales with plant debris		3
COAL	thin	
Pale mudstone-seatearth		6

Few exposures exist of the 300 to 400 ft of mainly argillaceous strata between the Cefn Cribbwr Rock and the Garw. The **Margam Marine Band** is known only at one locality [91138365], in the stream bed on the east side of Bryn-coch. The full thickness is not exposed, but 5 ft of rusty weathering dark shales yielded: *Dunbarella* aff. *elegans*, *Caneyella* cf. *multirugata* (? *sp. nov.*), *Posidonia* cf. *gibsoni*; a coiled nautiloid (cf. *Domatoceras*) and *Anthracoceras sp.*

Several sandstones, mostly impersistent and commonly quartzitic, are developed in the uppermost part of the sequence. They form definite though comparatively small features, colonized by bracken, and they stand out sharply against the featureless wet rushy land formed by the argillaceous strata. This contrast in vegetation is particularly well seen on the southern reaches of Hirwaun Common. One of the sandstones, 40 to 50 ft thick, lies close beneath the Garw and its crop seems to be more or less continuous from Kenfig Hill to Penprysg.

Elsewhere in the district, apart from a few old underground boreholes, which penetrated a little way below the Garw, these measures are unknown.   A.W.W.

B. GARW TO AMMAN RIDER

Fig. 27 shows generalized sections of these measures for each of the areas described below.

CYNON–TAFF AREA

The **Garw** and the measures upwards to the Gellideg are known only from the records of boreholes made before the present resurvey. The Garw is thin and has never been worked: at Fforchaman Colliery[1] it is recorded as being 18 in thick, at Lower Duffryn 14 in and at Lady Windsor 27 in. The measures between the Garw and the Gellideg are about 100 ft thick and consist mainly of mudstone with numerous ironstone bands and a few thin sandstones.

The **Gellideg** varies from about 2½ ft (in the northern part of the Penrikyber Colliery area) to 3½ ft, the thickening taking place, in general, in a westerly and south-westerly direction. Usually the seam consists of a single clean coal, but in the south-west of the area, at Lady Windsor Colliery, and in the north-west, at Fforchaman and Cwmneol collieries, a coal about 5 in thick is recorded near the base. No fauna is recorded.

In all collieries, except those of the Aberaman and Cwmaman areas, there is little separation between the Gellideg and the Five-Feet, and the two seams together form one workable unit. At Merthyr Vale and Deep Duffryn collieries only the Five-Feet portion of the section is worked at the face, but the full section is removed in the roadways; in parts of Penrikyber colliery the Gellideg together with the bottom coal of the Five-Feet is removed, and the upper coal of the Five-Feet is left as roof. In the Aberaman and Cwmaman areas the separation between the two seams is very variable, being as much as 50 ft in the north-west. In these areas a massive sandstone is developed above the Gellideg, and this is locally associated with washouts which can be traced south-westwards into the Ferndale area.

In the northern part of the area the **Five-Feet** Seam consists of two leaves, each 24 to 30 in thick. In the Fforchaman and Lower Duffryn colliery areas the separation between the two coals may be 4 ft or more, but south-eastwards this lessens rapidly, and south-east of a line drawn from Merthyr Vale pits through Navigation Colliery, the seam consists of a single main coal about 4 ft thick. In the Fforchaman–Cwmaman area a thin coal is developed locally between the two main leaves, while a further thin coal is nearly always present at the base of the section. The roof of the seam consists usually of well-bedded blue-grey mudstone containing sporadic *Curvirimula*; a sandstone roof occurs locally in Fforchaman Colliery.

The measures between the Five-Feet and the Seven-Feet are very variable, both in thickness and lithology. In general, the thickness varies from 70 ft in the south at Penrikyber, to little more than 10 ft in parts of the Fforchaman area. However, in the north, there is considerable variation even within single colliery areas: 12 to 60 ft at Fforchaman, 15 to 80 ft at Cwmneol, and from 25 to 60 ft at Aberaman. In these northern collieries a thin coal, rarely more than 12 in thick, occurs locally 2½ in to 30 ft below the Seven-Feet; this coal appears to die out southwards. The strata consist predominantly of mudstones with much ironstone, but sandstone, locally up to 30 ft thick, is extensively developed below the thin coal in the northern collieries. This sandstone dies out southwards and eastwards into the Merthyr Vale and Deep Duffryn areas, but it reappears in some strength in the Penrikyber area.

The **Seven-Feet** Seam in Merthyr Vale, Navigation, Deep Duffryn, Lower Duffryn, Aberaman and the southern part of Cwmneol collieries is 5 to 5½ ft thick and consists of three leaves with well-defined partings. The middle coal is consistently the thickest, being 24 to 36 in, while both the top and bottom leaves vary from 12 to 20 in. To the

---

[1] The National Grid-references of the major collieries are given in the Appendix of shaft and borehole sections, pp. 300 ff.

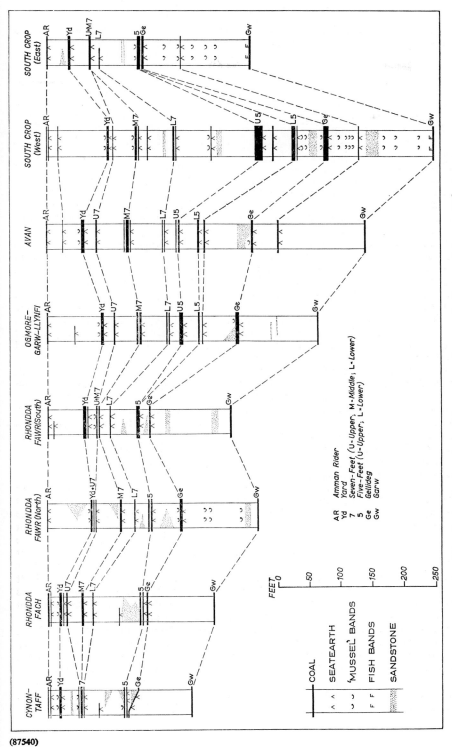

FIG. 27. *Generalized sections of the measures between the Garw and Amman Rider seams*

west and north-west, at Fforchaman and Cwmaman, the partings thicken, and the top coal, in particular, becomes separate. Over much of Fforchaman the middle and lower coals are worked together, but in the western parts of the old Cwmaman taking the splitting is so complete that only the middle coal was worked. Southwards from Mountain Ash the bottom leaf splits away and at Penrikyber Colliery the top and middle coals are worked together. The roof of the upper coal consists of grey mudstone, usually with small ironstone nodules. Fossils are rare, but poorly preserved 'mussels' including *Anthraconaia curtata* and *Naiadites sp.* have been obtained from Aberaman, Deep Duffryn, Merthyr Vale and Penrikyber collieries.

The strata between the top coal of the Seven-Feet and the Yard vary in thickness from 40 ft at Deep Duffryn to less than 2 ft in the northern parts of the Fforchaman and old Cwmaman takings. Southwards from Deep Duffryn they thin to about 16 ft in Lady Windsor Colliery. The beds normally consist of grey mudstones with numerous ribs and nodules of ironstone, and the Yard seatearth is usually well developed being as much as 12 ft in thickness. Where the cyclothem is thickest a thin sandstone is locally developed towards the middle.

The **Yard** Seam (locally called the No. 2 Yard) varies from 29 to 38 in, and is thinnest in the north and east of the area. Throughout the collieries of the Cwmaman–Aberaman area a coal up to 8 in thick lies 2 to 12 in below the main coal, but appears to die out southwards and eastwards. In parts of Merthyr Vale and Penrikyber collieries a few inches of rashings occur at the top of the seam. In the northern parts of Fforchaman and Cwmneol the top of the Seven-Feet is close enough to be included in the seam section, while in the southern parts of Aberaman, Cwmneol, and Fforchaman and the western part of Deep Duffryn, the Amman Rider is so close that the two coals have occasionally been worked together.

The roof of the Yard normally consists of blocky grey mudstone, usually containing scattered solid 'mussels' through a thickness of 1 to 3 ft. In the Merthyr Vale and Navigation areas the fauna is made up of variants of *Anthracosia regularis* associated with *Naiadites* aff. *triangularis* (J. de C. Sowerby), while at Deep Duffryn, Aberaman and Fforchaman these accompany large numbers of variants of *Carbonicola venusta*, together with *N. subtruncatus*. At Deep Duffryn *Lepidodendron ophiurus* and *Sigillariostrobus sp.* are also recorded.

The Yard to Amman Rider cyclothem is essentially argillaceous. It varies from a few inches of seatearth in the central parts of the area (see above) to 20 to 30 ft in the north and east, and from 10 to 15 ft at Lady Windsor. The seatearth beneath the Amman Rider is well developed, especially in the north where it may make up more than half the thickness of the cyclothem.

The **Amman Rider** (locally termed the Bute Rider) consists normally of two leaves, usually 5 to 10 ft apart, but separated by as little as 2 in. in the north (e.g. at Cwmaman). The top coal varies from 12 to 22 in and the bottom coal from 4 to 8 in.   A.W.W.

RHONDDA FACH AREA

The **Garw** has been proved only in Standard No. 2 Shaft and in a staple pit 300 yd W.S.W. of Mardy No. 3 Pit. In each case 21 in of clean coal was proved and the seam has nowhere been worked. The measures separating it from the Gellideg are 80 ft thick at Standard and 105 ft at Mardy; they consist of clift with numerous ironstone bands, 3 to 7 in thick; several sandstones, up to 10 ft, occur in the upper part.

The **Gellideg** varies from slightly less than 2 ft in the northern part of the Mardy area to rather more than $4\frac{1}{2}$ ft at National and Standard. It consists generally of a single clean coal, but in Mardy and National collieries a thin coal, usually less than 6 in, is present near the base, in the latter area often associated with bast. A well-defined washout crosses the northern portion of the Ferndale area from N.N.E. to S.S.W. Over most of the Ferndale ground the proximity of the Five-Feet gobs above has prevented the exploitation of the seam, and only small patches have been worked at National. At Standard Colliery it was formerly worked 5 to 14 ft beneath the Five-Feet. Northwards from Ferndale into the Mardy area the separation from the Five-Feet increases to about 40 ft, much of this thickness being made up of sandstone.

The **Five-Feet** is largely worked out in all the Ferndale collieries, as it is also in the National and Standard takings. Over most of this area the amount of separation between the two main leaves of the seam rarely exceeds 4 in, and this falls to nothing in the south-eastern parts of the National and Standard workings. In the Mardy area the parting thickens and is known to reach nearly 5 ft in the No. 3 Pit. The top coal varies from 28 to 39 in and the bottom coal from 22 to 36 in, the top coal being almost invariably the thicker. In the Mardy area a thin coal is present close below the top coal, while a thin leaf is developed in the base of the seam at Standard, National, parts of Ferndale No. 6 and at both Mardy pits.

The measures between the Five-Feet and the Seven-Feet range from 20 to 90 ft, being, in general, thinnest in the north at Mardy, and thickest in the south at Ferndale Nos. 8 and 9, and National. Over much of the Mardy and Ferndale Nos. 1 and 5 takings a massive sandstone, up to 30 ft thick, forms the roof of the Five-Feet, but southwards this splits up and its place is taken by mudstone with ironstone bands and nodules. The sandstone also dies out eastwards towards the Cynon Valley. A dirty coal, up to 11 in thick, usually lies above the sandstone at about 20 to 50 ft above the Five-Feet, but is not everywhere persistent.

The three coals, which in the Cynon Valley make up the full **Seven-Feet** Seam, are subject in this area to much splitting. They are always clearly distinct and are generally well separated. The Upper Seven-Feet varies from about 12 to 20 in; the Middle Seven-Feet (called in the Ferndale area the Two-Feet-Eight) 28 to 39 in; and the Lower Seven-Feet 10 to 20 in. The separation between the Lower and Middle coals varies from 4 to 20 ft, increasing fairly regularly in an east to west direction; that between the Middle and Upper coals from 1 in to 20 ft, being thinnest at Standard and National and in the north-east of the Ferndale area, and increasing to the west and south-west. The Middle and Upper Seven-Feet coals were worked extensively as a single seam at National and Standard, where it was called the No. 2 Seam. Over the rest of the area only the Middle Seven-Feet has been worked. Seatearth and mudstone with sporadic roots make up most of the ground between the split seams, but locally mudstone with ironstone occurs. At Mardy No. 1 Pit the mudstone roof of the Middle Seven-Feet yielded *Cyclopteris sp.*, *Neuropteris heterophylla* Brongniart and *N. obliqua* (Brongniart).

Over most of the ground the Upper Seven-Feet and Yard are separated by only 3 to 5 ft of seatearth. In the eastern parts of Ferndale and at National the separation increases, and up to 30 ft of mudstone with ironstone bands may occur. At Standard Colliery the Upper and Middle Seven-Feet and Yard coals are all close together.

The **Yard** Seam normally consists of a main coal 28 to 44 in thick, with a thin coal, 4 to 8 in, near the base. The seam is thinnest in the south and south-east of the area, at Ferndale No. 9 Pit and at National.

Over most of the Ferndale Nos. 1 and 5 and Nos. 6 and 7 areas the Yard and Amman Rider are separated by 2 to 6 ft of seatearth, and in parts of No. 5 Pit by as little as 5 in. Northwards into Mardy the Amman Rider splits away, and up to 40 ft of ground is developed between the two seams. Here the roof of the Yard consists of smooth grey mudstone with *Anthracosia regularis* and *Carbonicola venusta*, followed upwards by strong planty mudstone and striped beds. Southwards into Ferndale Nos. 8 and 9, the western parts of National and at Standard, 10 to 25 ft of strata intervene between the two seams, and the typical Yard fauna is no longer found. In Ferndale No. 9 the roof consists of fine-grained sandstone or strong silty mudstone with plants: *Alethopteris decurrens* (Artis), *A. lonchitica*, *Annularia radiata* Brongniart, *Asterophyllites sp.*, *Bothrodendron minutifolium* (Boulay), *Calamites sp.*, *Cyclopteris trichomanoides* Sternberg, *Mariopteris muricata* (Schlotheim), *Neuropteris heterophylla*, cf. *Renaultia gracilis* (Brongniart), *Sphenopteris trigonophylla* Behrend and *Sigillaria ovata* Sauveur. Large tree-boles, up to 2 ft in diameter, in position of growth, extend through the beds immediately above the coal, in some instances as far as the seam above.

The **Amman Rider** is everywhere a split seam: a top coal, 18 to 26 in, is underlain by a thin leaf, 6 to 18 in; in a few places a further few inches of coal is recorded below.

A.W.W.

NORTHERN RHONDDA FAWR AREA

The **Garw** has not been worked within the area. It proved to be 18 in thick in a trial pit at Fernhill Colliery, 27 in. in an underground borehole at Park Colliery, and 12 in. in another at Gelli. At Fernhill it lay 122 ft below the Gellideg, the intervening measures being mainly argillaceous, with numerous bands and nodules of ironstone and a few thin beds of sandstone. The log of a recent borehole, which reached 114 ft below the Gellideg, is as follows:

	Thickness		Depth below Gellideg	
	Ft	In	Ft	In
GELLIDEG	—	—	—	—
Silty seatearth and quartzitic siltstone	9	3	9	3
Striped beds; sandy at top, silty below	8	0	17	3
Sandstone	7	0	24	3
Silty mudstone	3	0	27	3
Striped beds, mainly siltstone; plant fragments; *Planolites montanus* near top	9	0	36	3
Dark grey mudstone; plant fragments and attached *Spirorbis sp.*; *Carbonicola* cf. *cristagalli* Wright, *C.* cf. *polmontensis* (Brown), *Curvirimula sp.*, *Naiadites sp.*; *Geisina arcuata* (Bean); *Belinurus sp.* The *Carbonicola* shells are preserved in thick calcite in the bottom 9 ft 4 in	15	0	51	3
Silty mudstone; shell fragments, *Geisina arcuata* and Palaeoniscid fish scales near top	17	6	68	9
Grey mudstone; *Curvirimula* cf. *trapeziforma* and cf. *Planolites ophthalmoides* near top	2	8	71	5
Grey mudstone, striped in parts, with a few ironstone bands	18	10	90	3
Evenly bedded mudstone; scattered fish	7	0	97	3
Quartzitic sandstone	17	0	114	3

Other boreholes at Fernhill proved similar sequences: in one a thin seatearth was noted 18 ft below the Gellideg, and in another *P. ophthalmoides* was obtained from silty mudstone at about the same horizon. At Tydraw Colliery the shaft proved a 3-in coal in the same position, and at Park what is presumably the same coal also carried *P. ophthalmoides*. At Fernhill *Neuropteris pseudogigantea* Potonié and *N. obliqua* were obtained just below this horizon.

At Park Colliery the Garw has been proved 117 ft below the Gellideg; a recent borehole passed through two horizons of thick calcite-shelled 'mussels' between the two coals; only the upper band appears to be present at Fernhill.

The **Gellideg** has been worked extensively in the southern part of the area and on a smaller scale in the north. A major washout, running from north-east to south-west through Tynybedw Colliery and Park shafts, affects the seam, and between the latter colliery and the Maindy workings has been proved to be 600 yd wide. To the south of this washout the seam is 4 to 5 ft thick, in general thinning towards the north-east. At Pentre Colliery a thin stone parting is developed about 3 in above the floor of the seam, and in the Maindy and Eastern areas there is a distinct top leaf of 3 to 8 in. To the north-west of the area the seam thins rapidly, and north of the washout the thickest recorded section is 39 in at Park, though it is more usually 27 to 33 in. In this northern area a 'stone' parting, 1 to 3 in thick, lies 30 in above the floor, and is presumably the same parting as that noted at Pentre. Another major washout, running north-west to south-east, forms the western limit of the workings from Maindy and Eastern, and continues south-eastwards into the Cambrian area and beyond.

The measures between the Gellideg and the Five-Feet vary considerably both in thickness and lithology. In the south-east at Pentre, Gelli, and Eastern collieries, the two seams are separated by less than 10 ft of mudstone and mudstone-seatearth. Westwards and northwards the measures thicken, largely through the incoming of sandstone in the roof of the Gellideg. West of Park they are 60 to 80 ft thick and consist largely of sandstone, which dies out north of Treherbert. At Tydraw the separation is less than 50 ft and sandstones are absent. In the Fernhill area boreholes show that there is a progressive westwards change in lithology from sandstones, first to striped beds, and then to mudstones; several boreholes show *Curvirimula sp.* associated with plant remains and attached *Spirorbis sp.* extending upwards for about 15 ft above the Gellideg. The upper part of the cyclothem has also yielded abundant plants including *Alethopteris decurrens*, *?Annularia radiata*, *Mariopteris daviesi* Kidston, *Neuropteris microphylla* Brongniart, *N. obliqua*, *N.* cf. *schlehani*, *N. tenuifolia* (Schlotheim) and *Sphenophyllum cuneifolium* (Sternberg).

The **Five-Feet** has been worked on a large scale from most of the collieries of the area. Typically the seam consists of two principal leaves, the lower usually 21 to 27 in thick, and the upper 30 to 36 in. Over much of the area there is also a middle coal, about 6 in, and a bottom coal of 3 to 6 in. Locally, around Treorky, another thin coal of about 6 in joins the section from above, and may be the coal often recorded between the Five-Feet and the Seven-Feet in the area to the north-east. At Abergorky Colliery the section of the Five-Feet is further complicated by the close approach of the Lower Seven-Feet, and the so-called No. 5 seam of the colliery comprises in different districts, the combined Lower and Upper Five-Feet coals, the Upper Five-Feet alone, and the combined Upper Five-Feet and Lower Seven-Feet. The parting between the Upper and Lower Five-Feet leaves is less than 12 in in the south of the area, but thickens northwards and as much as 8 ft is recorded at Lady Margaret, Ynysfeio, the northern part of Park, and in much of Tydraw. Boreholes at Fernhill suggest that the separation there is again reduced to a foot or so.

Around Abergorky the Five-Feet and Seven-Feet are as little as 2 ft apart, but more generally the two seams are separated by about 20 ft of argillaceous beds. The roof of the Five-Feet is normally a silty mudstone containing poorly preserved plants, though in places sandstones and striped beds may occur. At Gelli and Bodringallt the shafts record a thin composite coal between the Five-Feet and the Lower Seven-Feet.

Throughout the area the **Seven-Feet** is split into three distinct seams. The **Lower Seven-Feet** thickens westwards across the area from 18 to 27 in. It was formerly unnamed except at Tydraw, where it was worked on a small scale as the Five-Feet Rider or Upper Five-Feet. Both here and at Park and Dare collieries a thin parting may be developed near the middle of the seam. Characteristically the immediate roof contains a $\frac{1}{4}$-in band of cannel, overlain by a few inches of slickensided mudstone with plant fragments; above this the mudstone is darker and more fissile, but has yielded only worm burrows and tracks, including *Planolites montanus* at Park. It passes upwards into grey silty mudstones with ironstone bands, again containing wormy beds near the base, and these continue for 10 to 30 ft to the **Middle Seven-Feet.** This seam has been worked under a variety of names: at Abergorky it was known as the No. 4 Seam; at Pentre, Tydraw, and Tynybedw as the Lower Seven-Feet; and at Park and Dare as the Phils Seam. Its overall section thins from almost 4 ft in the south of Dare to about 27 in at Tydraw. The section includes a main top coal of 30 in and a bottom coal which thins rapidly from 10 in to 3 in when traced north-westwards from Maindy, and which, at Tynybedw and Tydraw, carries an inch or so of cannel coal at the top. The two leaves are separated by mudstones up to 3 ft thick. Locally, as at Dare Colliery, a thin cannelly mudstone lies in the roof of the seam and it is probably from this that Moore and Cox (1943, p. 233) record *Naiadites* cf. *triangularis*. The seam is affected by washouts locally; at Abergorky they run north to south, and at Park north-east to south-west. At Fernhill Colliery the Middle and Upper Seven-Feet coals are almost united to the north of the shafts, but elsewhere in the area the two seams are distinct and separated by up to 50 ft of silty mudstones with a few thin bands of sandstone. The **Upper Seven-Feet** is normally 10 to 12 in thick. Where recorded, it is everywhere closely associated with the overlying Yard and at Gelli, Eastern, Tydraw, and parts of Park forms a bottom coal to that seam. Over much of the area from Ton Pentre to Treherbert it is not recorded and is presumably absent or very thin.

The **Yard** has been worked from most of the collieries. At Ynysfeio it was called the No. 2 Yard; at Tydraw, Tynybedw, and Pentre the Seven-Feet; at Abergorky, and Lady Margaret the No. 3 Seam; and at Park and Dare the Three-Feet-Ten. It maintains a constant section of 27 to 36 in of top coal, separated by a few inches of 'stone' from a bottom leaf of 3 to 9 in. At Park and Dare up to half an inch of cannel occurs at the top of the seam.

The roof of the seam is normally a dark mudstone up to 2 ft thick containing abundant ironstone bands and nodules and many 'mussels'. This is followed by paler sandy mudstone with ironstones and sporadic shells. Variants of *Anthracosia regularis* and *Carbonicola venusta* associated with rarer cf. *C. oslancis* Wright have been collected at all the collieries where the seam was adequately exposed; in addition *Anthracosphaerium cycloquadratum* (Wright) and *Euestheria sp.* were recorded at Park. At the top of the cyclothem fragmentary plants, including *Mariopteris daviesi*, *Neuropteris gigantea*, *N. heterophylla* and *Sphenophyllum cuneifolium* occur at Fernhill.

The **Amman Rider** coal lies some 20 to 60 ft above the Yard. The two coals are closest along a line running east and west through Treherbert; here the intervening measures consist mainly of mudstone, and the Rider is usually a double coal, the section at Abergorky being: top coal 20 in, rashings 2 in, coal 6 in. Both to the north and south of this line the measures thicken, siltstones and sandstones develop, and the Rider splits into two distinct seams, 1 ft 6 in apart at Tynybedw, 3 ft 3 in at Pentre, and 20 ft at Dare. Southwards the lower coal thickens to 14 in and the upper thins to about 12 in. Northwards from Treherbert the splitting of the Amman Rider is accompanied by further deterioration of the lower coal, which at Fernhill is represented only by a seatearth 20 ft below the upper coal.   W.B.E.

SOUTHERN RHONDDA AREA

The measures below the Gellideg have not been seen during the present survey. A trial pit at Great Western Colliery and the Upcast Shaft at Coedely Colliery were sunk to depths of 201 ft 3 in and 299 ft 3 in respectively below the Gellideg; both sections were generally similar, that at Coedely reading as follows:

	Thickness		Depth below Gellideg	
	Ft	In	Ft	In
GELLIDEG	—	—	—	—
Seatearth and clift	30	7	30	7
Bastard rock	13	2	43	9
Clift	21	6	65	3
Clift and rock with 6 in of clay at base	14	6	79	9
Clift and pins of mine	13	10	93	7
Bastard rock and mine	10	3	103	10

	Thickness		Depth below Gellideg	
	Ft	In	Ft	In
Clift and mine	23	11	127	9
GARW: coal 13 in	1	1	128	10
Seatearth and rock	7	6	136	4
Clift	3	6	139	10
Rashings	1	0	140	10
Clift	2	5	143	3
COAL		2	143	5
Seatearth and rock	30	10	174	3
Clift with mine	29	8	203	11
Rashings	1	6	205	5
Clift with mine in upper part	48	9	254	2
Mixed ground	10	1	264	3
Clift and rock	25	0	289	3

At Great Western the trial pit reached nearly 100 ft below the Garw, the section of which was: top coal 10 in, rashings and coal 32 in, clod 6 in, coal 1 in. The Garw Seam has also been proved by boreholes at Britannic, Cilely, and Cwm collieries. At Britannic it is recorded as 26 in of clean coal lying 190 to 200 ft below the Five-Feet (the Gellideg being absent in washout). At Cilely 24 in of coal were proved 103 ft below the Gellideg and at Cwm 14 in of coal were penetrated 110 ft below the same seam.

The strata between the Garw and Gellideg are predominantly argillaceous, but several sandstones, up to 12 ft thick, occur at intervals at Britannic, Great Western, Cwm, and Coedely, while at Cymmer Colliery over 40 ft of 'rock' are recorded close below the Gellideg. Except at Cymmer, where 6 in of coal occur about 11 ft below the Gellideg, there is no record in this area of the thin coal which elsewhere characterizes the upper part of the measures between the Garw and the Gellideg.

The **Gellideg** has been worked extensively only at Cambrian, Llwynypia and Naval collieries (at all of which it was known as the Lower Five-Feet), at Cymmer Colliery (where it was called the Prince of Wales Seam), and at Cilely; in addition a few small patches have been worked beneath the gobs of the Five-Feet at Lewis Merthyr. Where not affected by washouts the seam varies from 27 in at Cwm and Maritime collieries to 5 to 5½ ft at Cilely, Naval (including Ely), Llwynypia, and Cambrian. In the Naval and Ely areas, 12 in to 2 ft of bast occur at the base, while thin coals are developed at the top of the seam in parts of Cambrian and near the base over parts of Cymmer, Lewis Merthyr, and Great Western. The extensive washout, already noted on the west side of Eastern and Maindy continues south-eastwards into this area (see Fig. 11). It crosses the Cambrian area in the neighbourhood of the shafts and continues into Britannic Colliery. Despite several widely spaced attempts to locate the seam at this last-named colliery it has not yet been proved, and it would seem that the washout embraces virtually the whole of this taking. The seam has not been proved at Coedely and it is possible that this colliery is also affected by the same washout.

At Lewis Merthyr, Great Western, Maritime and the greater part of Cwm collieries, the Gellideg and Five-Feet seams are close together, the separation being usually less than 2 ft. Northwards and westwards, however, the two seams split apart rapidly: in Naval and Ely pits 10 to 12 ft of strata, largely seatearth, separate the coals and these expand to 50 ft or more in the Cambrian area. Virtually the whole of the thickening is due to the development of massive sandstone, which forms the roof of the Gellideg over much of the western part of the area. Turbidity currents, associated with the influx of this sandstone, were presumably responsible for the extensive washouts in the Gellideg already described.

The **Five-Feet** everywhere consists of one main coal, the parting which separates the two principal leaves to the north having apparently died away through the northern margins of Cambrian and Llwynypia. This main coal varies from about $3\frac{3}{4}$ to $6\frac{1}{2}$ ft thick, being thinnest in the south-east in the Cwm taking, and thickening fairly steadily westwards to $4\frac{1}{2}$ ft at Great Western, Lewis Merthyr, and Cymmer, to 5 to $5\frac{1}{2}$ ft at Coedely, Cilely, Britannic, Naval, and Llwynypia, being thickest on the west side of the Cambrian workings. Over much of Llwynypia, Naval, and Britannic this single clean coal makes up the whole of the seam, but a thin coal, 4 to 9 in, is developed at the base, underlying 3 to 12 in of rashings, in parts of Cambrian, Cymmer, and Lewis Merthyr. Locally in Lewis Merthyr and Great Western, where the seam is very close to the Gellideg, a further few inches of coal occur immediately below. Over most of Lewis Merthyr, a bast, 4 to 10 in thick, is developed near the top of the seam, and passes locally into clod or rashings. The latter are particularly noticeable at Cwm, where 15 to 19 in of coal and rashings overlie the main coal of the seam.

The strata between the Five-Feet and the Lower Seven-Feet consist mainly of mudstones and silty mudstones with ironstone nodules, particularly in the upper part; a sandstone, usually thin but reaching 25 ft in the Ely area, occurs at about the middle. Over most of the area there is rapid variation of thickness from about 30 to 90 ft. The greatest thicknesses are recorded in a zone extending from Naval Colliery through parts of Cymmer, Lewis Merthyr, and Great Western to the north-eastern part of the Cwm taking, and there is a progressive thinning to the west and south. No fossils have been noted during the present survey from the roof of the Five-Feet, but *Curvirimula subovata* was collected from the roof of a 4-in dirty rider to the seam, which is locally developed in parts of Cambrian and Naval collieries. At Great Western Colliery *Annularia radiata*, *Asterophyllites equisetiformis* (Schlotheim), *Neuropteris heterophylla* and *Sphenophyllum cuneifolium* were collected from about 25 to 30 ft above the Five-Feet, while some 40 ft higher *Cyclopteris trichomanoides* occurred associated with *N. heterophylla*.

The **Seven-Feet** group of coals shows considerable variation. The **Lower Seven-Feet** coal, which has nowhere been worked in this area, ranges in thickness up to 21 in. It is thickest in the west, at Cambrian and in parts of Britannic; it thins eastwards and is usually less than 12 in at Cymmer and Lewis Merthyr, and appears to die out altogether in parts of Great Western, Maritime, and Cwm. The separation from the Middle Seven-Feet varies from 5 to 35 ft, being least in western parts of Lewis Merthyr and greatest at Britannic.

The **Middle** and **Upper Seven-Feet** coals form one extensively worked seam, locally called the Middle Five-Feet, at Cymmer, Lewis Merthyr, Great Western, Maritime, and Cwm collieries. At Cymmer and Lewis Merthyr the seam thickness averages about $3\frac{1}{2}$ ft, the upper leaf being 12 to 18 in, and the lower leaf 24 to 30 in. The parting dies out eastwards into Great Western and Cwm, and the single coal shows a marked thinning in the extreme south-east of the area, being only 20 to 28 in over most of Cwm.

Westwards into Cilely, Britannic, and Cambrian the Upper Seven-Feet leaf splits away and becomes a bottom coal to the Yard Seam above. In these collieries the Middle Seven-Feet (locally termed the Yard) shows a distinct westwards thickening and reaches 38 in. in the western parts of Britannic and Cambrian.

The amount of separation of the Upper Seven-Feet from the Yard is small over most of the area, but it reaches 10 to 30 ft in portions of the Lewis Merthyr and Great Western workings. The only fossils recorded are some poorly preserved *Naiadites sp.* in the roof of the upper leaf at Lewis Merthyr.

The **Yard** Seam has been known by various names: Bute at Cambrian, Llwynypia, Naval, and Britannic; Upper Five-Feet at Cymmer, Lewis Merthyr, Great Western, Maritime, and Cwm; Jubilee at Cilely, and Four-Feet at Coedely. The main coal has a thickness of $3\frac{1}{2}$ to $5\frac{1}{2}$ ft; it is thinnest in the north, in parts of Cambrian, Llwynypia, Naval and the northern parts of Lewis Merthyr and Great Western, and thickest in the south at Cwm and Coedely. Over the north-western half of the area a thin coal, 9 to 12 in, is typically developed at the base, separated from the main coal by a few inches of hard stone. Except in the south-west, and in the northern parts of Lewis

Merthyr and Great Western, the Upper Seven-Feet Seam lies close beneath the Yard, and in some areas in the west, e.g. at Cambrian, is so close that it may be regarded as forming part of the seam. At Lewis Merthyr and Great Western a thin coal occurs at the top of the seam, separated from the main leaf by 2 to 12 in of rashings or bast; this thin coal appears to develop by splitting.

The strata overlying the Yard usually consist of planty mudstone with cannelly mudstone forming the roof in a number of places. At Great Western Colliery the following plants are recorded: *Calamites carinatus* Sternberg, *Carpolithus sp. nov.*, *Mariopteris sp.*, *Neuropteris gigantea* and *N. heterophylla*.

In the eastern part of the Cwm area the separation between the Yard and the Amman Rider is less than 1 ft. North-westwards, however, this increases to 20 to 30 ft in the Great Western–Lewis Merthyr area, 45 to 60 ft at Cymmer and Cilely, 50 to 90 ft in the Naval–Llwynypia area, and thins again to about 40 ft at Cambrian. Except where thin and mainly seatearth, the strata consist of mudstones with ironstone nodules, though a few thin sandstone layers are developed locally. A dirty and impersistent coal sometimes occurs about 10 to 25 ft below the Amman Rider and may represent a split from that seam; this has been noted at Cambrian, Naval, Cymmer, Cilely, and Great Western, while a prominant band of rashings occurs in this position in the Naval and Ely shaft sections.

The **Amman Rider** coal itself is everywhere persistent and varies from 14 to 24 in. in thickness. A.W.W.

OGMORE–GARW–LLYNFI AREA

The strata below the Gellideg are known only from a few underground boreholes drilled before the present survey. At Wyndham Colliery two holes proved the **Garw,** 13 in and 20 in thick, 125 ft below the Gellideg, the intervening strata being described as light and dark clift with a few thin rock bands. One of the holes penetrated a further 150 ft through light and dark clift ending in 6½ ft of 'rock'. A similar section was proved at Ffaldau Colliery, 24 in of coal lying 125 ft 5 in below the Gellideg. At Caerau Colliery a borehole reached 183 ft 3 in below the Gellideg: the Garw, consisting of 12 in of coal underlain by a further 18 in of coal and rashings, is here separated from the Gellideg by 124 ft 9 in of clift, containing a few sandy beds towards the middle and numerous ironstone bands in the upper part, passing into 6 ft 3 in of seatearth at the top. The lowest beds recorded are 50 ft of clift with a few ironstone nodules. A hole at Coegnant Colliery was drilled to a depth of 167 ft below the Gellideg; though no Garw Coal was proved, the sequence was similar to that at Caerau. At St. John's Colliery two boreholes record a coal, 6 to 7 in thick, about 14 ft below the Gellideg.

At the time of survey the **Gellideg** had been worked only at Wyndham, Ffaldau, Garw, St. John's, and Coegnant collieries, though it had been proved by boreholes at Western, International, and Caerau. At Western several boreholes to the east of the Glyncorrwg Fault have recorded varying thicknesses up to 2 ft 4 in, and it seems possible that the seam here is still affected by the major washout, the eastern margin of which was proved at Maindy, Eastern, and Cambrian (see pp. 87, 89). Over the rest of the area the seam averages 4½ to 5 ft of clean coal. It is thickest in the south, at Ffaldau and St. John's, where it reaches 5½ ft, and thins northwards to about 4 ft in the northern parts of Garw.

A shelly roof to the Gellideg is recorded only at St. John's and in the south of the Coegnant workings. Here some 3 ft of mudstone overlie the coal and abundant 'mussels' occur towards the top; these latter range up for about 1 ft into silty mudstones containing sandstone bands. The fossils occur as well-preserved internal moulds, and forms allied to *Carbonicola communis* make up the bulk of the assemblage. Species identified include: variants of *C. communis, C.* aff. *cristagalli, C.* cf. *martini* and *C.* cf. *rhomboidalis* Hind. Elsewhere the mudstone roof is not developed. In the Ogmore and Garw valleys the coal is succeeded by strong striped siltstones or fine-grained sandstones, and in the north of the area strong sandstones form the roof. In the northern part of the Coegnant taking, washouts and roof-rolls associated with this sandstone are common, trending a few degrees north of west. Similar washout conditions have recently been encountered at the northern end of the Wyndham workings, and it is interesting to note that both these isolated areas of seam failure are in line with the extensive washout extending north-eastwards from Park and Dare through Ferndale to the Cwmaman–Aberaman area (see pp. 82, 84).

Throughout the area west of the Glyncorrwg Fault about 40 to 60 ft of strata intervene between the Gellideg and the lower coal of the Five-Feet group. Much of this ground normally consists of strong silty mudstone with massive sandy and silty beds in the lower part.

In the extreme east of the area the **Five-Feet** has a complex section, not unlike that in the south-west part of the Park workings: thus in the neighbourhood of the Wyndham shafts it consists of top coal 52 in, clod 9 in, coal 11 in, clod 3 in, coal 24 in, rashings 3 in, coal 6 in, a thin parting being sometimes developed about 3 to 4 in from the top of the main coal. Westwards from this area a major split develops between the top coal, or **Upper Five-Feet,** and the lower coals, at presumably the same horizon as that previously described in the northern parts of the Rhondda and Cynon valleys (see pp. 82, 87). The Upper Five-Feet Coal, usually called the Five-Feet in the Garw Valley and the Lower Five-Feet in the Maesteg area, varies from 4 ft in the north of the area to $6\frac{1}{2}$ ft in the south-west, e.g. at St. John's, which implies a considerable thickening south-westwards of the top coal of the Five-Feet of the Rhondda Valley. A thin coal of 6 to 9 in usually occurs within 1 to 2 ft of the base. The separation of this from the Lower Five-Feet increases rapidly west and south-west of Wyndham and reaches 40 to 45 ft at Ffaldau and in the Maesteg area, the strata consisting mainly of strong silty mudstones. The **Lower Five-Feet** is itself rather complex in the Garw Valley and the following sections have been recorded:

	Garw		International		Ffaldau	
	Ft	In	Ft	In	Ft	In
Coal		7		$4\frac{1}{2}$		
Clod or shale . .		2		1	2	0
Coal	1	9	2	5		
Clod or shale . .		1		1		9
Coal		4		$9\frac{1}{2}$	1	0

In the Maesteg area the top and middle coals of these sections appear to be represented by 18 to 27 in of coal, which at Coegnant has a sandstone roof. The bottom coal thickens somewhat and splits away: thus, at Coegnant, it is 19 in thick with two thin partings and lies about 15 to 25 ft below the main part of the Lower Five-Feet.

Through virtually the whole of this area the roof of the Five-Feet consists of mudstone, usually somewhat soft and weak. No fauna has been collected during the present survey, but Evans and Simpson (1934, p. 450) record *Naiadites sp.* and fragmentary shells in the Maesteg area. The measures upwards to the base of the lowest coal of the Seven-Feet group consist of 12 to 20 ft of strong silty mudstones passing into seatearth.

Everywhere within the area the **Seven-Feet** group of seams comprises three distinct coals, covering in all some 27 to 100 ft of strata. These coals represent the same three splits as those described in the previous areas. The **Lower Seven-Feet** (locally called the Upper Five-Feet or Little Seam) shows a distinct thickening when traced from the east into this area, and a parting is developed a little below the middle of the seam. This parting makes its appearance between Park, Cambrian, and Britannic collieries on the one hand and Western on the other, and is present in all the collieries. At

Wyndham and Garw this dirt band may reach as much as 8 in, but elsewhere it seldom exceeds 1½ in and may be less than ½ in. In total thickness the coal averages 2¾ to 3¼ ft, the top leaf being 19 to 24 in and the lower about 14 to 18 in. The seam has been worked on a small scale in the Llynfi Valley and at Garw Colliery. The roof invariably consists of mudstone, which in the western part of the area may be dark, soft and slickensided. This passes upwards into strong mudstones and silty mudstones, with sandstone bands developed towards the top of the cyclothem. These beds vary from 20 to 50 ft, being thinnest in the east of the area and thickest in the collieries of the Llynfi Valley.

The **Middle Seven-Feet** also becomes more complex when traced westwards from the Rhondda Valley area: at least two, and possibly three, new partings develop and thicken as the seam is traced towards Maesteg. The following representative sections illustrate this:

	Western		Ffaldau		Garw		St. John's		Coegnant		Caerau	
	Ft	In	Ft	In	Ft	In	Ft	In	Ft	In	Ft	In
Coal		3		11		10						
Clod or shale		5	3	8		10						
Coal	3	6	2	5	2	5	2	6	2	6	2	6
Stone or seatearth		1		9		9	7	3	2	6	2	0
Coal	1	2	2	1	2	1	2	2	1	6	2	0
Seatearth ..							3	3	0			
Coal							5	1	1			

The single coal of the Rhondda Valleys has already developed partings near the top and bottom in the south-western parts of Park Colliery; both these partings are well marked in Western, Garw, and Ffaldau, though in Wyndham the upper appears to have disappeared again. The lowest of the resulting three coals thickens westwards, though it always remains subordinate to the leaf above. The detailed correlation of the seam between the Garw and Llynfi valleys is obscure; it would seem that the top leaf either dies out or rejoins the middle coal, while the lowest coal again splits near its base. Except at Western Colliery the seam has not been worked in this area. The partings developed in the coal consist normally of clod or seatearth while the roof of the uppermost coal is usually silty mudstone with plants and plant debris.

The separation between the Middle and Upper Seven-Feet seams is very variable: at Caerau it appears to be but 3 ft; in the Garw Valley it is about 12 to 15 ft but this increases eastwards to about 35 to 40 ft at Wyndham and Western, and westwards to 50 to 65 ft in the southern collieries of the Maesteg area. The strata consist mainly of mudstone and silty mudstone with numerous ironstone bands, and in the Maesteg area subordinate beds of sandstone appear.

The **Upper Seven-Feet** varies from 9 to 15 in and normally consists of a single coal. In the eastern parts of the area, at Western and Wyndham, the coal lies close beneath the Yard Seam, the separation being as little as 1 in. Elsewhere the two seams have split away from one another. In the Garw Valley 15 to 20 ft of mudstone and mudstone-seatearth separate the two seams, while in the Llynfi Valley area this is reduced to 7 to 15 ft.

The **Yard** (locally known as the Three-Feet-Ten at Western, the Seven-Feet at Wyndham and in the Garw Valley, the No. 8 Seam at St. John's, Caerau, and Coegnant, and the New Seam at Garth Merthyr) everywhere consists of two coals. The upper and main coal is 27 to 41 in thick, being thickest in the east of the area, while the lower coal is 8 to 21 in, the higher figure being again recorded in the Ogmore Valley. The parting between the two coals is usually a hard 'stone', varying in thickness from 2 in to about 12 in. Locally, as in Ffaldau and Garw collieries, the parting consists of clod or rashings. In the Maesteg area a ½-in band of cannel lies in the immediate roof of the seam.

The roof of the Yard everywhere consists of grey mudstone with thin ironstone bands and it contains the usual characteristic 'mussel' fauna, the principal members of which are variants of *Anthracosia regularis* and *Carbonicola venusta*. In addition the following species have also been identified from most of the collieries: *A. sp.* cf. *aquilina*, cf. *C. oslancis*, *Anthracosphaerium sp.* and *Naiadites spp.* belonging to the *productus/quadratus* group. At Ffaldau and Garw collieries *Belinurus bellulus* (König) has also been recorded.

The measures between the Yard and the Amman Rider vary from about 60 to 100 ft, being, in general, thinnest in the Ogmore Valley collieries and thickest in those of the Maesteg area. A thin coal of 6 to 12 in seems to be everywhere developed about midway between the Yard and the Amman Rider, and this may represent a split from the Amman Rider, as previously noted in the Rhondda valleys (pp. 86, 91). The strata intervening between these coals are made up mainly of mudstone and silty mudstone with numerous thin bands and nodules of ironstone in the lower cyclothem. Locally, as at Ffaldau and in parts of the Llynfi Valley, ribs of sandstone are developed in the higher cyclothem, while at Ffaldau a prominant band of 'white rock' is recorded not far below the thin coal.

The **Amman Rider** is about 22 to 30 in thick. In the Ogmore and Garw valleys there is usually a thin parting near the top. In the Maesteg area the coal is again composite, consisting normally of three leaves up to 10 in thick separated by partings of 1 to 2 in. It appears likely that the lower leaf splits away when the seam is traced westwards into the Bryn area.

A. W. W., W. B. E.

AVAN VALLEY

At the time of survey these measures had been proved only at Bryn and Avon collieries, and were accessible at only the former.

The **Garw** is known only at Bryn Colliery; it is 27 in thick and lies 190 ft below the Gellideg horizon. The intervening measures are predominantly argillaceous, but thin sandstones occur and bands of ironstone are numerous. At Cwmavon (in the Swansea district, one-inch Sheet 247) several of these ironstones were formerly worked, and some of these workings appear to have extended eastwards into the present district. As in other areas where these ironstones were worked, the individual productive layers were named by the old miners: the lowest horizon, the Ram's Head Mine, lies about 100 ft below the Gellideg, the Yellow Mine some 25 ft higher, and the Black Mine and Spotted Pins are associated with an 11-in coal lying 40 ft below the Gellideg.

The **Gellideg** was worked at Bryn as the Lower Vein; it varies from 4 ft in the north of the taking to about 6 ft in the south. In the northern parts of the colliery it has a sandstone roof which in places has cut out part or all of the seam. The general trend of such wash-outs and 'rolls' varies between west to east and south-west to north-east. In the south of the taking the seam is reported to have a mudstone roof containing abundant 'mussels', presumably the same fauna as that known in the Llynfi Valley (see p. 91).

The main part of the **Lower Five-Feet** lies between 25 and 70 ft above the Gellideg at Bryn, the intervening measures being mainly sandy mudstones. It has been called the Yard and is said to be up to 3 ft thick, although where seen during the present survey it comprised barely 12 in of dirty coal. Some 10 to 30 ft higher there is another 12 in of coal, of sulphurous appearance, as in the Llynfi Valley.

The **Upper Five-Feet** is about 120 ft above the Gellideg at Bryn, where it has been extensively worked under the name Middle Vein. It varies in thickness between $4\frac{1}{2}$ and 6 ft, while further thin coal horizons lie about 4 ft and 10 ft beneath it.

At Avon Colliery the 'K' Seam was formerly worked with the following section: top coal 28 in, mudstone 1 to 2 in, coal 4 in, mudstone 1 to 2 in, coal (with a thin parting locally towards the base) 18 in. This section is similar to that of the central parts of the Rhondda Fawr and it may represent the whole of the Five-Feet.

Both at Bryn and Avon the measures between the Five-Feet and Lower Seven-Feet usually comprise some 30 ft of unfossiliferous silty mudstones containing numerous thin ironstone bands. In the west of the Bryn taking, however, they are reduced to only a few feet and at Cwmavon are measurable in inches.

The three main coals of the **Seven-Feet** group, where proved, comprise separate split seams, and are likely to remain split in those northern parts of the area yet unexplored. The **Lower Seven-Feet** has been worked on a small scale at Bryn as the Little Vein, a coal 2½ to 3 ft thick. At Avon its equivalent appears to be the 'H' Seam, which is recorded as having a top coal of 19 in, rashings 11 in, and bottom coal 9 in. The roof at Bryn is slickensided mudstone yielding only worm-tracks. Some 30 to 45 ft of sandy mudstones separate it from the **Middle Seven-Feet** or No. 6 Seam of Bryn. This seam here comprises three leaves: top coal 18 to 24 in, seatearth 12 in, middle coal 24 in, seatearth 5 ft, rashings 6 to 12 in, bottom coal 12 in. The roof is a barren silty mudstone which continues upwards through 30 to 35 ft to the No. 5 Seam, the local equivalent of the **Upper Seven-Feet,** which is about 2 ft thick. At Avon the Middle Seven-Feet appears to separate into two seams some 36 ft apart: the lower, seam G, is 26 in thick, with a 4-in rider close below; the upper is 20 in thick. The Upper Seven-Feet is here a thin rider coal close below the 'F' Seam. The roof, where seen at Bryn, is a yellow-weathering mudstone with fragmentary plants; it is overlain by silty mudstones which continue to the seatearth of the Yard, the two seams being 8 ft apart at Avon and 15 to 20 ft at Bryn.

The **Yard** has been called the No. 4 Seam at Bryn and the 'F' Seam at Avon. At Bryn it is about 3 ft thick with a thin parting about 9 in from the base and up to 1 in of cannel in its roof. At Avon it is about 2¾ ft thick. Its roof, at Bryn, consists of dark mudstone with ironstone bands, the lowest foot of which yielded an abundant fauna of *Anthracosia regularis*, and variants of *Carbonicola venusta* and *C. oslancis*, together with *Planolites montanus*.

The sequence continues at Bryn through 70 to 75 ft of mudstones and sandy mudstones to the **Amman Rider.** Two 10-in coals occur; one about 30 ft and the other about 15 ft below the top. Where seen, the Rider coal was badly exposed in disturbed ground; a single coal 10 in thick was measured, but the seam has also been recorded as two thin leaves, 3 to 4 in thick, separated by 18 in of seatearth. At Avon also the coal was seen only in disturbed conditions and its measured thickness of 18 in is unreliable.   W.B.E.

SOUTH CROP

West of the Ogmore Valley: The strata from the Garw to the Gellideg are known in detail only from the recently drilled Margam Park boreholes. This portion of Nos. 1 and 2 boreholes has been described elsewhere (Woodland and others 1957b, pp. 45, 48). No. 3 Borehole [81618730], situated 1900 yd E. 35° N. of Margam Church, showed the following section:

	Thickness (not corrected for dip)		Depth below Gellideg	
	Ft	In	Ft	In
GELLIDEG	—	—	—	—
Strong grey mudstone, silty in parts, with roots	5	10	5	10
Massive grey mudstone	4	6	10	4
Pale grey micaceous sandstone with silty layers ..	5	0	15	4

	Thickness (not corrected for dip)		Depth below Gellideg	
	Ft	In	Ft	In
Grey mudstone with striped laminae in middle; sporadic ironstones up to 3 in; *Planolites* cf. *ophthalmoides* and *Carbonicola sp.* at 20 ft 1 in	16	9	32	1
Dark blue-grey silty mudstone, becoming more carbonaceous downwards, with sporadic ironstones. Thick calcite-shelled 'mussels' (including *C.* aff. *communis* and variants of the *C. communis/pseudorobusta* group) associated with *Curvirimula subovata* and *Geisina arcuata* at 33 ft, 35 ft 4 in to 3 ft 2 in, 38 ft 7 in to 40 ft 1 in, and 43 ft 1 in	14	0	46	1
Grey mudstone	1	7	47	8
Dark carbonaceous shale, cannelly in part; fish scales		5	48	1
Cannel COAL		2	48	3
Black carbonaceous shale; *Curvirimula sp.*; fish scales	1	6	49	9
Grey mudstone; scattered plant debris and roots becoming fewer downwards	6	10	56	7
Pale grey fine-grained sandstone, silty at base	21	8	78	3
Blocky grey mudstone with sporadic ironstone bands and nodules; *Planolites sp.* near top; thick calcite-shelled variants of *Carbonicola pseudorobusta*, associated with *Spirorbis sp.* and *Geisina arcuata* from 89 ft 8 in to 91 ft 5 in; fragments of *Curvirimula sp.* and fish scales scattered thoughout	14	11	93	2
Blocky grey mudstone with ironstone bands up to 4 in; sporadic plant debris; *Belorhaphe kochi* (Ludwig)	10	2	103	4
Grey mudstone with a few ironstone bands; *Spirorbis sp.*, cf. *Planolites sp.*; *Carbonicola* aff. *pseudorobusta*	7	4	110	8
Strong grey silty mudstone, showing turbulence near top, with a few ironstone bands up to 4 in; abundant plant debris	20	7	131	3
Grey mudstone, darkening downwards, with sporadic ironstone bands up to 4 in; pyritic granules at 155 ft 10 in, *Planolites spp.* including large and small varieties as well as typical *P. ophthalmoides* from 138 ft 11 in to 169 ft 4 in; *C.* cf. *communis*, associated with *G. arcuata* at 144 ft 1 in; fragments of *Curvirimula sp.* common in upper part; fish remains, including Acanthodian spines and *Rhadinichthys sp.* frequent in lower part	41	1	172	4
Black carbonaceous shale with coalified wood debris		9	173	1
GARW: coal 19 in	1	7	174	8

The **Garw** or **Cribbwr Fach** Seam was formerly worked from bell-pits near its crop in the vicinity of the Porthcawl Branch railway between Kenfig Hill and Cefn Junction [85908340]: farther east several drift mines worked the coal on a small scale, as for example, Bankershill Slant [86998323], 1070 yd W. 19° S. of Park Slip Colliery. The dark roof shales with fish-remains, noted in the above borehole, weather at surface to brittle paper-shales, and these are a characteristic feature of the spoil from the bell-pits. The seam is said to vary from 2 ft 3 in to 3 ft of clean coal.

As on the eastern parts of the North Crop (Robertson 1927, pp. 63–7; 1933, pp. 100–2) the strata between the Garw and the Gellideg (or Cribbwr Fawr) on the western part of the South Crop contain important developments of ironstone. The ironstone layers, which vary up to 15 in. in thickness, were formerly worked extensively from open patchworks in a strip south of Nantiorweth-goch from Cwm-ffos westwards towards Kenfig Hill. They were also mined at several of the collieries in the same area, for example, Bryndu Colliery; while an old slip-mine [84878338] 370 yd W.S.W. of the Old Cefn Slip Colliery was driven for the sole purpose of working the ironstones. At Bryndu Colliery the following section, which names the principal productive ironstone horizons, is recorded:

		Thickness		Depth below Gellideg	
		Ft	In	Ft	In
GELLIDEG (CRIBBWR FAWR)		—	—	—	—
Fireclay and bastard fireclay or rock		3	6	3	6
Hard rock		3	0	6	6
Strong cliff with mine		6	0	12	6
Ironstone			5	12	11
Mine ground		2	0	14	11
Strong rock		2	8	17	7
Ironstone	4 in to	1	0	18	7
Mine ground		2	0	20	7
Ironstone			3	20	10
Mine ground		2	9	23	7
Ironstone .. .. ⎫			6	24	1
Mine ground .. ⎬ Black Vein Ground		1	6	25	7
Ironstone .. .. ⎬			3	25	10
Mine ground .. ⎬		2	0	27	10
Black Pin Ironstone.. ⎭			4	28	2
Black Pin Mine Ground with irregular balls of mine ..		12	0	40	2
Rock		3	0	43	2
Yellow Vein Mine Ground		4	6	47	8
Strong rock		6	0	53	8
Strong mine ground		11	0	64	8
Red Mine Ground with six courses of ironstone and two of balls		15	0	79	8
Mine ground with balls of ironstone		9	0	88	8
Ironstone			2	88	10
Mine ground		3	9	92	7
Ironstone .. .. ⎫			5	93	0
Mine ground .. ⎬ Double Pin Ground		3	0	96	0
Ironstone .. .. ⎭			6	96	6
Clod or inferior cannel		1	0	97	6
Strong cliff		13	0	110	6
Bastard rock		6	0	116	6
Shale		3	0	119	6
Ironstone			3	119	9
Strong shale		4	9	124	6
Ironstone			$1\frac{1}{2}$	124	$7\frac{1}{2}$
Mine ground		2	0	126	$7\frac{1}{2}$
Ironstone .. .. ⎫			4	126	$11\frac{1}{2}$
Cliff .. .. .. ⎬		2	0	128	$11\frac{1}{2}$
Ironstone .. .. ⎬ Blue Vein Ground			4	129	$3\frac{1}{2}$
Cliff .. .. .. ⎬		5	0	134	$3\frac{1}{2}$
Ironstone .. .. ⎭			3	134	$6\frac{1}{4}$

							Thickness		Depth below Gellideg	
							Ft	In	Ft	In
Strong cliff	..	..	..	..	..	..	4	0	138	6½
Ironstone	..	..	..	..	..	..		2½	138	9
Mine ground	..	..	..	..	..	..	1	9	140	6
Ironstone	..	..	..	..	..	..		4	140	10
Mine ground	..	..	..	..	..	..	2	9	143	7
Ironstone	..	..	..	..	..	..		7	144	2
Clod and cliff	..	..	..	..	..	..	8	8	152	10
Pin Garw Ironstone		..	..	..	..	..		6	153	4
Ground	..	..	..	..	..	..	5	0	158	4
Little Pin Garw Ironstone	..		..	..	..	..		3	158	7
Mine ground	..	..	..	..	..	..	4	0	162	7
GARW (CRIBBWR FACH)		..	..	..	..		2	10	165	5

The Garw Pins are reported as being the principal veins, and at Bryndu were taken when drawing back after the workings of the Garw coal. Most of the patch-workings are much degraded and grown over, and few exposures except of sandstones and strong silty mudstones are now to be seen. An old quarry [85268337], 1450 yd E. 22° N. of Kenfig Hill Church shows:

	Ft	In
Flaggy sandstones with thin beds of sandy shale	10	0
Greenish-grey silty shales with large ironstone septaria	11	0
Dark blue-grey and green-grey shales with a few ironstone bands and nodules	16	0
Gap, with GARW Coal near base	9	0
Grey shale seatearth, becoming silty downwards	2	6
Strong grey sandy mudstone	6	0

Apart from the ironstones obtained during the working of the quarry the associated mudstones were utilized to produce a good red brick.

The seams from the Gellideg upwards to the Yard have been worked fairly extensively in the old collieries between Pyle and Bryncethin. However, only at the western end of this tract, at Newlands Colliery, can the measures now be seen and a complete picture of the succession made out. All the other old drift mines and pits have long since been abandoned; in these, except for that portion of the sequence between the Gellideg and Five-Feet seams, the measures are subject to a great deal of disturbance, and since the only information now available consists of unreliable records of old cross-measures roadways, in which no attempt was made to unravel the structure or to record the true sections of the seams, exact correlation is seldom possible. In consequence, it is not thought worth while to reproduce the sections given in the plate facing p. 56 of the Second Edition of this memoir. Reliable sections of the measures on the South Crop between the Newlands–Cribbwr Fawr–Bryndu area in the west and Llanharan Colliery some ten miles to the east do not exist. Considerable expansion, accompanied by seam-splitting, occurs in this portion of the sequence from Llanharan to Newlands and little is known about the exact nature of the variations involved.

These measures crop out in a narrow east-west strip, varying from 150 to nearly 300 yd wide, extending from the neighbourhood of Bryndu Colliery to Evanstown, where they pass beneath the alluvial deposits of the Ogmore River. In the area between Old Cefn and Park Slip the surface is marked by numerous bell-pits and by extensive patchworks. In this western tract the strata between the Gellideg and Upper Five-Feet are relatively competent and the records in this portion of the sequence are generally much more reliable than those for the rocks above.

The crop of the **Gellideg** or **Cribbwr Fawr** between Kenfig Hill and Aberkenfig can be followed fairly easily, for numerous bell-pits were formerly sunk to win the seam, while, in addition, a number of adits or slips were driven on it, as for example Old Cefn Slip [85158350], Jenkin's Slip [85378346], Wain Arw Slip [85708342], Cwm-ffos [87048337] and New Cribbwr [88948370]. The seam was worked extensively at Cribbwr Fawr, Bryndu, Cefn, and Park collieries, at all of which it appears to have been relatively undisturbed. At Park Slip the lowest workings, in the so-called New Seam, are believed to have been in the Gellideg, a thrust-fault separating the two sets of workings. The seam consists everywhere in this area of a clean coal; it is thickest at Newlands Colliery, and the eastward thinning is indicated by the following sections: Newlands 9 ft 3 in to 8 ft 10 in, Cribbwr Fawr 8 ft to 7 ft, Bryndu 6 ft 6 in to 6 ft, Cefn 6 ft 4 in to 5 ft 6 in, and Park Slip 5 ft 6 in to 4 ft 6 in. At Ynysawdre the seam appears to be about 5 ft, while at the old Barrow Pit, Bryncethin, a seam called the Cribbwr, but by no means certainly the Gellideg, was said to be 3 ft 8 in. in thickness. In the Cefn area a thin bast, up to 4 in thick, is recorded at the top of the seam.

Over most of the area the roof of the Gellideg Seam consists of rather weak dark shaly mudstone, which in the Margam Park boreholes yielded the following fauna: *Anthraconaia* cf. *fugax* Eagar, *Anthracosphaerium sp.*, *Carbonicola* cf. *communis*, *C.* cf. *martini*, *C.* cf. *pseudorobusta*, *Curvirimula* cf. *subovata* and *Naiadites sp.* In parts of the Park Slip area this mudstone is cut out by massive sandstone (which normally lies a little way above the shell-bed) and this forms a massive roof to the seam.

The strata from the Gellideg to the Lower Five-Feet (or Five-Quarter) Seam vary in thickness from 30 ft in the Park Slip area to 70 ft at Newlands. They consist of strong mudstones or silty mudstones, with bands of sandstone, particularly in the middle of the cyclothem. In the Margam Park No. 2 Borehole there is an apparently abnormal development of *Curvirimula*-bearing mudstones as shown by the following section:

	Ft	In
LOWER FIVE-FEET bottom coal: **coal 25 in**	2	1
Grey silty mudstone-seatearth	5	4
Dark grey micaceous mudstone with layers of silty mudstone; non-marine lamellibranchs throughout, associated with plants; plants include *Alethopteris decurrens*, *A. lonchitica*, *Lepidostrobus variabilis* Lindley and Hutton, *Neuropteris hollandica* Stockmans, *N. obliqua*, *N. pseudogigantea*, *N. schlehani*, *N. tenuifolia*; shells are more numerous in lower half and consist chiefly of forms of *Curvirimula sp.*, including *C. subovata*, associated with *Spirorbis sp.* and *Geisina arcuata*	43	10
Massive pale grey quartzitic sandstone with silty layers near top and bottom	40	8
Grey shaly mudstone, darkening downwards, with irregular ironstone nodules; *?Anthracosphaerium sp.*, *Carbonicola sp.* (? *communis*), *Curvirimula subovata*; fish fragments	10	10
GELLIDEG: **coal 94 in**	7	10

The **Lower Five-Feet** (or **Five-Quarter**) consists everywhere of a top coal of 42 to 48 in and a bottom coal of 10 to 24 in separated by about 6 in to 3 ft of clod or fireclay. The coal usually has a strong roof and floor, and, in consequence, the main haulage slips at Cefn and Park collieries were driven in this seam. During 1924 Park Slip was reopened and exploratory work carried out in the seam in the neighbourhood of the main slant. The section proved to be very irregular, and this, associated with the massive sandstone roof, is suggestive of wash-out conditions.

The **Upper Five-Feet** (or **Nine-Feet** of the South Crop) usually lies about 50 to 70 ft above the Lower Five-Feet. The intervening measures are strong and consist mainly of massive mudstone with sandstone, and include a coal, locally called the Danllyd or Fiery, about 2 ft thick, in a position some 6 to 20 ft below the Upper Five-Feet. This coal appears to correlate with a thin coal near the base of the top coal of the Five-Feet in the north-east parts of the district. The Upper Five-Feet itself is always thick, but the available sections are now so few and variable that little can be said concerning its lateral behaviour. The following sections have been noted:

Newlands	Ft	In	Cribbwr Fawr	Ft	In	Cefn	Ft	In	Park	Ft	In
Coal ..	3	4	Coal ..	1	9	Coal ..	1	0	Coal ..		9
Brass ..		1	Clod ..		3	Clod ..	3	0	Shale ..		5
Coal ..	3	8	Coal ..		10	Coal ..	1	6	Coal ..	1	9
Rashings		1	Brass ..		7	Clod ..		6	Shale ..		3
Coal ..	2	4	Coal ..	1	0	Coal ..	7	0	Coal ..	4	6
Clay ..		1	Parting ..		1						
Coal ..	2	10	Coal ..	2	0						
			Clod ..		3						
			Coal ..	6	6						

The measures from the Upper Five-Feet to the Lower Seven-Feet, which are essentially argillaceous in character, show the usual east to west thickening, being about 60 ft at Park Slip and Cefn, and about 100 ft in the Cribbwr Fawr–Newlands area. At Newlands Colliery a 9-in coal, 19 ft below the Lower Seven-Feet, carries a roof of grey shaly mudstone with *Carbonicola* aff. *oslancis* and *Naiadites flexuosus* Dix and Trueman overlying a 10-in band of dark cannelly shale.

The **Lower Seven-Feet** (or **Slatog Fawr**) has a section similar to that in the Garw and Ogmore valleys. Recorded sections are reliable only in the Bryndu–Newlands area, where a top coal, 24 to 30 in thick, is separated from a bottom coal of 12 to 15 in by $\frac{1}{2}$ to 8 in of clod or rashings. At Newlands the roof consists of weak mudstone carrying plant debris including *Asterophyllites charaeformis* (Sternberg), *Diplotmema furcatum* (Brongniart), *Mariopteris muricata* and *?Neuropteris obliqua*.

The distance from the Lower Seven-Feet to the Middle Seven-Feet seems to be very variable, though few of the records are dependable. At Park Colliery the seams cannot be identified with certainty; at Cefn the two seams are said to be 90 ft apart, while at Bryndu only 22 ft appears to separate them; in Newlands and Cribbwr Fawr figures of 65 ft and 90 ft respectively seem reliable. The strata between the two seams are essentially argillaceous, and at Newlands a thin coal occurs about 10 ft below the Middle Seven-Feet.

The **Middle Seven-Feet** (or **Slatog Fach**) is a composite seam which, because of the large amount of dirt developed within it, has rarely been worked. At Newlands the following section was measured: roof of strong barren mudstone, carbonaceous shale with streaks of coal 2 in, coal 29 in, coal and dirt in bands 7 in, seatearth 3 to 20 in, coal 16 in, seatearth 4 in, coal with thin parting 4 in. At Cribbwr Fawr the top coal was 34 to 35 in, while at Bryndu a section reads: coal 30 in, clod 15 in, coal 15 in.

The ground between the Middle Seven-Feet and the Yard consists of rather silty mudstone with sandstone bands and varies from about 10 ft at the Park Slip opencast site to about 55 ft at Newlands. A thin coal, usually found a short distance below the Yard, is presumed to represent the Upper Seven-Feet.

The **Yard** (or **Six-Feet** of the South Crop) consists of 4 to $4\frac{1}{2}$ ft of clean coal at Newlands and Cribbwr Fawr, and it appears to thicken eastwards, 6 ft being recorded at Cefn and 7 ft at the Ironworks Colliery nearby. A roof of smooth grey mudstone with ironstone nodules yielded scattered solid 'mussels' at Newlands: *Anthracosia regularis, Carbonicola venusta, C.* cf. *oslancis* together with squashed *Naiadites sp.* intermediate between *productus* and *quadratus*.

At Park Slip opencast site the Amman Rider Seam was separated from the presumed Yard by only 1 ft of seatearth. Elsewhere, however, a fully developed cyclothem intervenes between the two seams. At Bryndu 87 ft of clift with a prominent sandstone band in the middle passes upwards into a thick seatearth immediately beneath the lower split of the Amman.

The **Amman Rider** is usually split into two distinct coals, both of which are composite and variable, separated by 15 to 20 ft of strata. The upper seam, known locally as the Wythien Fach or Little Seam, normally consists of two or three leaves having an overall thickness of 2 to 3 ft.

East of the Ogmore Valley: On Hirwaun Common the crop of the Garw can be followed by means of a double line of closely spaced bell-pits which extends for over a mile from about 100 yd N.E. of Heol-y-llan [93748332] to about 500 yd S. of Caeaucerrig [95298359]. The seam, about 2 ft 3 in thick and containing much disseminated pyrite, was exposed by the Opencast Executive in a trench [91848371] near the western end of Hirwaun Common, about 750 yd S.E. of Bryncethin Church, where the characteristic roof of dark blue-grey shale with fish remains was observed. The measures exposed by the trench were badly disturbed and it was impossible to determine the true succession of strata between the Garw and Gellideg seams. However, three distinct bands of dark slightly silty shale and mudstone, carrying the thick calcite shells typical of these beds, were seen. The lowest, composed of very thinly bedded black shale, occurred about 70 to 80 ft above the Garw and carried: *Spirorbis sp.*, *Carbonicola* cf. *centralis* (J. de C. Sowerby), *C.*aff. *communis*, fragmentary *Curvirimula sp.*; *Geisina arcuata* and fish remains. A dark silty mudstone, some 40 ft higher, contained *Carbonicola* aff. *communis* and *Curvirimula* cf. *candela*, while 20 to 30 ft above lay a dark micaceous silty mudstone with *Spirorbis sp.*, *Carbonicola pseudorobusta*, *C.* cf. *rhindi* (Brown), *Curvirimula subovata*, *Naiadites* cf. *flexuosus*, *Geisina arcuata* and fish remains.

At Llanharan a coal, 9 in thick, was encountered about 60 ft below the Cribbwr (Gellideg and Five-Feet) Seam in the Nos. 2 and 3 horizons main cross-measures drifts in the south-east sector of the colliery. This coal, which presumably correlates with the horizon of the 2-in cannel in the Margam Park No. 3 Borehole (see above), carries a much-slickensided roof of grey mudstone with *C.* cf. *subovata* and fish remains. A prominent shelly bed 20 ft higher marks the highest of the '*pseudorobusta*' bands: variants of *Carbonicola communis*, *C. martini* and *C. pseudorobusta* well preserved in dark shaly mudstone, and this is overlain by grey mudstone containing *Spirorbis sp.*, *C. rhindi*, *Geisina arcuata* and fish remains.

At Ynysawdre the correlation of the coals above the Five-Feet is doubtful. The following section from the Gellideg to the supposed Amman horizon was recorded in the North Drift from pit-bottom:

	Thickness		Depth below Amman Marine Band	
	Ft	In	Ft	In
?AMMAN RIDER: **coal 27 in**	2	3	2	3
Cliff and mine	23	0	25	3
Seam called Slatog, but possibly YARD with UPPER SEVEN-FEET: **coal 39 in**, clod 3 in, **coal 11 in**, rashings 3 in, **coal 8 in**, soft fireclay 18 in, **coal 4 in** ..	7	2	32	5
Fireclay	2	10	35	3
Shale and cliff	23	0	58	3
Rock	18	0	76	3
?MIDDLE SEVEN-FEET: **coal 46 in**, fireclay and cliff 54 in, **coal 18 in**	9	10	86	1
Cliff with bands of rock	16	6	102	7
?LOWER SEVEN-FEET: **coal 10 in**		10	103	5
Cliff	33	5	136	10
UPPER FIVE-FEET: **coal 17 in**, clay and shale 14 in, **coal 20 in**, shale 1 in, **coal 19 in**	5	11	142	9

	Thickness		Depth below Amman Marine Band	
	Ft	In	Ft	In
Fireclay, shale and cliff	17	0	159	9
COAL		6	160	3
Cliff	8	7	168	10
LOWER FIVE-FEET: **coal 42 in**, clod 6 in, **coal 13 in** ..	5	1	173	11
Bastard rock	3	0	176	11
COAL		3	177	2
Cliff and bastard rock	19	6	196	8
GELLIDEG: **coal 60 in**	5	0	201	8

The marked thinning of the measures between the Gellideg and Upper Five-Feet foreshadows the coalescence of the coals to form a single seam at Llanharan and in the area to the east. These lower coals were undoubtedly proved at Bryncethin (Barrow) Pits, but the sections which survive are not capable of interpretation. The seams must crop out on the southern reaches of Hirwaun Common, but no natural exposures exist and during the various stages of opencast activity in this area the strata exposed were always so highly disturbed that no true stratigraphical sections could be made.

Eastwards of Penprysg (at the eastern end of Hirwaun Common) the seams pass beneath an extensive cover of glacial deposits and it is only at Werntarw and Llanharan collieries that they can be studied again. The Llanharan sequence shows marked attenuation together with considerable re-uniting of the coals accompanied by reduction of the total coal thickness. The following section was taken in the No. 1 Conway cross-measures drift:

	Thickness		Depth below Amman Marine Band	
	Ft	In	Ft	In
AMMAN RIDER: **coal 20 in**	1	8	1	8
Fireclay	5	0	6	8
Rock	2	0	8	8
Clift with ironstone bands and nodules	26	0	34	8
Soft dark shale	1	0	35	8
YARD: **coal 4½ in**, rashings 1½ in, **coal 67 in**	6	1	41	9
Soft fireclay		9	42	6
Strong fireclay with ironstone nodules	12	11	55	5
Strong clift with ironstone nodules	13	11	69	4
UPPER and MIDDLE SEVEN-FEET: **coal 30 in** ..	2	6	71	10
Rock		9	72	7
Fireclay with ironstone nodules	5	2	77	9
Clift with large ironstone nodules	2	7	80	4
Rashings		8	81	0
?LOWER SEVEN-FEET: **coal 8 in**		8	81	8
Fireclay with ironstone nodules and streaks of coal ..	6	1	87	9
Strong sandy clift with irregular layers of ironstone ..	21	9	109	6
Strong sandy shale with rock bands	19	2	128	8
Dark slippery shale with bands and layers of ironstone ..	13	0	141	8
Dark shale and clod	1	4½	143	0½
Rashings		0½	143	1
FIVE-FEET ⎱ (CRIBBWR): **coal 48 in**, clod and rashings GELLIDEG ⎰ 9 in, **coal 52 in**	9	1	152	2

The coals of the **Upper** and **Lower Five-Feet** of the ground to the west have joined to form a single seam, whereas the **Gellideg** lies very close below. These two coals together were formerly known as the Cribbwr or No. 12 Seam. The **Seven-Feet** group of coals have deteriorated considerably; the 2 to 2½ ft of coal (No. 11 Seam), which occur some 1 ft to 30 ft below the Yard (No. 10), probably represent the combined Upper and Middle Seven-Feet, while the Lower Seven-Feet is developed only locally in the colliery. The **Yard** is 5 to 6 ft thick and frequently has a strong sandstone roof; under these conditions the seam is subject to wash-outs.

At Werntarw Colliery, about 2½ miles W.N.W. of Llanharan, the **Yard** (originally called the Six-Feet) has been worked fairly extensively; 5 to 5½ ft thick, it is separated from the Amman Rider by 44 ft of mudstone with ironstone bands and nodules. The Rider consists of a clean coal 23 in thick. Nothing is known with certainty of the sequence below the Yard.

There is no further evidence concerning these lower coals on the Sheet to the east of Llanharan, but sections are recorded on the abandonment plans of the old South Cambria Colliery, on the south-west margin of the Newport district. Here the Cribbwr Seam is reported to be a clean coal 5 ft 6 in thick, representing presumably both the Five-Feet and the Gellideg together. The Brass Seam above (with a section given as: coal 6 ft, clod 3 to 8 in, coal 18 to 24 in) presumably represents the Yard together with the Upper and Middle Seven-Feet. These sections, therefore, show further attenuation of the measures, a process which continues when these strata are traced farther eastwards to Nantgarw and beyond.   A.W.W.

REFERENCES

EVANS, W. H'. and SIMPSON, B. 1934. The Coal Measures of the Maesteg District, South Wales. *Proc. S. Wales Inst. Eng.*, **49,** 447–75.

MOORE, L. R. and COX, A. H. 1943. The geological sequence in the Taff Valley, and its correlation with the Rhondda Valley sequence. *Proc. S. Wales Inst. Eng.*, **59,** 189–304.

ROBERTSON, T. 1927. The Geology of the South Wales Coalfield. Part II. Abergavenny. 2nd edit. *Mem. Geol. Surv.*
——1933. The Geology of the South Wales Coalfield. Part V. The Country around Merthyr Tydfil. 2nd edit. *Mem. Geol. Surv.*

WOODLAND, A. W., ARCHER, A. A. and EVANS, W. B. 1957. Recent boreholes into the Lower Coal Measures below the Gellideg–Lower Pumpquart horizon in South Wales. *Bull. Geol. Surv. Gt. Brit.*, No. 13, 39–60.

CHAPTER V

MIDDLE COAL MEASURES: DETAILS

A. Amman Marine Band to Cefn Coed Marine Band

Fig. 28 shows generalized sections of these measures for each of the areas described in the following text.

CYNON–TAFF AREA

The measures between the Amman Rider and the Bute are thinnest in the Penrikyber and Lady Windsor colliery areas, where they vary from 12 to 16 ft; over most of the rest of the area they range from 20 to 40 ft. The greatest thicknesses occur at Fforchaman and Cwmneol, where 64 and 58 ft are recorded, but even here as little as 20 ft may separate the two seams locally. The **Amman Marine Band** is present almost everywhere, and its failure in the Navigation Colliery North Shaft is presumably due to wash-out at the base of a 40-ft sandstone. The band is up to $4\frac{1}{2}$ ft thick, and usually consists of blue-grey blocky mudstone with abundant pyritic granules, 'fucoids', and nodules in the lower part, and blue-grey micaceous silty shale with ironstones in the upper part. The pyritic mudstone carries small *Lingula mytilloides*, associated with rare *Orbiculoidea sp.*, *Donaldina?*, *Dunbarella sp.* and *Nuculana sp.*, and the silty shale is rich in larger specimens of *Lingula*. The marine strata are directly followed by smooth mudstone which everywhere contains poorly preserved 'mussels' including *Anthraconaia* cf. *williamsoni*, *Anthracosia* cf. *concinna*, *A.* aff. *ovum*, *A.* cf. *phrygiana*, *A.* cf. *regularis* and *Naiadites quadratus*; *Spirorbis sp.* has also been recorded. The rest of the cyclothem is mainly argillaceous, though one or more sandstone bands commonly occur in the middle part.

The **Bute** is fairly constant in character throughout most of the area: two coals, each 21 to 30 in thick, are separated by a well-developed clod or seatearth. The seam has been worked at Lady Windsor and the collieries of the Aberaman–Cwmaman area, where the clod is less than 2 ft. Elsewhere the clod thickens and the individual coals are not considered worth working; at Merthyr Vale, the separation is as much as 15 ft. Little is known about the seam at Penrikyber Colliery, where only one coal appears to be represented in the shafts, and it is possible that wash-out conditions affect the seam here as they do at Aberaman and Cwmneol collieries, where the top coal is locally absent.

At Merthyr Vale the bottom coal carries a fauna of small solid 'mussels', including *Anthracosphaerium exiguum* and *Naiadites* aff. *subtruncatus*, in blocky grey mudstone. Few fossils are found in the roof of the top coal: at Fforchaman poorly preserved squashed shells, including *Anthraconaia* aff. *williamsoni*, occur in a thin layer immediately above the coal, and *N.* cf. *quadratus* has also been found.

The strata between the Bute and the Nine-Feet consist of mudstone and silty mudstone with numerous ironstone bands and nodules, with, locally, a few thin sandstones in the middle. The thickness ranges between about 10 and 80 ft, and the variation is very rapid and local: thus in the Aberaman–Cwmaman area figures of 17 to 50 ft are recorded; at Deep Duffryn 18 to 75 ft; at Penrikyber 55 to 80 ft; at Lady Windsor only 20 ft separate the two seams.

104

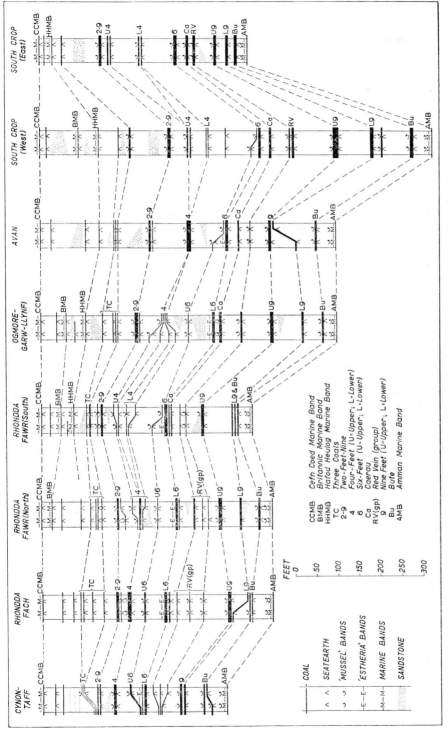

Fig. 28.   *Generalized sections of the measures between the Amman and Cefn Coed marine bands*

In much of the area, and more particularly in the north, the **Nine-Feet** is very disturbed. The disturbance is largely confined to the plane of the seam and is commonly associated with the development of thick structural rashings, especially in the roof. The coal varies in thickness from 4 ft to 15 ft or more, and is so mixed with dirt and rashings that the original section is difficult to make out. North of Mountain Ash a parting, locally associated with cannel, occurs about $3\frac{1}{2}$ to 4 ft from the top of the coal, and locally, as in Deep Duffryn and Merthyr Vale, may thicken to 12 ft. In the Cwm Cynon, Penrikyber and Lady Windsor areas, a parting appears in the bottom coal, while at the same time the top separation thins considerably. This lower split develops westwards and south-westwards to produce the distinct Upper Nine-Feet and the Lower Nine-Feet seams. The following sections illustrate the variations.

	Aberaman				Deep Duffryn				Cwm Cynon		
		Ft	In			Ft	In			Ft	In
Coal	..	3	6	Coal	..	3	7	Coal	..	3	6
Bast	..		6	Clift, etc.	..	11	1	Fireclay	..	1	0
Coal	..	5	0	Bast	..		5	Coal	..	2	7
				Coal	..	4	0	Clod	..		3
								Coal	..	4	0

At Merthyr Vale and the collieries in the south of the Cynon Valley the roof of the Nine-Feet consists of dark carbonaceous shale with abundant 'mussels': *Anthraconaia* aff. *salteri*, *Anthracosia* aff. *beaniana*, *A. sp.* cf. *disjuncta*, and variants of the group represented by the species *A. aquilina*, *A. ovum*, *A. phrygiana* and *A. retrotracta*. At Aberaman the shelly roof fails, and in the strong clift above the disturbed coal the following plants were found: *Asterophyllites equisitiformis*, *Calamites suckowi* Brongniart, *Diplotmema furcatum*, *Neuropteris heterophylla*, *Palaeostachya sp.*, *Pinnularia capillacea* Lindley and Hutton, *Sphenophyllum cuneifolium* and *Sphenopteris trigonophylla*.

Between the Nine-Feet and the Six-Feet the measures are irregular in character. This is mainly due to the variable development of the individual coals of the **Red Vein** group. In the Cwmaman area these are concentrated into a single compound seam, usually with three distinct leaves, e.g. coal 15 in, fireclay 31 in, coal 27 in, clod 3 in, coal 22 in. Eastwards and south-eastwards into the Aberaman and Mountain Ash areas the coals split apart and thin markedly. Thus at Lower Duffryn Colliery four coals, each 4 to 8 in thick, are recorded through 25 ft of strata; at Merthyr Vale three coals, 3 to 12 in thick, cover 40 to 50 ft. At Navigation Colliery no coals are present in the shafts, though a thick fireclay horizon 40 ft or so above the Nine-Feet may indicate the general position of the group. At Penrikyber and Cwm Cynon two thin coals associated with much seatearth occur between the Nine-Feet and the Six-Feet. Towards Lady Windsor the coals improve in thickness, for at that colliery 21 in of coal are recorded 74 ft above the Nine-Feet, as well as a coal 12 in thick 27 ft still higher and near the base of the Six-Feet.

Between the Nine-Feet and the Red Vein group there are usually some 40 to 70 ft of strata, mainly mudstones with ironstones together with a few intercalated sandstones in the upper half of the cyclothem. The measures associated with the Red Vein coals and extending to the Six-Feet are normally wholly argillaceous with a great deal of seatearth. In the Aberaman–Cwmaman area, where the Red Vein forms a distinct seam, 10 to 50 ft of measures separate it from the Six-Feet, the thickness increasing from east to west, while in the Mountain Ash area the corresponding strata measure 30 to 45 ft.

The only fossils associated with the Red Vein group in this area were found at Cwm Cynon and Penrikyber collieries. At both localities *Anthraconaia* cf. *curtata* and *Naiadites* cf. *productus* occur 6 to 24 in above a thin coal about 50 ft above the Nine-Feet.

In the south and east of the area the **Six-Feet** is a compound seam consisting of three main coals:

		Lady Windsor		Penrikyber		Cwm Cynon		Navigation	
		Ft	In	Ft	In	Ft	In	Ft	In
Coal ..	..	1	10	2	8	2	10	2	9
Dirt ..	..		0½		1		0½		6
Coal ..	..		9	1	0	1	0	1	3
Dirt ..	..		0½		2		7		3
Coal ..	..	3	9	4	0	2	9	3	3

Locally, as at Merthyr Vale, the middle coal is associated with bast, and a further thin coal occurs between it and the top coal. Northwards and westwards from Deep Duffryn Colliery the top coal splits away and becomes the Upper Six-Feet (No. 1 Yard); the lower coal—the Lower Six-Feet—is similar to the middle and lower coals of the above sections, with the difference that its upper leaf thins considerably northwards:

	Deep Duffryn		Lower Duffryn		Aberaman		Fforchaman	
	Ft	In	Ft	In	Ft	In	Ft	In
UPPER SIX-FEET: Coal ..	2	0	2	0	1	10	2	0
Measures	6	6	28	0	33	0	47	0
LOWER SIX-FEET: ⎰Coal..		9		5		4		2½
Dirt ..		10		10		4		0¼
⎱Coal..	3	0	3	3	3	4	3	6

Where the two seams are apart the separating measures, which may be as much as 70 ft thick in the extreme north-west of Cwmaman, consist of mudstones with ironstones, and with some local sandy developments in the middle part.

The Lower Six-Feet, where it develops a mudstone roof, as at Aberaman and Fforchaman, usually yields *Lioestheria striata* associated with *Naiadites obliquus* Dix and Trueman. At Merthyr Vale, Penrikyber and Lady Windsor, the roof of the Six-Feet contains abundant 'mussels', including *Anthracosia* cf. *aquilina*, *A*. aff. *atra*, *A*. cf. *lateralis*, *A. sp.* cf. *phrygiana*, *A. planitumida*, and *Naiadites* cf. *alatus* Trueman and Weir. *Lepidodendron ophiurus*, *Neuropteris heterophylla* and *Trigonocarpus parkinsoni* Brongniart have also been recorded.

About 12 to 55 ft separate the Six-Feet (or Upper Six-Feet) from the Four-Feet. The seams are closest together in the Aberaman–Cwmaman area, while the greatest separations occur in parts of Deep Duffryn and in the Penrikyber–Lady Windsor area. The strata are made up mainly of mudstones with ironstones, but sandy intercalations are noted in Deep Duffryn and Navigation.

In the Fforchaman–Aberaman–Deep Duffryn–Merthyr Vale region the **Four-Feet** consists of 5½ to 6 ft of clean coal. Formerly it was worked extensively but is now largely exhausted. In other parts of the area three partings develop, giving the four-leaf section which becomes characteristic of the seam in the Rhondda Fach Valley. In the extreme north-west, in the old Cwmaman workings, both upper and lower partings are present, giving a section: coal 19 in, clod 2 in, coal 48 in, clod 10 in, coal 10 in. South of Deep Duffryn in the Cynon Valley only the lower parting appears, e.g. at Cwm Cynon: coal 66 in, clod 4 in, coal 16 in. In the extreme south of the area all three partings are present at Lady Windsor: coal 15 in, stone 1 in, coal 15 in, stone 1 in, coal 23 in, rashings 3 in, coal 19 in.

The roof of the Four-Feet normally consists of blocky rather weak mudstone with ironstone nodules, and non-marine lamellibranchs often occur in the bottom 1¼ ft. These are particularly abundant at Cwmneol and Lady Windsor and include: *Anthraconaia sp. nov.* (cf. *curtata*), *A. sp.* (cf. *lanceolata*), *Anthracosia* aff. *atra*, *Anthracosphaerium* cf. *exiguum*, *A.* cf. *propinquum*, *A. sp.* intermediate between *propinquum* and *turgidum, Naiadites obliquus,* and *N. sp.* cf. *alatus. Spirorbis sp.* and the plants *Diplotmema sturi* Gothan, and *Lepidodendron rimosum* Sternberg have also been recorded during the present survey.

20 to 50 ft of strata normally separate the Four-Feet and Two-Feet-Nine, though thicknesses of 70 ft occur at Cwmneol and Merthyr Vale. The two seams are closest together in the Mountain Ash–Aberaman area, where the measures are wholly argillaceous; where the separation is markedly thicker, sandstones appear in the middle and lower portion of the cyclothem.

Northwards and westwards of Deep Duffryn Colliery the **Two-Feet-Nine** forms a single seam, 24 to 34 in thick, with one or two thin coals at or near the base. Typical sections are:

		Lower Duffryn		Aberaman		Fforchaman		Cwmaman	
		Ft	In	Ft	In	Ft	In	Ft	In
Coal..	..	2	10	2	8	2	5	1	4
Dirt ..	..		4	1	0	10	0	2	7
Coal..	..		5		4		6		9
Dirt ..	..		10				6		
Coal..	..		3				4		

To the south and east of Deep Duffryn the seam section is complicated by the fact that the compound **Three Coals**, which is clearly discrete in the ground to the north, here rests close upon the Two-Feet-Nine: at Cwm Cynon the section reads: coal 19 in, fireclay 27 in, coal 8½ in, fireclay with streaks of coal 6 in, coal 5 in, fireclay 7 in, coal 13 in, rashings and fireclay 24 in, coal (Two-Feet-Nine) 30 in; and at Penrikyber: coal 20 in, fireclay 26 in, coal 10 in, fireclay 27 in, coal 14 in, fireclay 33 in, coal (Two-Feet-Nine) 20 in.

Only in this last described area have the measures above the Two-Feet-Nine been seen during the present survey. Elsewhere they are known only from shaft records, from which correlation of the various horizons must be inferred. North and north-west of Deep Duffryn the Three Coals forms a distinct though compound and inferior seam, lying about 30 to 50 ft above the Two-Feet-Nine. It consists usually of three coals varying from 3 to 10 in thick separated by varying amounts of fireclay or rashings; locally as at Deep Duffryn and Cwmneol five coals may be present.

Fossils have been collected from the roof of the Three Coals only at Cwm Cynon Colliery. The following forms, concentrated in a thin layer 2¼ ft above the top coal, have been identified: *Anthraconaia sp. nov.* cf. *cymbula, Anthracosia* aff. *atra, Anthracosphaerium* ct. *affine, A. sp.* cf. *exiguum,* variants of *A. propinquum,* together with *Carbonita humilis.*

The strata between the Three Coals and the presumed position of the Cefn Coed Marine Band total about 85 to 95 ft in thickness. They vary a good deal in character, but much sandstone and seatearth are usually present. A coal about 3 in thick appears to underlie the Cefn Coed Marine Band in most cases, but this is not always recorded in the shaft logs, e.g. at Penrikyber and Lady Windsor, where the marine band may rest directly on a seatearth. In the Abercwmboi shaft a 2-in coal, 33 ft above the Three Coals, may carry the Hafod Heulog Marine Band, and a cannel and blackband horizon 14 ft still higher may represent the position of the Britannic Marine Band. At Cwmneol these two horizons are 63 and 76 ft respectively above the Three Coals. At Lady Windsor no coals are recorded between the Three Coals equivalent and the Gorllwyn horizon, seatearths carrying the marine strata having presumably passed unnoticed during the sinking of the shafts.   A.W.W.

RHONDDA FACH AREA

The **Amman Marine Band** has yielded only *Lingula mytilloides* in this area during the present survey, though *Orbiculoidea sp.* and *Productus sp.* were recorded by Moore and Cox (1943, p. 226) at National Colliery. The marine strata vary from about 3 to 6 ft in thickness and are succeeded by mudstone containing 'ghost'-like 'mussels'. At Mardy Colliery the following section was measured: striped silty mudstone with plant debris and 'ghosts' of stunted *Anthracosia spp.* at base, 5 ft 6 in, underlain by 3 ft 2 in of grey silty mudstone with abundant *L. mytilloides*. The rest of the cyclothem varies in thickness from 15 to 50 ft, the higher figure being recorded at Ferndale No. 5. The strata are mainly mudstones with ironstone bands and nodules, though in the upper part a few sandy bands are present in the lower Ferndale collieries and in the National–Standard area.

The **Bute** Seam (called the Nine-Feet at National and Standard) consists everywhere of two coals separated by a clod 1 to 17 in thick: the top coal is 22 to 30 in thick in the Mardy and Ferndale areas, and 27 to 36 in thick at National and Standard; the bottom coal is 20 to 28 inches in the former area and 20 to 36 inches in the latter. Few fossils are recorded from the roof of the seam; poorly preserved crushed forms include *Naiadites* aff. *quadratus* in the Ferndale Nos. 1 and 5 areas.

At Standard Colliery, the Bute and the Lower Nine-Feet, separated by only 2 to 3 ft of seatearth, were sometimes worked together. Traced northwards the measures between them expand progressively; at Ferndale they average 20 to 30 ft, and over much of the Mardy area 40 to 60 ft. The strata are mainly argillaceous, though a little sandstone occurs at Mardy and National. In parts of Mardy a thin coal and rashings appear in the middle.

The **Nine-Feet** comprises a single seam at Mardy, where it is thick and often highly disturbed; a section in No. 3 Pit reads: coal 37 in, coal and rashing 13 in, coal 16 in, bast and rashings 9 in, coal 42 in, stone 2 in, coal 29 in. Near the southern margin of the taking the seam splits and southwards from this point the two resulting seams (the Upper and Lower Nine-Feet) progressively diverge. The separating mudstones are 15 to 25 ft thick in the northern Ferndale Pits, 40 ft at Ferndale Nos. 8 and 9, 60 ft at National and 70 ft at Standard.

The **Lower Nine-Feet** (formerly known as the No. 1 Bute at Ferndale, and the Red Vein at National and Standard) consists of a clean coal 3 ft thick in the northern parts of Ferndale, but to the south partings are developed in the top and bottom of the seam, e.g. at Ferndale No. 7: coal 3 in, clod 9 in, coal 30 in, rashings 10 in, coal 9 in; and at National: coal 2½ in, dirt 4½ in, coal 36 in, fireclay and rashings 23 in, coal 2½ in, rashings 1 in, coal 4 in. No fauna has been recorded from the roof.

The **Upper Nine-Feet** (Six-Feet of National and Standard) is subject to much internal disturbance and has not been worked to any extent. At Ferndale a small area was worked under the mistaken impression that it was the Six-Feet. In normal section the seam everywhere appears to consist of two coals, separated in the Ferndale area by 10 to 12 in of bast, and in the National-Standard area by 1 to 5 in of rashings. The two leaves vary in thickness: at Ferndale the top coal is 30 to 48 in, and the bottom coal 42 to 60 in; in National and Standard collieries the top coal is 36 to 60 in and the bottom coal 26 to 32 in.

At Mardy Colliery (No. 3 Pit) only plants were found in the roof of the Nine-Feet: *Neuropteris gigantea*, *N. heterophylla*, *Sphenophyllum cuneifolium*. At the Ferndale collieries and at National, black carbonaceous, highly slickensided shales form the roof of the Upper Nine-Feet Seam and these yield: *Anthraconaia* aff. *salteri*, *Anthracosia* aff. *beaniana*, *A.* cf. *ovum*, and *A. phrygiana*. These shales are succeeded upwards by blocky grey mudstone yielding small solid shells including *Anthracosia* aff. *nitida*, *A.* cf. *angulata*, and problematical forms which may belong to either *Anthraconaia* or *Anthracosia*. At Ferndale No. 9 Pit the supposed egg-case *Vetacapsula?* was also recorded.

The measures between the Upper Nine-Feet and the Lower Six-Feet are variable. In the northern and eastern parts of the Mardy area the **Red Vein** forms a single though inferior seam, e.g. coal 6 in, fireclay 17 in, coal 19 in, stone ½ in, coal 12½ in. Southwards the seam splits into its component coals and at National the following section of the ground between the Upper Nine-Feet and the Lower Six-Feet was measured:

	Ft	In
LOWER SIX-FEET: **coal 42 in,** rashing and shale with coal streaks 4 in, **coal 18 in**	5	4
Carbonaceous shale		6
COAL	1	2
Massive seatearth with ironstone nodules	7	0
Massive fine-grained sandstone	7	0
Strong grey silty mudstone	12	0
Grey mudstone with ironstone nodules: *Anthraconaia* cf. *pulchella,* 'Estheria' sp.		$4\frac{1}{2}$
COAL with shaly parting		8
Massive fireclay with ironstone nodules	5	0
Massive blue-grey mudstone and silty mudstone	30	8
Strong grey striped mudstone with *Naiadites* aff. *productus* ..	2	0
Dark carbonaceous shale with squashed *N. sp.*	1	0
COAL and dirt		3
Massive grey silty mudstone with plants and roots at top	12	5

The three thin coals in this section below the Six-Feet appear to correspond to the Red Vein at Mardy. In most of the southern part of the Ferndale area the coals are still split, but in the north, at Ferndale Nos. 2 and 4, a seam with section: coal 10 to 12 in, clod 1 to 7 in, coal 18 to 24 in occurs 25 ft below a thin coal, which in turn lies 10 ft below the Lower Six-Feet. In the Mardy area upwards of 85 ft of mudstone with some sandstone bands separate the Nine-Feet from the lowest coal of the Red Vein group, but this decreases to the south as shown in the above section. The distance between the top of the Red Vein group and the Lower Six-Feet grows from less than 1 ft at National to 12 ft or so at Mardy.

Apart from the fossils listed above, *Anthraconaia* cf. *wardi, Naiadites* cf. *productus* and *Euestheria sp.* were obtained from the roof of the second thin coal below the Lower Six-Feet, and 'solid' *Naiadites* including *N.* aff. *productus* and *N. sp.* intermediate between *productus* and *subtruncatus*, from the first rider coal above the Upper Nine-Feet at Ferndale No. 1 Pit.

The **Lower Six-Feet** consists of a main coal $3\frac{1}{2}$ to 4 ft thick with a thin coal, 7 to 18 in, close below. When traced into the north of Mardy and north-eastwards into the Cwmaman ground the bottom coal appears to die out. In the south of the area, at National and Standard, a further thin coal near the base is part of the Red Vein group: top coal 55 in, clod 4 in, coal 18 in, clod $5\frac{1}{2}$ in, coal 15 in.

In Mardy No. 1 Pit the roof of the Lower Six-Feet yielded *Lioestheria sp.* from the immediate roof, while at Mardy No. 3 the following plants were obtained: *Eupecopteris volkmanni* (Sauveur), *Neuropteris gigantea, Myriophyllites gracilis* Artis, *Sphenophyllum cuneifolium, ?Sphenopteris striata* (Gothan).

The measures between the Lower Six-Feet and the Upper Six-Feet are mainly argillaceous, though some sandy beds occur in the upper part of the cyclothem in the Ferndale area. They vary in thickness from about 15 ft at Standard to about 60 ft in the Mardy area.

The **Upper Six-Feet** consists everywhere of a single clean coal, 22 to 26 in thick; only at Mardy Colliery have fossils been recorded from its roof: these include *Anthracosia* cf. *'faba', Naiadites sp.* and cf. *Planolites ophthalmoides* [2 mm.].

Some 20 to 40 ft of mudstones with much ironstone and sporadic thin bands of sandstone usually intervene between the Upper Six-Feet and Four-Feet seams. They are thinnest in the north at Mardy, while in parts of Ferndale Nos. 8 and 9 pits and at National they may reach 70 ft.

Throughout the entire area the **Four-Feet**, which was called the Two-Feet-Nine at National and Standard, is remarkably constant in character. Typically it consists of four coals, between which the partings tend to become more prominent in the south of the area. In the north the two middle coals coalesce. Typical sections are:

		Mardy		Ferndale No. 6		National		Standard	
		Ft	In	Ft	In	Ft	In	Ft	In
Coal ..	..	1	9	1	4	1	0		10
Dirt ..	..		2		1		10		4
Coal ..	.. ⎫			1	4	2	0	3	3
Dirt ..	.. ⎬ 4 7				3	1	0	1	0
Coal ..	.. ⎭			2	4	2	1	2	10
Dirt ..	..		4		9	1	10	1	6
Coal ..	..	1	4		9	1	1	1	0

The roof of the Four-Feet is usually made up of friable blocky mudstone with numerous ironstone nodules, which yields an abundant fauna: *Anthraconaia sp. nov.* (cf. *curtata*), *Anthracosia* cf. *acutella*, *A.* aff. *aquilina*, *A.* aff. *concinna*, *A.* cf. *planitumida*, *A. sp. nov.* (figured Trueman and Weir, pl. 18, fig. 5), *Anthracosphaerium* cf. *exiguum*, *A. sp.* intermediate between *exiguum* and *affine*, cf. *A. propinquum*.

Throughout the Mardy–Ferndale area the Two-Feet-Nine Seam lies about 20 to 30 ft above the Four-Feet, the intervening strata consisting almost wholly of mudstone.

The **Two-Feet-Nine** has been worked throughout the Ferndale area, where it consists of a main coal $3\frac{1}{2}$ to 4 ft thick with one or two thin coals at or near the base. Northwards the main coal appears to split, giving in the Mardy area the following section: top coal 20 in, soft and dark shale 55 in, coal with brass 22 in, fireclay 30 in, coal 17 in. In places the separation between the top and middle coals is even greater. A mudstone roof to the middle coal has yielded: *?Anthraconaia librata*, *A.* cf. *rubida*, *Anthracosia* cf. *aquilina*, *A. atra*, *A. lateralis* and *A.* aff. *simulans*. Southwards from Ferndale into the National and Standard takings there is a sharp deterioration of the section, not more than 22 to 24 in of coal being present in these areas. No fauna was recorded from the roof of the Two-Feet-Nine.

As in the Taff–Cynon area, the strata between the Two-Feet-Nine and Cefn Coed Marine Band are nowhere visible in the Rhondda Fach. In the north of the Mardy area the Three Coals forms a distinct compound horizon some 38 ft above the Two-Feet-Nine: coal 22 in, fireclay 58 in, coal 10 in, shale and fireclay 28 in, coal 8 in, rashings 6 in, coal 10 in. A coal 13 in thick 27 ft above may carry the Hafod Heulog Marine Band; a 7-in coal 54 ft above the Three Coals may mark the horizon of the Britannic Marine Band, while the Cefn Coed Marine Band presumably occurs above a 2-in coal some 51 ft 9 in still higher. These three horizons were seen at isolated places in the new cross-measures roadways driven at Mardy Colliery. *Lingula mytilloides* and 'fucoids' were recorded from the Cefn Coed Marine Band; the Hafod Heulog Band was presumably represented by 'sulphury' mudstones, 1 in thick, succeeded by dark mudstone with abundant ironstone nodules and sporadic *Planolites ophthalmoides*; while the intermediate coal carried a sandstone roof, stained yellow at its base.

The Three Coals horizon splits when traced southwards, and at Ferndale Nos. 8 and 9 four coals occur over a thickness of 64 ft: (top) coal 14 in, measures 8 ft 10 in, coal 7 in, measures 1 ft 7 in, coal 8 in, measures 21 ft 4 in, coal 10 in, measures 27 ft 11 in, coal 15 in. The top coal of the section may carry the Hafod Heulog Marine Band, and the assumed position of the Cefn Coed Marine Band is 90 ft higher. The sections at National and Standard indicate that the Three Coals group of coals has deteriorated still further in the south of the area.   A.W.W.

NORTHERN RHONDDA FAWR AREA

The **Amman Marine Band** has been identified at Fernhill, Lady Margaret, and Dare collieries. It is about 1 to 2 ft thick and contains abundant *Lingula mytilloides*. At

Tydraw and certain parts of Fernhill and Dare the marine strata fail, and planty mudstones with abundant macrospores and a few fish scales rest directly on the coal: this failure is related to wash-out conditions in the underlying coal. Immediately above the marine band abundant 'ghostly' non-marine shells are everywhere present in dark mudstone: *Anthracosia* cf. *aquilina*, *A.* cf. *concinna*, *A.* cf. *ovum*, *A. sp.* intermediate between *ovum* and *phrygiana*, *A. regularis* and *Naiadites sp.* The rest of the cyclothem normally consists of 15 to 30 ft of mudstones; at Pentre about 50 ft of beds are present, the thickening corresponding to the incoming of sandstones. A similar thickening occurs north of Fernhill associated with the development of 35 ft of sandstone.

The **Bute** has been worked extensively from most of the collieries. At Fernhill it was known as the New Seam; at Bute, Lady Margaret, Ynysfeio, and Rhondda Merthyr as the Seven-Feet; at Tydraw, Pentre, Tynybedw, and Gelli as the Yard; at Park, Dare, Eastern, and Maindy as the Brunts; and at Abergorky as the No. 2 Seam. Throughout much of the area the coal is divided into two leaves by a thin mudstone; this parting dies out westwards from Fernhill, the thickness of the seam decreasing from 4½ ft to under 3 ft. Elsewhere the overall thickness varies from about 4 to 4½ ft. The immediate roof consists of about 8 in of dark, well-bedded mudstone containing many 'mussels'; this is succeeded by paler silty mudstones which yield only rare shells; locally, as at Dare, these rest directly on the coal. The fauna includes variants of *Anthraconaia modiolaris* (J. de C. Sowerby), *A. williamsoni*, *Anthracosphaerium affine*, *A. bellum*, *A. exiguum*, *A. fuscum* (Davies and Trueman), *A. turgidum* and *Naiadites sp.*; *Spirorbis sp.* is also recorded.

The Bute and Nine-Feet are about 35 ft apart in the Ynysfeio area and this interval increases both northwards and south-westwards to about 60 ft. In the extreme south-east, at Gelli, the two seams close to within 3 to 15 ft. For the most part the beds are argillaceous, but in the north a well-developed sandstone up to 20 ft is present. Sandstones are also present in the south of the area, and locally, as at Park, are known to cut out the Bute. A 1- to 2-in coal is sometimes found about 10 ft below the Nine-Feet.

The **Nine-Feet** splits southwards within the valley, the line of split running between Fernhill and Tydraw collieries, and the separation increases to about 60 ft at Eastern and Gelli. The full Nine-Feet is found only at Fernhill where it comprises 8 to 9 ft of coal with a 6 to 12 in cannelly parting about 3 to 3½ ft from the top. In the workings north of the shafts the immediate roof of the seam is silty mudstone containing a few plant fragments; above this an abundant fauna of large crushed 'mussels' is found in about 3 ft of dark cannelly mudstone containing numerous ironstone bands. The 'mussels' include the usual Nine-Feet assemblage: *Anthracosia beaniana*, *A.* cf. *ovum*, *A.* aff. *phrygiana*, *A. sp.* intermediate between *phrygiana* and *aquilina* and *A. sp.* intermediate between *phrygiana* and *ovum*. Paler grey mudstones above yield only worm tracks.

The **Lower Nine-Feet**, usually called the Bute, but at Abergorky the No. 1 Seam, has been worked from several collieries particularly on the east side of the valley. The section varies little: a top coal 22 to 30 in thick is separated from a bottom coal 4 to 10 in by 1 to 10 in of clod or rashings. In places in the north-east a 2-in coal is recorded in the top of the seam. Only plant debris has been observed in the silty mudstone roof.

The **Upper Nine-Feet**, extensively worked under the name Nine-Feet, maintains a section of 5½ to 6½ ft throughout the area. In general, it comprises a double coal: the top leaf 3½ to 4½ ft thick is separated from the lower leaf 1 to 1½ ft thick by about 12 in of cannel, which often forms the floor of the worked section. On the western side of the valley another thin parting is often present 3 to 6 in below the roof. At Tydraw and Dare the roof is a planty mudstone.

The 70 to 130 ft of measures between the Upper Nine-Feet and the Lower Six-Feet are poorly exposed and the mining records show several anomalies. At Fernhill only one major coal—called the **Red Vein**—is recorded; this has a complex and variable section: a bottom coal of 12 to 15 in is overlain by up to three thin coals varying between 2 and 10 in. In the shaft a further thin coal lies between this seam and the Nine-Feet. At Tydraw a 2-in coal about 20 ft above the Nine-Feet carries a cannelly mudstone roof containing shells similar to those in the Nine-Feet roof at Fernhill: forms collected include *Anthraconaia* cf. *robertsoni* (Brown), *Anthracosia* cf. *phrygiana* and *A. sp.* intermediate between *phrygiana* and *ovum*. It is possible that this thin coal represents a split from the top of the Nine-Feet. Some 40 ft above, the 'Red Vein' consists of a top coal 3 to 12 in thick and a bottom coal 30 in thick; its roof consists of planty mudstone. Similar conditions exist further south. The thin coal carrying the Nine-Feet fauna is up to 22 ft above the Upper Nine-Feet at Park Colliery, and the Red Vein is about 30 ft below the Lower Six-Feet. At both Park and Dare the lower coal of the Red Vein often has a distinct parting near the middle, while the top coal splits away and may be several feet above the rest of the seam. At Tynybedw and Pentre this upper coal is near the floor of the Lower Six-Feet and farther south it forms a bottom coal to that seam.

The **Lower Six-Feet,** worked extensively as the Six-Feet, is made up of a main coal 4 ft thick and a bottom leaf, which is up to 22 in. in the south but thins to only 5 in at Rhondda Merthyr. At Park, Dare, Pentre and Tynybedw the bottom coal lies close beneath the main coal, but northwards it splits away and lies a few feet below. At Park and Dare a thin coal 2 to 6 in thick, and in part cannel, lies in the roof of the seam; this everywhere carries mudstone with abundant *Lioestheria striata* and *Naiadites sp.*, and northwards at Tydraw and Glenrhondda lies 3 ft above the Lower Six-Feet.

The **Upper Six-Feet,** 2 to 3 ft of clean coal, is everywhere distinct and has been worked on a small scale in the north of the area as the Upper Yard or No. 1 Yard. At Dare it is up to 60 ft above the Lower Six-Feet, but northwards this decreases, and to the north and north-west of Fernhill the two seams unite to form a composite Six-Feet. Throughout the area its roof carries a fauna of comparatively small *Anthracosia* including *A. atra*, *A.* cf. *planitumida* and *A. sp.* cf. *phrygiana*, together with a few *Naiadites obliquus* preserved as films; associated with the 'mussels' is a small form of cf. *Planolites ophthalmoides*. Above, mudstones and sandy mudstones continue for 25 to 40 ft to the Four-Feet.

The **Four-Feet** forms a single complex seam in the north and east of the area. At Fernhill a typical section reads: top coal 12 in, shale $1\frac{1}{2}$ in, coal 10 in, clod 12 in, coal 26 in, plane parting, coal 26 in, shale 12 in, coal 12 in. This is similar to the typical four-leaf section of the Rhondda Fach with the top coal of that area having split into two leaves. A similar section is recorded at Gelli, where this top coal appears to suffer further splitting into three distinct leaves: top coal 6 in, stone 2 in, coal 6 in, parting $\frac{1}{4}$ in, coal 6 in, clod 2 in, coal 22 in, bast and fireclay 12 in, coal 25 in, rashings 12 in, coal 12 in. Traced northwards from Fernhill the two top leaves of the above section split away and have been proved to lie 5 ft above the rest of the seam. In the central and south-western parts of the area the complex top coal splits even farther away, while in the extreme south-west the complete splitting of the seam in Cwm Clydach and the Ogmore Valley is foreshadowed at Eastern Colliery. The following sections illustrate this behaviour:

	Dare		Eastern		Abergorky		Rhondda Merthyr	
	Ft	In	Ft	In	Ft	In	Ft	In
Coal ..		8 ⎫						
Dirt ..		7 ⎬	1	3	1	10		7
Coal ..	1	1 ⎭						
Measures ..	30	6	16	10	11	7	13	10
Coal ..	1	5	2	3	1	6	1	8
Parting ..	1	4	10	8	thin		4	4
Coal ..	2	2	3	6	3	3	3	8
Parting ..	2	10	1	3	thin			6
Coal ..	1	0	1	6		6	1	0

At Fernhill the roof of the combined seam is a barren sandy mudstone. At Glenrhondda the uppermost coal, split 40 to 50 ft above the rest of the seam, carries plant fragments and film-like impressions of *Naiadites sp.* in its immediate roof, overlain by sandy mudstones containing *Anthracosphaerium* cf. *exiguum*. Nowhere else was the roof of the top coal accessible. Planty mudstones form the roof of the split lower seam.

The **Two-Feet-Nine** usually lies 15 to 30 ft above the top coal of the Four-Feet. South of Treorky it consists of a main coal 4 to 5½ ft thick with an underlying coal 12 to 27 in, sometimes in two leaves, up to several feet below. In the Park workings a parting develops about 18 in above the base of the main coal and this appears to increase to a true split in the collieries between Lady Margaret and Fernhill. In this latter area the top split has been worked sporadically with a section varying from 2½ to 4 ft. The following sections illustrate the variation of the seam.

	Maindy		Dare		Tylecoch		Bute		Glenrhondda	
	Ft	In	Ft	In	Ft	In	Ft	In	Ft	In
Coal					2	8	3	8	2	9
Measures	5	3	4	2		1	24	0	22	0
Coal					1	6	1	6	1	4
Measures	5	0	3	6	4	3	8	7	8	0
Coal	1	4						11½		
Dirt		1½	1	5	1	2	1	1		5
Coal		11						11½		

The roof of the lower split of the main coal has been examined at Fernhill and Glenrhondda. A fauna, poorly preserved in irony mudstone, includes: *Spirorbis sp.; ?Anthraconaia librata, Anthracosia atra, A.* cf. *planitumida, Naiadites productus* and scales of *Rhizodopsis sp.* At Park Colliery an abundant fauna was obtained 2 to 3½ ft above the main coal: *Gyrochorte carbonaria* Schleicher, *Planolites montanus; Anthraconaia* cf. *cymbula, Anthracosia* cf. *acutella, A.* cf. *atra, A.* cf. *concinna, A.* cf. *planitumida, A.* aff. *simulans* and *Naiadites* cf. *obliquus.* Small poorly preserved 'mussels' ranged upwards for several feet. At Fernhill a sparse fauna included *Anthraconaia* cf. *rubida, Anthracosia* cf. *planitumida* and *Naiadites sp.*

The succeeding measures to the Cefn Coed Marine Band are 140 to 200 ft thick. At Gelli Colliery the Three Coals horizon is represented by a complex group of four coals occurring about 25 ft above the Two-Feet-Nine: top coal 20 in, fireclay 55 in, coal 14 in, fireclay 38 in, clod 12 in, coal 11 in. Elsewhere in the area these seams split completely. At Maindy, for example, five thin coals occur over 54 ft. These are succeeded by three or four thin coals which presumably carry marine strata at two or three levels. The most complete section of these strata was examined at Fernhill. The coals of the Three Coals group comprise four thin leaves spread through about 45 ft and lying about 55 ft above the Two-Feet-Nine. The supposed Hafod Heulog horizon has yielded no fossils, but the **Britannic Marine Band** occurs on a 7- to 9-in coal about 40 ft below the Cefn Coed Band, and has yielded abundant *Planolites ophthalmoides* and rare *Lingula mytilloides.* The **Cefn Coed Marine Band** is about 2 to 3 ft thick at Fernhill and has yielded: macrospores; crinoid columnals, *Archaeocidaris sp.; Chonetes* cf. *granulifer* Owen, *C.* cf. *skipseyi* Currie, *C.* (*Tornquistia*) *diminutus* Demanet, *Crurithyris carbonaria* (Hind), *Orbiculoidea* cf. *nitida, Productus* (*P.*) *carbonarius, P.* (*Dictyoclostus*) *craigmarkensis* Muir Wood, cf. *P.* (*Levipustula*) *rimberti* Waterlot, *Rhipidomella* cf. *carbonaria* (Swallow); *Straparollus sp.; Nuculana?, Posidonia?; Ditomopyge sp.* Lenses of 'cank' occur within the band and the richest fauna is closely associated with these.

Locally sandstones are developed below the horizons of both the Hafod Heulog and Britannic bands; these thicken towards Park, where a third sandstone occurs close below the Cefn Coed.

W.B.E.

SOUTHERN RHONDDA AREA

The beds between the Amman Rider and the Bute vary in thickness between 13 and 40 ft. They are thickest in the central part of the area, over 30 ft being recorded at Britannic, Naval, Cymmer and Lewis Merthyr collieries; and they are thinnest at Llwynypia and Coedely. Mudstones generally predominate, but thin sandstones occur in the middle where the cyclothem is thickest. The **Amman Marine Band** consists of 10 in to 3 ft of mudstone, more or less silty and micaceous, everywhere overlain by paler smoother mudstone with the usual 'ghostly' non-marine lamellibranchs. *Lingula mytilloides* is ubiquitous in the marine band, and at Lewis Merthyr and Great Western collieries was the only fossil recorded. At Coedely *Orbiculoidea* cf. *nitida* was also found, while at Naval, Cambrian, and Maritime pyritized sponge spicules were abundant. In the south, at Coedely and Cwm collieries, more varied faunas occur: in addition to the forms mentioned above the following forms were collected: *Ammonema?*; *Emmonsia parasitica* (Phillips); *Sphenothallus?*, *Conularia sp.*; crinoid columnals; polyzoa; *Productus* (*Levipustula*) *piscariae* Waterlot; *Pleuroplax attheyi* (Barkas). Among the 'mussels', dwarfed *Anthracosia* predominate; they include *A. sp.* cf. *concinna*, *A. sp.* intermediate between *concinna* and *planitumida*, *A.* cf. *ovum* and *Naiadites sp.*, associated with *Spirorbis sp.*

Over most of the area the **Bute** and **Lower Nine-Feet** are close together, forming one seam, called the Nine-Feet at Llwynypia, Naval, Cymmer, and Cilely collieries, and the Two-Feet-Nine at Coedely. In the western parts of Cambrian and Britannic, and again in the northern and eastern parts of Lewis Merthyr and Great Western, and in the eastern parts of Cwm, the two seams split apart. Where separate, the Lower Nine-Feet was known as the Red Vein, and the Bute as the Nine-Feet. Typical sections of the combined seam are as follows:

		Llwynypia		Cambrian, east		Naval		Cilely	
		Ft	In	Ft	In	Ft	In	Ft	In
LOWER NINE-FEET {	Coal ..			1	7	3	2		
	Dirt ..	2	4		2		2	2	9
	Coal ..			1	6	1	3		
	Dirt ..	1	4	1	0		1		3
BUTE ..	Coal ..					2	11		
	Dirt ..	4	10	2	0		1½	4	2
	Coal ..						6		
	Dirt ..		2		2	1	4		0¼
	Coal ..	2	3	2	4	4	1	2	3
	Dirt ..		4			2	0		
	Coal ..		3				6		

In parts of Lewis Merthyr and Great Western collieries the separation between the two seams may reach 30 ft, while at Cwm 15 to 20 ft is locally recorded. Typical sections are:

		Lewis Merthyr		Great Western		Cwm		Britannic	
		Ft	In	Ft	In	Ft	In	Ft	In
LOWER NINE-FEET {	Coal ..		8				3	1	8
	Dirt ..		1½	3	0	—	—		2
	Coal ..	2	9			2	10	3	2
	Measures	17	0	30	0	15	0	24	0
BUTE ..	Coal ..	2	10	3	5½	2	9	2	8
	Dirt ..	1	1		0¾		4	1	2
	Coal ..	1	6	1	5	2	0	1	11
	Dirt ..				1¾		2		6
	Coal ..				8½		6		3

The Bute is usually a double coal, but thin leaves are locally developed in the middle of the seam (e.g. at Naval) and, more frequently, at the base. The Lower Nine-Feet consists of one main coal in the eastern part of the area, but this splits westwards into Cambrian and Britannic. In the northern parts of Lewis Merthyr a thin coal occurs at the top of the seam. On the west side of Cambrian Colliery an airbridge driven 30 ft above the Bute struck the Lower Nine-Feet with the following section: roof, dark grey mudstone with thin ironstone bands and nodules and sporadic coal streaks 8 in, cannelly shale with coal streaks and a layer of cannel 4½ in, coal 1 in, black cannelly shale 1 in, inferior coal 2 in, cannel, shaly at top, 11¾ in, coal 5 in, 'brass' 0 to 1½ in, coal 13 in, rashings 1 in, coal 18 in, hard seatearth floor. The presence of cannel in the top of the coal is characteristic of the Cambrian area.

Only plants have been recorded from these two seams. At Lewis Merthyr the roof of the Bute yielded: *Alethopteris decurrens, Asterophyllites charaeformis, A. equisetiformis, A. grandis* (Sternberg), *?Bothrodendron sp., Calamites sp., Diplotmema furcatum, Lepidophyllum morrisianum* Lesquereux, *Mariopteris muricata, Myriophyllites gracilis, Neuropteris heterophylla, Pinnularia capillacea, Trigonocarpus parkinsoni*. At Ty Mawr No. 1 Pit (Great Western Colliery) the following species were obtained from the roof of the Lower Nine-Feet: *Alethopteris decurrens, Boulaya fertilis* Renault, *Calamites undulatus* Sternberg, *Lepidodendron sp., Lonchopteris bricei* Brongniart, *L. eschweileriana* Brongniart, *Lycopodites sp. nov., Mariopteris muricata, Myriophyllites gracilis, Neuropteris gigantea, N. heterophylla, N. obliqua, Pinnularia capillacea, Sphenophyllum cuneifolium, Trigonocarpus parkinsoni, Urnatopteris tenella* (Brongniart).

In the north-eastern portion of the Lewis Merthyr taking the separation between the Lower and Upper Nine-Feet seams may be as little as 3 ft; this marks the beginning of the split in the Nine-Feet Seam from the direction of Lady Windsor Colliery, where the Nine-Feet coals form a single seam (see p. 106). Southwards and westwards the split develops rapidly, and over most of the area 50 to 70 ft, and locally, as at Llwynypia and Coedely, 100 ft or more of strata may be present between the two coals. Mudstones with ironstone bands and nodules make up the bulk of the cyclothem, but thin sandy beds are present in the middle or lower part in the Naval and Cambrian areas, while several beds of sandstone up to 25 ft thick occur at Coedely.

Except at Coedely Colliery the **Upper Nine-Feet** Seam was known throughout the area as the Six-Feet. At Great Western and Lewis Merthyr its section is similar to that described for the Rhondda Fach area: top coal 39 to 51 in, parting 1 in, bottom coal 18 to 36 in. Traced southwards to the southern parts of Lewis Merthyr, a second parting locally develops in the bottom coal: coal 48 in, clod 1 in, coal 18 in, clod 1 in, coal 12 in. In the Maritime and Cwm areas the partings die out and the seam consists of a clean coal 39 to 67 in thick, averaging about 54 in. In Cambrian and Llwynypia the parting between the two main coals often consists of bast. A thin leaf, 4 to 6 in thick, at the base of the seam is a constant character throughout the collieries west of Porth. On the south-west side of Naval, in Cambrian, and Britannic a further parting appears in the upper part of the top coal:

	Llwynypia Ft In	Cambrian Ft In	Naval (SW side) Ft In	Britannic Ft In
Coal	⎫	10	9	1 3
Dirt	⎬ 4 0	1	1	2
Coal	⎭	2 2	2 5	2 5
Dirt	6 (bast)	1 (bast)	2	1
Coal	2 3½	2 5½	2 3	2 6
Dirt	6	1	4	5
Coal	6	6	6	4

At Coedely the seam was formerly regarded as the bottom coal of the Three Coals horizon. Its section (coal 15 in, fireclay 32 in, coal 36 to 44 in) indicates that the main parting has died out, as in Cwm, and the upper leaf may be one of the Red Vein group coals.

At Llwynypia, Naval, and Cambrian collieries a massive sandstone, 25 to 50 ft thick, overlies the Upper Nine-Feet, and in the last-named area is associated with wash-outs in the underlying seam. Elsewhere the roof of the coal consists of blocky grey mudstone, usually carrying small solid 'mussels'. These are particularly well represented at Lewis Merthyr, Cwm, and Coedely collieries, and the following species have been identified: cf. *Anthraconaia pulchella*, *A. pumila* (Salter), *Anthracosia* cf. *angulata*, *A. nitida*, *A. sp. nov.* aff. *nitida*, *Naiadites* cf. *productus*. At Ty Mawr No. 1 Pit an abundant fauna of solid *Naiadites* was collected; species include: *N.* cf. *productus*, *N. subtruncatus* and *N. sp. nov.* (figured Trueman and Weir, pl. 31, figs. 44, 45), associated with *Anthraconaia* cf. *curtata*. In strong mudstones 7 to 10 ft above the coal at Cilely Colliery the following plants were collected: *Asterophyllites equisetiformis*, *Asterotheca miltoni* (Artis), *Calamites undulatus*, *Diplotmema furcatum*, *Lepidodendron sp.*, *Lonchopteris rugosa* Brongniart, *Mariopteris nervosa*, *Neuropteris heterophylla*, *Sphenophyllum cuneifolium*, *S. myriophyllum* Crépin.

The **Red Vein** group of coals consists of a well-defined and fairly consistent **Caerau** Seam, lying everywhere close to the floor of the Six-Feet, together with one or more thin coals not far below. The distance from the Upper Nine-Feet to the Caerau Seam varies from 10 to 15 ft at Coedely and 30 to 40 ft at Cwm in the south of the area, to 110 to 120 ft at Naval and Llwynypia in the north. Apart from the thick sandstone developed at the base in the Llwynypia, Naval, and Cambrian areas, these measures consist of mudstones with ironstones together with varying amounts of sandstone. In the Nantgwyn Hard Heading at Naval Colliery the following section was measured:

	Ft	In
SIX-FEET and CAERAU	—	—
Silty mudstone-seatearth	3	0
Strong silty mudstone, striped, with thin ironstone bands at base; plant debris	4	9
COAL	1	2
Massive mudstone-seatearth, passing down into blocky grey mudstone with ironstone bands and nodules	15	0
Striped grey mudstone; sporadic plants	5	0
Striped grey mudstone; *Naiadites sp.* intermediate between *productus* and *quadratus*; *Belinurus* cf. *bellulus*; *Vetacapsula sp.*	1	0
Black carbonaceous shale		2½
COAL		2½
Massive seatearth, carbonaceous, with coal streaks at top	3	0
Strong grey silty mudstone with roots	3	6
COAL, inferior, with seatearth bands	1	8
Shaly seatearth, passing down into silty mudstone	10	0
Irregularly bedded, fine-grained, sandstone with coal streaks	2	0
Position of UPPER NINE-FEET Seam	—	—

A similar section occurs in the shafts at Cymmer and Lewis Merthyr. At Great Western only one coal, 16 to 23 in, is recorded, about 11 ft below the Caerau, and at Cwm Colliery 8 to 9 in of coal occur in a similar position. At Cilely Colliery, where 70 to 80 ft of ground lie between the Upper Nine-Feet and the Six-Feet–Caerau group, a 7- to 8-in band of coal and bast occurs in the middle; the mudstone roof of this coal yielded *Anthracosia sp. nov.* aff. *nitida*, *A.* cf. *angulata* and *Naiadites sp.* cf. *productus* at 6 to 18 in above the seam.

The **Caerau** portion of the Red Vein group normally lies close beneath the Six-Feet. Only on the east side of Cwm Colliery is there sufficient ground between the two coals to allow the lower coal (called the Four-Feet Bottom Coal) to be worked, and even here the separation is little more than 3½ ft. The Caerau consists of two leaves, with an overall thickness of 2¼ to 2½ ft in the north and east and 3½ to 4½ ft in the west and south. The clod or rashings parting varies from a mere film to 2 ft. Representative sections are given with those of the Six-Feet below.

The **Six-Feet** (formerly known as the Four-Feet at all the collieries of this area except Llwynypia and Cambrian) splits into its component Upper and Lower coals. The line of split passes through Naval Colliery, thence southwards through the western part of Cilely before turning westwards. In the eastern workings of Cwm Colliery the parting between the two coals dies out. To the south-east of the line of split the Six-Feet has fairly constant characters: it consists of a top coal 15 to 22 in thick (the coal which forms the Upper Six-Feet), a main coal of 42 to 54 in and a bottom coal of 12 to 18 in, which does not seem to have a separate existence in the east and south of the area. The following sections of the combined Six-Feet and Caerau seams illustrate the variation:

		Cilely		Lewis Merthyr		Great Western		Cwm (east)	
		Ft	In	Ft	In	Ft	In	Ft	In
SIX-FEET	Coal ..	1	8	1	10	1	6		
	Dirt ..	2	0	1	0		1	5	7
	Coal ..								
	Dirt ..	5	0	4	2	4	4	1	9
	Coal ..							1	0
	Dirt ..	parting			5½		9	2	9
CAERAU	Coal ..	2	2	2	0½	1	2	1	2
	Dirt ..	parting			3		3		2
	Coal ..	1	5		7	1	1	3	2

Northwards and westwards the mudstones separating the Lower and Upper Six-Feet increase from 10 ft over much of the Naval area, to 30 ft at Llwynypia, 30 to 45 ft at Cambrian and 15 to 20 ft at Britannic.

Typical sections of the Lower Six-Feet together with the associated Caerau Vein in the areas of splitting are as follows:

		Naval		Llwynypia		Cambrian		Britannic	
		Ft	In	Ft	In	Ft	In	Ft	In
LOWER SIX-FEET	Coal ..	1	0						
	Dirt ..		3	5	4	5	0		
	Coal ..	2	6					5	3
	Dirt ..		3		2		2		
	Coal ..	2	0	2	0	1	9		
	Dirt ..		6	3	3	1	10		8
CAERAU	Coal ..	1	8	1	3	1	2	1	6
	Dirt ..		9	10	10	2	0	1	10
	Coal ..	1	8	1	6	2	6	2	2

In the western and southern parts of Naval Colliery and in parts of Llwynypia a persistent 'brass' development a few inches thick occurs 12 to 18 in from the top of the Lower Six-Feet Seam, while in Cambrian Colliery 7 ft of clean coal have been recorded locally.

The separated Upper Six-Feet usually consists of 20 to 28 in of clean coal, and is thinnest in the west of the area. In parts of Cambrian a thin clod is locally recorded about 5 in from the bottom of the seam.

No fossils have been recorded from the roof of the Lower Six-Feet in this area, out the roof of the Upper Six-Feet (or Six-Feet where the two coals are united) carries the characteristic 'mussel' fauna. The shells occur as solids or semi-solids in ironstone-shale (blackband) close above the coal, though some specimens occur in the grey mudstone above. Variants of the following species make up the bulk of the fauna collected: *Anthracosia acutella*, *A. aquilina*, *A. sp. nov.* cf. *aquilinoides*, *A. atra*, *A. caledonica* Trueman and Weir, *A. concinna*, *A. elliptica*, *A. sp.* cf. *phrygiana*, *A. planitumida*, *A. sp.* intermediate between *atra* and *lateralis*, *A. sp.* intermediate between *atra* and *barkeri* and *Naiadites sp.* ; *Carbonita humilis* is also associated.

The measures between the top coal of the Six-Feet and the lowest coal of the Four-Feet group vary from about 25 ft to about 90 ft. They are thinnest in the Cambrian area and thickest at Cwm, Cilely, and Britannic; elsewhere thicknesses of 50 to 65 ft are common. The strata consist mainly of mudstones and silty mudstones with ironstone bands and nodules, but thin sandstones are developed in the middle of the cyclothem at Naval and Great Western and, near the top, at Cymmer.

In the north of the Llwynypia taking the **Four-Feet** forms a single complex seam composed of four principal coals, as in the Rhondda Fach area (see p. 111): top coal 12 in, stone 1 in, coal 3 in, rashing 4 in, coal 28 in, 'bunker' 48 in, coal 24 in, rashing 15 in, coal 12 in. Southwards into the Naval and Dinas colliery areas a split develops in the middle of the section, giving two distinct seams, here called the **Upper** and **Lower Four-Feet.** At the main Naval shafts the separation is 10 to 10½ ft, and in Ely pit the section reads: Upper Four-Feet (called Lower Two-Feet-Nine), coal 5 in, rashings 9 in, coal 5 in, rashings 8 in, coal 20 in; separation 12 ft; Lower Four-Feet, coal 10 in, rashings 3 in, coal 22 in, fireclay 30 in, coal 24 in. At the collieries of the lower Rhondda, where the combined seam carried the name of Two-Feet-Nine, a similar behaviour is to be observed. Over much of the northern part of Lewis Merthyr and Great Western a single four-leaf section prevails, e.g. coal 13 in, fireclay 20 in, coal 29 in, fireclay 19 in, coal 26 in, rashings 8 in, coal 13 in; but locally the top coal splits away by as much as 30 ft. In Cwm and Coedely collieries two distinct seams are developed by the same splitting as that described at Naval Colliery:

		Cwm		Coedely	
		Ft	In	Ft	In
	Coal ..	1	11	2	5
UPPER FOUR-FEET	Dirt ..		4		2
	Coal ..	2	0	1	6
	Separation 10 ft to	20	0	20 ft to 25	0
	Coal ..	1	6	2	1
LOWER FOUR-FEET	Dirt ..		2		6
	Coal ..	1	2	1	0

At Cilely Colliery the Lower Four-Feet has the typical two-leaf section, but the Upper Four-Feet has split into its component parts. The following section was measured on the main roadway just west of the shafts:

	Ft	In
Strong silty mudstone	6	0
Shaly mudstone with 'mussels'	1	9
Hard cannelly shale passing into cannel		10
COAL	1	1½
Seatearth with many reddish ironstone nodules in middle part ..	9	7
Strong fine-grained sandstone	2	9
Soft clay		1½
COAL		5½
Seatearth	1	0½
COAL and rashings		4
Fireclay with numerous small nodules	3	10
Strong silty mudstone	11	0
COAL	1	6
Seatearth and mudstone	10	0
Rashings		3½
LOWER FOUR-FEET: **coal 22 in**, rashings 2 in, **coal 16 in**	3	4

Westwards from Llwynypia, Naval and Cilely, complete splitting of the various leaves takes place and the seam is represented by a series of single coals spread through a considerable thickness of strata. At Cambrian Colliery the section in No. 2 Shaft reads (from top): coal 11 in, shale 7 ft 6 in, coal 15 in, clift and shale 17 ft 2 in, coal 5 in, fireclay and clift 16 ft 5 in, coal 17 in. At Britannic the details are unreliable owing to structural complication; the section in the South Pit is given as: coal and rashes 5 in, measures 4 ft, coal and rashes 20 in, measures 15 ft 7 in, coal 12·in, measures 10 ft 7 in, coal 13 in, measures 25 ft 3 in, coal 13 in.

At Cilely, Coedely, and Cwm the immediate grey mudstone roof of the Lower Four-Feet carries a poor fauna of non-marine lamellibranchs, including *Anthraconaia obscura, A.* cf. *subcentralis* and *Naiadites sp.* At Naval, Cilely, and Cwm collieries the typical Four-Feet forms are abundant in the roof of the Upper Four-Feet: *Anthraconaia* cf. *cymbula, A.* cf. *librata, A. rubida, A. sp. nov.* (cf. *curtata*), *Anthracosia* cf. *atra, A. concinna, A.* aff, *fulva, A.* cf. *lateralis, A. planitumida, A. simulans, Anthracosia sp.* intermediate between *aquilina* and *lateralis, A. sp.* intermediate between *atra* and *fulva, A. sp.* intermediate between *atra* and *planitumida, Anthracosphaerium* cf. *exiguum, A.* aff. *propinquum, A. sp.* intermediate between *propinquum* and *turgidum, Naiadites obliquus.* At Lewis Merthyr, where the top coal splits away, a poor fauna of *Anthracosia* aff. *atra, Naiadites* cf. *alatus* and *N.* cf. *obliquus* occurs.

The measures between the Four-Feet group and the Two-Feet-Nine are thinnest in the east and south-east of the area, where they are as little as 10 ft in Cwm and Great Western, and in the north, being 20 to 30 ft thick in the Naval–Llwynypia area; they thicken south-westwards and westwards to 40 ft at Coedely, 45 to 50 ft at Cambrian, and 65 to 70 ft at Ely. They consist mainly of mudstones with ironstone bands, but some sandstone is recorded in the thicker sections.

The **Two-Feet-Nine** is a workable seam only at Llwynypia, Cambrian, and Coedely. At Llwynypia the general section is: coal 46 to 51 in, fireclay 2 to 3 ft, bottom coal 3 to 8 in. Southwards and south-westwards it thins rapidly and in the collieries of the Porth–Pontypridd–Cwm area the main coal is seldom more than 27 in thick and the seam is un-named. The coal has been extensively worked at Cambrian Colliery where the thickness varies from 5½ ft in the extreme west to 34 in about ½ mile E.S.E. of the shafts. In this latter area a thin rider coal, 7 to 8 in thick, is separated from the main coal by 3 to 4 ft of seatearth. In the neighbourhood of the shafts at Britannic Colliery the seam is 22 in thick and the thin rider occurs about 5 ft above. It is not known whether or not this rider coal is a split from the main coal, but it is significant that the fauna carried in its roof (see below) is very different from that immediately overlying the Two-Feet-Nine in the adjoining Ogmore Valley and on the South Crop. At Coedely Colliery the seam (formerly known as the Pentre) has been extensively worked: over most of the taking the section consists of 50 to 54 in of clean coal, but in the north it splits into two leaves, each of which thins quickly. On the east side, bordering on Cwm, the seam again thins rapidly to some 18 to 24 in of coal.

At Coedely a few shells occur above the normal seam, including *Anthraconaia sp. nov.* (cf. Hind, pl. xix, fig. 21) and *Anthracosia* cf. *atra,* but plants are more abundant: *Annularia radiata, Calamites sp., Mariopteris nervosa, Neuropteris gigantea, N. heterophylla, N. obliqua, Sphenophyllum cuneifolium.* At Naval Colliery *Leaia sp.* was found associated with *Naiadites sp.*

The thin rider coal close above the Two-Feet-Nine at Cambrian and Britannic carries a distinctive non-marine fauna in dark blackband: in the immediate roof *Anthracosia atra, A.* aff. *barkeri, A.* aff. *fulva* and *Naiadites sp.* were collected, and from 3 ft above *Anthracosia* cf. *atra, A.* aff. *barkeri, A.* aff. *fulva, A. sp.* intermediate between *atra* and *barkeri* and *A. sp.* intermediate between *atra* and *fulva.*

The strata between the Two-Feet-Nine and the Cefn Coed Marine Band thicken markedly when traced from east to west: at Cwm Colliery they are about 60 ft, and 77 ft is recorded in a trial pit between the Llanwonno and Daren-ddu faults on the east side of Great Western Colliery; in the west thicknesses of 130 to 150 ft are indicated at

Llwynypia, Naval, Cambrian, and Britannic collieries. The presumed positions of the marine bands and their correlation in these shafts is shown in plate IV. In the north and west of the area all the coals of the **Three Coals** group appear to be represented. At Llwynypia the following section occurs some 25 ft above the Two-Feet-Nine: coal 24 in, clift 5 ft, coal 10 in, clift 2½ ft, coal 6 in, rashings 4 in, coal 4 in; a sandstone, 25 ft thick, occurs above. Similar sections are recorded at Cambrian. At Naval the Three Coals consists of three coals, 9 to 21 in thick, spread over 25 ft of strata and associated with much sandstone. South-eastwards the coals deteriorate and tend to die out.

The marine bands of the Hafod Heulog–Cefn Coed group have been proved only at Naval, Britannic, and Cwm collieries. The Cefn Coed Marine Band, together with a second marine horizon beneath, was examined in the Nantgwyn Hard Heading at Naval Colliery, and the following section measured:

	Ft	In
Cefn Coed Marine Band:		
Yellow-weathering, grey, slightly silty shale with a few fossils; *Lingula mytilloides* and turreted gastropods	1	5
Ankerite-mudstone, impersistent .. .. 0 to		4
Dark, blocky mudstone with abundant pyrite nodules, streaks and 'fucoids'; crinoid columnals; *Productus (Dictyoclostus) craigmarkensis*, *L. mytilloides*; *Euphemites* cf. *anthracinus*, *Nuculana* cf. *attenuata* (Fleming), *?Nuculopsis gibbosa* (Fleming), *Pleurophorella* cf. *sesquiplicata* Price; *Huanghoceras?*, *Anthracoceras sp.*; *Hollinella sp.*; *Rhabdoderma sp.*	1	0
Ankerite-mudstone, hard .. .. 5 in to 1		4
Yellow-weathering, grey, slightly silty mudstone; *L. mytilloides*, *Dunbarella sp.*, *Pernopecten* [*Syncyclonema*] *carboniferus* (Hind); *Nuculopsis?*; *Anthracoceras* cf. *aegiranum*	1	4
COAL (with much ferrous sulphate)		5
Massive silty mudstone-seatearth with ironstone nodules	10	0
Grey mudstone, mostly massive	25	0
Compact ganister-like sandstone	3	0
Hard buff-grey silty mudstone-seatearth	2	6
Grey mudstone: *L. mytilloides*, *Orbiculoidea* cf. *nitida*	3	0
Shaly siltstone with abundant plant debris		1
COAL		6
Seatearth with streaks of coal	5	5

Because of major faulting immediately above and below this part of the section it is not possible to relate the sequence exactly to the general succession; the upper marine horizon is certainly the Cefn Coed, and it is likely that the lower one is the Britannic. At Britannic Colliery all three marine bands are developed. The following extract from the No. 4 Borehole sunk in 1952 is given in detail since it provides the type section for the Britannic Marine Band:

	Ft	In
Cefn Coed Marine Band: Dark, slightly silty mudstone with sporadic ironstone bands; pyritic in lower half; crinoid columnals; *Chonetes* cf. *skipseyi*, *L. mytilloides*; *Donaldina?*, *Straparollus sp.nov.*, *Nuculana* cf. *attenuata*; *Hollinella* cf. *bassleri* (Knight); *Paraparchites sp.* ..	2	6
COAL		3
Sandy mudstone-seatearth passing into sandstone	2	10
Grey silty mudstone with sphaerosiderite near top; *Planolites ophthalmoides*	18	7
Smooth carbonaceous mudstone with poor ?non-marine lamellibranchs	1	10
COAL, rashes and clod	1	2
Grey silty seatearth passing down into silty mudstone with siltstone laminae; plant debris	36	10

	Ft	In
Grey mudstone; *Planolites ophthalmoides*	6	8
Britannic Marine Band: dark mudstone with abundant pyrite granules:		
foraminifera, *P. ophthalmoides*; *Lingula?*	2	0
COAL		8
Seatearth, passing into silty mudstone with frequent siltstone laminae	17	1
COAL		3
Pale soapy seatearth, passing into grey mudstone	6	4
Hafod Heulog Marine Band: slightly silty mudstone with sporadic		
pyritic streaks: *L. mytilloides*, *Orbiculoidea* cf. *nitida*	10	11
Carbonaceous mudstone, with plant fragments	2	0
Coal 9 in, seatearth 21 in, **coal 8 in**	3	2

The No. 3 Borehole succession was similar, and *Sphenothallus sp.* was additionally recorded from the Hafod Heulog Band.

Eastwards the Britannic Marine Band deteriorates, and at Cwm seems to have died out, only one cyclothem being recognized between the Hafod Heulog and Cefn Coed bands.

A.W.W.

OGMORE–GARW–LLYNFI AREA

The **Amman Marine Band** has been examined at Wyndham, Garw, Ffaldau, St. John's, and Coegnant collieries. *Lingula mytilloides* was virtually the only fossil found: usually it occurs in the immediate roof of the Amman Rider, but the shells may be scattered through nearly 3 ft, as at St. John's. The marine strata are succeeded by $2\frac{1}{2}$ to 5 ft of grey mudstones carrying the usual small 'mussels' with 'ghostly' preservation, including *Anthraconaia?*, *Anthracosia* cf. *concinna*, *A.* aff. *ovum*, *A. sp.* of *ovum/aquilina* group and *A.* cf. *regularis*. At Coegnant *Spirorbis sp.* was attached to some of the shells.

The rest of the cyclothem consists of 20 to 55 ft of mudstones and silty mudstones, which in general thicken from east to west. In the south-west, at St. John's and Celtic, a rider coal, 6 to 8 in thick, lies about 10 ft below the Bute.

The **Bute** has been extensively worked; it was named the Yard at Penllwyngwent, Wyndham, and Ffaldau, the Brunt's at Western, Garw, and International, the Lower New at Caerau, Coegnant, and St. John's, the Five-Feet at Oakwood and Celtic and the Lower Three-Feet at Duffryn Madoc. Throughout the Ogmore and Garw valleys the seam has two leaves: top coal 16 to 24 in, stone or clod 0 to 2 in, bottom coal 30 to 42 in. In the Llynfi Valley the seam consists of a single coal varying from 3 ft in the north to 5 ft in the south, though the thin parting is present locally. At Coegnant and in the Garw Valley 3 to 8 in of inferior coal occur in the base of the seam, 3 to 12 in of rashings intervening, and this may be the thin coal recorded 10 ft below the seam in the south-west of the area. Typical sections are:

	Wyndham		Ffaldau		Garw		Caerau		St. John's	
	Ft	In	Ft	In	Ft	In	Ft	In	Ft	In
Coal ..	1	9	1	6	1	6	1	2		
Dirt ..		1		1		1		$\frac{1}{4}$	4	5
Coal ..	2	8	2	11	3	6	2	7		
Dirt ..			1	0						
Coal ..				7						

The roof of the Bute carries its distinctive fauna throughout the three valleys. At St. John's the lowest 5 ft of the roof are dark mudstones, increasingly cannelly towards the base and containing numerous ironstone bands. The basal layers abound in crushed shells, but in the upper part 'solids' are also present. The succeeding silty mudstones carry abundant solid forms. At other collieries the cannelly mudstone with crushed shells is thinner: 3 in at Ffaldau, 12 in at Garw and 4 in at Coegnant. At Caerau about 12 in of silty mudstone with a 'solid' fauna may intervene between the cannelly mudstone and the top of the coal. The crushed shells belong to the same species as the 'solids' and include variants of *Anthraconaia sp.* (of *modiolaris* group), *Anthracosphaerium affine, A. bellum, A. exiguum, A. turgidum, Naiadites quadratus* and *N. sp. nov.* cf. *modiolaris* Hind (*non* J. de C. Sowerby).

The Bute and the Lower Nine-Feet are mostly 40 to 50 ft apart, though thicknesses up to 80 ft are recorded in parts of Penllwyngwent and Wyndham, and in the north of the Llynfi Valley they may be somewhat closer. The strata consist mainly of mudstones and silty mudstones, with a few thin impersistent sandstones developed locally.

The **Lower Nine-Feet,** formerly known as the Bute Seam throughout the Ogmore and Garw valleys, the Upper New Seam at Caerau, Coegnant, and St. John's, and the Lower Four-Feet at Oakwood, Celtic, and Duffryn Madoc, has also been extensively worked. Typically the seam has a two-coal section with a separation between the leaves varying from 0 to 3 in. The top coal varies from 29 to 39 in. in the Ogmore Valley and in the north of the Garw and Llynfi areas, to 42 to 48 in. in the area south of Maesteg. At Western a cannel, up to 17 in thick, occurs at the top of the coal. Locally in Wyndham and Ffaldau collieries a parting develops 14 to 18 in from the bottom of the leaf, while in the Llynfi area thin and impersistent partings occur near the top, and pyrite lenses about 15 in above the base. The bottom coal is 34 in thick at Western, but thins westwards to 17 in and less in the Garw Valley, and to less than 12 in. in the north-west of the Caerau taking; in the Llynfi Valley the coal again thickens southwards to more than 2 ft in places in St. John's Colliery, where up to 15 in of cannel may form the top of the leaf. At Garw Colliery and south of the anticline in the Maesteg area there are no records of partings, the seam consisting of 40 to 63 in of clean coal. The following sections illustrate the variation:

		Western		Wyndham		Ffaldau		Caerau		St. John's	
		Ft	In	Ft	In	Ft	In	Ft	In	Ft	In
Coal	.. ⎫			2	0	2	6	⎫			
Dirt	.. ⎬	2	10		0½	thin		⎬ 2	11	3	6
Coal	.. ⎭			1	2	1	5	⎭			
Dirt	..		thin		0½		3		0½		1
Coal	..	2	11	1	8	1	2	1	4	2	5

The immediate roof of the Lower Nine-Feet is usually a carbonaceous mudstone which passes upwards into smooth shaly dark mudstone. It has yielded no definite fossils. At Penllwyngwent about 40 ft of strong mudstones separate the Lower and Upper Nine-Feet seams and in the Garw Valley these thicken to about 75 ft. In the Llynfi Valley the strata become more arenaceous, and at Coegnant up to 30 ft of sandstones occur in the middle of the cyclothem.

The **Upper Nine-Feet** has always been known as Nine-Feet except in the Llynfi Valley; at Caerau, Coegnant, and St. John's it was called the Harvey, at Duffryn Madoc the Upper Four-Feet, and it appears to be the seam known as the Two-Feet-Ten or Lower Two-Feet-Nine at Celtic and Oakwood. In the east of the area the seam consists of a main coal 4½ to 5¼ ft thick with an inferior or 'engine' coal lying close above. The main coal frequently has an impersistent 'brass' layer near the top and a thin stone parting about 1½ to 2 ft above the base. At Western the latter parting is replaced by 6 to 12 in of cannel; and cannel is also present at the top of the main coal section in many places. Traced westwards the main coal thins to about 3¾ ft in the Maesteg area. The position of the thin 'engine' coal at the top of the seam is variable, being sometimes separated by as little as 5 in of cannel from the main coal, as in Wyndham, or lying several feet above, as in parts of the Llynfi Valley. Locally a bottom coal, 3 to 6 in, has been recorded, separated from the base of the main seam by 3 to 12 in of clod or rashings. The following sections illustrate the variations exhibited by the seam in full section:

	Penllwyngwent Ft	In	Wyndham Ft	In	Ffaldau Ft	In	Caerau Ft	In
Coal		4		9		3		2
Clod	1	4		2		4	2	0
Bast				8		10		
Coal				6				
Brass				0½			1	11
Coal	4	6	1	6				
Dirt			1	0	5	3		0½
Coal			2	0			1	4
Rashings ..		4		7				5½
Coal		3		4				6½

At Penllwyngwent typical species of the 'Red Vein fauna' occur about 9 in above the seam. Many variants of the *Anthracosia nitida* series have been recognized, including *Anthracosia angulata*; also present are shells which approach *Anthraconaia pulchella*. At Western Colliery many shells occur above thin coal streaks 3 ft above the main coal: the fauna, which includes cf. *Anthraconaia pulchella, Anthracosia* cf. *nitida, Naiadites* aff. *productus, N.* cf. *subtruncatus*, is noteworthy for the abundance of well-preserved 'solid' *Naiadites*. This fauna has been followed eastwards and is the same as that recorded on the bottom coal of the Red Vein group in the Rhondda Fach.

In the Ogmore and Garw valleys the Caerau lies 80 to 120 ft above the Upper Nine-Feet, the mainly argillaceous strata thickening westwards. Within these, one thin coal is usually recorded, lying, in general, 20 to 30 ft below the Caerau. This coal evidently represents some part of the lower portion of the Red Vein of the Aberaman area; it is normally less than 12 in thick, but at International Colliery the shaft shows: top coal 18 in, fireclay 4¼ ft, coal 15 in. Immediately above this coal at Western occur 18 in of smooth dark carbonaceous, ferruginous mudstone with the typical 'Nine-Feet fauna': *A. sp.* cf. *beaniana, A.* cf. *ovum, A. phrygiana, A.* aff. *retrotracta* and *Naiadites sp.* intermediate between *productus* and *quadratus*. This is succeeded 3 to 4½ ft higher by a good example of the small 'Red Vein fauna' comprising variants of cf. *Anthraconaia pulchella, Anthracosia* aff. *nitida* and *Naiadites sp.*

The measures between the Upper Nine-Feet and the Caerau are not well known in the Llynfi Valley. Within the 70 to 100 ft of strata one seam is usually developed, lying 30 to 40 ft below the Caerau. At Coegnant Colliery it is about 13 in thick and carries a thin cannelly mudstone in its roof. Grey mudstone above yielded crushed 'mussels' of the *Anthracosia phrygiana* group at about 2 to 3 ft above the coal, and this was overlain by small 'solids' of 'Red Vein' type. Farther to the south-east the coal is 25 in thick with 7 in of cannelly mudstone in two bands in its roof; it is 26 in thick at Celtic Colliery and 33 in at Duffryn Madoc. Various old records in the valley show other coals between the Upper Nine-Feet and the Caerau, but the presence of these is believed to be the result of structual repetition.

The Caerau has been worked extensively in those areas where the separation from the overlying Lower Six-Feet is sufficient to permit adequate roof control. The seam, known as the Red Vein in the Ogmore and Garw valleys and the Seven-Feet in the Maesteg area, correlates with the upper part of the Red Vein of the Aberaman–Cwmaman area, though the details are obscure. It consists typically of two leaves separated by an inch or so of rashings, as shown by the following sections:

	Western Ft	In	Wyndham Ft	In	Ffaldau Ft	In	Garw Ft	In	Coegnant Ft	In	Caerau Ft	In
Coal ..	1	4	2	5	1	9	1	10	2	6	1	3
Rashings		3		3		0½		2		1		1
Coal ..	2	11	3	3	3	3	3	5	2	9	2	9

At Wyndham Colliery 3 to 6 in of cannel occur 5 to 8 in from the top of the seam, while at Coegnant a further thin coal is recorded 2 to 3 ft above. The Caerau is the lowest seam known to crop out in the Llynfi Valley, and was formerly worked in 'patches' in the area to the north-east of Maesteg Railway Station [85409157].

Around the shafts at Wyndham Colliery as little as 6 in of seatearth separate the Caerau from the Lower Six-Feet; at Ffaldau the separation varies from 2 to 30 ft. Westwards and north-westwards the two seams are 15 to 30 ft apart and the intervening measures are made up largely of seatearth or mudstone with roots and ironstone nodules. At Coegnant the immediate roof of the Caerau consists of seatearth which yielded a few 'mussels', identified as the internal moulds of *Carbonicola?*, and appear to represent a unique occurrence of this genus at so high a level. In the Llynfi Valley a thin dirty seam is usually recorded between the Caerau and Lower Six-Feet. At Coegnant this has a complicated section totalling 2½ ft; at St. John's 7 in of coal lies within 2 ft of the base of the Lower Six-Feet.

In the Ogmore and Garw valleys the Lower and Upper Six-Feet seams are 50 to 70 ft apart. The **Lower Six-Feet** has been virtually exhausted at all collieries except Penllwyngwent. The section varies little: 4 to 5½ ft of top coal is separated from a bottom coal of 16 in to 2½ ft by up to 2 in of dirt. In many parts of Wyndham and Ffaldau the seam has been recorded as 5½ to 7 ft of clean coal, but a smooth plane parting appears to be always present 1½ to 2 ft from the base. At Penllwyngwent up to 12 in of cannel occur in the top of the seam, and lenticular 'brass' is locally present in the middle of the top coal.

At Western Colliery *Lioestheria sp.*, associated with *Naiadites sp.* and *Planolites montanus*, has been collected from the 8 in of grey mudstone immediately overlying the Lower Six-Feet, the first-named being more common in the bottom 2 in. The strata up to the Upper Six-Feet consist mainly of strong mudstone with ironstones, and sandstone bands are recorded at Ffaldau and International and, near the top of the cyclothem, at Western. At the last-named colliery cannelly blackband shale lying about mid-way between the Lower and Upper Six-Feet seams yielded: *Spirorbis sp.* [on *Naiadites*], *Anthraconaia sp.* of *librata/oblonga* group, *N. productus* and *Carbonita humilis*.

The **Upper Six-Feet** (normally an un-named seam) has not been worked in any of the collieries of the Ogmore and Garw area; it consists everywhere of a clean coal varying in thickness from 20 to 26 in. The roof carries the typical Six-Feet fauna and the following species have been identified: *Anthracosia* cf. *aquilina*, *A.* cf. *atra*, *A. sp.* intermediate between *atra* and *lateralis*, *A. sp.* intermediate between *acutella* and *concinna*, *A. sp.* cf. *ovum*, *Anthracosphaerium* aff. *propinquum* and *Naiadites sp.*

The Lower and Upper Six-Feet seams, locally called the Six-Feet and Red Vein respectively, exist as two separate seams in the east of the Llynfi Valley area, but to the west they unite to form a single seam. The line of union follows the valley bottom, and to the west the full section, about 9 ft, has been worked in parts of Duffryn Madoc and Coegnant. This composite seam was formerly known as the Nine-Feet or Furnace Vein. Although the **Lower Six-Feet** has been worked extensively in the eastern part of the area, few working sections have been preserved and the seam is now only rarely accessible. In general it appears to be about 6 ft thick, but it is commonly much affected by structures due to incompetence, which give the seam a variable section. A few records show a top coal of about 5 in, and this may well represent the thin coal at or near the top of the seam which elsewhere carries *Lioestheria*. This conclusion is supported by the presence at St. John's Colliery of an inch or so of cannel between the thin coal and the main part of the seam, an association similar to that in the Treorky area (see p. 113). The roof of the Lower Six-Feet has been seen at only a few points: at St. John's 'slips' in the immediate top of the seam frequently cut out the true roof. At Coegnant the mudstone roof has yielded *Anthracosia* cf. *aquilina*, *A. planitumida* and *Naiadites* cf. *obliquus*. At Caerau the roof carries fragmentary plant remains, suggesting that the thin coal with its fauna may have split away from the main seam.

The Lower and Upper Six-Feet seams are separated by up to 60 ft of strong silty mudstones. The **Upper Six-Feet** has been worked on a small scale in most of the collieries in the valley and it has been examined at Caerau, Coegnant, and St. John's. At Caerau Colliery it consists of clean coal 26 to 40 in thick, while at Coegnant it is about 29 in. At all localities a rich fauna, well-preserved in ferruginous mudstone in the south of the area, and in a more cannelly one in the north, has been obtained. Among the forms identified are variants of: *Anthraconaia confusa?* (Trueman), *Anthracosia atra*, *A. aquilina*, *A. caledonica*, *A. concinna*, *A. fulva*, *A. lateralis*, *A. planitumida* and *Naiadites sp.* (*?alatus* group).

Over most of the Ogmore and Garw area some 30 to 40 ft of strata usually intervene between the Upper Six-Feet and the lowest coal of the Four-Feet group, though in parts of Wyndham and Ffaldau they may be reduced to as little as 11 ft. These measures consist of mudstones with a few sandy layers; thin ironstone layers are frequent in the lower part of the cyclothem and large nodules in the upper part.

In the Ogmore Valley four coals spread over 60 to 80 ft represent the **Four-Feet** group of coals. At Western the section is (from top): coal 7 in, rashes 2½ ft, coal 7 in, fireclay and rashes 5 ft, coal 2 in, strong mudstone with balls of mine 24 ft 11 in, coal 21 in, mudstone and fireclay 22 ft 11 in, coal 14 in; and at Wyndham: rashes with coal bands 6 ft 3 in, fireclay, clift and rock 42 ft 11 in, coal 18 in, mine ground 14 ft 8 in, coal 11 in, fireclay and rock 22 ft 5 in, coal 13 in. These sections are similar to those recorded at Cambrian and Britannic collieries (see p. 120), and the four coal horizons presumably correlate with the individual leaves of the typical Four-Feet section in the Cynon and Rhondda Fach valleys. The top coal of the group at Western Colliery yielded the typical Four-Feet fauna: *Anthraconaia* cf. *rubida*, *A. sp. nov.* (cf. *curtata*), *Anthracosia* cf. *acutella*, *A.* cf. *aquilina*, *A. sp.* intermediate between *aquilina* and *lateralis*, *A. sp.* intermediate between *atra* and *barkeri*, *Anthracosphaerium* cf. *exiguum*, *A.* aff. *propinquum* and *Naiadites obliquus*.

In the northern part of the Garw Valley all the coals of the Four-Feet group are much closer together; the following section of the ground between the Lower Six-Feet and the Two-Feet-Nine measured in a cross-measures roadway at Garw Colliery shows that the several coals are spread over only 16 ft 11 in of strata:

		Ft	In
TWO-FEET-NINE: **coal 57 in**, fireclay 42 in, **coal 14½ in**		9	5½
Streaky carbonaceous shale with coal layers		1	0
Strong seatearth passing down into strong siltstones		35	0
Strong silty mudstone with *Anthraconaia* cf. *rubida*, *A. sp. nov.* (cf. *curtata*), *Anthracosia sp.* intermediate between *atra* and *planitumida*, *Anthracosphaerium* cf. *exiguum*, *A.* aff. *propinquum*, *Naiadites sp.* ...		1	3
FOUR- FEET GROUP	Coal 8 in, dirt 1 in, **coal 13 in**	1	10
	Fireclay with thin streaks of coal	4	3
	Coal with large 'brass' nodules in upper part **18 in** ..	1	6
	Fireclay and rashes with dirty coal layers..	2	4
	Strong fireclay	5	7
	Coal with parting near top **17 in**	1	5
Fireclay passing down into grey mudstone; large ironstone nodules ..		25	0
Dark blue-grey shale with *Anthracosia* cf. *aquilina*, *A.* cf. *atra*, *A. sp.* cf. *ovum*, *A. sp.* intermediate between *atra* and *lateralis*		1	3
UPPER SIX-FEET: **Coal 21 in**		1	9
Fireclay passing down into mudstone		25	0
Cannelly shale passing down into shaly cannel		3	0
COAL			2½
Grey mudstone		12	0
LOWER SIX-FEET: **coal 57 in**		4	9

At International Colliery the Four-Feet coals are still closer together and the group is represented in the shaft by a single compound dirty seam 94 ft above the Upper Six-Feet: top coal 9 in, clod 2 in, coal 12 in, clod 46 in, coal 13 in, clay 13 in, coal 3 in.

In the Llynfi Valley the main part of the Four-Feet group lies about 90 to 135 ft above the Upper Six-Feet; here again it has a very variable behaviour. In the north, at Caerau Colliery, a single compound seam has a section similar to that at International: top coal 6 in, bast 2 in, coal 9 in, bast 1 in, coal 15 in, bast 1 in, coal 21 in. At Coegnant a parting of about 3 ft develops within the seam and foreshadows the splitting southwards into two distinct seams, formerly known as the Lower Six-Feet (lower split) and Truro (upper split), and which represent the Lower and Upper Four-Feet seams respectively. The individual coals are thicker than in the areas to the north and east, the shaft record showing the following sections: Lower Four-Feet, top coal 36 in, clod 9 in, coal 27 in, rashes 3 in, coal 4 in; and the Upper Four-Feet, top coal 13½ in, clod 3 in, coal 15½ in, rashings and coal 30½ in. Two further thin coals, each about 6 in, are recorded about 10 and 20 ft below the Lower Four-Feet, and these apparently equate with two thin streaks of coal developed within the seatearth of the main seam at Duffryn Rhondda Colliery to the north. At St. John's Colliery the position is obscure. The only details available are those of the shaft sections, where thrusting may be present. At Duffryn Madoc and Gin collieries the Lower and Upper Four-Feet are reputed to be 6 ft apart, and at Bryn Rhyg, 1200 yd east of Maesteg Station, they were worked separately, the former being said to have been 6 ft thick and the latter 3 ft. At Oakwood and Celtic where both seams were worked they are 30 ft apart: the lower seam is said to be 4 to 6 ft thick with an 18-in parting about 2 ft from the base; and the upper seam 3 ft thick in three leaves.

The roof of the Upper Four-Feet at Coegnant and the highest coal of the group at St. John's yielded the diagnostic fauna, including variants of *Anthraconaia sp. nov.* (cf. *curtata*), *Anthracosia planitumida*, *Anthracosphaerium exiguum* and *A. propinquum*. The lowest few inches of the roof mudstones contained crushed poorly preserved shells referred to *Anthracosia?* and *Naiadites obliquus*.

The strata from this horizon up to the Two-Feet-Nine vary in the Ogmore and Garw valleys from 15 to 40 ft, being thickest in the northern part of the Garw. In the Llynfi Valley they range from 25 ft in the south to 50 ft in the north. Mudstones make up the bulk, but sandstones occur towards the top in the thicker sections.

The **Two-Feet-Nine** is renowned for its quality throughout the Ogmore and Garw valleys, and it has been worked extensively. The seam varies little: a top coal, 4 to 5 ft, is separated from a bottom coal of 1 to 1½ ft by 6 in to 3½ ft of seatearth. Variable amounts of cannel up to 12 in are usually present in the top of the seam, and the bottom coal has a thin stone parting in the middle in parts of Western and Wyndham. An abundant 'mussel' fauna, which occurs everywhere about 3 to 5 ft above the top of the seam, is made up of variants of the following species: *Anthraconaia cymbula*, *A. sp. nov.* cf. *lanceolata*, *A. sp. nov.* cf. *wardi*, *Anthracosia acutella*, *A. aquilina*, *A. sp.* intermediate between *acutella* and *concinna*, *A. atra*, *A. sp.* intermediate between *atra* and *planitumida*, *A. concinna*, *A. planitumida*, *A. simulans*, *Anthracosphaerium propinquum*, *Naiadites obliquus*. *Planolites montanus* preserved in nodular ironstone is a constant associate of these forms, while at Ffaldau Colliery '*Estheria*' was also found.

In the Llynfi Valley the Two-Feet-Nine lies 25 to 50 ft above the Four-Feet, the distance increasing northwards. In the south the intervening strata are mainly mudstones containing numerous ironstone bands and nodules, but as they thicken they become more sandy. The Two-Feet-Nine has been worked at all collieries except Caerau. Throughout most of the area the main part of the seam is about 5 ft thick with a further coal, 1 to 2 ft thick, lying about 2 to 5 ft below, and it may be this lower leaf that crops out [85529187], 300 yd W.S.W. of Maesteg Deep Colliery. Over much of the area a plane parting, which locally thickens to an inch or so, is developed 9 to 18 in from the top of the main coal. In the north-west of the valley another parting develops about 2 ft above the floor of the main coal; this thickens rapidly and has limited the working of the seam, hardly any being extracted to the north-west of a line joining Coegnant and Caerau shafts. The roof of the seam has been examined at Coegnant and St. John's, where it consists of highly slickensided grey mudstone containing worm tracks and sporadic shells, including *Anthracosia sp.*, *Anthracosphaerium* cf. *exiguum* and *Naiadites obliquus?*. Spoil at Maesteg Deep Colliery yielded a well-preserved and varied fauna of solid shells, all of small size: *Anthraconaia librata*, *A.* cf. *obscura*, *Anthracosia* aff. *acutella*, *A.* cf. *aquilina*. *A. concinna*, *A.* aff. *lateralis*, *A.* cf. *simulans*, *Anthracosphaerium* cf. *affine*, *A. exiguum*, *A.* cf. *propinquum*, *A.* cf. *turgidum*. There are certain differences from the normal Two-Feet-Nine fauna, notably in the occurrence of the several species of *Anthracosphaerium*. In this connexion it should be noted that a small amount of a seam variously described as the 'Four-Feet' or 'Seven-Feet' has also been worked from this colliery; it seems likely that the various species of *Anthracosphaerium* came from this horizon, which presumably equates with the upper part of the Four-Feet.

In general the measures from the Two-Feet-Nine to the Cefn Coed Marine Band vary from about 200 to 250 ft in the Ogmore and Garw valleys. Except for an isolated occurrence of the Cefn Coed band, brought in by faulting, at Glengarw Colliery, these measures can be examined only at Penllwyngwent Colliery, where they were penetrated by a cross-measures drift from the No. 3 Rhondda to the Two-Feet-Nine through the Aber Fault. The strata in this old roadway are much obscured, but both the Cefn Coed and Hafod Heulog marine bands yielded fossils and the horizon of the Britannic Marine Band was identified. The **Hafod Heulog Marine Band** occurs as a band about 3 ft thick above a thin coal some 120 ft above the Two-Feet-Nine, and yielded foraminifera (including *Glomospira sp.*), *Lingula mytilloides* and large *Orbiculoidea* cf. *nitida* from dark blue shale. The horizon of the **Britannic Marine Band** occurs above an 8-in coal 45 ft higher in the sequence, and although it yielded no marine fossils other than doubtful foraminifera, the lithology was that of a typical marine sediment; furthermore *Planolites ophthalmoides* was obtained some little way above the seam. The **Cefn Coed Marine Band** is well developed and occurs above a 3-in coal about 100 ft above the Hafod Heulog horizon and 100 ft below the Caedavid (Gorllwyn). The marine strata are about 2 ft thick, consisting of smooth dark mudstone at top and bottom, with *Lingula mytilloides* and indeterminate gastropods, and about 6 in of blocky mudstone in the middle carrying the following varied fauna: crinoid columnals; *Archaeocidaris* aff. *acanthifera* Trautschold; *?Crurithyris carbonaria*, *Chonetes (Tornquistia) diminutus*, *L. mytilloides*; *Straparollus sp.*, turreted gastropods; *Aviculopecten sp.*, *Dunbarella* cf. *macgregori* (Currie), *Nuculana* cf. *acuta*, cf. '*Nucula*' *luciniformis* Phillips, *?Nuculopsis gibbosa*, *Posidonia sulcata* (Hind); Orthocone nautiloid, *Metacoceras?*, *Anthracoceras* cf. *aegiranum*; *Hollinella sp.*

At Wyndham and Western collieries the shaft sections show a complex coal horizon about 50 ft above the Two-Feet-Nine which is clearly the equivalent of the Three Coals of the Aberaman area. In Wyndham No. 2 Shaft the section of this coal reads: top coal 9 in, clod 44 in, coal 9 in, clod 22 in, coal 2 in; the Hafod Heulog Marine Band presumably lies on a coal some 50 ft higher in the sequence and the Cefn Coed Marine Band on a 1½-in coal 90 to 100 ft still higher. Between these two latter horizons occur two further coals, one of which presumably carries the Britannic Marine Band. The sequence is similar in the Garw Valley: at Ffaldau, for example, the Three Coals horizon, separated from the Two-Feet-Nine by 75 ft of strata, consists of: top coal 10 in, fireclay and clift with balls of ironstone 11 ft, coal 9 in, clift and fireclay 3 ft 9 in, coal 9 in, and the presumed positions of the Hafod Heulog and Cefn Coed marine bands lie 42 ft and 154 ft above respectively.

At Glengarw Colliery the **Cefn Coed Marine Band** is exposed in a drivage through a fault to the Caedavid workings. It occurs above a 1-in coal about 100 ft below the Caedavid, and the following fauna has been collected: *Chonetes (Tornquistia) diminutus*, Productid fragments, *Orbiculoidea* cf. *nitida*, *L. mytilloides*; *Glabrocingulum sp.*,

Nuculid lamellibranchs, *Palaeoneilo?*, cf. *Scaldia fragilis* de Koninck; *Metacoceras sp.*, *Anthracoceras* cf. *aegiranum*; *Hollinella sp.*

This part of the sequence between the Two-Feet-Nine and the Cefn Coed Marine Band in these areas contains a good deal of hard quartzitic sandstone in beds up to about 30 ft thick, those associated with the marine bands corresponding in position with the Cockshot Rocks of the Maesteg area.

In the Llynfi Valley the measures from the Two-Feet-Nine to the Cefn Coed Marine Band crop out in the area around Maesteg and certain of the shafts have been sunk through them. The measures thicken from north to south, from about 225 ft at Caerau to upwards of 300 ft at Celtic, Oakwood, and Duffryn Madoc, though the strata at these latter collieries appear to be highly disturbed. The sequence is similar to that in the Garw and Ogmore valleys, except that the quartzitic sandstones are more prominent, and clay-ironstone bands attain a local importance. About 50 ft above the Two-Feet-Nine occurs a thin coal, which at Maesteg Deep comprises 13 in of coal underlain by 4 in of inferior coal and 5 in of cannel. A short distance above, a coal, about 18 in thick and having a roof of dark cannelly shale with fish scales and spores, appears to be the seam formerly referred to as the Fireclay. A coal said to be 2 ft thick at Maesteg, but only a few inches at Caerau, lies some 30 ft higher in the sequence; this has been called the Coal Tar seam at Maesteg, though it does not appear to be the same as that so called at Oakwood (see below).

The **Lower Cockshot Rock**, developed within the overlying cyclothem, is 15 to 25 ft thick. The massive quartzitic sandstone lies immediately above the Coal Tar at Coegnant and Caerau while 20 to 30 ft of mudstones intervene to the south. Near Maesteg Deep Colliery its base is sharp and rests on a strongly eroded surface. Quarries are located [84639195, 85509210] to the north-west and north-east of Maesteg.

A short distance above the sandstone is the seam referred to at Oakwood and Celtic as the Coal Tar or Upper Two-Feet-Nine. At Oakwood two leaves, each 11 in thick, are separated by 9 in of dirt. Comparison with the closely similar sequences at Penllwyngwent and Britannic suggest that the roof of this seam should carry the Hafod Heulog Marine Band. The coal seems to deteriorate rapidly to the north, being insignificant at Caerau and not recorded at Coegnant.

The **Upper Cockshot Rock** overlies a 6-in coal some 40 ft higher in the sequence. The sandstone is quartzitic on the east side of the valley, though not on the west. It is thickest, over 50 ft, in the Caerau shafts, but it thins southwards to about 30 ft. It has been quarried [84709207, 85489226] in the Duffryn Madoc area, while in Maesteg itself it is exposed along Garn Road [85659096].

The Britannic Marine Band is assumed to lie in the roof of a 3- to 8-in coal almost immediately above the Upper Cockshot. At Oakwood an 8-in coal 35 ft above is thought to be the lower split of the Cefn Coed Marine Band coal, the upper split lying 25 to 40 ft still higher. Sandstone developed to the south between the two coals has been wrongly termed the Upper Cockshot at Oakwood.

The **Cefn Coed Marine Band** overlies a coal usually about 3 in thick. The band itself has been seen at only one locality—the stream [85759328], 200 yd E. of Coegnant Colliery—where it rests directly on seatearth. The lower part consists of 15 in of calcareous mudstone containing an abundant and varied fauna, and this is overlain by 3 ft of smooth dark mudstone with sporadic *Lingula*. The following forms have been identified from the calcareous mudstone: *Zaphrentites postuma* (S. Smith); *Archaeocidaris sp.*; crinoid columnals; *Chonetes (Chonetinella) sp.*, *Chonetes (Tornquistia) diminutus*, *Crurithyris carbonaria*, *L. mytilloides*, *Productus (Dictyoclostus) craigmarkensis*; *Straparollus (S.) sp. nov.*, turreted gastropods; cf. *Nuculopsis gibbosa*, *Pernopecten carboniferus*, *Posidonia* cf. *sulcata*, *Schizodus carbonarius* (J. de C. Sowerby); orthocone nautiloid; *Anthracoceras* cf. *aegiranum*, *Gastrioceras sp.*; *Hollinella* cf. *bassleri*, *Paraparchites sp.*; *Hindeodella sp.*, platformed conodonts; fish debris.

A.W.W., W.B.E.

AVAN VALLEY AND BRYN AREA

These measures crop out in a strip about half a mile wide on the north side of the Moel Gilau Fault in the neighbourhood of the village of Bryn. Surface exposures, however, are poor and provide little information additional to that obtained by mining.

The **Amman Marine Band** has been seen only at Bryn Colliery. Here *Lingula mytilloides* was obtained from 2 in of slickensided mudstone in the roof of a thin coal between the Nos. 3 and 4 seams. Above this several feet of mudstone yielded 'mussels', including *Anthracosia* cf. *ovum* and *A. sp.* cf. *aquilinoides*, with the 'ghostly' preservation typical of the horizon. At Avon Colliery a similar fauna occurs about 8 in to 3 ft above an 18-in coal which appears to be the first below the Bute: *Anthracosia* cf. *beaniana*, *A.* cf. *ovum* and *Naiadites sp.* have been identified. The mudstone in the immediate roof of the coal was highly disturbed and no marine fossils were found, despite the favourable lithology.

The **Bute**, 30 to 40 ft above, has been worked at Glyncorrwg, Duffryn Rhondda, and Bryn, while the workings from Glyncastle extend into the district. The seam has been named the Peacock at Glyncastle and Glyncorrwg, Seam 'E' at Avon, New Seam at Duffryn Rhondda, and No. 3 Seam at Bryn. Over most of the area it consists of a single clean coal, about 3 ft thick, but at Bryn a 2-in leaf occurs up to 12 in above the top of the main coal, and 6 to 18 in of inferior coal and rashings are present in the floor. At Duffryn Rhondda, Glyncastle, and Glyncorrwg the roof to about 6 in above the coal is planty and 'mussels' are abundant in the overlying 1 to 2 ft. At Avon and Bryn up to 18 in of cannelly mudstone containing many crushed shells are overlain by grey silty mudstone with ironstone bands and solid 'mussels'. The same species are found in all localities: *Anthraconaia williamsoni*, *A.* aff. *insignis* (Davies and Trueman), *Anthracosphaerium affine*, *A. exiguum*, *A. fuscum*, *A. turgidum* and *Naiadites sp. nov.* cf. *modiolaris*.

The Bute is succeeded by 30 ft of strata, mainly mudstone with up to 10 ft of sandstone, locally unconformable, in the middle. In places a thin coal is present midway in the cyclothem.

The **Nine-Feet** splits southwards between Glyncorrwg and Glyncastle in the north and Avon and Duffryn Rhondda in the south. To the north the seam averages 9 ft in thickness and it has been worked extensively at Glyncorrwg. At Glyncastle the roof consists of black cannelly mudstone with an abundant fauna, including *Anthracosia beaniana*, *A.* aff. *ovum*, *A. sp.* intermediate between *phrygiana* and *ovum*, *A. sp.* cf. *phrygiana*. At Glyncorrwg the immediate roof is planty.

To the south the **Lower Nine-Feet** is called the Seven-Feet at Duffryn Rhondda, the No. 2 Seam at Bryn, the Bute or 'D' Seam at Avon and the Lower Four-Feet at Eryl, south of Mynydd Bychan. It has been worked on a limited scale at all these collieries and it crops at Eryl. Characteristically the seam consists of an upper coal, $3\frac{1}{4}$ to 4 ft thick, and a bottom leaf, about 1 ft thick. At Avon, Duffryn Rhondda, Cwmavon, and Eryl the two leaves are 6 to 24 in apart, but the separation thickens south-westwards and is 2 to 10 ft at Bryn. Here a thin parting develops 4 in from the top of the seam, and a band of pyrite is frequently found 12 in from the base of the upper leaf. No fauna has been reported from the roof, which locally cuts down into the coal at Duffryn Rhondda and Bryn.

The separation of the Lower and Upper Nine-Feet increases southwards to 45 ft at Duffryn Rhondda, 70 ft at Avon, and from 70 to 100 ft at Cwmavon. Sandy mudstones and siltstones with a few thin sandstones make up the bulk of these strata.

The **Upper Nine-Feet** has been worked at outcrop from levels near Eryl under the name of Five-Feet, and attempts to work it have been made at Avon and Duffryn Rhondda. Few sections are reliable: it appears to vary from $2\frac{1}{2}$ to 4 ft thick with thin leaves present in floor and roof locally. At Nantewlaeth it is probably the coal called the New Seam, and was reported to be about $3\frac{1}{2}$ ft thick. At Duffryn Rhondda the roof yields a sparse 'Red Vein fauna': *Anthraconaia* aff. *pulchella* and *Naiadites* cf. *productus*; '*Estheria*' *sp.* was also noted.

The measures from the Nine-Feet to Six-Feet are among the least well-known. In Glyncorrwg the two seams are 100 to 120 ft apart, and two coals have been proved in the intervening measures. The **'Red Vein'**, equivalent to the Caerau of the Llynfi and Garw areas, lies some 80 ft above the Nine-Feet and has a section: top coal 7 in thin parting, coal 27 to 30 in. About 10 to 15 ft below the Six-Feet a 6- to 9-in coal, with a mudstone roof containing plants and fragmentary shells, appears to equate with the bottom leaf of the Lower Six-Feet in the Rhondda Fawr. At Avon Colliery this coal, about 12 in thick, lies 1 to 2 ft below the Six-Feet. The 'Red Vein', 20 to 30 ft below, has been worked on a small scale: top coal 10 in, clod 4 in, coal 32 in. A further dirty coal, 16 in thick, here lies 20 to 25 ft below the 'Red Vein' and 35 ft above the Nine-Feet. The roof of this coal carries small solid 'mussels' 10 to 16 in above the roof: a typical 'Red Vein fauna' includes *Anthraconaia sp. nov.* cf. *pulchella* and *Anthracosia* cf. *nitida*. A few hundred yards away the same coal carries a cannelly mudstone with ironstone bands and 'wormy' layers. A crushed fauna of characteristic 'Nine-Feet' type includes such forms as *Anthracosia* aff. *beaniana*, *A. sp.* intermediate between *lateralis* and *ovum*, *A.* aff. *phrygiana* and *A. sp.* intermediate between *phrygiana* and *aquilina*.

At Duffryn Rhondda the Upper Nine-Feet and Six-Feet are 150 ft apart, and only one coal has been proved between them. This is called the 'Red Vein', and is 24 to 27 in thick with an inch of cannel in its roof, and lies 30 to 40 ft above the Nine-Feet. *Anthracosia* cf. *phrygiana* and *A. sp.* cf. *ovum* were collected from the roof; 20 ft of mudstones overlain by 60 to 80 ft of sandstone and siltstone follow. It is possible that both here, and at Nantewlaeth, the 'Red Vein' (Caerau) of Avon is washed out by these sandstones.

At Bryn this portion of the sequence is much obscured by structural complications. In the workings a coal has been proved 5 to 15 ft below the Six-Feet. This 'No. 2 Seam', about 3 ft thick and disturbed, appears to be the Caerau Seam of the Llynfi area and the 'Red Vein' of Avon.

The sequence near Cwmavon is anomalous. The distance between the Nine-Feet (Five-Feet) and the Six-Feet (Big) is variously given as being as little as 108 ft and as much as 220 ft, but 175 ft is most often quoted. Three coals are recorded. The upper-most, called the **Little Vein,** is 15 in thick and lies 10 to 20 ft below the Six-Feet. About 30 ft lower is the **Clay** Seam, which varies from 2 to 2½ ft and appears to be equivalent to the Caerau. Some 40 ft below, a seam called the **Coal and Mine** is 2½ to 3½ ft thick and contains several bands of ironstone in its roof.

Except at Avon Colliery the **Six-Feet** forms a single composite seam. It consists of two main coals, with locally a third thin leaf developed in the middle by splitting of the bottom coal. The following sections are typical:

		Glyncorrwg Ft In	Glyncastle Ft In	Nantewlaeth Ft In	Duffryn Rhondda Ft In
Coal	..	2 3	2 11	2 0	2 3
Dirt	..	2	1	6	2
Coal	..	6 ⎫		9 ⎫	
Dirt	..	2 ⎬ 4 0	4 0	3 ⎬ 4 0	4 0
Coal	..	4 3 ⎭		4 0 ⎭	

At Glyncorrwg the immediate roof consists of 3 in of slickensided mudstone; above this 4 ft of dark mudstone carries crushed and solid 'mussels' associated with abundant ironstone nodules; forms include *Anthracosia* cf. *aquilina*, *A.* cf. *'faba'* (Wright), *A. planitumida* and *Naiadites* cf. *alatus* associated with a small form of *Planolites* cf. *ophthalmoides*. At Glyncastle a coal up to 6 in thick lies about 4 ft above the top of the seam; the mudstone roof carries plant debris, *Naiadites?* and *Lioestheria sp.* The last-named form is characteristically found above the Eighteen-Feet in the Neath Valley.

No reliable sections are known in the area between Bryn and Cwmavon, for the seam here is subject to intense disturbance. It is recorded as being between 5 and 40 ft in thickness, but 8 to 9 ft is probably normal. The roof consists of the ferruginous black-band mudstone distinctive of the seam along the South Crop. Spoil from Bryngyrnos Colliery [80519204] contains well-preserved *Anthracosia atra*, *A.* cf. *fulva* and *A.* aff. *lateralis*. At Eagle Brick-pit [79199253] and in bell-pit spoil from the adjacent Patch Mawr [79349225] the following fauna was collected: *Anthraconaia sp.* (of *subcentralis*/*pumila* group), *Anthracosia* cf. *lateralis*, *A. sp.* cf. *phrygiana*, *Naiadites* cf. *alatus*, *N.* cf. *obliquus* and *Carbonita* cf. *humilis*.

At Avon Colliery the top coal splits away to form the **Upper Six-Feet** or Yard; 27 in thick, it lies about 40 ft above the Lower Six-Feet. Its roof is similar to that of the Six-Feet at Glyncorrwg. The fauna includes cf. *Planolites ophthalmoides* (2 mm. diameter), *Anthracosia* aff. *lateralis*, *A. sp.* cf. *planitumida*; 'wormy' beds are also common. The **Lower Six-Feet**, 4 ft thick, carries a rider about 5 ft above the seam: it consists of two 1-in bands of coal separated by 5 in of cannel, succeeded by dark mudstone containing *Naiadites sp.* and abundant *Lioestheria sp.*

The **Four-Feet** at Avon is a single composite seam comprising up to 15 ft of coal, rashings and clay partings, and lying some 25 ft above the Upper Six-Feet. At Duffryn Rhondda, where the seam has been worked on a small scale, the section is cleaner and includes four leaves: top coal 8 to 12 in, rashings 4 in, coal 12 to 20 in, clod 2 in, coal 5 in, clod 6 to 22 in, coal (with parting near the middle) 48 in. Two further thin coals, 2 in and 5 in, occur at 9 ft and 17 ft respectively below this composite seam. At Bryn the seam was formerly worked in the south-east of the taking, but no reliable sections exist: in the shaft was recorded $7\frac{1}{2}$ ft of coal with a thin parting 27 in from the top; a rider seam of 9 to 12 in lies up to 6 ft above. Old crop workings near the Eagle Brickworks suggest that the seam is in two leaves about 10 ft apart, but no details are available. The seam was worked at Nantewlaeth with a section: top coal 8 in, parting 3 in, coal 3 to $4\frac{1}{2}$ ft.

At Nantewlaeth and Avon the typical 'Four-Feet fauna' was collected, including *Anthraconaia sp. nov.* (cf. *curtata*) and *Anthracosphaerium* aff. *propinquum*. At Duffryn Rhondda and Bryn similar forms occur together with *Anthracosia lateralis*, *A. planitumida*, *A.* cf. *simulans* and *Anthracosphaerium* cf. *exiguum*; the immediate roof contains *Naiadites obliquus* preserved as films. At Eagle Brick-pit the typical 'solid' fauna occurs, but at Glyncastle only the crushed fauna was obtained.

The **Two-Feet-Nine** was formerly worked at Bryn and Cwmavon as the Finery. The seam thins westwards across Bryn from $4\frac{1}{2}$ to $1\frac{1}{2}$ ft. This thinning is associated with the oncoming of a sandstone roof, which can be seen at Eagle Brick-pit overlying 2 ft of coal. Some 10 ft lower is the Sulphury Seam, an inferior coal 18 in thick. At Avon the Two-Feet-Nine has been worked south-east of the shafts where it has a section: top coal 36 in, clod 10 in, coal 19 in. At Duffryn Rhondda the whole seam appears to be represented by 15 in of coal, at Nantewlaeth by 9 in of 'rashings' and at Glyncorrwg possibly by 12 in of coal. Only at Avon has the seam yielded an abundant fauna—a typical Two-Feet-Nine assemblage—which includes: *Anthracosia acutella*, *A. atra*, *A.* aff. *concinna*, *A.* cf. *planitumida* and *Naiadites obliquus*.

The Two-Feet-Nine crops out on the south side of Nant Cwm-farteg, dipping south, and has been worked from old levels adjoining the ruined Farteg-fach farm [82159181]. It crops again in the disued Bryn brickyard [82669181], on the north side of the stream, dipping to the north-north-east at 15° on the northern limb of the Maesteg Anticline. The sequence continues upwards in mudstone about 70 ft thick to two thin coals, the **Balling** and the **Silver**. These are about 20 ft apart and probably equate with the Fireclay and Coal Tar seams of Maesteg. A few feet above the Silver, a persistent quartzitic sandstone, presumably the **Lower Cockshot Rock** of Maesteg, has been worked on a small scale south of the Bryn–Maesteg road [82309208], while it makes a prominent feature on Mynydd Bychan [796923]. Above this the strata are poorly exposed. The **Upper Cockshot Rock** of Maesteg is seen east of Bryn [82519213] but dies out near the village. Other impersistent sandstones occur in this part of the sequence, one of which has been quarried [80919212] near Bryngyrnos Colliery. The measures above the Two-Feet-Nine were penetrated in all the shafts of the Avan Valley collieries. The group as a whole varies from about 200 ft at Glyncorrwg to 225 ft at Nantewlaeth. Three thin coals appear to be developed between the Two-Feet-Nine and the Lower Cockshot Rock. The latter does not persist as far as the northern parts of the valley, but the Upper Cockshot is well developed, being up to 45 ft thick at Nantewlaeth. Of the marine bands only the **Cefn Coed** has been seen—at Duffryn Rhondda Colliery, where it rests on 3 in of coal, the first below the Caedavid Seam. Marine fossils were found throughout 18 in of mudstone, the most abundant fauna occurring in pyritic mudstone, 8 in thick, near the base of the band. The following forms have been identified: crinoid columnals; *Chonetes* (*C.*) cf. *skipseyi, Lingula mytilloides*; cf. *Nuculana attenuata, Pleurophorella?*, Pectinid fragments; *Straparollus* (*S.*) *sp. nov.*; *Anthracoceras sp.*; *Hollinella* cf. *bassleri* and fish debris. The top 10 in of the band carried only *L. mytilloides* and *Orbiculoidea* cf. *nitida*. Lenses of 'cank' occur in the pyritic mudstone. A specimen of this is described by Dr. K. C. Dunham as follows: 'A black compact finely crystalline chalybite-siltstone composed of interlocked well-developed rhombs of chalybite, probably containing some CaMg $(CO_3)_2$ in solid solution. The carbonate rhombs average 0·05 mm. long and are very even-grained. Irregular small streaks of pyrite are present and there is a sparse matrix of carbonaceous and clay material'.   W.B.E.

SOUTH CROP

The strata between the Amman and Cefn Coed marine bands crop out along the southern margin of the district in a strip varying in width from about 200 yd near Rhiwsaeson to about 1¾ miles in the Margam–Pyle area, where the measures are not only very much thicker but are also repeated several times by thrusts. Natural sections in this mainly argillaceous sequence are rare and knowledge of the stratigraphy is derived almost entirely from present day workings at the widely separated collieries in the extreme west and centre of the tract, from open-cast works to the west of the Ogmore Valley, and from deep exploratory boreholes in the Margam area. As is also usual with the beds of the upper part of the Lower Coal Measures these measures are everywhere much affected by compressional structures, and continuous undisturbed sections are almost non-existent. Many of the coals were worked, to some extent at least, at all the old collieries along the Crop. The records of these collieries include many sections of cross-measures roadways, all of which show such disturbance, both of the general strata and of the individual coals, that in no instance can satisfactory correlations be made.

West of the Ogmore Valley: In the extreme west the measures are concealed beneath morainic drift up to 100 ft thick, and the outcrops can be inferred only by calculation from mining and borehole information. The great width of the outcrop, more than 3000 yd, between Water Street and Margam Abbey is explained by the existence of major thrusting which apparently repeats the crop of the upper part of the sequence at least three times. In this area the southern limit of the exposed Coal Measures is formed by the down-faulted outcrop of the Keuper deposits. The British Industrial Solvents No. 1 Borehole [80028344], 1 mile 1100 yd W. 15° N. of Pyle Church, after penetrating about 70 ft of gravelly drift, entered the Middle Coal Measures just above the Upper Nine-Feet; No. 2 Borehole [79968316] some 300 yd S. 12° W. of No. 1, proved 65 ft of gravel and 161 ft of Keuper Marl before entering Coal Measures. It is probable, there-fore, that in the area between Pyle and the coast the lower beds of the Middle Coal Measures as well as the Lower Coal Measures 'crop' beneath the Trias to the south of the boundary fault.

Newlands and Cribbwr Fawr collieries have worked extensively the main coals of this group to the south of the Newlands Thrust. Newlands Slant [81048349] and Cribbwr Fawr No. 1 Slant [82268342] were both driven on the Upper Nine-Feet (South

Fawr) Seam, while the No. 2 Slant at Cribbwr Fawr descended along the Lower Nine-Feet (Ail). The only other seam located from the surface in this tract is the Red Vein (North Fawr) on the sub-drift crop of which a small working, Coalbrook Slant [82608353], 950 yd N. of Pyle Church, was formerly located. It is inferred that the coals from the Lower Nine-Feet to the Two-Feet-Nine crop out in a belt of country 500 to 700 yd wide situated, in general, to the north of a line drawn through Newlands and Cribbwr Fawr collieries. Although Cribbwr Fawr workings are now inaccessible the records show the sequence and coals to be very similar to those at Newlands which have been examined in detail during the present survey.

The **Amman Marine Band** is $3\frac{1}{2}$ ft thick at Newlands and yielded the following fauna: pyritized sponge spicules; pentagonal crinoid columnals; *Lingula mytilloides*, *Spirifer pennystonensis* T. N. George; *Planolites ophthalmoides*, pyritized 'worm burrows'; and fish fragments. The band proved to be 9 ft 2 in thick in the Margam Park No. 1 Borehole and provided the following additional forms: foraminifera including cf. *Hyperammina* and *Rectocornuspira?* and small gastropods.

Some 40 to 60 ft of argillaceous strata separate the Amman Marine Band from the **Bute** (or **Drydydd**) Seam, which is normally 5 to $5\frac{3}{4}$ ft thick with an impersistent layer of 'brass' about 10 to 12 in from the top. The roof is highly characteristic over large areas, the following being typical:

	Ft	In
Grey mudstone with many small nodules	–	–
Grey striped shaly mudstone with solid 'mussels': *Anthraconaia williamsoni*, *Anthracosphaerium affine*, *A. exiguum*, *A. turgidum* and *Naiadites* aff. *productus*	1	6
Black cannelly shale with numerous crushed shells: cf. *A. turgidum* and *N.* aff. *productus*		10
Black shaly rashings (on coal)		4

About 30 to 70 ft above the Bute a coal, 6 to 12 in thick, is separated from the lower Nine-Feet by 12 to 20 ft of mudstone and seatearth. Scattered shells, including *Anthraconaia williamsoni* and *N.* cf. *productus*, are recorded between 4 and 12 in above this coal.

The **Lower Nine-Feet** (or **Ail**) is characteristically a dirty seam consisting of three variable leaves: top coal 10 to 16 in, rashings and fireclay $\frac{1}{2}$ to 34 in, coal 12 to 14 in, stone and rashings $\frac{1}{2}$ to 17 in, coal 12 to 22 in. The roof has always proved to be barren of animal fossils.

Lying some 40 to 90 ft higher in the sequence, the **Upper Nine-Feet** (or **South Fawr**) was the principal seam worked at both Cribbwr Fawr and Newlands. In normal section the seam varies from about 10 to 13 ft and contains several distinct though thin partings; it is however subject to much structural variation, being in places as much as 30 ft, when it is soft and difficult to work. Under these conditions large quantities of coal were left behind and gob fires in the old workings were not infrequent. It is difficult to reconcile this great thickness of coal with the 3 to 5 ft which is normal in the nearest part of the main coalfield to the north. In the North Drift exploration through the Newlands Thrust a seam was proved in a tophole about 200 yd north of the disturbance which is almost certainly the Upper Nine-Feet: the section, top coal with 'brass' 22 in, rashings $1\frac{1}{2}$ in, coal 14 in, rashings $\frac{1}{2}$ in, and coal 26 in, is markedly different from that of the workings and is much more like that of the Maesteg and Garw valleys.

The true roof of the Upper Nine-Feet is a massive blocky mudstone, which immediately above the coal carries a sparse 'Red Vein fauna' with cf. *Anthraconaia pulchella*, *Anthracosia* cf. *nitida* and *Naiadites sp.* cf. *productus*.

The thickness of measures between the Upper Nine-Feet and the Red Vein is very variable, being as little as 50 ft in the North Drift at Newlands and as much as 126 ft in the Margam Park No. 2 Borehole. Strong mudstones with a few sandstone layers in the lower part make up the bulk of these strata and the Red Vein is usually underlain by a thick seatearth.

At Cribbwr Fawr and in much of the Newlands area the **Red Vein (North Fawr)** consists of two leaves: top coal 18 to 30 in, rashings 6 to 12 in, bottom coal 46 to 52 in. Locally, however, as in the No. 2 Scheme district at Newlands, the top coal splits away from the main coal, from which it may be separated by upwards of 20 ft of strong mudstone with ironstone, a hard fine-grained sandstone often being present at the base. Under these conditions the top coal is markedly thinner, 6 to 16 in being recorded in the No. 2 Scheme. A sparse fauna, consisting of typical Red Vein forms, occurs above the top coal and includes *Anthracosia* cf. *angulata* and *A.* cf. *nitida.*

The Red Vein is separated from the **Caerau Vein (Caegarw)** by about 30 ft of mudstone with sandstone bands. The Caerau is the topmost seam proved with certainty at Newlands and its section of 5 to 6 ft of top coal with irregular layers of coal and rashings up to 8 ft thick in the floor, is typical of the relatively small areas so far worked. The roof consists of about 12 ft of soft mudstone with poorly preserved planty debris, followed upwards by strong grey silty mudstone with good plants, among which the following have been identified: *Alethopteris lonchitica, Lepidodendron obovatum* Sternberg, *L. ophiurus, Mariopteris nervosa, Neuropteris gigantea, N. obliqua, N. pseudogigantea* and *Sphenopteris sp.*

On the north side of the Newlands Thrust several of the coals of this group were formerly worked at Bryndu Colliery, about one mile north-east of Pyle Church. The site of this colliery was presumably determined by the extensive bell-pitting of coal which in early days took place in the rectangle of ground between Bryndu Slip and the road from Waterhall Junction to Pentre and Aberbaiden, that is, close to Bryn-du Fawr Farm [83408400]. In this area a window of solid rocks, overlain only by head deposits, appears from beneath glacial drift and includes all seams from the Yard to the Two-Feet-Nine. The crops of the Upper Nine-Feet, Red Vein and Caerau, in particular, are marked by lines of closely spaced bell-pits, and just to the north of Bryn-du Fawr spoil from crop workings yields blackband fragments carrying 'mussels' typical of the Six-Feet Seam. Bryndu Slip itself was driven on the Upper Nine-Feet. Westwards from Bryndu the crops of the coals of this group presumably lie beneath thick drift in a band several hundred yards wide extending through Caegarw [82208375] and Eglwys-Nynnid [80308470].

The few records still available of the coals at Bryndu Colliery do not seem to be wholly reliable. A seam, presumably the **Bute,** is given as 50 in thick, and the section of the **Lower Nine-Feet (Ail)** appears to be: top coal 14 in, parting 1 in, coal 11 in, clod 4 in, coal 20 in. The various sections given for the **Upper Nine-Feet (South Fawr)** indicate that the coal was frequently in a disturbed condition; according to one authority it was generally about 14 ft in thickness.

Between Bryndu and Cribbwr Fawr collieries the Newlands Thrust takes on a markedly west-north-west to east-south-east trend and abuts against the boundary fault in the Kenfig Hill area before running out of the Coal Measures beneath the Trias. The eastwards continuation of the crop from the Bryndu area through the ground north of Mill Pit [83968367] and Park Slip [87928352], therefore, constitutes the southernmost outcrop of these measures in this part of the coalfield. The coals of this group do not appear to have been worked to any great extent at Mill Pit, Cefn or Park collieries, presumably because of structural difficulties. However, in the area between Mill Pit in the west and Evanstown in the east the ground was extensively worked over to a depth of 20 to 30 ft in search of ironstone, which, in the mid-nineteenth century, was smelted at the Cefn Iron Works, just over a half-a-mile north-west of Cefn Cross. This patch-working for ironstone was associated with bell-pitting for coal on a large scale.

Before the recent extensive opencast mining the shales of the **Amman Marine Band** were to be seen as spoil from shallow workings at several points along the crop between Cefn and Park collieries: they consisted of dark blue micaceous paper-shales showing a tendency to yellow 'sulphury' weathering, and carrying abundant *Lingula mytilloides.* These were observed at the following localities: [86568351] 300 yd S. by W. of

Ysgubor-y-parc Farm; [87228358] 800 yd W. by N. of Park Colliery; and [87668368] 350 yd W. 30° N. of Park Colliery.

The characteristic blackband of the **Six-Feet** was observed in the roof of the seam proved in trial pit No. 4 of the Kenfig Opencast Site, sunk to a depth of 36 ft at a point [85058410] 1750 yd N. 40° E. of Kenfig Hill Church, and from the spoil of old bell-pit workings [85768393] 650 yd S.S.E. of Aberbaiden Farm. This seam was exposed in a small area near the plane of the Kenfig Thrust in the Kenfig Opencast Production Site [848841] where variants of the following forms were collected from the roof: *Anthracosia atra, A. aquilina, A. aquilinoides, A. sp.* cf. *elliptica* and *Naiadites sp.* At all these localities the Six-Feet exists as a single though usually disturbed seam apparently about 6 ft thick. On the Cefn Opencast Site about one mile to the east, a trial-cut approximately 400 yd north of Cefn House [86078385] showed the seam split into its usual two component parts separated by some 10 to 12 ft of seatearth. The Lower Six-Feet was about 3 to 4 ft thick; the Upper Six-Feet, 2 ft in thickness, carried a blackband fauna: *Anthracosia atra, A.* cf. *aquilinoides, A. sp.* intermediate between *atra* and *barkeri, Naiadites* cf. *angustus* Trueman and Weir and *N. obliquus.*

An exploration trench [approx. 864836] on the Cefn Opencast Site, about 200 yd S.S.W. of Ysgubor-y-parc, on the west of Ffordd-y-gyfraith, exposed the measures from the Amman Rider to the presumed Caerau Vein. The beds above the Nine-Feet were very disturbed (see Fig. 43), but the following section gives a probable interpretation of the sequence:

	Ft	In
Grey mudstone with layers of ironstone nodules; a few sandstones up to 18 in thick	40	0
?CAERAU VEIN: **coal 72 to 84 in**	6 ft to 7	0
Seatearth with ironstone nodules passing down into	3	0
Grey mudstone with ironstone nodules	14	0
Carbonaceous shale and rashings; streaks of coal and some ironstone nodules	3	3
Strong silty mudstone with plant debris and a layer of sandstone (contorted and faulted)	about 32	0
?RED VEIN: highly disturbed **coal** in numerous thin and discontinuous layers occurring in three distinct horizons indicative of repetition ..	—	—
Grey seatearth passing down into	2	0
Grey mudstone with sporadic ironstone nodules and thin streak of coal	12	0
Alternations of grey mudstone and fine-grained sandstone ..	9 ft to 11	6
Highly sheared black carbonaceous shales with rashings with coaly streaks	2 ft to 3	6
UPPER NINE-FEET: **coal about 54 to 84 in**	4½ ft to 7	0
(This seam was intensely disturbed with repetition of the coal in thick and thin layers over a thickness of about 60 ft.)		
Dark grey seatearth with nodules of ironstone passing down into ..	3	6
Strong grey mudstone with a few layers of ironstone	34	6
Grey mudstone with a few streaks of coal; plant debris	3	6
LOWER NINE-FEET: **coal 30 to 39 in,** clod 1 to 3 in, **coal 18 to 27 in** about	5	0
Dark grey seatearth with ironstone nodules	7	0
COAL, rashings and seatearth	1	6
Dark grey seatearth with ironstone nodules	2	6
Grey mudstone; plant debris	14	0
Alternations of strong grey mudstone and fine-grained sandstone in layers up to 3 ft	21	0
Blocky grey mudstone with sporadic ironstone nodules ..	11 ft to 14	0
BUTE: **coal 36 to 80 in**	3 ft to 6	8
Soft rashings	2 ft to 3	0
Seatearth with ironstone nodules	4	0

	Ft	In
Thinly bedded grey mudstone with ironstone layers and nodules; 'ghost'-like 'mussels' at base	23	0
Amman Marine Band: darker grey shaly mudstone with *Lingula mytilloides* and pyritic 'fucoids'	8	0
AMMAN RIDER: **coal 18 in,** seatearth 48 in, **coal 12 in**	6	6

The crop of the Two-Feet-Nine has been proved by drilling followed by opencast working almost continuously from a point [81678500] 300 yd S.S.W. of Hirwaun Farm as far east as [88038414] 200 yd S. by E. of Park Farm, a distance of 4 miles. The Upper and Lower Four-Feet seams below were also proved over much of this ground, the crops lying 50 to 100 yd south of that of the Two-Feet-Nine. The Opencast Production Sites were known (reading from W. to E.) as East Lodge, Bryndu West, Kenfig, Cefn, and Park Farm.

Measurements made at the Kenfig and Cefn Opencast sites give a good section from the Lower Four-Feet to some distance above the Two-Feet-Nine, measures which were remarkably constant in character over the entire length of opencast working.

Cefn Site, near Ffordd-y-gyfraith:		Ft	In
COAL, weathered	about	1	6
Dark hard band, inaccessible	about		6
Pale massive seatearth and grey blocky mudstone		8	0
Grey shaly mudstone with a few thin ironstone layers		9	0
Grey shaly mudstone with *Anthracosia* cf. *acutella*, *A.* cf. *aquilina*, *A.* cf. *atra*, *A.* cf. *planitumida* and *Naiadites sp.*			3
Cannelly shale			3
COAL, dirty and inferior			10
Grey seatearth, cannelly at top, with ironstone nodules		2	4
Grey shaly mudstone with 'mussel'-band 3 ft from base: *Anthraconaia sp.* cf. *lanceolata*, *A.* cf. *librata*, *A. sp.* cf. *obscura*, *Anthracosia acutella*, *A.* cf. *atra*, *A.* cf. *concinna*, *A.* cf. *lateralis*, *A.* cf. *planitumida*, *A. simulans*, *A. sp.* intermediate between *acutella* and *concinna*, *Anthracosphaerium* cf. *turgidum*, *A. sp.* cf. *affine*, *A. sp.* cf. *exiguum*, *Naiadites* cf. *obliquus*, *N. angustus?*		7	0
Hard nodular bed with *Mariopteris sauveuri* (Brongniart), *Neuropteris pseudogigantea*, *N. tenuifolia*, *Sphenopteris sp.*			6
LANTERN: **coal 28 in**		2	4
Black coaly rashings		1	10
Soft dark seatearth passing down into			6
Pale seatearth and grey striped siltstone		9	7
Alternations of grey sandstone and blocky grey siltstone		17	2
Strong blocky greenish-grey mudstone		11	10
Grey well-bedded mudstone with sporadic large nodules of ironstone; 'mussel'-bed 12 ft at base: *Anthracosia acutella*, *A.* cf. *atra*, *A.* cf. *lateralis*, *A. planitumida*, cf. *A. simulans*, *Anthracosphaerium propinquum*, *Naiadites* cf. *obliquus*		30	0
Black rashings: highly disturbed cannelly shale..	6 in to 1		0
TWO-FEET-NINE: **coal about 58 in**	about	4	10

Kenfig Site:	Ft	In
Seatearth passing down into rashings with ironstone nodules.. ..	8	6
Grey mudstone with ironstone layers and nodules; 'mussels' 3 to 4 in from base: *Anthraconaia sp. nov.* (cf. *curtata*), *A. sp.* of *cymbula/ librata* group, *Anthracosia* cf. *atra*, *A. planitumida*, *Anthracosphaerium* cf. *exiguum*, *A.* aff. *propinquum*..	12	10
UPPER FOUR-FEET: **coal 15 in,** rashings 5 in, **coal 18 in,** dirt ½ in, **coal 12 in**	4	2½

	Ft	In
Seatearth		9
Grey mudstone, silty downwards, with ironstone	8	0
Pale grey fine-grained sandstone; layers of grey mudstone	11	0
Dark grey mudstone with ironstone layers and nodules	10	1
Grey shale; a few poor plants	2	0
Dark shale; *Anthraconaia cymbula?, A. sp.* cf. *subcentralis, Anthracosphaerium* aff. *propinquum, Naiadites* cf. *productus*; *Carbonita* cf. *humilis*; and insect wing indet...		1
LOWER FOUR-FEET: **coal 1 to 5 in,** dark carbonaceous shales with abundant plant debris 7 to 42 in, **coal 48 in** .. .. .. about 6		6

A similar section was proved in the Oaklands Drift [84998413] which was sunk on ground immediately adjoining the Kenfig Opencast Site to the east. This trial followed the Lower Four-Feet from outcrop until it was lost in disturbed ground some 250 ft from the surface. After passing through the Kenfig Thrust the Caerau Vein was located in a disturbed condition and a level cross-measures roadway driven to the Two-Feet-Nine. The section below the Lower Four-Feet was as follows:

	Ft	In
LOWER FOUR-FEET	—	—
Hard seatearth with ironstone nodules	8	0
Strong mudstone with thin layer of sandstone	16	7
COAL and rashings	1	5
Strong seatearth	1	6
Blocky grey mudstone with ironstone layers and nodules and a few thin sandstones; plant debris	45	0
Blackband shale (highly sheared); *Anthracosia aquilinoides, A. atra* ..		6
SIX-FEET: **coal 42 in**	3	6
Seatearth and rashings	5	9
Strong grey mudstone with nodules of ironstone	30	0
CAERAU: **coal 84 in**	7	0

Throughout this area the **Two-Feet-Nine Rider** or **Lantern** Seam is about 28 in thick and it was worked to some extent during opencast operations. It normally lies about 70 to 100 ft above the Two-Feet-Nine and the roof carries 'mussels' similar to those above the Two-Feet-Nine. The **Two-Feet-Nine** was intensely disturbed throughout the length of the opencast workings, being particularly subject to 'overlapping'. Thicknesses of coal varied from 0 to 30 ft or more, and nowhere was a section observed which could be regarded as normal (see Fig. 44, p. 261). It is thought, however, that the true thickness of the coal is about 5 ft. The Upper Four-Feet also was always subject to much disturbance, the coal often proving to be thinner than normal due to small scale lag-faulting. An extensive wash-out in the Lower Four-Feet was proved in the Oaklands Workings.

Immediately to the west of Evanstown the measures from the Amman Rider to the lower part of the Four-Feet group of coals have been quarried on a large scale for the making of bricks by the Tondu Brick Works Co., Ltd. The quarry [890840] covers an area of about 12 acres and has been excavated to a depth of about 80 ft. The following section was measured in April 1957.

	Ft	In
Massive sandstone	10	0
Blocky grey mudstone; bands of fine-grained micaceous sandstone ..	15	0
Massive sandstone .. .. 3 ft to 5		0
Blocky grey mudstone, passing into shaly mudstone with ironstone layers; a few fragments of crushed and indeterminate lamellibranchs 6 in to 8		0

	Ft	In
Coal 10 in, soft grey seatearth 3 in, **coal 1 in**	1	2
Blocky silty seatearth	1	6
Hard pale grey sandstone; muddy partings	2	3
Blocky blue-grey mudstone, slightly silty in parts, with regular layers of sandstone up to 5 in; ironstone layers, nodular in places, up to 6 in	16	0
Massive striped sandy mudstone; plant debris	10	0
Blocky blue-grey mudstone	2	0
Hard fine-grained quartzitic sandstone .. .. 2 ft to	4	0
Blocky grey mudstone with a few ironstone layers	8	0
Pale grey fine-grained striped sandstone	1	0
Blocky grey mudstone with ironstone layers, shaly towards base 10 ft to	12	0
Dark rusty-weathering shales; a few shell traces		6
Blackband ironstone shale; *Anthracosia* cf. *aquilina, A.* cf. *atra* ..		5
UPPER SIX-FEET: rashings 1 in, **coal 29 in**	2	6
Grey seatearth with ironstone nodules passing down into blocky grey mudstone with roots	6	0
Blocky silty mudstone; sandy bands and layers of ironstone	12	0
Dark blue-grey shaly mudstone with ironstone nodules	6	0
Gap	4	0
LOWER SIX-FEET: **coal about 48 in** (full section not exposed) about	4	0
Gap (? mainly seatearth)	12	0
Black rashings and **coal** mixture	1	6
Grey seatearth .. .. 7 ft to	8	0
Black rashings		6
Grey shaly seatearth with ironstone nodules	2	0
Grey mudstone with plants	2	6
COAL with layer of pyrite nodules in middle	2	5
Grey silty mudstone-seatearth	5	6
Grey shaly mudstone with plant debris; **coal** ½ **in** in middle		4½
Alternations of blue-grey silty mudstone and striped siltstone; scattered ironstone nodules	4	6
Carbonaceous shale		3
COAL (inferior) .. .. 5 in to	2	0
Carbonaceous shale with coal streaks		9
Silty seatearth		6
Alternations of silty sandstone with silty mudstone-seatearth; ironstone nodules	6	0
Blocky grey mudstone with large ironstone nodules; scattered plant debris	6	0
COAL		10
Grey mudstone-seatearth with ironstone nodules; soft contorted clay 3 in at top	2	0
Grey mudstone with scattered large nodules	4	6
Massive sandstone	1	6
Gap, with blue-grey shale at base	12	0
Rashings 3 in, **coal 22 in,** cannel 7 in	2	8
Grey mudstone-seatearth with ironstone nodules	2	0
Blocky grey mudstone and silty mudstone with a few harder sandy layers; ironstone nodules; plant debris	15	0
Massive false-bedded, fine-grained sandstone wedging laterally into mudstone	7	0
Blocky grey silty mudstone; *Anthraconaia?* and *Naiadites* cf. *productus* at base	15	0
COAL		6

	Ft	In
Carbonaceous mudstone with coal streaks		4
Black rashings with coal streaks		7
Grey mudstone-seatearth		9
Dark blue-grey shale; numerous ironstone layers and nodules	20	0
Gap	about 15	0
UPPER NINE-FEET: **coal,** poorly exposed but at least	8	0
Grey mudstone-seatearth with numerous small ironstone nodules; two prominent bands of ironstone 5 in and 8 in thick	7	0
Grey shaly mudstone	3	0
Massive grey mudstone	30	0
Fine-grained sandstone in layers 3 to 4 in thick; mudstone partings 4 ft to	5	0
Grey mudstone with ironstone nodules	1	1
Black cannelly shale		2
LOWER NINE-FEET: **coal** (disturbed)	up to 8	6
Grey seatearth passing down into blue-grey mudstone	2	9
Ironstone	3 in to	5
Blocky blue-grey silty mudstone; a few sandy bands up to 4½ ft; sporadic septarian ironstone nodules	35	0
Cannelly shale		4
?BUTE: **coal 42 in,** seatearth with coal streak 17 in, **coal 2 in**	5	1
Strong silty seatearth with ironstone nodules in upper part	3	0
Gap, due to faulting	—	—
Blue-grey shaly mudstone; numerous ironstone layers up to 2 in	9	0
Dark blue-grey shale with 'ghost'-like 'mussels'; *Anthracosia* cf. *aquilina, A.* cf. *ovum, A. beaniana*	1	0
Amman Marine Band: Pale blue-grey micaceous silty shale, weathering yellow with ferrous sulphate encrusted surfaces; scattered ironstone nodules; *Ammodiscus sp.*; pyritized sponge spicules; *Lingula mytilloides* (scattered throughout); *Dunbarella sp.*	7	6
Ironstone	3 in to	5
Blue-black shale; cannelly at base with poorly preserved goniatite or gastropod		4
AMMAN RIDER: cannel 2 in, rashings and coal streaks 6½ in, **coal 22 in**	2	6½
Grey seatearth and silty mudstone passing down into strong sandy siltstone with sandstone layers (disturbed and poorly exposed) about 10	0	

The Red Vein–Caerau group of coals appears from this section to have become much split, with accompanying deterioration of coal sections. The split in the Six-Feet Seam, recorded as being 12 ft at the Cefn Opencast Site, has here increased to about 28 ft.

Apart from a very few exposures in the Aberbaiden area, knowledge of the Hafod Heulog, Britannic, and Cefn Coed marine bands is confined to the Margam Park boreholes (pp. 343–6). These marine horizons show a remarkably consistent pattern. The **Hafod Heulog Marine Band** lies about 200 ft above the Two-Feet-Nine, and consists of dark blue shaly micaceous mudstone with abundant pyrite, 3 to 8 ft thick, containing rare *Lingula mytilloides, Orbiculoidea nitida* and Bellerophontid gastropods, overlying black carbonaceous shale, up to 1 ft 8 in thick, which carried abundant foraminifera, fish debris and sporadic *Lingula*. Some 35 to 40 ft of mudstone separate this horizon from the **Britannic Marine Band** which normally rests on 1¼ to 2¾ ft of cannelly shale with cannel bands. The marine fauna is scanty, consisting of *Hollinella sp.* and rare *Lingula*, associated with fish debris and doubtful foraminifera in dark carbonaceous shales about 6 in thick; these are overlain by 8 to 17 ft of blue-grey shaly mudstone with pyritic granules and 'fucoids', carrying scattered *Planolites ophthalmoides*. The **Cefn Coed Marine Band** is well developed in all four boreholes and lies 87 to 149 ft above the cannelly shales at the base of the Britannic horizon. The lower half of the intervening cyclothem consists of mudstone with numerous ironstone layers and nodules, whilst in the upper part silty and sandy beds, together with sphaerosideritic seatearths, occur interbedded with mudstone. In one borehole (No. 1) a 30-ft band of hard fine-grained quartzitic sandstone, evidently one of the Cockshot Rock horizons, is well developed. The Cefn Coed Marine Band itself rests on 2 to 4 in of coal and is 4 to 5 ft thick. It consists of dark grey blocky mudstone with layers rich in pyritic granules and 'fucoids', and the varied fauna includes crinoid columnals, calcareous and horny brachiopods, turreted gastropods, goniatites, ostracods, trilobites, and conodonts. At the the base there is usually a thin layer of black carbonaceous shale with *Lingula mytilloides*.

The **Hafod Heulog Marine Band** is exposed in the Afon Kenfig at a point [84388447] 300 yd E. 27° S. of Hafod Heulog Farm. This is the type locality for the marine band and the section is as follows:

	Ft	In
Grey shales with ironstone	—	—
Blue-grey shales with ferruginous weathering; *Lingula mytilloides* and Orbiculoidea nitida	2	0
COAL, cannelly in top 3 in		9
Grey mudstone seatearth	—	—

This horizon was poorly exposed [85758421] in the small stream 400 yd E. 42° S. of Aberbaiden Farm. 340 yd upstream from this locality the **Cefn Coed Marine Band** crops out in the south bank of the stream at a point [86068421] 720 yd E.S.E. of Aberbaiden Farm. The section and fauna at this locality are described in detail by W. H. C. Ramsbottom (1952, pp. 8–32) and are summarized below:

	Ft	In
Soft grey shale, much weathered	1	3+
Cannelly shale		1½
Hard grey shale with *Lingula*		4
Dark grey shale with crinoid and echinoid remains, *Lingula*, Productids, Pectinids, ostracods, conodonts and fish fragments		4
Very soft grey shale with rare crinoids, Chonetids, gastropods, ostracods and fish scales		8
Hard dark grey-blue shale with boring sponges, Zaphrentoid corals, crinoids, echinoid spines, annelid jaws, calcareous brachiopods, including Athyrids, Chonetids, Productids and Spiriferids, of which the Chonetids are especially abundant and varied, various lamellibranchs including Pectinids, a gastropod, nautiloids both coiled and straight, goniatities, trilobites, ostracods, conodonts and fish debris		5½
Soft grey shale		2
COAL		5
Soft grey planty shale or seatearth	—	—

East of the Ogmore Valley: Werntarw and Llanharan collieries provide most of the information in this area about the measures between the Amman and Cefn Coed marine bands. The main coals of the group were undoubtedly penetrated at Ynysawdre [900842] and Bryncethin (Barrow) [918843] collieries, but individual seams cannot be identified with any certainty. The recorded thicknesses of the coals appear to be abnormal and it is not known to what extent any of them were worked. For the sake of completeness, sections at these collieries are given in Appendix I.

It is believed that Tyn-y-Waun Pit [93188444] was sunk to the Four-Feet group, but identification of the lower coals is not certain and there is no evidence that any of the main seams was ever worked. The section, with suggested correlation, is given on p. 363.

At Werntarw Colliery the No. 1 Pit [96638469] (the only one for which a record exists) appears to have been sunk to the horizon of the Six-Feet Seam, but the strata in the lower part of the pit are intensely disturbed and the seams can only be identified tentatively (see p. 363). The main haulage roads north of the pits were driven about 700 yd in extremely broken ground (in which no major development took place and where no seam was identified) before emerging in a relatively undisturbed seam, the so-called Six-Feet. The coal was worked under this name for many years, nothing being known of the measures above. During the present survey the seam was identified as the Yard and, as a result, the measures above were explored by means of level cross-measures roadways. The section proved as follows:

	Ft	In
Mudstone with ironstone bands and nodules	10	0
?LOWER FOUR-FEET: **coal 4 in,** parting 11 in, **coal 9 in**	2	0
Fireclay passing down into mudstone with ironstone nodules ..	15	3
SIX-FEET: **coal 2 in,** rashings 1 in, **coal 28 in,** brass 6 in, **coal 36 in,**		
clod 4 in, **coal 14 in**	7	7
Fireclay with rashings at top	4	4
CAERAU: **coal 28 in,** rashings 4 in, **coal 24 in**	4	8
Fireclay with rashings at top	6	10
Dark shale	2	2
COAL		8
Mudstone, shaly in lower part, with ironstone	14	6
RED VEIN: **coal 17 in**	1	5
Fireclay	2	1
Mudstone, shaly in upper part; sporadic ironstone nodules	39	6
UPPER NINE-FEET: stone 2 in, **coal 3 in,** stone 2 in, **coal 46 in** ..	4	5
Rashings and fireclay with ironstone nodules	4	8
Strong silty mudstone with a little ironstone	15	10
LOWER NINE-FEET: **coal 24 in**	2	0
Fireclay with small ironstone nodules	4	9
Sandstone		10
Mudstone, darker in upper portion; a few ironstone bands	20	9
Sandstone	1	0
Blocky mudstone with ironstone bands and large nodules	30	3
BUTE: rashings 6 in, **coal 20 in,** rashings 1 in, **coal 12 in,** rashings 2 in,		
coal 24 in, parting $\frac{1}{2}$ in, **coal 62 in**	10	7$\frac{1}{2}$
Fireclay	3	8
Mudstone with ironstone nodules	12	3
Amman Marine Band: shaly mudstone with *Lingula mytilloides* ..	4	1
AMMAN RIDER: **coal 23 in**	1	11

The sections of both the Bute and Lower Nine-Feet coals are anomalous, and it may be that the Lower Nine-Feet of this section represents only the top part of that seam elsewhere; the lower part being split away and resting directly on top of the Bute. The limited development in the Bute showed that the section as given above is normal for the area and not due to structural thickening. Apart from *Lingula mytilloides* in the Amman Marine Band, no fossils were obtained from above any of the coals in the exploratory cross-measures described above.

Owing to repetition by the Llanharan Thrust, the strata from the Amman to Cefn Coed marine bands crop out on Hirwaun Common in two strips. On the lower limb the crops of the coals extend in a narrow band about 200 to 300 yd wide, lying mainly south of the Pencoed Branch railway on the west of the Common, and which to the south of Heol-y-Cyw swings east-south-eastwards to occupy the ground immediately south of the Nant Crymlyn (Glam. 35 S.W.). From here, as far as the Penprysg road, the coals were worked very extensively near the outcrop, and the remains of many hundreds of bell-pits and shallow excavations can now be seen in the swampy waste ground.

The eastward continuation of this outcrop passes beneath thick morainic drift and in the area south of Brynna and Llanharan passes out of the district. Throughout this tract the crops of the coals can be inferred only by extrapolation from the workings at Llanharan Colliery.

On the north side of the Llanharan Thrust the outcrop of the group extends through Heol-y-Cyw in an east-south-east direction to the immediate south of Brynna, and thence to the area just south of the Llanharan pits before being cut off at crop by the thrust. Except for the area west of Wern Fach [96228362] this strip is concealed beneath morainic drift and there is little or no evidence at the surface of the position of the crops of the individual seams. A coal which is almost certainly the **Two-Feet-Nine** was proved by exploration by the Opencast Executive at two localities in the Heol-y-Cyw area. The first was to the west of the hamlet about 750 yd W. of where the road crosses the railway, and the second about 400 yd E. of the same point; the seam was $3\frac{1}{2}$ to $4\frac{1}{2}$ ft thick. In the neighbourhood of the old Brynna Colliery [986830] the crops of the Two-Feet-Nine and Six-Feet seams are closely fixed, the former seam, under the name of Lantern, having been worked extensively.

Well-developed sandy beds some little distance above the Two-Feet-Nine crop out in a window through the drift just north of Tre-nos-uchaf Farm [995825], and a line of bell-pits marks the presumed crop of the seam near Bryn-y-cae [99808226].

The South Pit at Llanharan Colliery was sunk in 1870 to just below the Six-Feet Seam at a depth of 144 yd. Only one coal was worked, and that on a limited scale—the Two-Feet-Nine (then called the **Lantern** on account of the well-developed cannel in the top of the seam). The colliery was abandoned about 1875 and remained derelict for nearly fifty years. In the early 1920's it was re-opened and, in addition to the deepening of the South Pit, the North Pit was sunk. It was not realized at the time that the shafts had passed through a large thrust (see pp. 245–9) in the vicinity of the Six-Feet Seam, and that the main coals were repeated about 750 ft lower down. This defeated all endeavours to identify the coals with those of the main part of the coalfield. Instead, the main coals beneath the thrust were numbered on a purely local basis, and this nomenclature remained in force until the present resurvey, when recognition of the fossil-horizons above most of the seams proved the existence of the thrusting, and showed that the undisturbed sequence was not markedly different from that of the collieries to the north.

Due to an extensive system of cross-measures roadways more information is available at Llanharan concerning the measures under discussion than at almost any other colliery in the district. Despite the existence of many small-scale structures in the measures, detailed examination of the main roadways reveals much minor variation in the sequence, and in the following general account of the strata between the Amman and Cefn Coed marine bands an attempt is made to summarize these variations.

The **Amman Marine Band** is usually thin being often less than 9 in and rarely more than 24 in. In one locality it was absent, non-marine shells resting directly on the coal below. The sparse though varied fauna includes the following species: sponge spicules [pyritized]; *Sphenothallus sp.*; crinoid columnals, both round and pentagonal; *Aviculopecten* cf. *scalaris* (J. de C. Sowerby); *Lingula mytilloides, Orbiculoidea nitida, Productus* cf. *carbonarius, P.* (*Levipustula*) *piscariae, Spirifer pennystonensis*; '*Cypridina phillipsi*' Corsin; and *Pleuroplax sp.* Immediately above the marine shale non-marine shells with the typical 'ghostly' preservation occur in smooth grey mudstone, though this phase is not always present. In the poorly preserved material a short form of *Anthraconaia sp.* and *Anthracosia* cf. *ovum* [?dwarfed] occur.

The marine band is separated from the Bute Seam by 8 to 20 ft of mainly argillaceous strata, all of which is seatearth where the development is thinnest.

The section of the **Bute** (No. 9 Seam of the old colliery nomenclature) is: top coal 6 to 10 in, rashings 3 to 12 in, coal 9 to 13 in, seatearth or clod 4 to 12 in, coal 54 to 70 in, rashings 2 to 3 in, and coal 6 in. It is not clear whether this section correlates in detail with the Bute of the main collieries to the north: it may be that the two thin coals in the top of the seam are the lower part of the Lower Nine-Feet which has split into two distinct portions, the lower part resting on the Bute and the upper part tending to occur close to the Upper Nine-Feet. The roof of the Bute consists of cannelly shale crowded with indeterminate plant fragments followed upwards by 10 to 20 ft of striped silty mudstone with ironstone nodules and layers.

The **Lower Nine-Feet,** possibly only the upper part of the standard seam, (No. 8 Seam), is a dirty coal usually in three distinct leaves, e.g. top coal 7 in, rashings 2 to 21 in, coal 15 to 27 in, rashings and coal streaks 1 to 9 in, coal 12 to 24 in. The seam is separated from the Upper Nine-Feet by rarely more than 5 ft of seatearth; in the No. 1 Conway Drivage the separation consists merely of 1 in of rashings, and the two coals make one complex seam, with five distinct partings, nearly 12 ft in thickness.

The **Upper Nine-Feet** (No. 7 Seam) has the following characteristic section: top coal 5 to 8 in, rashings or stone 1 to 15 in, coal 36 to 48 in, rashings 1 to 3 in, coal 12 to 24 in. The roof of the seam is formed typically of a weak shaly mudstone with small ironstone nodules and sporadic shells, passing upwards into silty mudstone. Among the fauna, which is of the dwarfed 'Red Vein' type, the following forms have been identified: *Anthraconaia* cf. *curtata, A. sp.* (of *pumila* group), *A. sp.* (cf. *williamsoni*), *Naiadites subtruncatus.* Among plants obtained at one locality *Alethopteris decurrens, Neuropteris heterophylla, N. pseudogigantea* and *N. tenuifolia* have been named.

About 30 to 55 ft of strong silty mudstone, often with sandy beds in the middle and upper portions, intervene between the Upper Nine-Feet and the **Red Vein** (No. 6 Seam). The latter is a variable seam which has never been worked and the section usually approximates to: top coal 16 to 18 in, rashings 3 to 9 in, coal 10 to 16 in. The roof consists of cannelly shale, about 4 in, followed upwards by striped shales with small ironstone nodules and small solid 'mussels' which occur through a thickness of about 12 in. This is followed by 3 ft of barren striped silty mudstone with small nodules, in turn overlain by 5 in of cannelly mudstone with poorly preserved shells. The fauna consists of typical dwarf 'Red Vein' forms: *Anthraconaia* cf. *pulchella, Anthracosia* cf. *angulata, A.* aff. *nitida, A. sp.* cf. *ovum, A. sp.* intermediate between *ovum* and *phrygiana* and *Naiadites sp.*

The Red Vein and the Caerau (No. 5 Seam) are rarely separated by more than 12 ft of dark shale with ironstone and seatearth. The **Caerau Vein,** where undisturbed, is uniform in character consisting of two leaves, the upper one averaging 14 to 20 in, and the lower 15 to 24 in, separated by 8 to 21 in of rashings and seatearth. Frequently a third coal about 6 in thick is recorded lying $2\frac{1}{2}$ to 10 ft below the main seam. The roof, where examined, consisted of 4 in of grey shaly seatearth with poorly preserved shells, mainly flattened smears, followed upwards by 1 ft 8 in of smooth shale, with many shells both solid and flattened, overlain by striped siltstone. The fauna—a typical 'Red Vein' suite—represents the only well-developed assemblage collected from this horizon during the present survey and the following species have been recognized: *Spirorbis sp.*; *Anthraconaia* cf. *pulchella, A. sp.* aff. *williamsoni, Anthracosia sp. nov.* aff. *aquilinoides, A.* cf. *nitida, A. sp.* cf. *simulans, Naiadites productus*; *Carbonita sp.*

The **Six-Feet** (No. 4 Seam) normally lies some 20 to 30 ft higher in the sequence, the intervening strata consisting mainly of mudstone and silty mudstone. The seam has been extensively worked, and together with the Yard and the Two-Feet-Nine has provided the bulk of the output from the colliery. The section varies little: main coal 75 to 84 in, rashings 14 in, bottom coal 13 to 15 in. The roof is formed by up to 18 in of hard, massive, black, laminated though non-fissile blackband, which contains a well-preserved *Anthracosia* fauna typical of the horizon. The forms include *A.* aff. *aquilina, A.* aff. *atra, A. concinna* and variants, *A.* cf. *fulva, A.* cf. *lateralis* with associated *Naiadites* cf. *productus* and *Carbonita sp.* The strong competent nature of this roof has undoubtedly protected the coal which is not so disturbed as many of the other seams.

60 to 90 ft of strong silty mudstone with ironstone layers and nodules, passing upwards into sandstone at the top of the cyclothem, separate the Six-Feet from the **Lower Four-Feet** (No. 3 Seam). The latter is subject to much disturbance, but appears typically to have a three-coal section: top coal 6 to 15 in, seatearth or rashings 6 to 30 in,

coal 10 to 18 in, seatearth or carbonaceous shale 4 to 12 in, coal 19 to 24 in. 'Mussels' occur in the roof to 6 in above the seam, in blocky grey shaly mudstone with sporadic ironstone nodules, followed upwards by thinly bedded sandstone with much plant debris. *Anthraconaia obscura, Anthracosia aquilinoides* and *Carpolithus membranaceus* Goeppert have been identified. The Lower and Upper Four-Feet seams may be close together (as little as 3 ft 10 in of seatearth separating them) or 30 to 50 ft of silty and sandy mudstone with sandstone bands may be developed between them as in the vicinity of the shafts. The **Upper Four-Feet** (No. 2 Seam) is typically a double coal, two leaves each about 14 to 24 in being separated by 6 to 30 in of rashings. 10 to 12 in of thinly laminated blackband with a few mudstone partings overlie the coal and are succeeded by grey shale carrying shells in the bottom ½ in. A typical 'Four-Feet' fauna is present with the following species occurring in some abundance: *Anthraconaia sp. nov.* (cf. *curtata*), *Anthracosphaerium* cf. *exiguum, Anthracosia planitumida,* cf. *A. simulans, Naiadites sp.*

The **Two-Feet-Nine** (No. 1 Seam) normally lies about 5 to 14 ft above the Upper Four-Feet, most of this thickness being made up of seatearth. Everywhere 5½ to 6¼ ft of clean coal are overlain by 6 to 12 in of cannel and up to 3 ft of cannelly shale. These are followed upwards by grey mudstone with ironstone nodules and layers, a characteristic feature of which is the common occurrence of 'worm-casts' on the surface. An abundant and typical fauna occurs in the lower part of the mudstone: *Anthraconaia sp. nov.* cf. *lanceolata, A. sp. nov.* cf. *wardi, Anthracosia* cf. *acutella, A. aquilina, A. atra, A. concinna, A. planitumida, A.* cf. *simulans, Anthracosphaerium propinquum, Naiadites* cf. *angustus.* A second and similar shell-horizon was located in similar lithology about 10 to 14 ft above.

The sequence from the Two-Feet-Nine to the Cefn Coed Marine Band contains fewer cyclothems at Llanharan Colliery than elsewhere in the district. A rider coal 16 to 27 in thick lies 60 to 85 ft above the Two-Feet-Nine and the next coal, 4 to 5 in thick and 20 ft or so above, carries the Hafod Heulog Marine Band. Sandy beds are well developed in the middle and upper portions of the cyclothem above the Two-Feet-Nine and they also occur beneath the seatearth of the Hafod Heulog coal. The strata from the Hafod Heulog Marine Band to the Cefn Coed are also much condensed, 30 to 50 ft normally covering the group.

The **Two-Feet-Nine Rider** coal carries badly preserved *Anthracosia* in mudstone with many ironstone nodules. The **Hafod Heulog Marine Band** is characterized by relatively abundant specimens of *Orbiculoidea* cf. *nitida,* up to ½ in across, associated with subordinate *Lingula mytilloides* in medium to dark grey, slightly silty, mudstone. Other fossils which occur rarely are: sponge spicules [pyritized]; *Productus (Levipustula) rimberti; Schizodus?;* and fish debris including *Rhizodopsis sp.*

The Britannic Marine Band has not been identified at Llanharan: it is presumably represented by ferruginous mudstone containing *Naiadites spp.* and *Samaropsis sp.* which occurs above a thin cannelly coal normally lying rather nearer to the Cefn Coed horizon than to the Hafod Heulog. The **Cefn Coed Marine Band** rests directly on a buff-grey blocky seatearth, and carries abundant *L. mytilloides* in smooth grey-blue shaly mudstone. The blocky mudstone with abundant pyritic material, which characterizes the Cefn Coed horizon elsewhere, is absent at all localities where the band has been identified in the colliery.

The shaft section of Cardiff Navigation (Llanelay) Colliery, abandoned over fifty years ago, is difficult to correlate and there is undoubtedly much structural disturbance in the neighbourhood of the shaft. Apart from the so-called 'Drydydd or Upper Four-Feet' Seam, 5 ft 6 in thick at 450 ft 7 in. in the shaft (see p. 310), which is almost certainly the Two-Feet-Nine Seam, it is not possible to identify any of the seams of the present group. The characteristic cannelly shale from the roof of the Two-Feet-Nine and the equally diagnostic shelly blackband of the Six-Feet both occur abundantly in the spoil heaps, and from this it would seem that these two seams provided most of the output of the colliery.

Numbers 1 and 2 pits at Llantrisant Colliery were undoubtedly sunk to the main coals, but the measures in this part of the sequence appear to have been very disturbed and correlation of the seams in the shaft sections is difficult. Coal working stopped in 1941 and the shafts had been abandoned following an explosion before the present survey was undertaken. The only coals of the present group that can be tentatively identified are (in No. 1 pit): Two-Feet-Nine, coal 83 in at 1631 ft 7 in; Upper Four-Feet, coal 35 in at 1655 ft 7 in; Lower Four-Feet, top coal 19 in, rashings 9 in, coal 20 in, fireclay 11 in, coal 2 in, at 1678 ft 9 in; and the Six-Feet, coal (dirty) 11 ft at 1738 ft.

The Llanharan Thrust probably dies out not far east of the River Ely. Eastwards from Mwyndy Junction it is inferred that the measures occupy a single narrow outcrop. In the area from south-west of Pont-y-parc [037822] to Llwynmilwas [070823] the coals are mostly concealed beneath the Dolomitic Conglomerate and, except for some old bell-pits 300 to 400 yd W.N.W. of the latter farm, the coals in this tract are totally unexploited. Nothing is known concerning the behaviour of the sequence but the eastwards attenuation, so marked from Newlands to Llanharan, evidently continues, for near the eastern margin of the district less than 200 yd covers the width of the group at outcrop.

South Rhondda [99208488], 1 mile N. by E. of Brynna, and Meiros [00408400] just over half-a-mile N. of Llanharan, can both be described as South Crop collieries; both have been abandoned for many years. The former was sunk to the No. 3 Rhondda Seam and the latter to the Pentre, and these were the principal coals worked at both collieries. At both collieries cross-measures drifts penetrated to the main coals, but such is the nature of the records now in existence that little can be made of the correlation at either. At South Rhondda the Two-Feet-Nine, Four-Feet and Six-Feet seams appear to have been proved, but all the recorded sections appear to be abnormally thin. The Two-Feet-Nine is recorded as 39 in, and the Six-Feet as 57 in. At Meiros details of the section below the Pentre are not clear and the correlation given on p. 347 is only tentative.

A.W.W.

B. Cefn Coed Marine Band to Upper Cwmgorse Marine Band

Generalized sections for these strata in each of the areas described below are given in Fig. 29.

TAFF–CYNON AREA

These strata have only a limited outcrop in the Cynon Valley near the northern margin of the district, and even here they are almost completely obscured by drift. The following account, therefore, is based mainly on mining records, which, in certain instances, are nor entirely reliable.

The 60 to 70 ft of measures between the Cefn Coed Marine Band and the Gorllwyn are almost wholly argillaceous, and are apparently rich in ironstone layers and nodules, since the expression 'mine ground' appears in most records.

The seams from the Gorllwyn to the Pentre Rider form a compact group, but since they have never been worked the individual coals have not been named. An attempt has been made in most collieries to identify the Gorllwyn, but this has not always been successful. The measures between these seams thicken steadily northwards and north-westwards from about 85 ft at Lady Windsor to 120 to 150 ft in the Mountain Ash and Merthyr Vale areas, and 150 to 170 ft in the upper part of the Cwmaman Valley.

The **Gorllwyn** presumably crops beneath drift in the middle of the valley near Lletty Shenkin Lower Pit and Old Duffryn Colliery. It varies a good deal in thickness, being best developed in the north and north-west where it is 18 to 26 in. Here it is normally separated from the **Gorllwyn Rider** above by 9 to 16 ft of seatearth and clift. In the south, in the Mountain Ash, Penrhiwceiber and Ynys-y-bwl areas, the two seams are

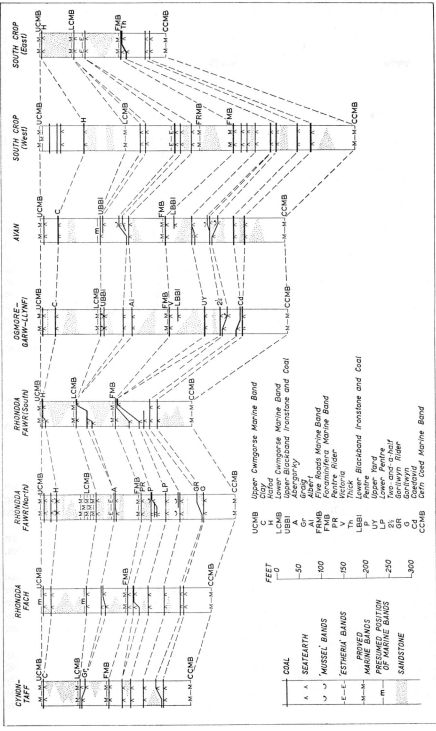

Fɪɢ. 29. *Generalized sections of the measures between the Cefn Coed and Upper Cwmgorse marine bands*

usually close together, being separated by 2 to 7 ft of seatearth. The following sections illustrate the behaviour of these coals.

	Cwmneol Ft	Cwmneol In	Abercwmboi Ft	Abercwmboi In	Navigation Ft	Navigation In	Lady Windsor Ft	Lady Windsor In
Coal	1	0		7		7½	1	5½
Seatearth, etc. ..	16	0	9	0	7	4	2	5
Coal .. .. ⎫						1 0		3½
Seatearth .. ⎬	2	2	2	6	{	4		
Coal .. .. ⎭						8		

The exact correlation at Lady Windsor is obscure, and it may be that the coals recorded represent the Gorllwyn, the Rider being absent in wash-out. Such an interpretation is supported by the evidence at Deep Duffryn and Merthyr Vale where both coals appear to be washed out by sandstone.

The Gorllwyn Rider is overlain by about 25 to 50 ft of strata, the highest figures being recorded in the Aberaman–Cwmaman area, and the lowest at Penrhiwceiber and Lady Windsor. In the former area a well-developed sandstone, known locally as the 'Big Rock', is present in the upper part of the cyclothem, while in the south mudstones predominate. The succeeding coal, which has not been worked, appears to correlate with the **Eighteen-Inch** of Maesteg, and consists of 9 to 14 in of clean coal.

Some 21 to 37 ft of mudstone, locally with a few bands of sandstone, separate the Eighteen-Inch from the Lower Pentre, which is underlain by a thick seatearth. The **Lower Pentre** is everywhere a double coal: top coal 8 to 19 in, 'stone' or seatearth 28 in, coal 5 to 11 in. It is separated from the Pentre Seam by 10 to 14 ft of mudstone and seatearth in the Cwmaman area, and by 18 to 25 ft of similar beds elsewhere.

In the northern part of the area the **Pentre** consists of two coals but in the south only one is usually recorded. Here it is not clear whether the two coals have united or whether one has died out, since there is an overall thinning of the seam. The following sections are typical:

	Trewen Ft	Trewen In	Abercwmboi Ft	Abercwmboi In	Merthyr Vale Ft	Merthyr Vale In	Navigation Ft	Navigation In	Lady Windsor Ft	Lady Windsor In
Coal ..		9		7	1	0				
Clod ..		4		3		2½	1	2	1	0
Coal ..	1	8		7		6½				

The separation between the Pentre and Pentre Rider is greatest in the north-west and north-east of the area. In the Cwmaman Valley 45 to 50 ft of clift with a few sandstone bands occur between the two seams; in Merthyr Vale Colliery 60 ft of clift are recorded. Elsewhere there is a gradual southward thinning from about 30 ft north of Mountain Ash to 20 to 24 ft at Cwm Cynon and Penrikyber, and only 6½ ft at Lady Windsor. The **Pentre Rider** consists of 7 to 20 in of clean coal, a more or less gradual thickening of the seam being evident as it is traced from Cwmaman southwards to Penrikyber. The **Foraminifera Marine Band,** which should occur in the roof of the seam, has been proved only at one locality. In an inset into the wall of No. 2 Pit at Cwm Cynon Colliery 18 in of coal at a depth of 1024 ft were exposed, and from the roof blue-grey mudstone was obtained which contained abundant foraminifera, a few *Lingula mytilloides* and *Planolites ophthalmoides*.

Throughout most of the area the Pentre Rider coal is succeeded by 50 to 60 ft of measures, though, locally, greater thicknesses are recorded, e.g. 74 ft at Cwmneol and 67 ft at Lletty Shenkin. In the north and north-west the strata are almost wholly argillaceous, but in the south up to 20 ft of sandstone occur in the top part of the cyclothem.

The succeeding **Graig** Seam correlates with the Abergorky of the Rhondda Valley, but '*Estheria*' which is characteristic in the roof there and elsewhere has not been found in the present area. The absence of '*Estheria*' can be explained, for the seatearth of a thin overlying coal here forms the roof of the Graig. The Graig crops near the floor of the valley on the west side of Aberaman, where it is reported to have been worked on a small scale. The seam has also been worked extensively from Abergorki Colliery at Mountain Ash, where it varies in thickness from 22 to 32 in, a steady thinning taking place towards the north and north-west. In the extreme north-east at Trewen and Cwmaman collieries, two coals in the general position of the Graig are separated by 6 to 15 ft of seatearth and mine ground; these suggest the typical double-coal section of the Abergorky in the Rhondda Fawr. Elsewhere the seam usually consists of a single coal 14 to 32 in thick, though locally thin coals are recorded in the floor as at Abercwmboi Pit: top coal 30 in, clod 2 in, inferior coal 5 in, clod 8 in, coal 4 in.

A rider coal, 1 to 3 in thick, overlies the Graig at no great distance. The separation is least in the workings at Abergorki Colliery where as little as 3 ft is recorded; elsewhere up to 18 ft normally intervene and increase in the north-west to 42 ft at Trewen shaft. The **Lower Cwmgorse Marine Band** is presumed to occur above the rider, but it has been seen only in the workings at Abergorki Colliery. In a cross-measures drift from the Clay Seam to the Graig a considerable thickness of marine strata was proved, though the fauna is not abundant. In the immediate roof of the rider coal *Lingula sp. nov.* was found; 2 ft above, foraminifera, including *Glomospira sp.*, occur, associated with *Planolites ophthalmoides*; *Orbiculoidea?* or '*Estheria*'? are recorded $3\frac{1}{2}$ ft above, *Lingula mytilloides* and *Dunbarella sp.* 12 ft above, and *Nuculana* cf. *stilla* (McCoy) 35 ft above the coal.

The thickness of strata between the Lower Cwmgorse rider coal and the Clay Seam usually varies from about 70 to 85 ft, but higher figures are noted in a few places, e.g. Cwmaman Pit 134 ft, Cwmneol Pit 90 ft, Deep Duffryn 128 ft, and Navigation 96 ft. Except at Lady Windsor Colliery massive sandstones are well developed in the middle or upper part of the cyclothem, 30 to 60 ft being commonly recorded.

The **Clay** Seam, which correlates with the Hafod of the Rhondda Valley, crops in the bottom of the Cynon Valley on both sides of Aberaman. It was worked from the Clay Level [02200010] together with its underlying seatearth, which formed the basis of a flourishing brick and sanitary-pipe industry. It was also worked at Abergorki Colliery (where it was known as the Gorllwyn). Over most of the area the seam varies from 24 to 27 in, but to the south of Mountain Ash it thins to 19 in at Penrikyber Colliery and 14 in at Lady Windsor. The immediate roof of the seam comprises the **Upper Cwmgorse Marine Band,** which is well seen in the workings at Abergorki Colliery, where the band is about 8 ft thick. The lowest 3 ft consist of blocky, micaceous, silty mudstones, which weather readily to a soft clay with a heavy yellow 'sulphury' discoloration; fossils are relatively rare, but *Lingula mytilloides*, *L. sp. nov.* and *Productus* (*Dictyoclostus?*) *sp.* have been collected. This blocky mudstone passes upwards into grey, somewhat more shaly, mudstone, relatively resistant to weathering, which has yielded the following fauna: echinoderm fragment; crinoid columnals; *L. mytilloides*, cf. *Productus* (*P.*) *carbonarius*, *P.* (*Dictyoclostus*) *craigmarkensis*, *P.* (*Linoproductus*) *sp.*; *Aviculopecten?*. In addition *Huanghoceras* [*Metacoceras*] of the *costatum*/*postcostatum* group was identified from material submitted by Mrs. J. V. Harrison in 1935; it is not known from what level above the coal these specimens were collected. In spoil at the old Clay level at Aberaman, *L. mytilloides* and *P.* (*D.*) *craigmarkensis* were found.   A.W.W.

RHONDDA FACH AREA

These measures do not crop out anywhere within the valley and except at Mardy Nos. 3 and 4 Pits, there are no opportunities for their examination. None of the seams has been worked, but during the recent re-organization of the Mardy–Bwllfa area, the uppermost (Yellow) horizon was driven through the lower part of this sequence, and it was possible to examine limited portions.

The Cefn Coed Marine Band is overlain by 55 to 90 ft of clift with ironstone layers and nodules, the lower figure being recorded in the northern part of the Ferndale area and at Mardy Nos. 1 and 2 Pits, and the higher one at the southern Ferndale collieries.

It is only in the northern part of the Mardy taking that the **Gorllwyn** Seam is workable. Here it consists of 32 to 35 in of coal with a thin parting about 12 in from the floor. Traced southwards the coal thins rapidly and a thickening dirt parting develops, as shown by the following sections:

		Ferndale No. 3		Ferndale No. 8		National		Standard		Lady Lewis	
		Ft	In	Ft	In	Ft	In	Ft	In	Ft	In
Coal	..		10		6		6		5		
Dirt	..		1		3	3	4	5	0		
Coal	..	1	0	1	9[1]	1	4		8		6

The distance between the Gorllwyn and the Gorllwyn Rider varies considerably. In the National and Ferndale Nos. 8 and 9 areas they are parted by only 3 to 5 ft of seat-earth, but both to north and south the separation increases to 22 ft in the Mardy area and more than 10 ft at Lady Lewis.

The **Gorllwyn Rider** normally comprises 10 to 19 in of clean coal, being in general thickest in the north. In the Mardy area the roof consists of 4 in of cannelly mudstone with fish remains and plant debris, including *Trigonocarpus parkinsoni*, succeeded by grey mudstone with ironstone layers carrying worm tracks. In parts of the Ferndale area blackband and mine is developed in the roof.

The **Eighteen-Inch** Seam rests on 20 to 40 ft of clift with intercalated sandy beds; in the National and Standard areas a well-developed sandstone occurs in the upper part of the cyclothem. The seam is 8 to 14 in thick and in the Ferndale area carries blackband in the roof.

The **Lower Pentre,** which follows 30 to 45 ft of clift and sandy clift, has a characteristic section throughout most of the valley: top coal 11 to 15 in, stone 2 to 3 in, coal 6 to 11 in; at National and Standard collieries a single coal, 12 to 15 in thick, is recorded. The roof was seen only at Mardy, where it consisted of 3 in of planty mudstone, overlain by silty mudstone with plant debris.

Over much of the area 24 to 40 ft of clift and fireclay intervene between the Lower Pentre and Pentre seams, but in the Ferndale collieries and at Lady Lewis only 9 to 12 ft are recorded.

The **Pentre** is very variable. In the southern parts of Ferndale its section is similar to that of the neighbouring parts of the Rhondda Fawr: top coal 14 to 16 in, stone 17 in, coal 6 to 8 in. In the northern part of the Ferndale area a further split develops in the top coal: top coal 9 in, clod $\frac{1}{2}$ in, coal $12\frac{1}{2}$ in, clod $1\frac{1}{2}$ in, coal 6 in. This split thickens rapidly in the Mardy area, where the two resulting seams are separated by about 15 ft. At Mardy No. 3 Pit the so-called **Bottom Pentre** (top coal 12 to 15 in, stone 2 to 3 in, coal 7 to 8 in) has a roof of dark grey silty mudstone with plant debris. The **Top Pentre** ($7\frac{1}{2}$ to 10 in of coal) is overlain by soft mudstone with indeterminate shell fragments, followed upwards by slightly silty mudstone with abundant worm tracks.

In the Mardy and Ferndale collieries the Pentre Seam is succeeded by 20 to 30 ft of clift with ironstone nodules and fireclay. In the south of the valley the separation is much less—10 to 12 ft at National and Lady Lewis and only $3\frac{1}{4}$ ft at Standard. The **Pentre Rider,** which carries the **Foraminifera Marine Band** in its roof, consists of 10 to 20 in of

---

[1] **coal 10 in, dirt 7 in, coal 4 in.**

coal, with some bast in the top of the seam at Ferndale. At Mardy the following section was measured:

	Ft	In
Mudstone with *Planolites ophthalmoides*	2	0+
Mudstone with many ironstone nodules; *P. ophthalmoides* and foraminifera	2	0
'Sulphury' weathering mudstone with a few foraminifera		1
COAL	1	0

The **Abergorky** lies about 60 to 75 ft above, the strata consisting of clift with sandy developments in the upper part. In the Ferndale and Mardy areas the seam is comprised of two leaves, the parting between them thickening markedly both northwards and southwards, as shown by the following sections:

	Mardy No. 3		Mardy No. 1		Ferndale No. 3		Ferndale No. 6		Ferndale No. 8	
	Ft	In	Ft	In	Ft	In	Ft	In	Ft	In
Coal ..	1	1	2	0½	2	2	1	5	1	5
Measures ..	18	6	1	3	2	0	12	2	11	1
Coal ..		11	1	1	1	2	1	2	1	0

At National, Standard and Lady Lewis only one coal is recorded, varying in thickness from 20 to 24 in. It is not clear whether this represents the full section, or whether one of the leaves has disappeared. The roof of the top coal was examined in workings at Mardy No. 1 Pit; it yielded '*Estheria*' *sp.*, *Naiadites?*, *Vetacapsula kidstoni* Crookall and the plants *Alethopteris sp.* and *Hexagonocarpus sp.*

The Lower Cwmgorse Marine Band is presumably developed in the roof of a 2- to 3-in coal which normally lies 20 to 30 ft above the Abergorky, but it has nowhere been seen in this area. The **Hafod** horizon apparently occurs some 70 to 95 ft higher, and a well-developed sandstone is usually present in the upper part of the intervening strata. Except in the extreme south of the valley the Hafod is split into two distinct coals, separated by upwards of 25 ft of clift and fireclay. The lower coal ranges from 15 to 24 in, and the upper from 6 to 12 in. The Upper Cwmgorse Marine Band almost certainly occurs in the roof of the upper coal, but it is nowhere seen in the Rhondda Fach Valley.   A.W.W.

NORTHERN RHONDDA FAWR AREA

The measures from the Cefn Coed Marine Band to the Gorllwyn vary from 60 ft at Fernhill to 85 ft at Gelli. There is, however, one exception to this general southerly thickening: between Treherbert, Ynys-wen and Cwm-parc these beds are thinner than would be expected from regional considerations, and are reputed to be only 37 ft at Ynysfeio. Apart from subordinate bands of sandstone the cyclothem is composed of mudstones and siltstones; ironstone nodules and ribs are common, especially at Fernhill and Rhondda Merthyr, where part of these measures was formerly called the Black Pins Mine-ground.

Both the **Gorllwyn** and **Gorllwyn Rider** are well developed throughout most of the area. The two seams are normally 15 to 20 ft apart, but north and north-east of the Fernhill shafts the parting between them is reduced to less than 12 in, and they have been worked together as one seam. The Gorllwyn was formerly worked extensively between Treorky and the head of the valley. Around Treherbert it has a three-coal section, each leaf being about 10 in thick, with thin partings. To the north and west the upper parting disappears and there is a tendency for the lower coal to deteriorate. Southwards the three leaves·thin markedly and at Gelli the overall thickness is only 14 in. A sandstone immediately above the seam appears to be restricted to the Fernhill–Glenrhondda area. More usually the roof is a silty mudstone containing abundant plant remains. The Gorllwyn Rider has been worked to a limited extent at Ynysfeio and

Abergorky collieries, where it consists of a single clean coal varying from 18 to 24 in. At Gelli a 1-in band of soft yellow-weathering mudstone overlies the coal and passes upwards into smooth barren mudstone.

The mudstones above the Gorllwyn Rider coarsen upwards and pass into sandstones, which are 20 ft thick at Glenrhondda and Tydraw. The overlying **Eighteen-Inch** usually lies 40 ft above the Rider; it has been seen only at Gelli Colliery, where the following section was measured:

	Ft	In
Mudstone with worm tracks; solid 'mussels', including *Anthraconaia hindi* variants, *A.* cf. *warei, Naiadites* cf. *hindi* in the bottom 6 in ..	4	0
Mudstone with worm tracks and abundant sphaerosiderite	1	6
Ferruginous mudstone with 'mussel' impressions, including *A.* cf. *hindi*		3
COAL, dirty		6
Mudstone, dark grey and cannelly at top and bottom..	4	0
COAL	1	0

At Tylecoch, Rhondda Merthyr, and Park collieries the two coals are recorded in the shafts, and at Tydraw they are noted as two bands of bast. In all the other shafts only a single coal, 2 to 13 in thick is recorded. Mine-ground is invariably present in the roof, and blackband is noted at Gelli and Park. This horizon, as well as the underlying Gorllwyn and Gorllwyn Rider, must crop beneath boulder clay to the west of the Dinas Fault at Cwm-parc; a trial on the hillside [94609737], 200 yd S.E. of Nantdyris Colliery, shows spoil containing seatearth and ferruginous mudstone, and is probably on the Eighteen-Inch.

The sequence is continued by 25 to 40 ft of mudstones, which contain an abundance of ironstone. The mudstones, which are thinnest around Treorky, coarsen upwards and a few feet of sandstone usually underlie the seatearth of the **Lower Pentre**. This seam, which has been worked on a small scale at Nebo and Gelli, normally consists of a top coal of 12 to 15 in, separated by a 2-in 'stone' from a bottom coal of 6 to 10 in. It crops out around Treorky; and there are trials on it close above Nantdyris Colliery [94439743], and also 250 yd to the S.S.E. [94609728]. Mudstone spoil from these levels carries abundant plant pinnules, as does the roof of the seam at Gelli.

The Lower Pentre and Pentre are about 20 to 40 ft apart, the intervening measures being almost wholly argillaceous, though in places a few feet of sandstone underlie the seatearth of the Pentre.

The **Pentre** Seam takes its name from the village 2 miles S.E. of Treorky. It has been extensively mined from levels in the neighbourhood and from Nebo Colliery at Ystrad Rhondda; it is also said to have been worked from levels [95789615] about 100 yd S. of Ystrad-fechan House at Treorky; and there are several old trials on the hillside S.W. and S.E. of Nantdyris. Around Pentre the seam comprises three leaves: top coal 10 to 14 in, stone ¼ to 15 in, coal 9 to 10 in, stone 1 to 2 in, coal 7 to 10 in. At Gelli the parting between the top and middle coals disappears, but northwards it thickens and, at Nebo, reaches 15 in. At Abergorky and Ynysfeio collieries the Pentre appears to have split into two distinct seams by the thickening of this parting: the top coal, 12 to 15 in, lies 12 to 17 ft above the lower coal, which is 17 in thick. Farther north the position is more obscure: the top coal is clearly recognizable, but the lower appears to deteriorate, partly as a result of further splitting. At Nebo the immediate mudstone roof of the Pentre contains worm tracks and *Naiadites* fragments.

The Pentre Rider usually lies 20 to 30 ft above, though locally, as at Pentre, the two seams are nearly 40 ft apart. Over the east of the area the measures are argillaceous, but westwards and southwards sandstones appear, and at Fernhill No. 2 Pit they form the roof of the Pentre. The sandstones vary from 5 to 9 ft at Fernhill to 29 ft at Pentre; at the head of the valley they are recorded as 'white rock', suggesting that they are quartzitic in character.

The overlying **Pentre Rider** is a single coal thickening north-westwards from 10 in to 21 in. The **Foraminifera Marine Band** in its roof has been seen at Glenrhondda and Gelli collieries, at a depth of 152 ft in a borehole from the surface [94079599] near Park Colliery, and in the stream [93739800] 300 yd W. by N. of Bute Colliery. At each locality dark mudstones in the roof of the seam contain many ironstone nodules. *Planolites ophthalmoides* is abundant and ranges to about 8 ft above the coal; foraminifera are scattered through the lower part of the band and are especially abundant in the bottom 1 to 2 ft. *Ammonema sp.* and *Glomospira sp.* have been identified and were associated with pyritized tubular wavy markings, possibly formed by a boring sponge. In Nant y Pentre, dark mudstone dug [97189624] about 150 yd N.E. of Pentre Church also contains abundant foraminifera, and is presumably from the same horizon. A thin coal exposed in Cwm Parc [93789590] 700 yd W.N.W. of Park Colliery and another proved by trials on the hillside [95069629] 400 yd N. of Dare Colliery are both probably the Pentre Rider.

Above the marine band the sequence is continued by mudstones, which coarsen upwards. In the south thin bands of sandstone and 'bastard rock' occur towards the top of the cyclothem, and farther to the north and west these are better developed, about 15 to 45 ft being recorded close below the seatearth of the Abergorky Seam. These measures are poorly exposed, but the sandstones have been quarried on a small scale [96409680] east of Treorky.

The **Abergorky** Seam, which lies 50 to 80 ft above the Pentre Rider, is best developed around Abergorky Colliery from which it takes its name. It crops on the hillside from Blaenrhondda to Gelli and has been extensively worked from the crop, as well as at Gelli Colliery. It consists of two coals, the upper 18 to 27 in thick, and the lower 10 to 18 in, usually separated by less than 24 in of seatearth. To the north and east of the Abergorky Colliery workings, however, the 'clod' between the coals thickens to 8 ft, and a similar thickening takes place to the west and south of Treorky. The seam has been examined at Glenrhondda and Gelli collieries: at both the immediate roof of 2 to 3 in of cannelly mudstone with fragmentary plants was succeeded by mudstones containing *Lioestheria vinti*. This fossil has also been obtained from spoil of a level [94989678] at Tylecoch Colliery, from small trials on the west side of the railway [94959715] S.E. of Nantdyris Colliery, from diggings in Coed Mawr [94689614], and from levels [93579881] near Rhondda Merthyr Colliery.

A thin rider coal lies 35 to 40 ft above the Abergorky, though at Park it is about 65 ft above. The intervening measures are mainly mudstone, but sandstones are developed in places; one such sandstone occurs 5 to 10 ft above the Abergorky and may be 12 ft thick; another up to 18 ft thick lies close below the rider coal. These were formerly worked from small quarries near Tylecoch [94409659] and north-east of Dare [94659622].

Exposures of the rider can be seen in several streams around Treherbert and Treorky, where it is usually 3 to 10 in thick. At Rhondda Merthyr Colliery a second thin coal of about 1 in is recorded $5\frac{1}{2}$ ft above, and the same two coals can be seen in Nant Orky [95879804], where the lower carries a thin blackband.

The cyclothem which follows carries the **Lower Cwmgorse Marine Band,** and is one of the thickest in this part of the measures, 85 to 110 ft. Throughout most of the area mudstones with ironstone bands predominate, but in the north sandstones are developed. At Rhondda Merthyr 'gritty shale' and 'bastard rock' appear in the middle; at Fernhill one of the shafts records 59 ft of 'bastard rock' in the middle, while the other two show virtually the whole of the cyclothem as sandstone. There are several exposures of these beds around Treherbert, Treorky, Ton Pentre, and Cwm-parc where they are dominantly argillaceous. The mudstones at the extreme base, although dark and marine in appearance, have yielded no fossils. They pass upwards into grey, rusty-weathering mudstones which contain sporadic *Dunbarella*, *Lingula* and *Hollinella* cf. *bassleri*. Near Rhondda Merthyr [93409908] these occur 4 ft above the underlying coal. On the other side of the valley [93429825; 93309834], near Bute Colliery, *Dunbarella* was obtained 15 ft above the coal. In Nant Orky [95609821] and in a tributary gully [95609764] similar fossils occurred about halfway between the underlying coal and the Hafod, whilst in Nant y Pentre [97289646] and at a locality [96470538] west of Ton Pentre a few specimens were collected from only 30 ft below the Hafod. Immediately beneath the Hafod is a thick seatearth, so leached as to be almost white.

The so-called **Hafod** of this area represents only the lower part of the Hafod of the southern parts of the valley. Several small trials lie on its outcrop, but its thickness is only 6 to 12 in, though locally it is reputed to be almost 2 ft. The seam is seen in an old quarry [96639676] N. of Tynybedw Colliery: coal smut 1 in, ferruginous shale and blackband with crushed 'mussels' 2 to 6 in, coal smut 1 in, mudstone 1 in, coal 10 in; the roof is mudstone with ostracods. A similar section is seen in an old quarry [94469624] N. of Park Colliery. In a quarry [95349645] S. of Tylecoch a single coal occurs, carrying blackband in its roof. The blackband with crushed 'mussels' is distinctive and can be seen in spoil wherever levels have been driven on the coal: it has been seen from a line of levels west of Treherbert railway station, and in a landslip scar [95989587] 700 yd S. of Treorky station. In spoil [97239619] 350 yd N.N.E. of Pentre Colliery the following forms were collected: *Anthraconaia* cf. *stobbsi*, cf. *Anthraconauta phillipsii*, *Naiadites* cf. *hindi*; *Geisina? subarcuata* and *Carbonita sp.*

Two thin coals overlie the Hafod, one about 15 ft above and the other 15 to 20 ft still higher. They rarely exceed 3 in, and are frequently squeezed out at crop; together they form the upper part of the Hafod of the lower Rhondda. At Nant Orky [95419825] the lower smut carries a mudstone roof with *Geisina?*.

The **Upper Cwmgorse Marine Band** occurs in the roof of the upper smut and has been identified at numerous localities on either side of the valley from Ton Pentre northwards to Treherbert: in gullies [93259826; 93369822] S.W. of Ystrad-ffernol; [93479921] N.W. of Rhondda Merthyr shafts; at two localities in Nant Orky [95189846; 95449822]; at several points on the spur between Nantdyris and Park collieries [94459706; 94549707; 94619703; 95099458]; on the hillside [95949584; 96329550] W. of Ton Pentre; in the stream [96549684] N. by W. of Tynybedw Colliery; and in Nant Coly [96149744]. In general the band varies from 2 to 5 ft, and *Lingula* is usually the commonest fossil, associated with small lamellibranchs and brachiopods. On the hillside [96329550] a particularly varied fauna was collected including several undescribed species of lamellibranchs. The faunal list is as follows: *Coleolus sp.*; *Crurithyris sp.*, *Lingula mytilloides*, *Orbiculoidea* cf. *nitida*; *Bucanopsis* cf. *undatus* (Etheridge), *Euphemites anthracinus*; *Aviculopecten* aff. *delepinei* Demanet, *Dunbarella sp.*, *Nuculana* cf. *stilla*, *Palaeoneilo sp.*, *Posidonia* or *Myalina*, *Prothyris?*, Gen. et sp. nov. aff. *Solenomorpha*, '*Yoldia glabra*' Demanet *pars*; Orthocone nautiloid, *Huanghoceras postcostatum*; *Anthracoceras* cf. *cambriense*, *Homoceratoides kitchini*. At other localities *Planolites ophthalmoides*, foraminifera including *Ammonema sp.*, *Patellostium sp.*, and *Platyconcha* cf. *hindi* Longstaff have also been recorded.   W.B.E.

SOUTHERN RHONDDA AREA

Of the coals in this group only the Pentre Seam has been worked to any extent and this only in a few areas. The measures as a group are therefore seldom seen. Several of the coals were examined in the main drivage from Margaret Pit (Cwm Colliery) to the Maritime area, and in Naval Colliery the strata from the Gorllwyn horizon to the Pentre were examined in the Nantgwyn Hard Heading on the west side of the Colliery.

Except in the south and south-east of the area the measures between the Cefn Coed Marine Band and the Gorllwyn horizon vary from about 95 to 120 ft, a gradual thickening to the west and north-west being observed. In the south a marked thinning occurs, 95 ft being recorded at Great Western, 75 ft at Coedely, and only 60 ft at Cwm Colliery. In the lower part the strata consist entirely of mudstone (clift), but in the upper part sandstone (in places mis-correlated with Cockshot Rock) is almost always developed, even when the beds are at their thinnest.

The correlation of the **Gorllwyn** horizons is far from clear. Except at Cambrian and Coedely collieries only one seam is developed in this part of the sequence and its exact relationship with the Gorllwyn and Gorllwyn Rider coals of the north Rhondda is not known. The single coal recorded in nearly all the shafts varies from 11 in at Lewis Merthyr to 33 in at Britannic though 17 to 20 in is more usual. At Cambrian a 3- to 4-in coal occurs 35 ft beneath this seam, and at Coedely a 9-in leaf lies 5½ ft below. This thin coal may represent the true Gorllwyn, its absence throughout the rest of the area being explained by washout by sandstone, as has previously been noted at Deep Duffryn and Merthyr Vale collieries (p. 148). The single coal above may represent the Gorllwyn Rider. In the Nantgwyn Hard Heading at Naval Colliery the roof of the latter, a cannelly mudstone, carried fish remains including Palaeoniscid and Platysomid scales and *Rhabdoderma sp.*

The measures extending up to the Eighteen-Inch Seam thin from about 50 ft in the north of the area, to about 35 ft in the Cambrian–Naval area, to about 25 ft at Lewis Merthyr and Great Western collieries, and to only 15¼ ft at the Mildred Pit of Cwm Colliery (although 29 ft 7 in is recorded in the Margaret Shaft section). The cyclothem everywhere consists of mudstone in the lower part and sandy beds are usually developed in its upper part.

The **Eighteen-Inch** is usually recorded as clean coal varying from 8 to 21 in, 12 in being average. Where examined in the Nantgwyn Hard Heading the roof yielded abundant *Anthraconaia* from grey mudstone about 21 in above the coal. Shells identified include many variants of *Anthraconaia hindi*, including elongate forms of the *A. wardi/ pruvosti* group, *Naiadites sp.* cf. *productus* together with associated *Planolites montanus.*

The **Lower Pentre** lies some 20 to 40 ft higher, overlying mainly argillaceous strata with a little interbedded sandstone in the upper part. The seam is variable: in the south only a single coal is recorded, varying from 12 to 15 in at Dinas and Cymmer to 21 to 25 in at Cwm and Great Western. In the south and west of the Naval area a parting develops about 4 in from the top of the seam, which has thickened considerably, while in the Cambrian and Blaenclydach areas a further parting, usually recorded as 'stone', is present in the middle or lower portion of the main coal. Northwards and westwards into Britannic and towards Maindy the lower stone parting persists but the upper one dies out. The following sections illustrate the variation:

	Nantgwyn		Cambrian		Britannic	
	Ft	In	Ft	In	Ft	In
Coal ..		4		6		
Clod ..	1	0		2	1	6
Coal ..			1	2		
Stone ..	2	2		2		2
Coal ..			1	0		10

At Naval Colliery the seam carries carbonaceous shale with coal streaks in the roof, above which grey mudstone yields plants including *Neuropteris obliqua, N. tenuifolia* and *Sphenophyllum cuneifolium.*

In the north of the area up to 55 ft of clift and sandstone separate the Lower Pentre and Pentre seams. Southwards these strata diminish in thickness and over much of the central part of the area 20 to 30 ft of mainly argillaceous measures is the rule. In the south-eastern sector extending from Cymmer through Lewis Merthyr and Great Western to Cwm 10 to 20 ft of fireclay and clift separate the two coals. Locally the thinning is very marked and at Penrhiwfer and Coedely only 3 to 4 ft of seatearth intervene.

The **Pentre,** the best developed seam in this group of measures, has been worked extensively in the central part of the area at Llwynypia, Naval, Blaenclydach, Cambrian, Britannic, and Cymmer collieries. Throughout this area it varies in thickness from 33 to 39 in. The seam consists of two leaves, an upper, 22 to 26 in, separated from a lower, 9 to 12 in, by 1 to 3 in of stone. In the south and south-east the coal deteriorates: normally only a single coal is recorded varying from 16 to 26 in.

Except in the extreme south 20 to 35 ft of clift with ironstone normally intervene between the Pentre and Pentre Rider seams. At Cwm Colliery this thickness is reduced to about 10 ft and at Coedely to only 1 ft 2 in.

The **Pentre Rider** is 10 to 22 in thick, the variation being somewhat irregular. The seam and its roof has been examined at two localities underground. At Naval Colliery, in a cross-measures drift towards Ely shaft, smooth grey mudstone 10 to 18 in above the coal yielded foraminifera including *Glomospira sp.* and *Hyperammina sp.*, *Lingula sp.*, *Edmondia?* [pyritized] and *Geisina? subarcuata.* Similar mudstones 1½ to 5 ft above the coal were characterized by more foraminifera associated with abundant *Planolites ophthalmoides.* An air bridge over the Pentre Main at Cambrian Colliery exposed the Rider; foraminifera including *Glomospira sp.*, *L. mytilloides* and *P. ophthalmoides* were obtained from the roof to 18 in above the coal.

Reference has already been made to the thinning of the beds between the Pentre Rider, Pentre, and Lower Pentre seams at Coedely Colliery. The section of these coals in the upcast shaft reads: top coal 23 in, clod 14 in, coal 36 in, rashings 11 in, fireclay 25 in, coal 24 in. This attenuation foreshadows the development of the so-called Thick Seam on the South Crop at Llanharan Colliery (p. 170).

The cyclothem above the Pentre Rider varies from 50 to 85 ft, being thickest in the western half of the area and thinnest in the east and south-east. It consists everywhere of clift or mudstone in the lower part with sandy beds becoming increasingly prominent in the upper part as the beds thicken westwards. The sandstone reaches 50 to 60 ft in the Coedely and Cambrian areas where it makes up the bulk of the cyclothem.

The **Abergorky** coals are very variable. Only in the north are the two leaves close enough to form a single seam; thus at Eastern Colliery on the edge of the area: top coal 17½ in, rashings and stone 14½ in, coal 17 in. Traced southwards the two leaves split apart rapidly and 10 to 20 ft normally separate them, as shown by the following sections:

		Llwynypia		Cambrian		Britannic		Naval	
		Ft	In	Ft	In	Ft	In	Ft	In
Coal ..	..	1	6	1	0	1	8	1	8
Measures	..	10	11	17	8	14	0	14	4
Coal ..	..	1	4	1	2	1	8	1	7

From Porth southwards only one leaf is recorded in the shaft sections. This coal, which varies from 14 in at Ty Mawr Pit to 22 in at Cymmer and Coedely, appears to be the top coal of the above sections. It is not clear what has happened to the bottom coal, but it seems certain that it does not re-appear as a separate seam to the south, where a single coal is recorded in several of the cross-measures road-ways at Llanharan on the South Crop. Though the Graig Seam at Mountain Ash (p. 149) consists of a single coal, it is thick enough to represent a combination of the two leaves. There is no evidence that the two seams re-unite southwards, and the most likely explanation is that the bottom coal is lost to the south in wash-out.

The single Abergorky coal was doubtfully recorded at outcrop in the crest of the Pontypridd Anticline on the east side of the Ely Valley [01188911] 500 yd S. of Cilely Colliery, where spoil of blue-grey paper shale with '*Estheria*'? was noted.

The **Lower Cwmgorse Marine Band** is presumed to occur everywhere above a rider coal which over most of the area lies some 30 to 50 ft above the upper Abergorky Coal. The intervening measures are thickest in the north and west; they are mainly mudstones with a little sandstone near the top in a few places. In the south-east the separation is much reduced; at Ty Mawr Pit, for example, only 4½ ft of fireclay intervenes. The rider is a single coal 4 to 18 in thick and is best developed at Britannic, Penrhiwfer, Cwm, and Coedely. The marine band has been recorded during the present survey only at Cwm Colliery. In the Margaret Main at a distance of 924 ft from the shaft, an 18-in coal was overlain by grey shale which yielded *Lingula mytilloides*, *Dunbarella sp.*, *Myalina?* and *Rhizodopsis sp.*

The thickness of strata between the Lower Cwmgorse Rider coal and the Hafod horizon is everywhere considerable. In the north and west at Llwynypia, Cambrian, and Britannic it approaches 100 ft, but this gradually diminishes to the south-east and east to 60 ft at Cymmer and Lewis Merthyr, to 50 ft in a trial pit on the east side of

Great Western Colliery (in the trough between the Llanwonno and Daren-ddu faults), and to about 50 ft also at Coedely and Cwm collieries. Over most of the area these measures consist of mudstone with interbedded sandy beds in the upper part, but in the south at Cwm and Coedely, despite the overall thinning, sandstone predominates.

The **Hafod** Seam was formerly worked from several collieries in the south central part of the area, more particularly at Cilely, Cymmer, and Hafod (adjoining Lewis Merthyr). Throughout this area the seam is 2½ to 3 ft in thickness and consists essentially of two leaves. Southwards and eastwards these unite, and a single coal is recorded, 28 in and 33 in thick respectively, at Cwm and Coedely. Northwards and north-westwards the seam deteriorates rapidly, partly through overall thinning, but more particularly because the top coal splits away. About 1¼ miles north of the shafts at Hafod Colliery the deterioration was very rapid and a typical section at the limit of the workings reads: top coal 4 in, rashings 3 in, coal 5 in, clod 4 in, coal 7½ in. The main split appears to be at the rashings bed, while the clod appears to be a local development in the bottom coal. A similar section is recorded in the northern part of the workings at Cymmer Colliery: top coal 11 in, rashings 4 in, coal 5 in, stone 3 in, coal 14 in. The seam was worked over a limited area near the shafts at Penrhiwfer Colliery with a section: top coal 20 in, clod 10 in, coal 18 in; but within 200 yd to both north and west of the shafts the clod had increased to 5 ft. The development of the split to the west and north-west of the area is illustrated by the following sections:

	Britannic North Pit		Naval		Cambrian		Llwynypia	
	Ft	In	Ft	In	Ft	In	Ft	In
Coal		5		10		2		5
Measures ..	36	0	24	2	30	0	25	0
Coal	1	5	1	2	1	0		9

The seam crops out along the crest of the Pontypridd Anticline in the Cilely–Cwm Cae'r-gwerlas area. Several levels were driven on the crop of the seam and from the spoil a few marine fossils have been obtained. One such level [01158904] on the east side of the track leading southwards from Cilely Colliery and 600 yd from the pits is driven on the strike of the seam dipping steeply south: from the spoil *L. mytilloides* and *Orbiculoidea nitida* have been collected. Another level occurs in a corresponding posi-tion on the west side of the valley [00908907], 650 yd S.S.W. of Cilely. A third level [00688940], 550 yd W.S.W. of the colliery, is driven on the flattish northern limb of the anticline; the seam section here is given as: coal 19 in, stone 2 in, coal 19 in; and the badly weathered spoil yielded *L.* cf. *squamiformis* Phillips. The Hafod is exposed in the axis of the anticline in a small area at the bottom of the Nant Cae'r-gwerlas [00158915], about 1200 yd W.S.W. of Cilely Colliery. Here are the remains of several old workings in which the spoil was too weathered to yield fossils. The seam was also worked from the old Caerlan Colliery [00288894], which was driven on the crop of the No. 3 Rhondda Seam on the steep southern limit of the anticline with cross-measures roads intersecting the Hafod Seam. Spoil of 'sulphury'-weathering ferruginous shale yielding *L. mytilloides* presumably came from these Hafod workings.

The Hafod Seam was also examined at Cwm Colliery. The seam was intersected in the Margaret Main; in the roof shales, weathered almost to a soft clay, a fragment of *Orbiculoidea sp.* was noted.   A.W.W.

OGMORE–GARW–LLYNFI AREA

Although a complete section of these measures is not known within this area, only a small portion at the top is missing at Western and Caerau collieries.

In the Ogmore and Garw valleys the measures between the Cefn Coed Marine Band and the Caedavid vary from about 90 to 105 ft. They consist mainly of mudstone, but sandstones are developed near the top in the Ogmore Valley: 33 ft are recorded at Western Colliery though Wyndham shafts, only 300 yd to the south-west, show less than 12 ft. In the Llynfi Valley the measures thicken south-westwards from about 90 to 120 ft: they are almost wholly argillaceous and carry numerous nodules and layers of ironstone. To the south, in the Celtic shafts, a sandstone is recorded between 8 and 32 ft below the top of the cyclothem.

The **Caedavid** is a compound seam which represents the Gorllwyn and Gorllwyn Rider of the north-eastern part of the district. It has been worked extensively in the Garw Valley, on the eastern side of the Llynfi Valley, and from Celtic and Oakwood collieries. At Penllwyngwent and Wyndham it is said to consist of two leaves: top coal 18 to 20 in, shale or fireclay 11 to 12 in, coal 5½ to 11 in. Elsewhere in the Ogmore and Garw valleys the top coal appears to be split and a three-leaf section prevails:

		Western		Glengarw		International		Ffaldau		Waun Bant	
		Ft	In	Ft	In	Ft	In	Ft	In	Ft	In
Coal	..	1	2	1	9	1	9	1	9	1	9
Dirt	..		3		4		1		5		4
Coal	..		8		10½		10		7		8
Dirt	..	1	10		9		9	1	3	1	0
Coal	..		6	1	0		7		9		10

The upper parting is recorded as 'stone' at Western, International, Glengarw, and Darren; as 'rashings' in some of the Ffaldau sections, where they reach 22 in, and as 'bast' in other parts of Ffaldau and at Waun Bant [90709161], a small drift colliery to the east of Pontycymmer. The lower parting is usually described as rashings. The three-leaf section continues into the Maesteg area where the top coal is 14 to 24 in thick, the middle 6 to 8 in, and the bottom 9 to 12 in; the floor is usually composed of dark coaly rashings which locally pass into inferior coal. In the north of the Llynfi the top coal splits away to form a separate seam, presumably the equivalent of the Gorllwyn Rider. The line of split runs approximately east to west just south of the Caerau shafts, thence turning south towards Duffryn Madoc; from here it runs between Oakwood and Celtic, where the two seams are separated by 18 ft of 'bastard rock'.

Spoil from the Caedavid at Glengarw Colliery includes a hard carbonaceous stone carrying closely packed specimens of a deep variety of *Anthraconaia adamsi*. These fossils were not observed underground, but it is thought that they come from the upper of the two partings and not from the true roof of the seam, which around Maesteg consists of dark mudstone yielding scattered plants.

30 to 50 ft usually separate the Caedavid and Eighteen-Inch. The beds, which are thickest in the Ogmore Valley and the northern Garw and thinnest in the Llynfi, consist of mudstones and silty mudstones with many sandstone bands. The sandy beds increase in frequency towards the east and in the Ogmore Valley they make up the bulk of the cyclothem.

The **Eighteen-Inch** Seam was hitherto named only in the Maesteg area. It varies from 12 to 26 in of clean coal. In the Garw–Glengarw area it is associated with shaly coal or bast, and at Ffaldau and Darren pits blackband, coal and shale up to 45 in thick are recorded in the roof of the seam. At Caerau 3 in of bast occur at the top of the seam, which has been worked on a small scale from levels [85709076], 200 to 400 yd W. of Oakwood Colliery.

In the Ogmore Valley the Eighteen-Inch is separated from the Two-and-a-Half or Lower Pentre Seam by 32 to 42 ft of mudstone with a little sandstone near the top. In the Garw Valley only 9 to 14 ft of beds occur and the thinning continues westwards so that at International and over much of the Llynfi Valley less than 5 ft of seatearth with ironstone bands separate the two seams. To the south of Maesteg the separation again increases to 8 ft under Garn Wen and 18 ft at Garth where ironstone bands and nodules are prominent in the mudstones. The **Two-and-a-Half** Seam has a complicated section at Penllwyngwent: top coal 3 in, clod 3 in, coal 14 in, coal and rashings 5 in, coal 6 in, rashings ½ in, coal 2 in. To the north at Western this is reduced to a two coal section similar to that recorded over much of the Rhondda Valley: top coal 19 in, stone 2 in, coal 10 in. Elsewhere throughout the area the seam is recorded as a single coal 18 to 30 in thick, there being a tendency for the section to thicken towards the south. It has been worked fairly extensively from Caerau Colliery, from numerous levels on the hillsides east of Nantyffyllon, and on a smaller scale from levels west of Oakwood and near Duffryn Madoc. Spoil from these old workings has yielded a rich fauna of non-marine lamellibranchs especially from levels south-east of Coegnant [85719300], from No. 9 Level [86339189] east-north-east of Maesteg Railway Station, from trials [85709077] W. of Oakwood, and from Ton-Hir Colliery [93759217] on the watershed west of Maesteg. Despite the occurrence of these fossils in spoil from Two-and-a-half workings it is more likely that they occur in the roof of the Eighteen-Inch close below, for, where seen at Caerau Colliery, only plants were noted in the roof of the Two-and-a-half. Moreover the shelly mudstones carry many roots, and near Oakwood where small diggings occur on both seams, the fauna appears to be restricted to the lower line of levels on the Eighteen-Inch. Additional support is found in the occurrence of an identical fauna on the Eighteen-Inch in the Rhondda Fawr (p. 155). Forms collected in the Llynfi Valley during the present survey include many variants of *Anthraconaia hindi*, including elongate forms approaching *A.* cf. *pruvosti*, *Naiadites hindi*, *N.* sp. cf. *productus*.

Some 15 to 45 ft of mudstones with ironstone layers and nodules separate the Two-and-a-half from the overlying Upper Yard or Pentre Seam. Sandstone bands occur locally in the thicker sections, for example, 10 ft at Caerau, and 21 ft at Celtic.

The **Upper Yard** has a very variable section. A three coal section is recorded at Wyndham and Western: top coal 16 to 18 in, dirt ½ in, coal 7 to 8 in, stone 1 to 7 in, coal 2 to 12 in. In the Garw Valley the seam has been worked from Glengarw Colliery, New Braichycymmer Drift and from Waun Bant and Dundee levels: the crop runs through the east side of Pontycymmer where levels occur on it behind Victoria Street, and in Cwmgelligron near the southern end of Adare Street. In the Pontycymmer area the two top coals of the Ogmore section appear to have united and a seam similar to that in the Rhondda Valley results: top coal 20 to 22 in, stone 1 to 2 in, coal 11 to 14 in. To the west and north a new split develops a few inches from the top of the upper leaf giving a section which persists in the northern part of the Garw Valley and throughout the Maesteg area: top coal 1 to 4 in, stone, fireclay or rashings 1 to 3 in, coal 14 to 18 in, stone 2 to 3 in, coal 6 to 16 in. In the Maesteg area the seam has been worked in several small levels surrounding the town and from Celtic Colliery.

In the Ogmore Valley the Yard Seam is succeeded by 24 to 40 ft of measures, predominantly mudstone, though some sandstone is recorded in the lower part. These thicken westwards to about 60 ft in the Garw Valley, and sandstone is still subordinate. In the Llynfi Valley the seam is overlain more or less directly by sandstone and for this reason it was once referred to as the Rock Vein. This sandstone is 70 ft thick at Caerau, and almost 90 ft at Celtic, but it thins to only 30 or 40 ft east of Coegnant. It has been quarried west of Oakwood [85809066], on the hillside east of Nantyffyllon [85869236], and alongside Nant Gwyn Bach [85949405] near Caerau. At each of these localities 20 to 30 ft of massive, false-bedded sandstone, closely resembling pennant, can be seen. The sandstone is overlain by 25 to 30 ft of mudstones near the base of which the impersistent **Lower Blackband Ironstone**, 15 in thick, has been worked around the sides of Mynydd Bach.

The **Victoria** or Pentre Rider has been worked extensively in the northern part of the Garw Valley, from Glengarw and International collieries, and from various levels at Darran Fawr; farther south it has been wrought at the New Braichycymmer Drift and from Gelligron and Oakwood levels on the east side of Pontycymmer. It has also been worked on the east side of the Llynfi Valley and from Caerau and St. John's collieries. The seam consists almost everywhere of a single coal, 16 to 20 in thick on the east side of the Ogmore Valley, thickening westwards to 28 to 35 in in the Garw Valley. In the Llynfi the coal is thickest in the St. John's area where it reaches 39 in, whence it thins both northwards and southwards to about 2 ft. The roof has been examined at Glengarw where it consists of blocky laminated mudstones with numerous small round nodules of ironstone: foraminifera are common in the lower 12 to 15 in, and abundant *Planolites ophthalmoides* occur in the beds above. Spoil from workings on the western slopes of Mynydd Bach [86229343; 86269330] has also yielded foraminifera including *Ammonema sp.*, *Glomospira sp.* and *Tolypammina sp.*

The strata above the Victoria floor much of the Garw Valley between Pontycymmer and Blaengarw, and they have extensive outcrops in the Llynfi Valley. Few good sections exist, though the positions of some of the seams are marked here and there by levels, and sandstones crop out in many of the streams or are indicated by features along the valley sides.

The cyclothem overlying the Victoria thickens from 40 ft at Penllwyngwent to 80 ft in the Garw Valley, and as much as 100 ft in the Llynfi. Its lower part is always argillaceous, while sandstones predominate above. In the Llynfi Valley the latter reach 40 ft.

The **Abergorky** Seam of the Rhondda Valley appears to be represented in this area by a single seam only at International Colliery. Elsewhere the characteristic two coals are always distinct and separate, the lower one being known as the **Albert.** The separation between the two coals varies from 10 to 20 ft in the Ogmore Valley, to 30 ft in the Llynfi Valley, while on the west side of the Garw Valley as little as 1½ to 4 ft intervene at International and New Braichycymmer collieries. The Albert varies from 15 to 22 in; the upper coal is usually 14 to 19 in, though it appears to be somewhat thinner in the Llynfi area. The latter presumably carries the '*Estheria*' Band in its roof, but it has not been possible to confirm this during the present survey.

The Lower Cwmgorse Marine Band has nowhere been proved in the Ogmore and Garw valleys; the coal beneath its presumed position lies 50 to 55 ft above the top coal of the Abergorky, the separating measures consisting of clift with sandstone bands in the upper part. In the Llynfi the top Abergorky coal is succeeded by 25 to 40 ft of mudstone overlain in turn by 10 to 20 ft of sandstone. Then follows a combined coal and ironstone horizon, usually referred to as the **Upper Blackband,** though in some old records it is called the Middle Blackband. In general there are two coals, the lower 6 to 9 in thick, and the upper 12 to 15 in. They are usually separated by 3 to 9 ft of mudstone, within which occurs the blackband 18 to 30 in thick. The ironstone was formerly dug in deep trenches following the outcrop, and it was mined from several levels along Mynydd Bach, at Garn Wen, and on the western slopes of Garth Hill. North and east of St. John's Colliery it has yielded ostracods, including *Geisina? subarcuata.* Spoil from the diggings on this horizon also includes ferruginous mudstone containing 'mussels' including *Naiadites* cf. *daviesi* and cf. *Anthraconauta phillipsii*, together with *G.? subarcuata.* These have been noted at localities [88429171; 87969105], and on the north-east slope of Garth Hill [87569052]; ostracods were also present in spoil west of Mynydd Bach [85969273]. The precise horizon of this fauna is not known, but it is probably derived from the immediate roof of the lower coal. Spoil from the Mynydd Bach locality mentioned above also yields foraminifera. Mine plans show no connexion between these workings and those of the Victoria below, and the foraminifera probably indicate the presence of the Lower Cwmgorse Marine Band, presumably in the roof of the upper of the two coals referred to above. The cyclothem above the Blackband horizon is everywhere about 100 ft thick, and contains a 20- to 35-ft sandstone in the middle.

The **Hafod** appears to be split into two distinct seams throughout the area. The lower coal is 18 in thick in the Ogmore Valley, but no details are known concerning it in the Garw. In the Llynfi Valley it is called the **Clay** Seam, and is a single coal 18 to 27 in thick: it has been worked from levels [86959025; 86829003] on Garn Wen and near Celtic Colliery, and there are small trials on it east of Coegnant [86979355; 86969335; 86778979]. The coal rests on a thick seatearth, which at Garn Wen is up to

6 ft thick and contains local stone partings; the detailed sections extant suggest that it was once worked as a source of fireclay.

The top split of the Hafod together with the **Upper Cwmgorse Marine Band** in its roof are exposed in the bottom of the Ogmore Valley at the northern end of Nant-y-moel, the following section being measured in the stream bed immediately north-east of Llywelyn Street [93279338]:

	Ft	In
Brittle dark blue-grey shales with small ironstone nodules; *Lingula mytilloides, Orbiculoidea nitida*; *Rhabdoderma sp.*	6	0
Dark shale with plant debris		3
Bast		2
COAL, said to be	1	6
Hard seatearth	3	0
Sandstone		8
Strong silty mudstone		6+

The marine shales are also seen in the side of the stream 130 yd south-east of Nant-y-moel Farm [93209320].

A thin seam recorded in the Cymmer Tunnel [85549477] as 13 in of dirty coal, overlain by 9 in of rashings, represents the same top split of the Hafod. Throughout the Llynfi Valley it lies some 20 to 50 ft above the Clay Seam, the intervening measures consisting of mudstone with a persistent sandstone 5 to 10 ft thick at the top. The two coals appear to be closest under Garn Wen and farthest apart in Nant Gwyn Bach [86979328], east of Coegnant. At the latter locality and also in Nant Cwmdu [87949274], north of St. John's, small trial levels have been dug in the rider coal and the spoil yields marine fossils. The locality at Nant Gwyn Bach was first noted by Evans and Simpson (1934, p. 463) who recorded: '*L. squamiformis, Orbiculoidea sp.*; *Pterinopecten papyraceus*; *Gastrioceras sp.*; *Productus sp.*; *Loxomena sp.*, lamellibranchs'. The following forms were collected during the present survey: echinoid tooth; *Crurithyris?*; *Dunbarella sp.*; and large ostracods. In Nant Cwmdu, spoil yielded *L. mytilloides*, *O. nitida* and *Productus (Dictyoclostus) craigmarkensis*. A.W.W., W.B.E.

AVAN VALLEY AND BRYN AREA

The measures of this group crop out on the north side of the Moel Gilau Fault in the area between the Llynfi Valley and Cwmavon. Although most of the coals have been worked in the past, little active mining is going on at present. Over the rest of the district knowledge is largely derived from the shaft sections of the principal collieries.

The strata from the Cefn Coed Marine Band to the Caedavid thicken from 70 ft at Glyncorrwg to 90 ft at Duffryn Rhondda and this thickness is probably maintained south-westwards. Sandstones are developed only in the uppermost part of the cyclothem; they thicken towards Cwmavon where up to 40 ft of quartzitic rock of 'Cockshot' type can be observed [79259290].

The **Caedavid** is split throughout the area—the two seams representing the Gorllwyn and Gorllwyn Rider of the upper Rhondda. The lower coal has been worked near Bryn [82359241] as the **Cockshot** Seam and is 2 ft thick. A seam at Avon Colliery, consisting of 26 to 33 in of coal in three leaves, is recorded at this horizon. North-westwards it deteriorates steadily: at Nantewlaeth a double coal 17 in thick is underlain by 10 in of coal and rashings; at Glyncorrwg 10 in of coal rest on 10 in of rashings; and at Glyncastle only a seatearth is recorded. To the west of Bryn the position is obscure but it seems likely that the Upper Cockshot is represented by two or three thin coals at Cwmavon. The upper split of the Caedavid, some 10 to 20 ft above, is usually 9 to 12 in thick, though locally at Cwmavon it is said to attain 2 ft.

The **Eighteen-Inch** overlies 35 to 40 ft of mudstones with sandy beds towards the top. The seam varies from 6 to 16 in, thinning in general to the west and north-west.

At Duffryn Rhondda, and east of Bryn the Eighteen-Inch and the overlying equivalent of the Lower Pentre are separated by only 2 to 5 ft of seatearth. In the north of the area, on the other hand, 35 ft of mudstones with ironstone layers intervene. West of Bryn, records are ambiguous and it is not clear whether the two seams are associated as the Golden or whether this name is applied to the Eighteen-Inch alone.

The Lower Pentre is known as the **Two-and-a-half** at Duffryn Rhondda and the **Golden** at Bryn. It has been worked on a small scale at the former locality, and more extensively along its outcrop from Garn Wen westwards to the Avan Valley, where some of these workings may be in the Eighteen-Inch. Under Garn Wen it is about 30 in, at Duffryn Rhondda 29 in, and at Cefn-y-Bryn Colliery [82359240] it is said to be 33 in. West of Bryn the seam believed to be its equivalent is recorded as varying from 18 to 26 in., in places with a thin bottom coal; to the north at Nantewlaeth and Glyn-corrwg it is only 12 in; and its probable equivalent at Glyncastle is only 2 in. At Duffryn Rhondda the roof is a sandy mudstone, and a rider, 3 to 5 in thick, lies some 3 ft above the main seam. The immediate roof of this rider consists of 8 in of irony silty mudstone containing drifted plants and fragmentary 'mussels': a single solid specimen referred to *Anthraconaia* aff. *hindi* was also found.

Mudstones 30 to 40 ft thick underlie the **Upper Yard** Seam. In the Avon shafts its section is similar to that in the Garw Valley: top coal 4 in, parting 4 in, coal 14 in, stone 2 in, coal 7 in. At Nantewlaeth the thin top coal appears to die out, and farther west and north the seam is apparently represented by a mere 7 in of coal at Duffryn Rhondda and 4 in at Glyncastle.

In the south of the area sections in the measures above the Golden are rare, but two thin coals, about 10 ft apart, are recorded near Cwmavon, the lower lying 10 to 30 ft above the Golden. The lower coal is 18 in thick and has been called the **Little** or **Golden Rider.** The upper coal consists of 9 to 24 in of dirty coal. In this area the Golden Rider may equate with the Two-and-a-half, and the upper with the Upper Yard, though this is by no means certain. It is possible that the seam is split at Glyncorrwg and Glyncastle, but the general deterioration of the seam there makes precise correlation difficult.

The Upper Yard is usually succeeded by sandstone. At Avon this is 8 ft thick, but thickens westwards and south-westwards to 60 ft at Duffryn Rhondda, and at outcrop forms a marked feature between Maesteg and Cwmavon. Along this southern tract and at Nantewlaeth a thin coal horizon closely overlies the sandstone, and at Cwmavon is said to be associated with a thin blackband ironstone, which may, therefore, represent the Lower Blackband of Maesteg. At Avon its likely equivalent lies 45 ft above the Upper Yard, but it does not appear to be developed at Duffryn Rhondda.

Mudstones, 15 to 30 ft thick, continue the sequence upwards to the **Victoria** or **Black** Seam, which comprises 18 to 24 in of clean coal. The roof of the seam has not been seen within the confines of the present district, but on the west side of the Avan [78949309], a short distance within the Swansea district, spoil has yielded abundant foraminifera, so confirming the occurrence of the Foraminifera Marine Band at this horizon.

The succeeding cyclothem is 75 to 90 ft thick. Mudstones with numerous ironstone bands pass upwards into sandstones which are 30 ft thick at Duffryn Rhondda, 10 ft at Glyncorrwg, and 4 ft at Glyncastle. The succeeding **Cwmmawr** Seam, lying close above, varies from 2 ft at Bryn to 10 in at Nantewlaeth, and 8 in at Glyncastle. Some 20 to 30 ft above is another thin coal called at Bryn the **Cwmbyr.** The intervening measures are usually argillaceous, though to the west they become sandy. At Cwmavon a thin coal between the Cwmmawr and the Cwmbyr is developed locally. The Cwmbyr varies from 15 to 26 in, though some of this includes dirty coal. It is likely to improve towards the north-west since it appears to be the equivalent of the Red Vein of the anthracite field. Near Bryn [80989260] the roof of the Cwmbyr carries a few inches of dark mudstone containing abundant large *Spirorbis sp.*, *Naiadites melvillei* Trueman and

Weir, *Geisina? subarcuata* and fish remains; but there is no sign of the '*Estheria*' normally characteristic of this horizon. At Glyncorrwg, the Cwmmawr and Cwmbyr are only 3 ft apart, foreshadowing the uniting of the two seams eastwards to form the Abergorky of the Rhondda Fawr.

Above the Cwmbyr 45 to 60 ft of strata continue the sequence upwards to the **Upper Blackband Coal and Ironstone.** The measures are mudstones, frequently passing upwards into siltstones and sandstones which may reach 20 ft. In the north and east, a single coal 6 to 9 in thick occurs: at Nantewlaeth, a rashings horizon 4 ft below, which passes into 9 in of coal at Duffryn Rhondda. Around Bryn the two coals are 5 ft apart, and 18 in of blackband lies in the roof of the lower. This ironstone was formerly dug in trenches between Maesteg and Bryn, and was mined from levels on the east side of the Avan Valley.

The overlying cyclothem is particularly thick: it varies from 75 ft at Glyncastle to as much as 130 ft at Cwmavon. Most of this thickness is made up of mudstone, but there is at least one significant sandstone, 15 to 20 ft thick, lying about 60 ft above the Upper Blackband. The top is marked by an especially thick seatearth, which gives its name to the overlying Clay Seam. No fossils have been seen, but the Lower Cwmgorse Marine Band presumably lies at the base of the cyclothem.

The **Clay** is usually 12 to 15 in thick over the northern part of the area, thickening southwards to 27 in between Bryn and Cwmavon, where it has been worked from numerous small levels. Its roof is a silty mudstone with plant debris. The **Clay Rider** lies above 30 ft of mudstones: it is insignificant in the north, but in the south it comprises two coals, each about 6 in thick, and separated by 3 in to 3 ft of seatearth. It has been seen only at Brynawel Colliery [81989280] where it has been intersected so near its outcrop that the roof is very soft and weathered. Though only doubtful specimens of *Planolites ophthalmoides* and cf. *Hyperammina sp.* were obtained, it is clearly the horizon of the Upper Cwmgorse Marine Band.   W.B.E.

SOUTH CROP

West of the Ogmore Valley: These measures crop out in a band 350 to 400 yd wide lying immediately south of the Pennant scarp, and extending from the general vicinity of the Fishpond at Margam Park through the area between Aberbaiden Colliery and the farm of that name eastwards to Tondu and Pwllandras. Little is seen of the strata at outcrop though several persistent sandstones are marked by well-developed features. The only reliable information for the sequence as a whole is contained in the records of the Margam Park Boreholes (see pp. 343–6), using No. 4 for the beds between the Cefn Coed Marine Band and the Pentre Rider and the No. 3 Borehole for those above; the correlation of the Pentre Rider horizon is certain since the Foraminifera Marine Band was clearly recognizable in both holes.

	Thickness		Depth[1]	
	Ft	In	Ft	In
Upper Cwmgorse Marine Band:				
Dark blue silty micaceous mudstone; *Rhabdoderma sp.*		10		10
Blue-grey smooth shaly mudstone: *Planolites ophthal-moides*; *Lingula mytilloides, Orbiculoidea sp.*; *Curvirimula sp.* (at top), *Nuculana?, Palaeoneilo sp.*; *Anthracoceras?*; *Megalichthys?, Rhabdoderma sp.* ...	13	1	13	11
Grey silty micaceous mudstone; plant debris	9	10	23	9

---

[1] below top of Middle Coal Measures.

	Thickness		Depth	
	Ft	In	Ft	In
Grey shaly mudstone with carbonaceous layers and coal streaks; abundant plant debris	1	8½	25	5½
COAL		½	25	6
Grey shaly mudstone; plant remains	2	3	27	9
Dark blue-grey shale; *Spirorbis sp.*; *Naiadites hindi*; *Carbonita?*		6	28	3
Grey silty mudstone with a few layers of pale siltstone; plant debris	11	4	39	7
Grey mudstone with coal streak in middle; plant debris	11	8	51	3
COAL		2	51	5
Blocky grey mudstone-seatearth, passing down into strong grey mudstone with silty layers	36	10	88	3
Blocky blue-grey smooth mudstone with sporadic ironstone bands; *Naiadites?* near base	15	5	103	8
Rashings		6	104	2
HAFOD (? lower split): **coal 24 in**	2	0	106	2
Pale grey mudstone-seatearth, with sphaerosiderite ..	3	5	109	7
Strong pale grey sandstone, coarsening downwards ..	16	0	125	7
Grey mudstone, silty in parts; plant debris	10	4	135	11
Quartzitic sandstone and siltstone; plant debris ..	7	5	143	4
Grey mudstone silty in upper part; numerous thin ironstone bands; *P. ophthalmoides* near base	12	11	156	3
Siltstone with ironstone nodules and roots		7	156	10
Grey mudstone with plant debris and sporadic roots, passing down into striped beds..	16	9	173	7
Grey mudstone; worm tracks; '*Estheria*' *sp.*; *Hollinella sp.*..	2	9	176	4
Grey silty mudstone; plant debris	3	3	179	7
Grey mudstone, silty in parts; *Naiadites sp.*; *Geisina? subarcuata*; *P. ophthalmoides* at base	6	6	186	1
Pale grey silty mudstone with quartzitic laminae ..	23	6	209	7
Lower Cwmgorse Marine Band:				
Grey silty mudstone with quartzitic layers; *L. mytilloides*; *Hollinella sp.*	17	10	227	5
Grey mudstone; *P. ophthalmoides*	5	8	233	1
Dark grey silty mudstone with ironstone layers; *Myalina?*; '*Estheria*' *sp.*	8	7	241	8
Dark blue-grey shale; fish including Palaeoniscid scales		3½	241	11½
COAL		10	242	9½
Grey mudstone-seatearth, carbonaceous downwards; sphaerosiderite	4	11½	247	9
Dark carbonaceous shale: *G.? subarcuata*	1	8	249	5
COAL		6	249	11
Mudstone-seatearth, carbonaceous at top	5	6	255	5
Dark blue-grey shale, cannelly at base; *Naiadites sp.*; *G.? subarcuata*	6	9	262	2
Grey silty mudstone with ironstone layers; plant debris	19	6	281	8
Grey smooth mudstone with ironstone layers and nodules; worm tracks; *Naiadites sp.*; '*Estheria*' *sp.*	22	4	304	0
Dark mudstone, smooth and carbonaceous at top, grey towards base; scattered plant debris	5	0	309	0
ABERGORKY (Top split): **coal 23 in**	1	11	310	11
Grey mudstone; '*Estheria*' *sp.*; fish debris	9	1	320	0

	Thickness		Depth	
	Ft	In	Ft	In
ABERGORKY (Top split): **coal and rashings 16 in** ..	1	4	321	4
Black carbonaceous and cannelly shale and mudstone; roots and other plant debris	4	10½	326	2½
COAL		½	326	3
Grey mudstone-seatearth, passing down into grey silty mudstone, interbanded with grey siltstone and sandstone	17	11	344	2
Grey smooth mudstone; *Naiadites sp.*	3	2	347	4
ABERGORKY (Bottom split): **coal 17 in**	1	5	348	9
Grey silty mudstone-seatearth, passing into silty mudstone and mudstone	11	10	360	7
Five Roads Marine Band:				
Massive grey silty mudstone, striped with siltstone layers; *Myalina sp.*	5	6	366	1
Grey silty mudstone, interbanded with siltstone and fine-grained sandstone	29	4	395	5
Blocky grey mudstone; scattered plant debris	39	4	434	9
Foraminifera Marine Band:				
Blue-grey smooth mudstone; *Curvirimula sp.* (at top); abundant *P. ophthalmoides* throughout	6	0	440	9
Blue-grey smooth mudstone with many small ironstone nodules; *P. ophthalmoides* throughout; foraminifera including *Ammonema sp.* and *Glomospira sp.* in bottom 3 ft	4	7	445	4
PENTRE RIDER: **coal 17 in**	1	5	446	9
Blocky grey mudstone-seatearth, passing into grey mudstone, seatearth-like, with sphaerosiderite	21	3	468	0
Black mudstone rashings with associated sheared grey mudstone-seatearth; coal streaks in lower part; ?horizon of PENTRE	19	11	487	11
Grey mudstone-seatearth with irregular ironstone nodules		9	488	8
Pale grey fine-grained sandstone	7	2	495	10
Grey mudstone; scattered plant debris	9	5	505	3
Carbonaceous mudstone and rashings		6	505	9
COAL		4	506	1
Grey mudstone-seatearth with irregular ironstone nodules; passing into grey mudstone with plant debris	21	0	527	1
?LOWER PENTRE: **coal and rashings 24 in**	2	0	529	1
Grey mudstone-seatearth passing into grey mudstone, silty in part; a few ironstone bands and nodules ..	21	6	550	7
?EIGHTEEN-INCH: **coal 22 in**	1	10	552	5
Grey silty seatearth with sphaerosiderite, passing into blocky grey mudstone, silty in part; scattered plant debris; *?Anthraconaia adamsi* at base	16	6	568	11
?GORLLWYN RIDER: **coal 15 in**	1	3	570	2
Grey mudstone-seatearth	2	3	572	5
Massive pale grey sandstone	1	5	573	10
Grey mudstone; plant debris	7	2	581	0
Pale grey sandstone, siltstone and striped beds	12	2	593	2
Grey mudstone with ironstone bands and layers ..	22	4	615	6
Dark carbonaceous shale and mudstone, ferruginous at base	4	5½	619	11½
?GORLLWYN: **coal 26 in**	2	2	622	1½

			Thickness		Depth	
			Ft	In	Ft	In
Mudstone-seatearth, carbonaceous at top	..	..	3	$3\frac{1}{2}$	625	5
Grey mudstone; plant debris	..	..	14	5	639	10
Pale grey sandstone with mudstone band	..	..	7	10	647	8
Buff-grey soapy mudstone-seatearth with sphaerosiderite			3	2	650	10
Massive fine-grained sandstone with carbonaceous part-						
ings and coal streaks in bottom 4 ft ..	..	..	32	10	683	8
Grey to dark grey mudstone with ironstone bands;						
'Estheria' sp. ..	..	..	8	6	692	2
COAL, inferior ..	..	..		1	692	3
Pale grey fine-grained sandstone ..	..	..	7	9	700	0
Grey mudstone, silty in parts, with numerous ironstone						
bands and nodules ..	..	..	104	2	804	2
Cefn Coed Marine Band ..	..	..	4	1	808	3

In Margam Park No. 1 Borehole only 80 ft separates the Cefn Coed Marine Band from the Gorllwyn, and the thick sandstone in the upper part of the cyclothem in No. 4 is scarcely developed. In No. 2 Borehole the strata between the Britannic Marine Band and the Pentre Rider, about 200 ft above, were very disturbed and no horizons could be identified with certainty. The Foraminifera Marine Band was well developed; *Lingula sp.* was associated with the foraminifera and abundant *P. ophthalmoides*, and ostracods including *Geisina?* and *Hollinella sp.* were also present in the top of the band. The Five Roads Marine Band was represented by a single specimen of *Myalina compressa* Hind in pale grey striped siltstone about 12 ft below the presumed bottom coal of the Abergorky. Further disturbance affected the strata above this level, but the Lower Cwmgorse Marine Band was indicated by the occurrence of Nuculids 40 to 50 ft beneath the Hafod.

The **Hafod** Coal has been worked at Aberbaiden Colliery. The occurrence of the **Upper Cwmgorse Marine Band** about 80 ft above the seam indicates that the worked coal represents only the bottom split of the true Hafod of the Rhondda Valley. During the years 1935–7 nine boreholes, examined by the late A. Templeman, were drilled within $1\frac{1}{2}$ square miles of the Pentre and Aberbaiden Colliery areas (seven of them down from the No. 3 Rhondda and two from a somewhat lower level at the surface). All proved the Upper Cwmgorse Marine Band, with recorded thicknesses varying from 16 to 42 ft. A typical section (here abridged) is that proved in No. 7 Borehole [83788609]:

						Ft	In
Dark grey shales; non-marine lamellibranchs and ostracods ..	..	..	2	0
Dark grey shales; marine fossils and ironstone bands ..	..	..	32	0
Grey to dark grey shales; poor plants ..	..	..	..	..	29	0
Grey shale; *Spirorbis sp.*; *Naiadites sp.* ..	..	..	..	..		6
Grey silty seatearth ..	..	..	..	..	..	4	6
Strong grey sandy shale and shaly sandstone ..	..	..	27	6
Grey shale with ironstone layers and nodules; *Naiadites?* at base	..	15	2
HAFOD (bottom split): **coal 34 in** ..	..	..	..	..	2	10

The fauna of the marine band is abundant and varied, and includes goniatites typical of the horizon. The following forms were collected during the present survey from man-holes on the main cross-measures drift to the Hafod workings: foraminifera; Archaeocidarid spine; *Crurithyris?*, *Lingula spp.* including *L. squamiformis*, *Orbiculoidea* cf. *nitida*, Orthotetid fragments; *Bucanopsis?*, *Coleolus sp.*, *Euphemites sp.*, turreted gastropods; *Aviculopecten* aff. *delepinei*, *Curvirimula sp.*, *Dunbarella sp.*, *Edmondia?*, *Myalina* cf. *compressa*, *Nuculana* cf. *stilla*, *Palaeoneilo sp. nov.*, *Solemya* cf. *primaeva* (Phillips), *Gen. et sp. nov.* aff. *Solenomorpha*; orthocone nautiloid, *Huanghoceras postcostatum, Anthracoceras cambriense, Homoceratoides kitchini; Geisina? subarcuata; Paraparchites sp.*; fish remains including *Rhadinichthys sp.* Immediately overlying the marine band a non-marine mudstone carried *Spirorbis sp., Anthraconaia?, Naiadites* cf. *melvillei* and *Geisina? subarcuata.* The following fossils, additional to those enumerated above, are taken from the lists of boreholes 2 to 9: *Conularia sp.*; crinoid columnals; *L. mytilloides, Productus (P.) carbonarius; Bucanopsis? navicula* (J. de C. Sowerby), *Coleolus carbonarius flenuensis* Demanet, *Euphemites anthracinus, Platyconcha sp.*; cf. *Anthraconaia pruvosti, Anthraconeilo?*, cf. *Edmondia sulcata* (Phillips), *E.* cf. *transversa* (Hind), cf. *'Nucula' wewokana* Girty, *Nuculana* cf. *acuta, Posidonia sp., Schizodus sp.*; *Coelogasteroceras dubium* (Bisat); *Hollinella* cf. *bassleri.* The roof of the Hafod carried poor specimens of *Naiadites sp.* and three other horizons with *Naiadites sp.* were recorded at intervals between the coal and the base of the marine band. In addition a well-marked plant horizon occurs about 50 to 60 ft above the coal, the following species being recorded in the boreholes: *Annularia radiata, Calamites undulatus, Cyclopteris sp., Lonchopteris bricei, Mariopteris muricata, M. nervosa, Neuropteris gigantea, N. heterophylla,* cf. *N. rarinervis* Bunbury, *N. scheuchzeri* Hoffmann, *N. tenuifolia, Sphenophyllum cuneifolium, Sphenopteris striata.*

Several coals have been proved over the years by pitting in the area to the south of Pentre and Aberbaiden collieries. They cannot now be identified with certainty, and the records contribute little to the knowledge of the sequence.

In Cwm gwyneu Colliery, to the south of the Moel Gilau Fault and about 3 miles north of Margam, the measures between the Cwmmawr (or bottom Abergorky split) and the top of the group have been proved in two boreholes, details of which are combined to provide the following section:

	Thickness Ft	Thickness In	Depth Ft	Depth In
Presumed **Upper Cwmgorse Marine Band**: Shale with coal 'joints'	3	5	3	5
Fireclay passing into shale with ironstone bands ..	51	2	54	7
?CLAY: **coal 39 in**	3	3	57	10
Fireclay passing into shale with sandstone bands in middle; **?Lower Cwmgorse Marine Band** at base ..	126	10	184	8
?UPPER BLACKBAND: **coal 7 in**		7	185	3
Fireclay passing down into shale	72	6	257	9
?CWMBYR: **coal 24 in**	2	0	259	9
Fireclay and shale	8	4	268	1
?CWMMAWR: **coal 27 in**	2	3	270	4
Fireclay	10	6	280	10

Apart from general thickening this sequence is similar to that proved around Cwmavon.

East of the Ogmore Valley: The strata from the Pentre Seam up to the Llynfi Rock is well exposed in the quarry [918844] adjoining Bryncethin Brickworks. In the following section, measured in 1950, the beds above the Upper Cwmgorse Marine Band are included for completeness.

	Ft	In
Llynfi Rock:		
Massive irregularly bedded, pennant-type sandstone	70	0
Sandstone with coal pebbles and irregular layers of detrital coal ..	5	0
TORMYNYDD: rashings and clay 3 in, **coal 8 in,** rashings 4 in ..	1	3
Mudstone-seatearth, carbonaceous at top, passing down into grey shale	15	5
WHITE: rashings and clay 12 in, **coal 22 in**	2	10
Seatearth	1	4
Grey silty mudstone with a few sandstone layers and ironstone nodules	20	2
Gap—presumed grey mudstone	17	0
Blocky grey shale	1	6

Ft In

Upper Cwmgorse Marine Band: lavender grey micaceous mudstone, weathering to soft clay: *Lingula mytilloides, Orbiculoidea nitida, Productus* (*Dictyoclostus*) *craigmarkensis*; *Dunbarella sp.*; fish debris including *Rhadinichthys sp.* and Acanthodian and Palaeoniscid spines ... 4

HAFOD (upper split): **coal and rashings 5 in,** clay 4 in, **coal 16 in** .. 2 1
Gap—presumably seatearth passing down into silty mudstone and shale 26 0
Thinly bedded sandy siltsone passing into sandstone 3 6
Grey shale 2 6
HAFOD (bottom split): **coal 24 in** 2 0
Mudstone-seatearth with ironstone nodules 4 8
Thinly bedded sandstone passing into grey silty mudstone 2 6
Irregularly bedded sandstone with pebbly layers 3 0
COAL and rashings 1 1½
Hard seatearth passing into irregularly bedded sandstone 9 0
Fine-grained sandstone with layers of siltstone 19 9
Gap—probably grey mudstone 15 0
Blocky grey mudstone with sporadic ironstone nodules 10 0

Lower Cwmgorse Marine Band: grey mudstone with pyrites; *Nuculopsis* cf. *gibbosa* 4
Coal 24 in, seatearth 12 in, **coal 3 in** 3 3
Grey mudstone-seatearth, passing into irregularly bedded sandstone .. 12 0
Grey silty mudstone; sandy layers in upper part 17 0
Grey shale with 2 in rashings at base 6 2
Dark grey shale with poor shells; *Naiadites?* 4
Grey shale with ironstone layers; '*Estheria*' *sp.* 9 5
ABERGORKY: **coal 32 in** 2 8
Seatearth passing into sandstone 4 0
Thinly bedded sandy mudstone with irregular layers of fine-grained sandstone 20 0
Blue-grey well-bedded mudstone 16 0

Foraminifera Marine Band: blue-grey mudstone with *P. ophthalmoides* in top 3½ ft and foraminifera in bottom 6 in 4 0
PENTRE RIDER: **coal 18 in** 1 6
Pale seatearth, carbonaceous at top 5 6
Grey silty mudstone with thin layers of fine-grained sandstone .. 20 0
Black rashy shale 7

PENTRE: **inferior coal and dirt 4 in,** banded seatearth with coal streaks 12 in, **coal with two thin rashings bands 33½ in** 4 1½
Seatearth .. .. 1 0+

The crop of the measures continues eastwards from the vicinity of the brick-pit to Heol-y-Cyw and the area south of Wern-tarw, to Brynna, and Dolau immediately to the south of Llanharan village. The outcrop is repeated to the south by the Llanharan Thrust in a tract extending eastwards and south-eastwards from the northern part of Hirwaun Common south-east of Heol-y-Cyw, through Brynnau Gwynion and the western part of Gwaun Llanharry to Coed Trecastell. Throughout these areas the crops of several of the upper coals are maked by the remains of old adits from which they were formerly worked. The section proved in the shaft at the old Tyn-y-Waun Colliery [93188445] is shown in Plate IV. The **Gorllwyn** is represented by 12 in of coal about 55 ft above the presumed position of the Cefn Coed Marine Band. The other seams can be readily identified in a normal sequence apart from the Eighteen-Inch. Four coals (4 in, 12 in, 6 in, and 18 in) spread over 34 ft of strata appear in the position of this seam, and repetition by faulting cannot be ruled out. The **Pentre** was called the **Lantern** and this name persisted throughout the Heol-y-Cyw–Wern-tarw area. The **Abergorky** was identified as the **'Wern-tarw Seam'**, and the **Hafod** is again split into two coals 34 ft apart.

In the Heol-y-Cyw area the **Pentre** Seam and the **Hafod** have both been worked, the former extensively, and the seams can be fixed at the surface by reference to several old adit mouths. The most important of these is the Raglan Colliery No. 4 Drift [95018422] driven on the Pentre or Lantern Seam. Other openings in this seam were located at Drefach Colliery [94268420], 850 yd. W of Raglan Colliery, and 550 yd to the east [95518420], and numerous bell-pits were located near the crop elsewhere. More recently the Pentre Seam has been located by drilling for opencast coal, the proved crop extending from 600 yd S.W. of Tyn-y-Waun Colliery [92648423] to Heol-y-Cyw [94478418]. Old drifts on the Hafod Seam are to be seen 350 yd N. by E. of Raglan [95098448], 600 yd E.N.E. of the same colliery [95508430] and about 100 yd S.E. of Wern-tarw Farm [96148432].

At Raglan Colliery the **Pentre** Seam consists of: top coal 3 in, clod 2 in, coal 17 in, plane parting, coal 37 in, fireclay 6 in, coal 9 in, clod 2 in and coal 7 in; the last two coals of this section probably represent the **Lower Pentre**, the two seams being close together as at Llanharan Colliery (see below). The **Pentre Rider** Seam, 28 in thick, is separated from the Pentre Seam by 7 to 27 ft of mudstone and seatearth. The **Abergorky** occurs some 50 ft higher in the sequence and the lower split of the **Hafod** some 150 ft higher still. About 30 ft of beds separate the two coals of the Hafod on the west side of the workings but these diminish to no more than 4 ft some 500 yd to the east. The occurrence of marine fossils in the spoil from the adit near Wern-tarw Farm indicates that the two leaves are together at this point.

At Werntarw Colliery these measures were penetrated by the shafts and the strata between the Pentre and the Hafod were examined in a cross-measures roadway during the present survey. The **Gorllwyn** horizon is poorly developed, a single coal 14 in thick being recorded about 100 ft above the Cefn Coed position. The **Eighteen-Inch** and **Lower Pentre** seams are also thin, being recorded in No. 1 Shaft section as 11 and 12 in respectively. The **Pentre**, although subject to wash-outs, has been extensively worked and the average sections reads: top coal 2 in, clod 3 to 5 in, coal 32 to 51 in. The **Pentre Rider**, carrying the **Foraminifera Marine Band** in its roof, is 19 in thick, and is separated from the Pentre by about 40 ft of strata, mostly massive sandstone. The beds separating the Rider from the **Abergorky** Seam also consist mainly of strong sandstones about 70 ft thick. The latter coal, 24 to 28 in thick, is overlain by about 7 ft of grey mudstone crowded with *Lioestheria* cf. *vinti*. The **Lower Cwmgorse Marine Band** is well developed above a 12-in coal 45 ft higher in the sequence, the following forms being identified from grey shaly mudstone up to 3 ft above the coal: foraminifera undetermined; *L. mytilloides*; *Myalina compressa, Nuculopsis gibbosa, Palaeoneilo ?*; and *Diplodus sp.* The **Hafod** or **Wern-tarw** Seam everywhere has a two-leaf section, the two splits of the areas to the west having come together to form a single seam. The upper coal varies from 17 to 24 in and the lower 14 to 18 in. The separation between the two leaves decreases from rather more than 2 ft on the west side of the workings to less than $\frac{1}{2}$ in on the east side. The seam is overlain by 2 ft of blocky lavender-grey mudstone containing *L. mytilloides, L. squamiformis, Orbiculoidea* cf. *nitida*; cf. *Platyconcha* and '*Estheria*' *sp.* This passes upwards into grey mudstone with *L. squamiformis,* '*Estheria*' *sp.* and *Pleuroplax sp.*; pyritic concretions extend to 6 ft above the seam. This is the **Upper Cwmgorse Marine Band.**

A section similar to that of the Raglan–Wern-tarw area was proved in the Blackmill Borehole [92428557], drilled in 1938; because of a deviation varying from 27° to 43° from the vertical the true inter-seam distances are difficult to determine. *Lingula sp.* was recorded from the Upper Cwmgorse Marine Band and *Myalina* from the horizon of the Lower Cwmgorse Marine Band.

At Llanharan Colliery an excellent section through these measures was examined in the No. 1 Horizon Main East roadway. The Upper Cwmgorse Marine Band was exposed at the bend where the lateral from the South Pit turns into the main cross-measures and the section to the Cefn Coed Marine Band was completed in about 200 yds. The section reads:

	Ft	In
Upper Cwmgorse Marine Band: blocky micaceous mudstone, weathering to yellowish soft clay; *Coleolus?*; *Dunbarella sp.*; *Euphemites sp.*; fish debris		9
HAFOD: coal 23½ in	1	11½
Fireclay passing down into light grey, false-bedded sandstone	90	0
Grey mudstone	3	3
Lower Cwmgorse Marine Band: grey mudstone; foraminifera undetermined; *Myalina compressa, Nuculopsis gibbosa, Palaeoneilo*; fish debris including *Rhabdoderma sp.* and *Rhadinichthys sp.*	3	3
COAL	1	9
Mudstone-seatearth	4	2
Strong grey shaly mudstone	18	0
Grey shale with ironstone bands; *Belinurus arcuatus* Dix and Pringle non Baily; *Lioestheria vinti* in bottom 6 in	6	10
ABERGORKY: coal 19 in	1	7
Mudstone-seatearth	8	2
Strong grey silty mudstone, sandy in parts	115	0
Foraminifera Marine Band: dark grey shaly mudstone with numerous small round ironstone nodules; abundant foraminifera including *Glomospira sp., Glomospirella?* and *Hyperammina sp.* near base	5	6
THICK SEAM = PENTRE RIDER, PENTRE and LOWER PENTRE: **coal 32 in**, parting 1 in, **coal 11 in**, pan 1 in, **coal 21 in**, clod 9 in, **coal 11 in**, clod 3 in, **coal 24 in**	9	5
Mudstone-seatearth	8	2
Grey sandy mudstone with ironstone nodules	30	6
EIGHTEEN-INCH: coal 15 in	1	3
Mudstone-seatearth	7	0
GORLLWYN: coal 16 in, clod 2 in, **coal 11 in**	2	5
Seatearth with ironstone nodules	5	0
Strong grey mudstone with **Cefn Coed Marine Band** at base	42	3

On the north side of the colliery the Thick Seam splits into its component coals and development has taken place along the Pentre Seam on the upper limb of the Llanharan Thrust.

Between Wern-tarw and the Ely Valley the measures are largely concealed by drift. North of the Llanharan Thrust the strata are completely obscured, and the crops of coals can only be determined by reference to underground workings and the surface positions of the higher coals. South of the Thrust, on the northern margins of the Bridgend district, the Thick Seam is marked by a belt of bell-pits, 50 to 100 yds wide, which line the crop of the seam extending from a point on Mynydd Hywel-Dio [99368205] through Brynawel and across Coed Trecastell to a point near the railway at Woodland Terrace [02608204]. The presumed Abergorky Seam crops in the stream about 150 yd W. of Cwm Graian Farm [01038193] while the Hafod was probably worked from bell-pits 150 yd N.E. of the farm [01288202]. Grey shales with *Lingula mytilloides* obtained from loose material alongside the stream 300 yd S.S.W. of Hendre-Owen [01638208] are thought to mark the proximity of an outcrop of the Upper Cwmgorse Marine Band.

The sections at Meiros and South Rhondda are closely similar to those seen at Werntarw. The exact position of the Cefn Coed Marine Band is not recognizable in the shaft section of either colliery: at Meiros it may occur above a thick seatearth about 140 ft above the Two-Feet-Nine. The Gorllwyn does not appear to be present at either colliery, a well-developed sandstone possibly being responsible for its

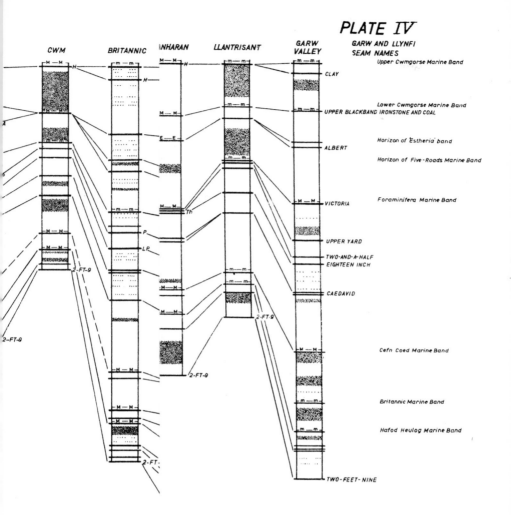

PLATE IV

CWM	BRITANNIC	ANHARAN	LLANTRISANT	GARW VALLEY	GARW AND LLYNFI SEAM NAMES

Upper Cwmgorse Marine Band

CLAY

Lower Cwmgorse Marine Band
UPPER BLACKBAND IRONSTONE AND COAL

Horizon of 'Estheria' band

ALBERT

Horizon of Five-Roads Marine Band

Foraminifera Marine Band

VICTORIA

UPPER YARD

TWO-AND-A-HALF
EIGHTEEN INCH

CAEDAVID

Cefn Coed Marine Band

Britannic Marine Band

Hafod Heulog Marine Band

TWO-FEET-NINE

tone

rth unaccompanied by coal
band proved
ned position of marine band
ia' band proved

the Garw Valley are repeat

two main seam-nomenclatur

ROM TWO-FEET

widespread wash-out. The first coal in this general position appears to be the **Eighteen-Inch,** which varies from 12 to 15 in. The **Lower Pentre,** recorded as 14 in at Meiros and 30 in at South Rhondda, is separated from the Eighteen-Inch by about 20 ft of mudstone. The **Pentre,** 20 to 30 ft above, was worked extensively at both collieries: it varies in thickness from 42 to 48 in. Some 25 to 35 ft, variously described as rock or clift, intervene between the Pentre and Pentre Rider which is 21 in at Meiros and 25 in at South Rhondda. The **Abergorky** Seam again appears to be a single coal lying 50 to 60 ft above the Rider. At Meiros it is recorded as 21 in thick, while at South Rhondda, where it was named the Wern-tarw, a top coal 21 in is separated from a lower leaf 6 in thick by 10 in of clod. The Lower Cwmgorse Marine horizon apparently lies above 23 to 27 in of coal some 25 ft higher in the sequence. At Meiros the **Hafod,** 28 in thick, lies 63 ft above, the intervening measures consisting of sandstone: at South Rhondda the separation has increased to 106 ft, the Hafod itself being 27 in.

Nothing is known of the measures on the South Crop of the Pontypridd district east of the Ely River for they are concealed beneath the alluvium and terrace deposits of the Afon Clun and Nant Myddlyn to the south of Llantrisant and Rhiwsaeson. They are presumably attenuated, for their outcrop is scarcely more than 200 yd wide.

A.W.W.

References

EVANS, W. H. and SIMPSON, B. 1934. The Coal Measures of the Maesteg District, South Wales. *Proc. S. Wales Inst. Eng.*, **49,** 447–75.

HIND, W. 1894–96. A Monograph on *Carbonicola, Anthracomya* and *Naiadites. Palaeontogr. Soc.*

TRUEMAN, A. E. and WEIR, J. 1946–60. A Monograph on British Carboniferous non-marine Lamellibranchia. *Palaeontogr. Soc.*

CHAPTER VI

PENNANT MEASURES: DETAILS

A. LLYNFI BEDS

TAFF–CYNON AREA

WHERE FULLY developed, this group varies from 250 to 310 ft, being thinnest in the east at Merthyr Vale and thickest in the south-west at Lady Windsor. Over most of the Cynon Valley 260 to 270 ft of beds are present. No coals of importance are developed nor have any been named.

Three thin coals spread over 20 to 30 ft occur some 40 to 60 ft above the base. The lowest, 10 to 18 in thick, appears to be the Blackband Coal of the Merthyr area, which probably equates with the White of Maesteg. The two thin riders above may correspond with the Tormynydd and No. 3 Rhondda.

In the upper part of the group two thin coals are usually developed, while at Merthyr Vale a third rider occurs not far beneath the Castell-y-Wiever or No. 2 Rhondda. The lowest of these coals, 17 to 24 in thick, corresponds to the **Taldwyn** of the Merthyr area; 15 to 20 ft above, the **Gilfach** equivalent varies from 13 to 15 in. These two coals may correspond to the Wernpistyll and Wernpistyll Rider of the western parts of the district. At Penrhiwceiber only one 9-in coal is present in the Taldwyn–Gilfach position.

At Lady Windsor the **Blackband** horizon is poorly developed, but in the upper part of the group eight fireclay horizons, associated with thin coals or bands of rashings, are recorded. These thin seams cannot be correlated precisely, but a 21-in coal 190 ft above the Clay is possibly equivalent to the Taldwyn or Wernpistyll.

At Merthyr Vale sandstones are prominent only above the Taldwyn, but in the Mountain Ash area they are well developed in the middle part of the group, associated with the riders of the Blackband, and in the measures immediately above. Throughout this area 'hard white rock', 20 to 40 ft thick, occurs close below the Castell-y-Wiever Coal. South of Mountain Ash the sandy facies becomes less evident while seatearths tend to be especially thick. Seatearths are also prominent at Lady Windsor, where sandstones, though developed at several levels, are not thick, although one occurring midway between the base of the group and the Blackband reaches 25 ft. The Llynfi Beds crop in the bottom of the Cynon Valley to the north of Mountain Ash. They are much obscured by drift and exposures are few. Two quarries on the lower part of the hillside near Aberaman Brickworks give fragmentary sections. The lower one [01840007] exposes strata about 50 ft above the Clay Seam:

	Ft	In
Sandstone	4	0
Silty mudstone and shale	15	0
Sandstone	4	0
Sandy mudstone	4	0
Black coaly shale with plant debris	3	0
COAL		6
Mudstone-seatearth	3	0
Sandstone	4	0

The coal is thought to be one of the thin riders to the Blackband. The higher quarry, situated some 300 yd E.S.E. [02119990], was worked as a source of shale and mudstone for use in the brickworks. The section is as follows:

	Ft	In
Head	5	0
Sandstone		7
Mudstone	1	0
Sandstone		3
Mudstone	4	0
Sandstone passing down into sandy mudstone	5	0
Well-bedded silty mudstone, mudstone and blue-black shale	30	0
Inferior COAL (not seen)	2	0

Old trials on this coal, which may be the Gilfach, are recorded just south of Cwm-du Farm [01539982], 1200 yd S. by W. of Aberaman Church.

Further fragmentary exposures occur on the west side of the valley at Aberaman, particularly in landslip scars in the vicinity of the Gadlys Fault on the north side of Craig Abergwawr. On the east side of the fault [00370122] strata in the lower part of the group are seen:

	Ft	In
Alternations of mudstone, silty mudstone and sandy mudstone	40	0+
Massive sandstone	15	0
Blue-black shale	6	0
COAL	smut	
Seatearth	3	0
Massive sandstone	12	0

This coal would appear to be at about the horizon of the No. 3 Rhondda and it is possible that the lower sandstone is the equivalent of the Llynfi Rock. The sandstone above the coal dies out when traced southwards, but the underlying sandstone is persistent. On the west side of the fault [00340120] the uppermost beds of the group are exposed:

	Ft	In
Massive sandstone	12	0+
NO. 2 RHONDDA: coal	smut	
Seatearth	2	0
Pale grey mudstone with ironstone nodules; thin sandstone near base	8	6
Ganister-like siltstone	7	0
Mudstone	8	0
COAL	smut	
Seatearth	1	6
Grey mudstone with spheroidal weathering	6	0
Hard sandstone	3	0
Strong sandy mudstones	10	0

A.W.W.

RHONDDA FACH AREA

Very few exposures of these rocks occur within this area. The uppermost strata underlie the lower parts of the valley between Mardy No. 3 Pit and the southern end of Ferndale, but they are almost completely concealed beneath thick scree, hillwash

or glacial drift. The only exposures of note occur in the bed and sides of the river
[00109696] immediately west of Ferndale Nos. 1 and 5 pits. Here several feet of dark
ferruginous shale with abundant non-marine lamellibranchs (consisting mainly of
Anthraconauta phillipsii but including also a form identified as cf. *Naiadites sp.*) are to
be seen at stream level, while on the west bank a little to the south a 12-in coal is
overlain by about 15 ft of weathered shale and mudstone. It is likely that these strata
lie close above the Blackband Coal.

Shaft records show that the measures vary in character considerably and they
range in thickness from 300 to 400 ft. They are thinnest at the northern end of the
valley and thickest at Cynllwyn-du (Ferndale Nos. 8 and 9), while in the southern part
of the area they vary from 350 ft at National to 365 ft at Lady Lewis.

Over most of the area the development of coals is so variable that correlation of the
widely spaced sections is difficult. At Mardy No. 1 Pit a coal, 18 in thick, lying 60 ft
above the base of the group, is probably the Blackband. A coal, 15 to 24 in, some
60 to 70 ft higher in the sequence, may represent either the Tormynydd or No. 3
Rhondda. The 260 ft of strata present at the top of the No. 3 Shaft are mainly
argillaceous, though several sandstone bands are prominent in the uppermost
50 ft.

The 150 to 200 ft of measures belonging to the lower part of the group at Ferndale
Nos. 1, 3 and 5 shafts are almost wholly argillaceous; fireclays (associated with
several thin coals) are reported to be especially prominent from 50 to 120 ft above
the Upper Cwmgorse Marine Band position.

Although the two sections available of Ferndale Nos. 6 and 7 shafts are conflicting
and unreliable, they agree in recording a considerable increase in the amount of
sandstone, which is developed fairly evenly throughout the group. Several seatearths
are shown, but only two coals, 8 in and 1 in respectively, which it is impossible to
identify.

At Cynllwyn-du numerous sandstones in beds 10 to 30 ft thick are again developed
throughout the group. The 6 or 7 recorded coals are all thin and difficult to correlate.
No coals are present beneath what appears to be the Llynfi Rock, about 30 ft thick,
but this is overlain by an 11-in coal which may be the No. 3 Rhondda. Several
seatearths with associated thin coals are developed in the middle part of the group.

National and Standard collieries, although less than 700 yds apart, show very
different sections. Numerous sandstones and thin coals are shown fairly evenly
developed through both sections, but they cannot be correlated with any degree of
certainty. At National a well-developed 'mine' above a fireclay 50 ft above the Upper
Cwmgorse position presumably represents the Blackband horizon. A 34-in coal with
4 in of dirt, some 30 ft above, may be the Tormynydd. At Standard this coal appears
to be represented by only 3 in of coal lying beneath 25 ft of sandstone which presumably
represent the Llynfi Rock. Only two coals, 5 in and 12 in respectively, occur above this
horizon at National, but the Standard section shows three well-developed seams in
the upper part of the group. The lowest of these (top coal 21 in, fireclay 2 ft, coal 6 in)
lies 200 ft above the base of the group and is probably the No. 3 Rhondda. About 60 ft
higher 36 in of coal with 6 in of rashings 6 in above the base, may equate with the
Wernpistyll, whilst a further 25 in of coal and rashings occur about 25 ft beneath
the No. 2 Rhondda.

In the Lady Lewis shaft section three coals, 3 to 6 in thick, situated between 60 and
120 ft above the Hafod, are clearly representatives of the White–Tormynydd sequence
farther west. The uppermost is overlain by 32 ft of hard conglomeratic 'rock' and
bedded sanstone; these were originally referred to as the 'Lower Cockshot', but they
clearly equate with the Llynfi Rock. They are in turn overlain by the **No. 3 Rhondda,**
and the $25\frac{1}{2}$ in of coal here recorded thickens towards areas west and south where
the seam is extensively worked. Above this level correlation is not so certain. The
strata consist of about 200 ft of mudstones interspersed with numerous sandstones
up to 15 ft thick and several thin coals. The principal of these (top coal 5 in, clod 1 in,

coal 16 in, clod 36 in, coal 9 in) 60 ft above the No. 3 Rhondda appears to correlate with the seam at Standard which may be the Wernpistyll.

No coals have been worked from the Llynfi Beds within the Rhondda Fach area.

A.W.W.

NORTHERN RHONDDA FAWR AREA

The Llynfi Beds crop out on the lower slopes of the valley north of Blaenrhondda and south-eastwards from Ton Pentre, and form the higher slopes in the intervening anticlinal areas. They are 400 to 450 ft thick, and contain between eight and eleven cyclothems.

The measures below the Tormynydd are 120 to 140 ft thick and have been proved by the shafts at Fernhill Colliery and by Park Upcast. The section of Fernhill No. 2 Pit is:

	Thickness		Depth	
	Ft	In	Ft	In
TORMYNYDD: coal 21 in	1	9	1	9
Underclay and shale	9	6	11	3
Sandstone with fireclay band	28	9	40	0
COAL		1	40	1
Underclay and sandstone	32	9	72	10
COAL		6	73	4
Underclay and shale with ironstone nodules	20	1	93	5
Sandstone	12	0	105	5
Shale; ironstone nodules towards base	26	3	131	8
Sandstone	8	0	139	8
Carbonaceous shale		1	139	9

At outcrop the beds are usually obscured by hill-wash, landslips and scree, but parts of the sequence are exposed locally, notably in Nant Ystrad-ffernol [934991] at Tynewydd, in gullies near Treherbert [932982], in Cwm Parc [941961] and on the hillside west of Ton Pentre [963955]. At all these localities the sequence is broadly similar. In exposures near Ton Pentre [966954] *Naiadites sp.* and *Spirorbis sp.* were collected from mudstones resting immediately on the Upper Cwmgorse Marine Band. These mudstones continue upwards for 50 ft to an impersistent seatearth, which is in turn overlain by 20 to 30 ft of siltstones and sandstones, including a particularly persistent sandstone post at the top. A coal, 3 to 6 in thick, rests on the upper surface of this sandstone. The cyclothem overlying this coal is wholly argillaceous and barely 10 ft thick at Cwm-parc. Northward it is much siltier with, at the base, a ferruginous mudstone, almost a blackband, yielding fragmentary 'mussels', suggesting a possible correlation with the White Seam of Maesteg and the Avan Valley. In the extreme north, at Fernhill [926005], the entire cyclothem is composed of 30 ft of sandstone.

Overlying this cyclothem is another thin coal, in the roof of which sandstones and siltstones are well developed. In Nant Ystrad-ffernol and west of Ton Pentre sandstones up to 20 ft thick are present near the base of the cyclothem; elsewhere the sandstones are thinner and confined to the upper part. These sandy beds are overlain by a thin coal which is separated from the Tormynydd by about 10 ft of sandstone.

Throughout most of the area the **Tormynydd** has been known as the No. 3 Rhondda. It was also worked under the name of Hafod from Church Level [97359562], Ystrad Rhondda, where it is said to be 38 in thick. Nowhere else is it known to exceed 24 in, and at Ynyswen Level [951979] was worked at 22 in. Numerous small trials have been driven on it on the east side of the valley between Treorky and Pentre, to the south of Cwm-parc, west of Treherbert and north-east of Blaen-y-cwm; at all these places the thickness varies from 12 to 21 in.

In the immediate roof of the Tormynydd, or within a few inches of it, a persistent sandstone equates with the **Llynfi Rock**. Its thickness varies considerably. In the three shafts at Fernhill it ranges from 10 to 30 ft. On the east side of the valley southwards to Pentre it rarely exceeds 10 ft, but from Pentre to Ystrad Rhondda it thickens rapidly and in Church Quarry [97319550], where a face of 30 ft has been worked, its total thickness must be about 50 ft. Eastwards it splits up and is replaced by striped beds, and around Bodringallt no significant sandstone appears in this position. On the west side of the valley it is 10 to 15 ft thick as far south as Nant Saerbren, but on the south side of the stream [93209765] it is 50 ft thick; traced southwards and westwards it reaches 70 ft in Cwm Parc. Old quarries for building stone are located on the north side of Mynydd Maendy [95339551], near Cwm Dar [94079526], south of Gelli-goch [94399706], and within the large Treherbert landslip [93209765].

In the south-east, near Bodringallt, the **No. 3 Rhondda** lies close above the Llynfi Rock. It has been worked from Ty-fry levels [97629572], north of Ystrad Rhondda, where it was about 3 ft thick. To the north these workings were limited by a belt of 'thin coal' running N.E. to S.W. a short distance east of Pentre Colliery. This line appears to represent the margin of a regional washout which can be traced to the South Crop. Its width in the Rhondda Fawr is unproved, but the seam has not been recognized to the north.

The measures from the No. 3 to No. 2 Rhondda are 200 to 250 ft thick but poorly exposed. Their lower part can best be seen in the streams that drain the eastern and southern slopes of Mynydd Tyle-coch [93909651; 93509663]. Here about 80 ft of sandy mudstones and striped beds overlie the position of the No. 3 Rhondda. They include two sandstone ribs 50 ft and 80 ft above this horizon. A coal 10 ft above the lower sandstone has the following section on the southern side of the hill: coal 2 in, carbonaceous shale 6 in, coal 2 in. A second thin coal is seen 20 ft above the upper sandstone. Although exposures are bad on the east of the valley two lines of old trials, presumably on these two thin seams, can be seen north-east of Treherbert [94639860; 94659860] and suggest that both coals thicken in this direction. A massive sandstone, up to 50 ft thick, containing numerous conglomeratic layers, lies in the roof of the upper coal and has been quarried extensively on Mynydd Ynys-feio [94879850]. It is probably the same sandstone that closely overlies a thin coal in the landslip above Treherbert [93769728], where 35 ft are exposed, conglomeratic at the base and flaggy towards the top.

The sequence upwards to the No. 2 Rhondda is well exposed in several streams near the head of the valley. All the sections are similar: that in Nant y Gwair, [91079872] is given below, the lowest item apparently being the upper coal of Mynydd Tyle-coch, noted above:

	Thickness Ft	Thickness In	Depth Ft	Depth In
NO. 2 RHONDDA	—	—	—	—
Seatearth and sandy mudstone	5	6	5	6
Silty and massive sandstone	6	0	11	6
Mudstone and sandy mudstone	2	0	13	6
?WERNPISTYLL RIDER: **coal 14 in**	1	2	14	8
Seatearth and mudstone with thin bands of sandstone in middle	8	6	23	2
Sandstone passing into sandy siltstone with thin ironstone bands in lower part	11	0	34	2
Seatearth with nodules of ironstone	6	0	40	2
?WERNPISTYLL: **coal 12 in**	1	0	41	2
Seatearth passing into sandstone	9	6	50	8
Shaly mudstone	5	0	55	8
Silty sandstone with a few mudstone partings	22	6	78	2
Striped beds	12	0	90	2
Flaggy sandstone	4	0	94	2
Mudstone	4	0	98	2
COAL		7	98	9

The presumed **Wernpistyll Seam** lies 40 to 60 ft below the No. 2 Rhondda and the **Wernpistyll Rider** 14 to 30 ft below the same horizon. The measures between them vary lithologically, the lower part tending to become more arenaceous northwards. Neither coal has been worked, with the possible exception of workings on the upper seam north-east of Tynewydd [94329912]. It is, however, possible that these workings are in a lower split of the No. 2 Rhondda.   w.b.e.

SOUTHERN RHONDDA AREA

The thickness of the group varies from about 330 ft to 360 ft in the east and north-east of the area (at Gelli, Llwyn-y-pia, Hopkinstown and Pontypridd) to 460 ft to 480 ft in the western part (at Penrhiwfer and Gilfach Goch). Although the strata occupy extensive tracts at the surface, mainly on the lower slopes of the valleys, they are much covered by drift.

The crop may be followed from the middle and lower slopes of the hillside south of Cwm-parc around the Maendy spur into Cwm Ian above Maindy Colliery. Here the full thickness of the group occupies the ground approximately between the 700 and 1100 ft contours. In Nant Gelli above Eastern pits the measures from the No. 3 Rhondda to the No. 2 Rhondda are largely covered by boulder clay and only sporadic exposures, not capable of precise correlation, are seen in the streams. The No. 3 Seam itself crops at pit-top level at Eastern.

East of the Dinas Fault the higher strata of the group underlie the main valley floor between Gelli and Llwyn-y-pia, which is completely obscured by drift. South of Llwyn-y-pia the beds dip beneath the overlying Rhondda Beds, the uppermost horizons emerging again only in the valley bottom near Naval Colliery. Here the rocks are again obscured by glacial drift and alluvium as they are in the neighbouring Clydach Vale to the west, where the top 200 ft of strata occupy the ground below the 1000 ft contour to the west of the Dinas Fault. Traced downstream from the Naval area in the main valley the group again disappears beneath the Rhondda Beds midway between Dinas and Porth and remains concealed throughout the rest of the valley. The group has two other areas of outcrop. At Gilfach Goch the highest beds occupy the lower slopes of Cwm Ogwr Fach between Scotch Row [97959055] and the point of confluence with Nant Abercerdin [97808900], where they are almost completely covered by boulder clay. The beds also come to the surface in an elongated oval belt, largely faulted on the northern side, flanking the crest of the Pontypridd Anticline, extending from the east side of Mynydd-y-Gilfach in the west, through Nant Cae'r-gwerlas and the area south of Cilely Colliery to the neighbourhood of Glyn [02408890] in the east. The full sequence of beds must crop in the Nant Cae'r-gwerlas area since the Hafod Seam, which lies beneath the Upper Cwmgorse Marine Band at the base of the group, has been located in the stream [00158915] near Caerlan Colliery. A few exposures, mainly of mudstones with a few thin sandstones, occur on the banks of the stream not far above the base of the group, but in the main the measures are obscured by great spreads of stony boulder clay.

In the Ely Valley around Cilely the full sequence is again present at crop. The Llynfi Rock does not appear to be developed, but the **No. 3 Rhondda** is overlain by 30 to 40 ft of pennant-type sandstone, which make a feature easily followed on either side of the valley. A little higher in the sequence a white quartzitic sandstone (referred to locally as the Cwar Gwyn) was formerly quarried at a point [01308955] 200 yd E.S.E. of Cilely Colliery. Immediately behind the No. 1 Pit winding house this white quartzite is faulted against the No. 3 Rhondda Coal. Further exposures of interest are seen in the small stream running approximately along the strike midway between the No. 3 and No. 2 Rhondda seams [01168892], about 750 yd S. of Cilely No. 1 Pit: a $\frac{1}{2}$-in band among bedded mudstones yielded *Anthraconauta* cf. *phillipsii*, *Carbonita* cf. *humilis* and fish remains.

In the Glyn Valley the No. 3 Rhondda emerges in the stream bottom at a point [02308895] 100 yd N.W. of Glyn Farm, brought to the surface in the crest of the Anticline. The seam was formerly worked on a small scale and the sandstone overlying the coal is well seen, dipping away steeply towards the south and more gently towards the north.

Since the outcrop of the beds is so drift-covered detailed knowledge of the group is derived from the numerous shafts within the area, most of which penetrated at least the lower part of the sequence. The measures conveniently fall into two main divisions at the No. 3 Rhondda Coal, the only seam of economic significance in the group. The lower division is thinnest in the north and south-east of the area (being about 140 ft at Gelli and Llwyn-y-pia and 130 to 135 ft at Maritime and Cwm) and thickest in the west (180 ft at Cambrian and 190 to 200 ft in the Penrhiwfer–Britannic area). Two, sometimes three, coal horizons are developed. The lowest, presumably equating with the Blackband or White seams, occurs about 40 to 90 ft above the base of the group, the distance varying with the overall thickness. The **Tormynydd** Seam varies from a few inches to 2 ft, usually of clean coal, and it has been worked from crop on the northern slopes of Maendy and in the upper reaches of Cwm Ian [95309500]. A thin coal is developed between the Blackband and the Tormynydd in the south-eastern parts of the area, while at Gilfach Goch and Lewis Merthyr further thin coals are developed not far beneath the No. 3 Rhondda.

The occurrence of sandstone between the Tormynydd and the No. 3 Rhondda is very variable. Typical Llynfi Rock is developed only in the Maendy–Cwm Ian areas where it forms a readily traceable feature overlying the Tormynydd, and at Dinas, Cymmer and Great Western collieries where sandstones 30 to 50 ft thick are recorded immediately below the No. 3 Rhondda. Elsewhere sandstones are reduced to relatively thin bands, while at Britannic, Penrhiwfer, Coedely and Cwm they are virtually absent.

The **No. 3 Rhondda** Seam was extensively wrought as a source of gas coal throughout the Southern Rhondda Valley in the early days of mining. At Gelli it was worked between the Dinas and Cymmer faults: it is 3 to $3\frac{1}{2}$ ft thick with, locally, two stone partings in the middle. At Clydach Vale extensive working took place to the south-west of the Dinas Fault from drifts driven from the bottom of the valley some 200 to 300 yd S.W. of the church. Blaenclydach Colliery [98029289] was situated on the north bank of the river and Dinas Clydach [98039284] to the south. The workings in the seam, said to be 32 in thick, were limited to the west side by a line of 'complete washout' running approximately N.N.E. to S.S.W. about $\frac{1}{2}$ mile north-west of Cambrian Shafts. Perche's Pit [96449267], 650 yd W. by S. of Cambrian Colliery, connected with the workings from Dinas Clydach. Farther down the main valley Ty-newydd Pit [02349162], 700 yd W.N.W. of Porth Railway Station, intersected the seam at 270 ft and workings were extensive on the north-east side of the Cymmer Fault. To the north-east marked thinning of the coal takes place along a N.W.–S.E. line running beneath the watershed in the neighbourhood of Mynydd Troed-y-rhiw.

Throughout the area of six-inch Sheet Glamorgan 27 S.E. (underlain by the seam everywhere except in the crest of the Anticline in the Nant Cae'r-gwerlas–Cilely area) the coal is almost exhausted; some of the workings were very old and no plans are now known to exist. The coal was wrought from shafts at Llwyncelyn [03429101], at Cymmer Colliery, from Dinas Lower Pit [01289172], at Dinas Isaf [00579033] and from the old Penrhiwfer Pit [99708973]. The seam was also worked from surface drifts on the north side of the axis of the Pontypridd Anticline at Penygraig [99649125], Caerlan [99908920], Cilely [01308950] and Rhiwgarn [01568924] collieries; on the steep southward dipping limb of the Anticline were situated the lower drifts at Caerlan [00258906] and those of Collena [01108896] and Glyn [02538878] collieries. Throughout this area the seam consisted of clean coal 31 to 36 in thick, and the roof was usually a strong micaceous silty mudstone with only a little plant debris.

The seam is exhausted south of the Rhondda River on Sheet 28 S.W., the main workings being those of Coedcae [03759111], Hafod No. 2 [04159128] and Gyfeillon [05369097] pits, while south of the Anticline the workings from Pen-y-rhiw [06408894] extended well on to Sheet 36 N.W. and virtually linked up with those from Glyn Colliery to the west. In most of these areas the seam is 2½ to 3 ft thick, but pronounced thinning is recorded on the south side of the Penrhiw area where a stone or clod develops locally about 6 to 9 in from the top of the seam.

In the Gilfach Goch area the coal was extensively worked from pits near Scotch Row [97969048], from Dinas Main Colliery [98059012] and from drifts in the bottom of the valley [98128970; 98178965] to the S.W. of Britannic. In the same area Glynogwr Colliery [97748907] exploited the seam from cross-measures drifts driven on the south limb of the Anticline. Throughout the Gilfach Goch area the coal varied from 24 to 34 in and was usually overlain by 3 to 7 in of cannel.

At Coedely Colliery the No. 3 Rhondda is intersected at 1133 ft in the Downcast Shaft and the seam was extensively worked to the west, north and east of the shafts. It varies from 12 to 30 in, and up to 12 in of cannel are usually developed in the top of the seam. A limited amount of extraction took place at Cwm Colliery where it was encountered at a depth of 1691 ft; the section varied from 19 to 43 in.

The measures between the No. 3 and No. 2 Rhondda seams, which vary from 190 ft in the south-east to 290 ft in the west of the area, are very variable both in coal and sandstone development. Usually some three to six coals, all thin, are present. In the area from Porth southwards a coal up to 18 in thick, lying about 40 to 60 ft above the No. 3 Rhondda, seems to occupy much the same position as the Rock Fach of the south-west of the district, and another, commonly 14 to 18 in thick, about 120 to 200 ft above the No. 3, may correlate with the Wernpistyll.

At Cambrian, in the west, and Cwm and Coedely, in the south, sandstone is prominent above the No. 3 Rhondda, but elsewhere the seam is overlain by mudstone. In the latter area sandstones are prominent throughout the sequence, but at Britannic, Penrhiwfer and Great Western they are subordinate. Elsewhere sandstones in beds about 10 to 30 ft thick are developed sporadically throughout.

The sequence from the No. 3 to No. 2 Rhondda seams was examined in the main hard heading to the latter seam at Coedely Colliery, the following section being recorded:

	Thickness		Depth	
	Ft	In	Ft	In
Massive sandstone with quartz pebbles, rolled ironstone and irregular coal streaks and 'pebbles'	7	0	7	0
NO. 2 RHONDDA: **coal 48 in**	4	0	11	0
Mudstone-seatearth	3	0	14	0
Striped, fine-grained sandstone passing down into coarse quartzitic grit	17	0	31	0
Strong muddy sandstone in beds about 12 in	5	10	36	10
Grey shale with fragmentary plants		5	37	3
COAL, inferior		2	37	5
Strong seatearth with ironstone nodules	3	0	40	5
Hard, massive, fine-grained striped sandstone	20	0	60	5
Strong silty and sandy mudstone	18	0	78	5
Rashings		1	78	6
Blocky mudstone with ironstone nodules	7	0	85	6
COAL		8	86	2
Hard sandy seatearth	4	0	90	2
Pale quartzitic sandstone passing down into fine-grained sandstone or siltstone	22	0	112	2
Strong silty mudstone	7	0	119	2
Black shale and rashings		5	119	7

	Thickness		Depth	
	Ft	In	Ft	In
Hard siltstone with streaks of coal and small ironstone nodules	3	0	122	7
Fine-grained mudstone	7	0	129	7
COAL (irregular)		2	129	9
Seatearth	3	0	132	9
COAL		7	133	4
Seatearth passing down into massive striped sandstone	9	0	142	4
COAL (irregular) .. .. ½ in to		3	142	7
Sandy seatearth		10	143	5
Striped fine-grained sandstone in thick beds	22	0	165	5
Massive sandy mudstone	7	9	173	2
Soft sulphury shale and sandy mudstone with ironstone nodules	1	11	175	1
Rashings		1	175	2
COAL		10	176	0
Hard seatearth	2	6	178	6
COAL		1	178	7
Hard seatearth	1	6	180	1
Pale quartzitic sandstone passing down into massive sandstone	39	0	219	1
COAL and rashings		6	219	7
Sandy seatearth passing down into massive sandstone with silty beds, coal streaks and 'pebbles' in bottom 6 ft	52	3	271	10
NO. 3 RHONDDA: bast 8 to 10 in, coal 34 in ..	3	7	275	5

A.W.W.

OGMORE–GARW–LLYNFI AREA

No shafts or boreholes penetrate the beds which are fully exposed in all three valleys, though good sections are few. In thickness the group varies from rather less than 500 ft in the north-east of the Ogmore Valley to more than 600 ft to the south and south-west of Maesteg, a steady increase in the Llynfi Rock being mainly responsible.

Virtually nothing is known of the strata below the Llynfi Rock in the Ogmore Valley, and only a few incomplete sections are to be seen in the main Garw Valley. In the Blaengarw area a strong feature marks the crop of a sandstone not far below the White Seam; this may be seen in Cwmgwineu [89369320], 400 yd W. of International Colliery, where 20 ft of sandstone with bands of sandy mudstone may be seen at the site of a waterfall. What appears to be the **White Seam** has been worked on a small scale from levels [89509334] above International, and just south of Darren Fawr [89699254], 700 yd S. of the same colliery. Trial levels at the same horizon also occur [90169118] above the abandoned New Braichycymmer Colliery, while on the south side of Fforch-wen [91479087] a line of crop-workings probably exploited the same coal.

Just south of New Braichycymmer the measures below the Llynfi Rock are exposed on the west side of the Bridgend Road [90409080]:

	Ft	In
Llynfi Rock: massive sandstones	30	0+
Shales, mudstones and silty mudstones with a few sandy bands ..	27	0
?WHITE: coal not seen	—	—
Seatearth passing down into siltstone and sandstone	4	9
Silty mudstone with a few sandstone bands	50	0
?HAFOD (top split): **coal 20 in**	1	8

The mudstones at the base of the section are poorly exposed and the presence of Upper Cwmgorse Marine Band was not confirmed. Similar strata are to be seen in the banks of the Garwfechan stream. The following section occurs at the confluence with Nantcwmdu [89569080]:

	Ft	In
Llynfi Rock	—	—
TORMYNYDD:	smut	
Seatearth, mudstones and shales	18	0
WHITE: **coal** (not well seen) **3 in** + ...		3+
Seatearth, passing down into mudstones	20	0

At the northern end of the Garwfechan exposures of the Llynfi Rock and the beds below are well seen. A section in a side stream entering from the north-east [89059240] was measured as follows:

	Ft	In
Llynfi Rock: strong flaggy and false-bedded sandstone; at the base many impersistent coal streaks and rolled pellets of clay-ironstone ..	20	0+
TORMYNYDD: **coal 1 in**, seatearth 4 in, **coal 6 in**		11
Seatearth passing into grey silty mudstone and blue-grey shale ..	26	0
?WHITE: **coal 12 in**	1	0
Seatearth	2	0
Strong sandy mudstone with thin beds of sandstone	3	0
Blue-grey silty mudstone and shale with bands of blocky darker shale	40	0
COAL	smut	
Hard seatearth passing into grey silty mudstone with sandstone bands	19	6
Mudstone and shales	12	0+

Though no fossils were found it seems likely that the dark shales overlying the lowest coal mark the horizon of the Upper Cwmgorse Marine Band. 450 yd S.S.E. of this locality the presumed White Seam is overlain by 20 ft of silty mudstone with sandstone bands up to 2 ft thick. Beneath lie 12 to 14 ft of sandy mudstones with sandstone layers, passing down into 20 ft of silty mudstone and shale carrying a few poorly preserved *Naiadites?*.

In the Llynfi Valley the Upper Cwmgorse Marine Band is succeeded by mudstones, well exposed in Nantyffyllon north of Maesteg, where they continue up to the White Seam about 80 ft higher in the sequence. On Mynydd Bach [845927] these measures are only about 50 ft thick with 5 ft of silty sandstone directly underlying the White. Farther south, at Garth Hill, they thicken to 90 or 100 ft.

The **White** Seam has been extensively worked under Garn Wen and also beneath Garth Hill. At Garn Wen [84489265] it has a variable but complex section, e.g. top coal 3 in, fireclay 18 in, coal 6 in, parting 3 in, coal 10 in, parting 1½ in, coal 12½ in, fireclay up to 24 in, coal 16 in, parting 1 in, coal 15 in. This section appears to represent both the White and Jonah seams of Cwmavon, the two seams being separated by the 2-ft fireclay of the above section. Spoil from the seam contains abundant ferruginous mudstone yielding *Anthraconaia* aff. *stobbsi*, *Naiadites* aff. *daviesi*, associated with *Carbonita sp.* and *Geisina? subarcuata*. Evans and Simpson (1934, p. 464) also record *Euestheria* from this horizon, while the holotype of *Naiadites daviesi* came from spoil material from this same locality. The exact position of the fossiliferous material is uncertain; the roof of the seam is a silty mudstone carrying plant debris and is quite unlike the ferruginous mudstone containing the shells. It is possible that this shelly blackband comes from the floor of the seam since several inches of 'inferior blackband' are recorded from this position in the workings. The White Seam is largely unproved around the head of the valley, though its smut can be seen here and there, and trial levels occur north-east of Caerau. The seam has been worked east and south-east of Garth [86879000; 86898956] with the following section: top coal 13 in, rashes 3 in, coal 18 in. This seems to represent only the lower part of the Garn Wen section. Spoil from these workings carries ferruginous mudstones with shells, but they are too poorly preserved for positive identification; they belong either to *Naiadites* or *Anthraconauta*. The seam was also worked with a similar section at Hafod Deep Colliery [87779015] in Cwmdu, under the name of 'Havod'.

In the Maesteg area some 15 to 35 ft of strata intervene between the White and the Tormynydd, the Jonah, when present as a separate seam, lying within these measures. Little is known concerning this portion of the sequence as it is generally obscured by scree and head from the Llynfi Rock escarpment above.

The **Tormynydd** Seam is everywhere thin and unworkable and indeed may not always be present. Small trials have been driven on it east of Caerau [87229442].

The **Llynfi Rock** is the lowest thick sandstone of Pennant type developed within the central part of the coalfield. In the Ogmore Valley it increases in thickness rapidly from about 70 ft in Craig Talgarth east of Nant-y-moel to about 160 to 180 ft on the west side of the valley in Coed Nant Dyrys. Almost its full thickness is exposed in Cwm Nant-y-moel where the lower beds were quarried near the south end of Garon Street [93159320]. Here 30 ft of massive sandstones with intercalations of flaggy beds and irregular lenses of silty mudstones are to be seen. Other old quarries are located on the hillside west of Nant-y-moel [93129293; 93149244], and on the east side of the valley along Craig Talgarth, north-west of Western Colliery. In Cwm-y-ffosp, immediately behind Price Town [94209235], 50 to 60 ft of irregularly false-bedded, massive and blocky sandstones, with impersistent lens-like developments of silty mudstone and streaks and rafts of coal, were formerly quarried.

The Rock again forms a prominent feature around the sides of the northern Garw and Garwfechan valleys where it varies in thickness from about 60 ft (north-west of International) to about 150 ft (south-west of Pontycymmer). On the spur immediately north of Blaengarw an old quarry [90179347] shows:

	Ft	In
Dark blue shale (loose)	—	—
COAL	smut	
Seatearth passing down into blocky blue-grey, silty mudstone with spheriodal weathering	10	0
Faulted against:		
Massive sandstones with irregular lenses of silty mudstone	15	0
NO. 3 RHONDDA: **coal**..	smut	
Seatearth and silty mudstone	3	0
Llynfi Rock: massive false-bedded sandstone with irregular silty developments	30	0+

The landslip scar at Darren Fawr [89609270], west of Blaengarw, shows 40 to 50 ft of sandstones. Extensive quarrying also took place from a scar [89709210], 650 yd N.W. of Braichycymmer Pit, where massive false-bedded sandstones 50 ft thick are split by up to 10 ft of grey-blue and dark blue barren shales. In the Pontycymmer area the sandstone was quarried west and north-west of New Braichycymmer Colliery and on the spur extending from just east of St. Theodore's Church [90649119] southward to the junction of Oxford Street and Penton Place. Here faces 35 to 40 ft high in massive false-bedded sandstone are to be seen.

In Cwm Garwfechan where, in the crest of the Maesteg Anticline, the Rock caps the hill-tops of Tarencwmdu on the west and Wauntynewydd on the east, the beds immediately underlying the No. 3 Rhondda Seam consist of hard fine-grained quartzites which weather white and contrast sharply with the normal pennant-type sandstone. The surface of the hillside is strewn with blocks of this conspicuous rock which can be clearly seen from considerable distances. In this area, as on Garn Wen to the west, the basal beds of the Rock carry impersistent streaks and rafts of coal and many pellets of rolled clay-ironstone.

Throughout most of the Llynfi Valley the Llynfi Rock varies between 120 and 160 ft in thickness but north and east of Caerau it thins to little more than 70 ft. Exposures are especially good in extensive old quarries on the west of the valley from Caerau to Nant-y-ffyllon, on the southern side of Mynydd Bach and around Garth Hill. In all these quarries alternating beds of massive and false-bedded sandstone with sporadic developments of flagstones are seen. In general the lower part of the Rock is more massive while flagstones are more common towards the top.

The No. 3 Rhondda was formerly worked extensively on the east side of the Ogmore Valley at Ogmore Vale. Access to the seam, which normally comprises 28 to 36 in of clean coal, was by means of a cross-measures slant at Aber Colliery [93469054] situated 500 yd N.N.E. of the church. These workings were sharply limited on the north side by an east-west line of washout. On the west side of the valley the seam was exploited from levels at Fron-wen [92949131], and Tynewydd [93049095], 500 yd W.N.W. and 400 yd S.W. respectively of Penllwyngwent Colliery. Farther north there were limited workings in the seam, 28 in thick, at Wyndham Level [93099267] and in Cwm Daren-fawr [92819334], north-west of Nant-y-moel. Between these two points numerous old strike-workings indicate the crop of the seam.

The seam was also worked to a considerable degree on either side of the Garw Valley, especially south of Pontycymmer. The principal points of exploitation were: Lluest Colliery [90558966] near the confluence of Garwfechan with the main valley, where the coal varied from 22 to 42 in; Braichycymmer [90359070] and Duchy [90359071] collieries; Garwfechan Colliery [89929047] with numerous openings in the crop of the seam on the hillside above; on the crop above New Braichycymmer [90179097]; and on the south side of Cwmfforchwen above Pant Street [90729072]. Further trials and strike-levels occur higher up along the south and south-east sides of Cwmfforchwen. Farther north the seam was worked from Gelli-wern Level [91409210], 1100 yd E.N.E. of Pontycymmer Church with a section varying from 21 to 32 in. On the south-west side of the Tableland Fault, Nant-hir Level [91359320], 900 yd E. by N. of Garw Colliery, followed the coal in a south-eastward direction. The coal here was generally about 34 in thick, but about one mile south-west of the level-mouth the section is broken by stone partings: top coal 18 in, stone 24 in, coal 4 in, stone 20 in, coal 16 in.

There is no evidence that the coal has been proved to the north of Cwm Nant-hir or on the western side of the valley at Blaengarw, and it is not unlikely that the seam is in washout in this area. Neither is there any sign of the coal north of the anticline in the Maesteg area. It has been worked, however, on the eastern side of the Cwmdu Valley north-east of Pont Rhyd-y-cyff. In this area levels are located at Gelli-hir [87518942] and northward along the crop towards Nant-y-fforest [87748992; 87909024; 88489057], the seam varying from 28 to 36 in.

Throughout the Ogmore, Garw, and Cwmdu valleys the No. 3 Rhondda is everywhere overlain by sandstone 15 to 50 ft thick, and over much of the area little more than a foot or so of hard ganister-like seatearth intervenes between the seam and the top of the Llynfi Rock proper.

In the Ogmore Valley the measures between the No. 3 and No. 2 Rhondda seams are about 250 to 300 ft thick. Exposures are only sporadic. At the northern end of the valley about 180 ft of strata are exposed in the steeply falling stream [92659470] immediately west of the hairpin bend in the road over Bwlch-y-Clawdd:

	Ft	In
Very massive sandstone	20	0+
Sandstone with numerous rolled clay-ironstone pellets and streaks, rafts and 'pebbles' of coal	3	0
NO. 2 RHONDDA: coal (not seen)	—	—
Mudstone-seatearth passing down into shales and silty mudstones with thin sandstone bands	15	0
Massive, false-bedded sandstones, silty at base	15	6

	Ft	In
COAL	smut	
Silty mudstones with a few thin sandstone layers	80	0
Blue shale	3	0
?WERNPISTYLL: **coal** thin, seatearth with ironstone nodules 36 in,		
coal thin .. .. say	3	6
Seatearth and rashings	1	6
Silty and sandy mudstones interbedded with sandstones up to about		
2 ft thick	56	0

Farther south in Cwm-y-ffosp a persistent sandstone up to 90 ft thick is developed about 60 to 100 ft below the No. 2 Rhondda. This sandstone has been quarried at several localities: at the head of Cwm-y-ffosp [94989225], where 25 ft of massive and flaggy sandstones are overlain by 10 ft of silty mudstones; 250 yd N.E. of Penllwyngwent Colliery [93689125], where 10 ft of blocky silty mudstones rest on 40 ft of massive false-bedded sandstones; on the west side of Wyndham above Fairy Glen [93059165]; 650 yd W. of Penllwyngwent [92779117], where a 50 ft face in massive sandstone may be seen; on the north-west side of Ogmore Vale [92909085; 93029080]. One mile farther south at Rhondda Main Colliery, Anne Pit [93648895] proved the following section below the No. 2 Rhondda:

	Thickness		Depth below No. 2 Rhondda	
	Ft	In	Ft	In
NO. 2 RHONDDA	—	—	—	—
Fireclay and clift	82	2	82	2
COAL		3	82	5
Fireclay and rock	1	3	83	8
Clift	32	7	116	3
?WERNPISTYLL: **coal 11 in**		11	117	2
Clift	9	6	126	8
Rock	39	6	166	2
Clift and rock	52	0	218	2
COAL		4	218	6
Clift	20	6	239	0
Rock, very jointy	18	6	257	6
NO. 3 RHONDDA: **coal 21 in**	1	9	259	3
Fireclay and clift	42	6	301	9
Rock	31	9	333	6
?TORMYNYDD: **coal 4 in**		4	333	10
Clift	6	5	340	3

In the Garw Valley the measures between the No. 3 and No. 2 Rhondda seams total about 350 ft. Apart from the bed which everywhere overlies the No. 3 Rhondda or the position of the seam, sandstones up to 20 ft or so are developed at several horizons, the most persistent and prominent being that above the Wernpistyll position. This thickens southwards and reaches about 100 ft to the south of Pontycymmer. It has been quarried in the neighbourhood of Pant-y-gog [90409040; 90609050], and on the west side of the Rhyl in the Garwfechan [90139026].

The strata below the No. 2 Rhondda may be seen in the small stream rising in Ffynnon Daren-Goch [90759381] about ½ mile N.N.E. of Blaengarw:

	Ft	In
NO. 2 RHONDDA position	—	—
Measures, mainly argillaceous, not well exposed .. .. .. c.	40	0
Massive false-bedded sandstones	20	0
Silty and sandy mudstones	25	0

	Ft	In
?WERNPISTYLL: **coal 10 in**		10
Seatearth passing down into shale and mudstone	15	0
Strong blocky silty mudstones	15	0
Dark shale	1	0
COAL		9
Seatearth passing down into silty mudstones with dark shale bands ..	25	0
COAL	smut	
Silty mudstones	20	0+

The same part of the sequence may be examined on the west side of the valley in and above a quarry [89909378] on the spur 650 yd N. by E. of International Colliery:

	Ft	In
Massive false-bedded sandstones (in crags)	—	—
NO. 2 RHONDDA (slipped ?)	smut	
Silty mudstones	10	0
COAL, dirty	smut	
Strong sandstone	1	6
Blue-grey silty mudstone (with ferruginous weathering) passing down into blue silty mudstone	10	0
Strong grey sandy mudstone	4	0
Massive slightly false-bedded sandstone (worked)	20	0
Grey-blue silty mudstone	2	0
Blue-black paper shales .. .. c.	1	0
?WERNPISTYLL	smut	

Beneath the Wernpistyll are some 40 ft of measures containing thin sandstones, at the bottom of which there is a thin coal, proved by old trial levels. The dark paper shales overlying the ?Wernpistyll horizon are again exposed at the head of Nant Gwineu [89409380] 800 yd W. by N. of International Colliery, where they yield *Anthraconauta phillipsii*. The presumed Wernpistyll Coal has been proved a short distance below thick sandstones at several localities on the south side of Pontycymmer [89439062; 90349050; 90629058; 91079068], but nowhere does it appear to have been worked commercially.

In the Llynfi Valley the sequence from the top of the Llynfi Rock to the Wernddu (No. 2 Rhondda) is very variable. On the west side of the valley and around Moel-cynhordy (1½ miles S.E. of Maesteg) it is essentially similar to that known around Bryn (pp. 187-8), but northwards many of the individual sandstones thin out and the succession is transitional to that of the northern Rhondda Valley where the measures are mainly argillaceous.

On Garn Wen and Foel y Duffryn, to the west of Nantyffyllon and Caerau respectively, the general sequence is as follows:

		Ft	In
WERNDDU		—	—
Mudstones with subordinate sandstones and two coals (WERN-PISTYLL RIDER and WERNPISTYLL)		140	0
Sandstones		80	0
COAL, impersistent		—	—
Seatearth, mudstone, etc.	up to	20	0
Sandstone		70	0
ROCK FACH: **coal 18 in**		1	6
Seatearth, mudstones, etc.		30	0
Siltstone and sandstone		30	0
Mudstones, etc. (horizon of No. 3 Rhondda in 'washout' presumably near base)		40	0
Llynfi Rock		—	—

The coal, here equated with the **Rock Fach,** lies at the base of a sandstone which has been quarried on the west side of the stream west of Nantyffyllon [84669317]. At this point a small mine has proved the coal to be locally more than 2 ft thick, but it appears to deteriorate rapidly when traced northwards. At Garn Wen the overlying sandstone is about 150 ft thick. Northwards a mudstone develops within this sandstone and on the southern slope of Foel y Duffryn small trials have proved a thin coal near the top of this 'slack'. Mainly argillaceous measures underlie the Wernddu and include two coals, each probably less than 2 ft thick: the **Wernpistyll** near the base and the **Wernpistyll Rider** about midway to the Wernddu. These two seams lie at the base of two low features indicating the presence of siltstone or striped beds. A small trial west of Caerau [84909406] has proved the Wernpistyll to have a roof of dark fissile mudstone.

South of the Maesteg Anticline a similar but thinner sequence has been proved at Moelcynhordy to the north-east of Pont Rhyd-y-cyff, the Wernddu lying about 300 ft above the No. 3 Rhondda.

East of Bryndefaid Farm [88249160] the watershed between the Cwmdu and Garw-fechan valleys is formed by the Llynfi Rock which arches over the Maesteg Anticline. When the succeeding measures reappear to the north of Foel Gwilym Hywel [883924] they are predominantly argillaceous, but traced still farther northwards the sandstones gradually re-develop. On Bryn Siwrnai [877931] the highest sandstone, underlying the presumed Wernpistyll, is about 75 ft thick. Two underlying sandstones also reappear, but cannot be followed far. Between here and Blaencaerau Farm [86959476] the Llynfi Rock is succeeded by 130 to 150 ft of strata which contain no sandstones of note and which are capped by a coal proved to be 18 in thick on Mynydd Caerau. To the west a major sandstone develops within these beds and has been quarried north of Blaencaerau [86409495]. Close beneath the Wernpistyll a persistent sandstone reaches 150 ft in the area north of Caerau. The Wernpistyll has been worked [87339496] just below Blaencaerau Level, dark shale appearing in the spoil. About 50 ft of mudstones separate this seam from the Wernddu.   A.W.W., W.B.E.

AVAN VALLEY AND BRYN AREA

Apart from two small inliers at Blaen-gwynfi and Glyncorrwg, exposures of these beds are restricted to the area between Garn Wen and Pont Rhyd-y-fen. They have, however, been penetrated by colliery shafts, which reveal a marked thickening sonth-westwards across the area: at Avon and Glyncorrwg they are 450 to 500 ft thick; at Duffryn Rhondda nearly 600 ft; and near Cwmavon they reach about 750 ft.

On the southern outcrops, the basal 60 to 80 ft of beds consist of mudstones with a few thick posts of fine-grained sandstone. These are succeeded by the **White** and **Jonah** seams, both of which have been worked from crop. The following sections illustrate the development of these coals:

			Brynawel [81119281]		Drysiog [83169276]		Garn Wen [84489265]	
			Ft	In	Ft	In	Ft	In
JONAH	Coal	⎫						3
	Fireclay	⎬	2	1		11	1	3
	Coal	⎭					2	0
	Measures		18	0	18	0		9
WHITE	Coal	⎧	2	7	2	0	1	9
	Parting	⎨		5		11		2
	Coal	⎩	2	0	1	7	1	2

These sections illustrate the westwards splitting of the 'White' of Garn Wen into the White and Jonah of Bryn. In the extreme west near Cwmavon the White may be only the lower leaf of the Brynawel section. Spoil from workings of the White at Drysiog Colliery yielded *Anthraconaia stobbsi, Naiadites daviesi* and ostracods including *Carbonita?*. The Jonah lies 40 ft beneath the Tormynydd, the intervening measures consisting largely of mudstones.

At Duffryn Rhondda the measures between the base of the group and the Llynfi Rock are about 140 ft thick; at Glyncastle they are about 120 ft, at Avon about 110 ft, and at Nantewlaeth about 75 ft. At Glyncorrwg the true sequence is complicated by faulting. The White-Jonah coal horizon is not developed in any of these collieries except possibly Glyncastle, where it may be represented by 15 in of coal and dirt 30½ ft below the presumed Tormynydd. All the shafts record a sandstone 25 to 60 ft thick in the general position of the coals and it is possible that they are affected by washouts.

The **Tormynydd** has been worked from a number of adits along the crop. North of Bryn it is said to be 22 in thick, and between here and Cwmavon the following section has been recorded: top coal 22 to 28 in, parting 3 to 6 in, coal 3 in. It thins to the north, being 15 in at Duffryn Rhondda, and 18 in at Nantewlaeth. At Glyncastle it is thicker; known as the No. 3 Rhondda, it is 27 in thick, with a further 6 in of coal lying about 5 ft below.

North and north-east of Bryn the sequence between the Tormynydd and Wernpistyll is as follows:

	Thickness Ft	In	Depth below Wernpistyll Ft	In
WERNPISTYLL ..	—	—	—	—
Mudstones, etc. ..	20	0	20	0
Sandstone ..	90	0	110	0
COAL, impersistent	0 to 1	0	111	0
Mudstones, etc. ..	0 to 15	0	126	0
Sandstone ..	60	0	186	0
ROCK FACH: **coal 18 in** ..	1	6	187	6
Mudstones, etc. ..	30	0	217	6
Siltstones and sandstones ..	40	0	257	6
Mudstones, etc. with No. 3 Rhondda (Rock Fawr) in washout near base ..	30	0	287	6
Llynfi Rock: sandstone ..	180	0	467	6
TORMYNYDD ..	—	—	—	—

On Garn Wen the uppermost impersistent coal finally disappears as do the underlying mudstones. Still farther west, on the south side of Mynydd Pen-hydd [80899323], the Rock Fach and its associated mudstones, together with the mudstones in the Rock Fawr position, abruptly pinch out. They appear for a short distance on Moel y Fen [79909308], but as the Avan Valley is approached the whole sequence from the Tormynydd to close below the Wernpistyll, about 400 ft, is arenaceous. The Llynfi Rock has been quarried near Drysiog Colliery [83149283]; north of Bryn [81979298], where up to 80 ft of massive false-bedded sandstone containing flaggy partings can be seen; and near Pwll-y-gwlaw [79709320; 79609340] where quarrying has extended upwards almost to the Wernpistyll.

To the north the Llynfi Rock is well developed at Duffryn Rhondda, where it is 140 ft thick, and at Nantewlaeth, where it is 185 ft. It thins northwards to rather more than 90 ft at Glyncastle, and eastwards to barely 80 ft at Avon. Except at Avon, a thin coal is developed in the general position of the No. 3 Rhondda, and the mudstones between this and the Rock Fach can be distinguished.

The measures between the Wernpistyll and the Wernddu thin considerably when traced from south-west to north-east, as shown by the following sections:

	Cwmavon	Parc-y-Bryn [82079332]	Glyncymmer [85679593]	Glyncastle
WERNDDU	—	—	—	—
Measures	75 to 100 ft	50 ft	33 ft	35 ft
WERNPISTYLL RIDER ..	—	—	—	—
Measures	55 to 65 ft	87 ft	50 ft	28 ft
WERNPISTYLL	—	—	—	—

The Wernpistyll, the Rider and the Wernddu all lie at or near the top of mudstone 'slacks', each 15 to 30 ft thick. Between the 'slacks' siltstones and lenticular sandstones are developed, the latter being particularly prominent south of Pont Rhyd-y-fen. The **Wernpistyll** is thickest in the west, and is reputed to have been 34 in at Oakwood (Pont Rhyd-y-fen) Colliery [79929411]. It has also been worked at Cwmavon (top coal 24 in, parting 2 in, coal 2 in), and from small collieries in Cwm-yr-Argoed and on Pendisgwylfa. In Cwm-yr-Argoed spoil yielded abundant *Anthraconauta* cf. *phillipsii* accompanied by *Carbonita pungens* (Jones and Kirby). At Glyncymmer the coal is 22 in thick, and at Glyncastle the section reads: top coal 22 in, dirt 8 in, coal 8 in.

The **Wernpistyll Rider** usually shows a two-leaf section: top coal 3 to 7 in, rashings 3 to 8 in, coal 15 to 22 in. It has been worked at intervals south-westwards from Nant yr Hwyaid [828941]. In the Neath Valley it appears to be the equivalent of the Pantrhydydwr at Glyn Merthyr Colliery, the workings from which extend southwards beneath Melin Court.

In the south-west of the area a further thin coal lies close beneath the Wernddu; it is called the **Wernddu Rider,** and, near Pont Rhyd-y-fen, is locally of workable thickness. It is a split from the No. 2 Rhondda, but for convenience of description has been retained within the Llynfi Beds.

Near Blaen-gwynfi a thin coal (probably the Wernpistyll), overlain by about 12 ft of sandstone, can be seen in the railway cutting east of the village [89919612]. What appears to be the same seam is exposed on the east side of the Glyncorrwg Fault, in Nant Gwyn [89709663] and near the source of the Avan [90969575]. Although some 200 ft of Llynfi Beds must crop out on this side of the fault, they are almost wholly concealed by boulder clay, landslips or mine-spoil.

W.B.E.

SOUTH CROP

Cwm Wernderi area: The Llynfi Beds have been proved on the south side of the Moel Gilau Fault by boring from the Wernddu (No. 2 Rhondda) Seam at Cwmgwyneu Colliery, 1½ miles S.W. of Bryngyrnos. The borehole was located [80339048] about 100 yd E. by S. of the shafts and the following record summarizes the journal for the upper part of the hole:

	Thickness		Depth	
	Ft	In	Ft	In
WERNDDU (NO. 2 RHONDDA)	—	—	—	—
Fireclay passing down into pennant sandstone	36	1	36	1
WERNPISTYLL RIDER: **coal 17 in**	1	5	37	6
Fireclay passing down into sandy shale with sandstone bands	36	7	74	1

	Thickness		Depth	
	Ft	In	Ft	In
Pennant sandstone with a band of fireclay	49	5	123	6
WERNPISTYLL: coal 16 in	1	4	124	10
Fireclay passing down into sandy shale with sandstone bands	28	8	153	6
Pennant sandstone with a few shaly partings	122	3	275	9
Sandy shale with interbedded sandstone	12	5	288	2
ROCK FACH: coal 15 in	1	3	289	5
Fireclay passing down into sandy shale with a few sandstone bands	45	0	334	5
?NO. 3 RHONDDA horizon	—	—	334	5
Fireclay passing down into sandy shale and sandstone ..	54	8	389	1
Pennant sandstone	69	8	458	9
TORMYNYDD: coal 7 in		7	459	4
Fireclay, shale and pennant sandstone	30	5	489	9
Shaly fireclay, sandstone and bands of shale	26	10	516	7
Shale	45	0	561	7
WHITE: coal 57 in	4	9	566	4
Shale	60	3	626	7

This sequence is closely similar to that proved in the lower part of the Avan Valley and in the Llynfi Valley. In the colliery a small area of 'New Seam' was worked about 500 ft below the Wernddu. From its section (top coal 12 in, rashes 2 in, coal 12 in, rashes 8 in, coal 30 in) it appears to be the White Seam possibly combined with the Jonah.   W.B.E.

West of the Ogmore Valley: Pennant-type sandstones make up the bulk of the 650 to 700 ft of measures which comprise the group along this western part of the South Crop. The southern limit of the Pennant plateau is marked by the bold escarpment of the **Llynfi Rock** rising above the almost featureless low-lying ground occupied by the Middle Coal Measures and the upper part of the Lower Coal Measures to the south. This escarpment stretches from Graig Fawr on the north-west of Margam Park eastwards along Craig y Lodge, Craig Goch, Coed Tonmawr to Aberbaiden Colliery, Ton Phillip and Cwmrisca, to the north side of Ton-du and Brynmenyn Common in the east. The thinning of the sandstone is reflected by the gradual decline of the feature as it is traced eastwards. The outcrop of the group occupies a belt some 200 to 500 yd wide.

In the extreme west the 300 to 400 ft high escarpment of Graig Fawr is formed by massive sandstones. The lower part of the slope, comprising the subordinate bluff of Craig y Capel [80168645], is formed by the Llynfi Rock, which also gives rise to the isolated feature of Mynydd y Castell, the site of ancient earth works, immediately to the north of the modern house of Margam Abbey. Above the Llynfi Rock a complex 'slack' represents the outcrop of about 100 ft of measures carrying the horizons of the **No. 3 Rhondda (Rock Fawr)** and overlying **Rock Fach.** These strata, which consist of mudstones and silty mudstones with one prominent rib of sandstone, are overlain by massive sandstone, nearly 150 ft thick, well exposed in the great southward-facing landslip scar of Graig Fawr. At Cwar y Graig Fawr [79128727], on the north-west of the scarp, about 80 ft of massive, false-bedded pennant with irregular siltstone developments can be seen. A thin coal occurs at the base of this sandstone while a somewhat thicker coal, possibly the Rock Fach, about 50 ft below, is exposed at intervals along the slip scar. There is no evidence along this hillside or in Cwm Phillip that the No. 3 Rhondda is present, and the seam may be subject to extensive washout in this area.

A well marked 'slack' extending from Cwm y Brombil to Crugwyllt-fawr [79788706] and Cwm Maelwg [80528715] marks the crop of the uppermost beds of the group. These strata include a coal, presumably the **Wernpistyll** of the area to the north, lying about 90 ft below the No. 2 Rhondda. This coal was proved by opencast pitting immediately to the north-east of Crugwyllt-fawr as follows: top coal 5 in, clay 2 in, coal 19 in, clay 5 in, coal 7 in. It was formerly worked on a small scale from levels on the crop near the bottom of Cwm Cae-Treharne [81058747] about 1200 yd N.N.E. of Margam Park Fishpond. A feature immediately overlying this coal varies in strength along the crop suggesting that sandstone is developed impersistently.

Almost the whole of the group was proved by two boreholes on the same site in Cwm Philip [81618730], about 1 mile N.E. of Margam Abbey. The first borehole was drilled in 1946 in an attempt to prove the No. 3 Rhondda Seam on the N.W. side of the Pentre Colliery workings; it was stopped at a depth of 409 ft 6 in. Margam Park No. 3 Borehole was open-holed to 389 ft 3 in and then cored into the Lower Coal Measures. The bores evidently started at an horizon close below the No. 2 Rhondda Seam, and only one thin coal (at the top of the hole) was proved above the Llynfi Rock. The combined record for the Llynfi Beds reads as follows:

	Thickness		Depth	
	Ft	In	Ft	In
Drift	8	0	8	0
COAL		9	8	9
Pennant sandstone	49	3	58	0
Dark and hard shaly fireclay (?WERNPISTYLL horizon) ..	3	0	61	0
Pennant sandstone with beds carrying streaks of coal and ironstone pellets	142	0	203	0
Hard fireclay passing into dark clift	37	6	240	6
Pennant sandstone with coal streaks	40	0	280	6
Sandy clift and mudstone with sandstone bands ..	125	4	405	10
Massive pale grey pennant, strong grey silty mudstone and striped beds	59	6	465	4
?NO. 3 RHONDDA: **coal** and mudstone **5 in**.. ..		5	465	9
Massive pale grey pennant sandstone with a few siltstone and mudstone layers (?TORMYNYDD cut out by fault at base)	153	8	619	5
Dark grey mudstone with coal streaks	1	$2\frac{1}{2}$	620	$7\frac{1}{2}$
Strong mudstone-seatearth passing down into mudstone and silty mudstone with bands of siltstone	22	$11\frac{1}{2}$	643	7
COAL		6	644	1
Buff-grey seatearth, passing into strong grey mudstone ..	8	5	652	6
COAL with carbonaceous shaly bands	2	0	654	6
Grey seatearth, passing into grey mudstone	9	2	663	8
COAL		7	664	3
Carbonaceous shale and grey mudstone	4	4	668	7
COAL		3	668	10
Buff-grey seatearth	3	8	672	6
COAL and rashings		10	673	4
Grey mudstone	5	11	679	3
WHITE: **coal 24 in**	2	0	681	3
Seatearth and grey mudstone with ironstone bands ..	25	0	706	3
Upper Cwmgorse Marine Band	—	—	—	—

The absence of coal in the upper part of the sequence may have been due to poor recovery, for the **Wernpistyll** Coal was extensively proved by opencast exploration in the area between Cwm Philip and Cwm Kenfig. Several pits located the coal, one 500 yd E.S.E. of Ton-mawr [82838606] showing: sandstone roof, shale and clay 13 in, coal 10½ in, clay ½ in, coal 34 in. The **No. 3 Rhondda** (Rock Fawr) was also proved by opencast drilling immediately overlying the Llynfi Rock escarpment of Craig y Lodge, Craig Goch and Coed Tonmawr. The seam was extensively worked over about 5½ square miles from the combined Pentre and Aberbaiden collieries [84138545; 85008485]. Over the south and east of the Aberbaiden workings the seam varied from 40 to 54 in. in thickness, but a parting, usually of rashings, is present in the north and west, and the general section over much of the taking reads: top coal 15 to 26 in, rashings 1 to 3 in, coal 19 to 36 in. The parting thickens on the western side of the Pentre workings to 12 in or more, while an extensive washout was proved in the general area of Nant Cwm Philip.

At Aberbaiden the strata below the No. 3 Rhondda were proved in seven boreholes from the workings, and the following generalized section summarizes the results.

								Ft	In
NO. 3 RHONDDA	..	..	..	..	..	..	..	—	—
Llynfi Rock: massive pennant sandstone			..	..	150 ft to 165		0
TORMYNYDD: **coal 12 to 22 in**	..	..	..	..	..		1	10
Measures, mainly argillaceous, with two or more seatearths	..	42 ft to 52		0
WHITE: **coal 12 to 23 in** ..	..	..	..	..	..	..	1	11
Measures, mainly argillaceous	..	..	..	..	..	62 ft to 80		0
Upper Cwmgorse Marine Band	..	..	..	..	..	..	—	—

The measures between the White and the Tormynydd carried well-developed plant-bearing beds, usually in two distinct horizons. Among the species recorded are: *Alethopteris lonchitica*, *Annularia microphylla* Sauveur, *A. sphenophylloides* (Zenker), *Bothrodendron minutifolium* (Boulay), *Cyclopteris orbicularis* Brongniart, *Lepidodendron aculeatum* Sternberg, *Linopteris munsteri* (Eichwald), *Mariopteris muricata*, *Neuropteris heterophylla*, *N. rarinervis*, *N. tenuifolia* and *Sphenophyllum cuneifolium*.

The strata above the No. 3 Rhondda have also been drilled but full details are lacking. At Pentre Colliery the section up to the Rock Fach was proved as follows: Rock Fach coal 17 in, clift 37 ft 9 in, coal 5 in, rock with a thin band of clift 43 ft, No. 3 Rhondda. At Aberbaiden where the Malthouse (No. 2 Rhondda) occurred 370 ft above the No. 3, the Wernpistyll (coal and shale 15 in, coal 17 in) was struck 53½ ft below the No. 2.

The crop of the **Wernpistyll** has been proved by opencast exploration on the hillside extending from Ty-fry [84248563] in an east-south-east direction past the north side of Pentyla [85108523] towards Craig-yr-aber Valley. The coal may be seen in the mouth of an old trial level [85318524] 220 yd E. of Pentyla with the following section: silty mudstone 6 ft, soft grey clay 15 in, rashings with layers of inferior coal 24 in, coal 22 in, seatearth. Features suggest that sandstone makes up much of the sequence between this seam and the No. 2 Rhondda.

The crop of the **No. 3 Rhondda** can be followed eastwards from Aberbaiden marked by a number of collieries and minor levels. Several of these are sited on the western side of the valley opposite Ty-talwyn Colliery [85718496]. Ton Phillip Colliery consisted of a main slant in the seam [86528469], and a cross-measures adit [86268458] from a point well below the base of the Llynfi Rock. The section of the coal in this area varied from 31 to 48 in, a plane parting being sometimes recorded. During recent years a small licensed mine was located on the crop of the seam [88598466] about 300 yd S.W. of Ton-du House.

Several quarries on the west side of the Llynfi Valley north of Ton-du formerly provided stone for local building. Several of these worked the Llynfi Rock [89178474; 89298478; 89388485], while a major excavation [89138500] is located in the thick sandstone between the Rock Fach and Wernpistyll seams, 400 yd N.E. of Ton-du House. Here a 60 to 70 ft face in very massive pennant may still be seen.

Coytrahan Colliery, north of Ynysawdre, exploited the **No. 3 Rhondda** from slants [89588502; 89788498] driven on the crop of the seam. The seam varied from 3½ to 4 ft near the crop but thinned to 2 to 2½ ft at the limit of the workings about three-quarters of a mile to the north. A cross-measures roadway proved the strata to the **Malthouse (No. 2 Rhondda)** about 330 ft above. The section, measured from an old plan, reads:

	Ft	In
MALTHOUSE (NO. 2 RHONDDA)	4	1
Rashes and fireclay	4	2
COAL	1	0
Rashes and fireclay	7	1
COAL		6
Fireclay passing down into clift	23	6
Sandstone	21	0
Clift	12	0
WERNPISTYLL: **coal and rashes 45½ in**	3	9½
Fireclay passing down into sandstone with a few thin bands of clift ..	163	0
COAL		6
Fireclay passing down into shale and clift	10	0
COAL	3	0
Fireclay	1	0
COAL	1	0
Fireclay and rashes passing down into clift	24	0
Sandstone		9
COAL and rashes		9
Bastard fireclay and clift	12	0
Sandstone	32	0
NO. 3 RHONDDA (ROCK FAWR): **coal 38 in**	3	2

East of the Ogmore Valley: In the Bryncethin area the Llynfi Rock is still thick though it appears to thin rapidly when traced to the east. As the sandstones of the group generally become less prominent there is a gradual thinning of the whole group to about 500 ft at Tynywaun and 450 ft at Raglan and in the South Rhondda–Meiros area.

At Bryncethin Brickpit (see p. 167) the **White** Seam, 22 in thick, is separated from the Upper Cwmgorse Marine Band by about 40 ft of argillaceous strata, mainly silty and sandy mudstones. A thin coal 17 ft above, at the base of the Llynfi Rock, represents a poorly developed **Tormynydd.** The **Llynfi Rock** was formerly quarried in the north-western part of the pit, where the face shows at least 75 ft of sandstone, of which the bottom 5 ft are characterized by numerous coal 'pebbles' and irregularly developed layers of detrital coal.

The **No. 3 Rhondda** was worked at Heol-laethog Colliery [92998453] near Ty'n-y-waun, where 34 in of coal were underlain by 5 ft of fireclay. The Llynfi Rock does not appear to be developed in this area but the No. 3 Rhondda is overlain by well-developed sandstone. Between this and the No. 2 Rhondda only one significant sandstone, 50 ft or so thick, is recognized, presumably below the Wernpistyll horizon.

At Ty'n-y-waun Pit [93188444] the White Seam, 2 ft thick, is said to occur about 57 ft above the top split of the Hafod, with the Tormynydd, 1½ ft thick, 12 ft higher in the sequence. The No. 3 Rhondda is 190 ft above the base of the group and the No. 2 Rhondda about 315 ft above the No. 3.

About one mile north-east of Bryncethin the Black Mill Borehole (see p. 303) penetrated the full thickness of the Llynfi Beds. Apart from the Llynfi Rock, about 95 ft thick, and a well-marked sandstone above the No. 3 Rhondda position, no massive pennant developments are recorded. Though no coals were recovered between the No. 2 Rhondda and the marine shales at the base of the group, their positions are indicated by seatearths. The Tormynydd horizon, at the base of the Llynfi Rock, lies about 90 ft above the marine band; between this and the No. 2 Rhondda Seam, 360 ft above, five or six other seatearth horizons were noted, in a sequence of silty mudstones with numerous sandy bands.

The crop of the **No. 3 Rhondda** was traced by opencast exploration between Ty'n-y-waun Colliery and Tynywaun Farm, where a pit [93728460] proved the coal to be 34 in thick. The coal was worked on a moderate scale from two pairs of levels at Raglan Colliery on either side of the Cwm-llwyd stream [94678469; 94788467]. Records of the Llynfi Beds at Raglan are unreliable but the No. 3 Rhondda is reported to lie 165 ft above the Upper Cwmgorse Marine Band and the No. 2 Rhondda about 280 to 300 ft higher. Eastwards of Raglan the crop of the No. 3 Seam, everywhere at the base of a prominent sandstone feature, has been traced by drilling for opencast coal nearly as far as Pen-y-lan Farm [96088484]; several trial pits along this tract proved the coal, varying in thickness from 26 to 34 in. The crop of the seam is marked by occasional levels which, for the most part, were worked only on a small scale.

Workings from levels [96628488] at Werntarw Colliery extended northwards almost as far as the Ogwr Fach river. The coal varied from 33 to 50 in. in thickness, and in the north up to 12 in of bast occurred at or near the top of the seam, apparently replacing true coal. A hard heading connecting the No. 3 workings with those of the Hafod proved the following section which is notable for the complete absence of Llynfi Rock:

	Ft	In
NO. 3 RHONDDA: **coal 48 in** ..	4	0
Fireclay passing down into clift with two bands of sandstone ..	60	0
TORMYNYDD: **coal 17 in**	1	5
Clift ..	10	0
COAL		8
Clift ..	35	0
WHITE: **coal 10 in**		10
Clift and dark shale with sandstone band in middle; Upper Cwmgorse		
Marine Band at base	72	0
HAFOD: **coal 36 in**	3	0

The measures between the No. 3 and No. 2 Rhondda seams have not been proved but mapping of the slopes above the colliery indicates that in addition to the sandstone above the No. 3 Rhondda, which still maintains a thickness of about 50 ft, at least three well-defined sandstones occur in the upper part of the group.

To the east of Wern-tarw the outcrop of the Llynfi Beds is much obscured by drift, and reliable details are scanty. The **No. 3 Rhondda** was worked extensively from levels at Cwm-ciwc [98208430] and from the shafts at South Rhondda and Meiros where 2½ ft of coal were generally overlain by several inches of cannel. At Cwm-ciwc a small area of 'New Seam' was worked 60 to 70 ft below the No. 3 Rhondda, the section being, top coal 7 in, black holing 2 in, coal 29 in, rashings 6 in; the seam presumably equates with the White.

At South Rhondda the No. 3 Rhondda occurs about 190 ft above the Hafod and 260 ft below the No. 2 Rhondda. Only three thin coals, spread over 35 ft of strata, are recorded in the shaft section in the middle part of the sequence between the No. 2 and No. 3 Rhondda seams. The underlying measures are predominantly argillaceous but sandstones comprise nearly two-thirds of the strata above. On the hillside south of the colliery trial levels indicate the existence of thin coals about 50 ft [99158406] and 110 ft [99248436] below the No. 2 Rhondda.

The crop of the group is thrown southwards by the Meiros Fault extending in an east-south-easterly direction from just north of Brynna towards Dolau. Because of extensive drift cover few details of the sequence are known. The No. 3 Rhondda was formerly worked from adits on the crop [99368311] between Brynna and Llanharan, the coal being 25 in thick with cannel in the roof. The seam is overlain by massive sandstones, in part white and quartzitic, which may be seen in quarries near the level.

The lower part of the Llynfi Beds appears on the south limb of the Llanharan Thrust between Tre-nos-isaf [98758262] and the area south of Hendre-Owen [01808237]. No good sections are seen but at least two coals appear to have been worked from bell-pits.

The group was proved in the sinkings at Llantrisant Colliery, where the total thickness is little more than 400 ft, and marked thinning appears to have taken place in the sequence between the No. 3 Rhondda and the Wernpistyll. The following abridged abstract is taken from the journal of the No. 1 Pit [03238409]:

	Thickness		Depth below No. 2 Rhondda	
	Ft	In	Ft	In
NO. 2 RHONDDA	—	—	—	—
Hard sandy fireclay, passing into very hard rock ..	29	0	29	0
Rashings 3 in, on strong sandy fireclay with ironstone balls	17	0	46	0
Hard bedded rock and strong sandy shale	39	11	85	11
COAL	1	0	86	11
Strong fireclay passing into hard grey rock	33	6	120	5
?WERNPISTYLL: coal 22 in, fireclay 10 in, coal 2 in ..	2	10	123	3
Strong sandy fireclay passing into strong sandy shale with rock	21	0	144	3
Hard rock	20	8	164	11
Shale and fireclay with ironstone balls	3	0	167	11
COAL and rashings	1	6	169	5
Strong fireclay with ironstone balls passing into very hard brownish grey rock, grey sandy shale and fireclay ..	20	11	190	4
COAL	1	2	191	6
Strong sandy fireclay passing into very hard grey rock ..	49	5	240	11
NO. 3 RHONDDA: coal, dirty with pyrite, 23 in ..	1	11	242	10
Very strong fireclay passing to hard rock	11	4	254	2
Strong fireclay and shale	46	5	300	7
TORMYNYDD: coal 22 in	1	10	302	5
Strong sandy fireclay passing to bluish-grey rock ..	17	2	319	7
Dark shale with ironstone bands and balls	7	0	326	7
Black rashings with coal veins		10	327	5
Strong sandy fireclay and shale	6	10	334	3
COAL		4	334	7
Strong grey fireclay and dark shale	12	10	347	5
WHITE: coal 20 in	1	8	349	1
Sandy fireclay passing to strong shale with a thin rock-band	67	0	416	1
Upper Cwmgorse Marine Band: black shale	4	10	420	11

The measures below the No. 3 Rhondda are almost wholly argillaceous while sandstones predominate in the upper part of the group.

Eastwards from the Ely Valley the Llynfi Beds are concealed beneath the drift deposits of the Clun and Myddlyn streams. The narrowness of the outcrop suggests that further attenuation takes place in this direction but no details of the sequence are known. Several coals of the group were worked, all except one on a small scale, at Tor-y-coed Colliery [06768269], Rhiwsaeson. The old plans are obscure and the seams cannot be identified with certainty. The main coal worked is said to have been

3 ft thick with a plane parting 12 in from the top. It appears to be the No. 3 Rhondda though it lies only 120 ft below the No. 2 Rhondda. If this correlation is correct the two seams proved close below, 27 and 30 in thick respectively, presumably represent the Tormynydd and White seams.   A.W.W.

B. RHONDDA BEDS

TAFF–CYNON AREA

Throughout this area the group is of fairly uniform thickness varying only between about 550 and 650 ft. The upper beds occupy the lower slopes of the Taff Valley in the vicinity of Merthyr Vale; but in the Cynon Valley the full thickness crops on the middle and lower slopes in the Mountain Ash area and on the hillsides above Cwmaman.

The **No. 2 Rhondda** at the base has not hitherto been so-named in this area; in the Taff Valley it was formerly worked as the **Castell-y-Wiever** or **Saron** Coal. The seam crops out on the east side of the Cynon Valley towards the top of Coed Tir Estyll and Craig y Duffryn and it can be traced into the Cwmpennar Valley where it was formerly worked from levels [04050040; 04160037] with a section reputed to be 33 in thick. On the opposite side of the valley the crop can be traced along the slopes immediately above Capcoch, where it was also worked on a small scale. In the Cwmaman Valley the crop between the Gadlys and Cwmneol faults is everywhere obscured by scree, head and landslip, but to the west of the latter fault it is marked by numerous levels extending from Ffyrnant to the head of the valley above Blaenaman Fawr [98409957]. Along this tract the coal is said to have varied from 1½ to 2 ft. The smut, resting on plastic fireclay, can also be seen on either side of the Gadlys Fault on the north and north-east sides of Rhos-gwawr and Craig Rhiw-ddu to the north-west of Aberaman.

The coal is 25 in thick in the Merthyr Vale shafts and 30 in at Deep Duffryn. When traced southwards in the Cynon Valley the section becomes more complex; at Naviga- tion Pit a three-coal section is recorded: top coal 10 in, clod 4 in, coal 12 in, clod 18 in, coal 10 in. South of this the coal appears to split into two distinct seams, as shown by the following pit sections:

Abergorki		Ft	In	Cwm Cynon			Ft	In	Penrikyber		Ft	In
Coal ..	..	2	0	Coal ..	..		1	2	Coal with stone		1	4½
Fireclay and				Fireclay, clift					Fireclay, rock			
clift	..	25	2	and rock	..		33	5	and clift	..	28	11½
Coal ..	..		9½	Coal ..	..		2	8	Coal ..	..		9½
Clod ..	..		4½	Black rock and					Stone ..	..		2½
Coal ..	..		11	rashings	..		3	2	Coal ..	..		11½
Black shale	..		4	Coal ..	..			5	Clod, rock and			
Coal ..	..		3						clift ..	..	5	7
									Coal ..	..		6

At Lady Windsor only one coal, 25½ in thick, is recorded at the base of thick sandstone and the exact correlation is uncertain.

The No. 2 Rhondda is succeeded by 130 to 300 ft of massive pennant, the lowest figures being recorded in the Merthyr Vale and Lady Windsor areas. To the north and west of Cwmaman it forms the prominent hills and rolling moorlands of Rhos- gwawr, Cefnrhos-gwawr, the watershed between the Aman Fawr and Fach streams and the upper reaches of the Rhondda Fach above Maerdy. It also forms the bold spur between Cwmpennar and the Cynon Valley as well as the lower slopes of that valley to the north and west of Mountain Ash. The base of the sandstone, where seen, for example, in the Blaenaman Fawr area and at Hafod-wen [983008], consists of several feet of coarse conglomerate in which quartz pebbles, rolled clay-ironstone pellets and coal 'pebbles' and rafts abound. The sandstone, more particularly its lower portion, has been extensively worked in former times for building stone, particularly at Cwmaman, Aberaman and Cwmpennar.

In much of the area this massive sandstone is interrupted about 100 ft from the top by a development of mudstone and seatearth, associated, in the central part of the area, with a thin coal. This coal, which may be the equivalent of the Daren Rhestyn horizon of the Rhondda, has been proved in the base of a quarry [03680073] at Cefnpennar, and by trials in Gelli-ddu-fawr Plantation [04240047] and above Capcoch [02209967]. A borehole near the old pond in Gelli-ddu-fawr [04380044] proved 20 in of coal overlain by 16 in of mixed coal and clod lying about 90 to 100 ft above the horizon of the No. 2 Rhondda. At Navigation, Cwm Cynon, and Penrikyber collieries up to 17 ft of fireclay and clift lie 125 to 165 ft above the base of the sandstone, but no coal has been recorded. Correlation at Lady Windsor is doubtful: the No. 2 Rhondda is overlain by 130 ft of sandstones, which are succeeded by 90 ft of mainly mudstone measures overlain by a coal 12 in thick which may be the Daren Rhestyn.

The 400 ft or so of Rhondda Beds lying above the No. 2 Rhondda or Castell-y-Wiever Sandstone consist of alternations of mudstones and sandstones with several thin coals, the sequence in the main being argillaceous. No detailed sections are available along the outcrop; individual sandstones can be followed by features and some coals can be located. Shaft sections show that the measures are variable and difficult to correlate in detail.

In the Taff Valley, just north of the district, the **Cwmdu 2** Coal has been equated with the No. 1 Rhondda, and the **Cwmdu 1** with the No. 1 Rhondda Rider. These seams cannot be identified with certainty in the Merthyr Vale shafts, where four thin coals spread over 100 ft of measures occur close above the Castell-y-Wiever Sandstone. The highest is probably the No. 1 Rhondda Rider and the seams below presumably include the No. 1 Rhondda. The Cwmdu 2 Coal, 18 in thick, was worked from levels on the hillside [06900148] west of Troed-y-rhiw, where the roof consisted of black papery shales with *Anthraconauta phillipsii* and ostracods. Some 80 ft above, a coal, known locally as the **Berry** Seam, has been worked from a trial in the same area; where worked in Nant Cwmdu, some 1000 yd to the north-west, this coal was 24 to 28 in thick. Two further thin coals are recorded in the Merthyr Vale shafts; one 14 in thick 330 ft above the Castell-y-Wiever, and another 18 in thick some 64 ft higher. One of these seams, possibly the lower, is exposed on the west side of the valley behind Nant-y-maen Farm [06850105], and lies 200 ft below the Brithdir. Here a mudstone roof carries variants of *Anthraconaia pruvosti* and *Anthraconauta phillipsii* associated with abundant *Euestheria simoni*. A further smut is seen in the stream about 70 ft above.

In the Cynon Valley the measures above the Castell-y-Wiever Sandstone were penetrated in varying degree in the shafts south of Mountain Ash. At Cwm Cynon a coal, 16 in thick, formerly called the 'Havod' is presumed to represent the No. 1 Rhondda. 80 ft above, the mis-named 'No. 3 Rhondda' represents the No. 1 Rhondda Rider, while two further coals, 53 in (with rashings) and 6 in thick respectively, lie 100 and 140 ft higher in the sequence.

At Penrikyber these measures are markedly argillaceous. A coal, 17 in thick, 45 ft above the top of the Castell-y-Wiever Sandstone, is presumed to be the No. 1 Rhondda, while the Rider or 'No. 3 Rhondda' lies 80 ft above. The 270 ft of strata up to the Brithdir (locally called 'No. 2 Rhondda') include several sandstones and three coals. The thickest of these, consisting of 3 ft of coal and rashings, lies about 90 ft above the No. 1 Rhondda Rider, and 50 ft or so above it 'mottled blue and red marl, clift and rock' are recorded through 70 ft, succeeded by 'white rock'. These coloured beds represent one of the few developments of 'Deri Beds' noted west of the Taff. The 'white' sandstone presumably thickens westwards and becomes the massive pennant between the No. 1 Rhondda Rider and Brithdir horizons.

The **No. 1 Rhondda Rider** (or No. 3 Rhondda as it was known locally) has been worked extensively in the Mountain Ash and Penrhiwceiber areas. The principal collieries were: Fforest Level [05249913], Glyngwyn Level [04949857] and Penrikyber No. 3 Pit. In Gelli-ddu-fawr Plantation [04480063] 34 in of coal overlain by at least 16 ft of massive sandstones were seen in an old adit mouth. The coal rarely exceeded 28 in throughout the main workings, which were limited in almost every direction by thinning to less than 2 ft. The Fforest Level workings were restricted to the east by a washout running N.N.W.–S.S.E. beneath Twyn Brynbychan.

At Lady Windsor Colliery the presumed No. 1 Rhondda, 19½ in of coal with two rashings bands, lies 20 ft above the presumed Daren Rhestyn, with the No. 1 Rhondda Rider, 23 in thick, 62 ft higher in the sequence. The measures above the latter show a greatly increased development of sandstone, and include thin coals 90 ft above the Rider and 50 ft below the Brithdir. The Mynachdy Borehole (see p. 348) proved most of the Rhondda Beds.

The highest strata of the group are well exposed in the landslip scar of Taren Pwllfa [00859810], about a mile south of Cwmaman, where the following section was measured:

	Ft	In
Massive pennant	—	—
BRITHDIR: weathered **coal** with numerous dark shale and fireclay partings	5	2
Pale mudstone seatearth	3	0
Blocky ganister-like siltstone passing into bedded sandy mudstone ..	16	0
COAL		3
Soft seatearth passing into hard silty seatearth	2	10
Shaly micaceous silty mudstones with several bands of sandstone ..	23	2
Grey-blue shales with *Anthraconauta phillipsii*	2	6
COAL		1
Black carbonaceous shales	2	0
Pale greyish green seatearth	2	3
Blue-grey shaly sandy mudstone; sporadic ironstone nodules ..	8	6
Well-bedded sandstone	7	6
Thinly bedded silty sandstone passing into blocky sandy and silty mudstone	20	0+

Nothing was seen of the 'vari-coloured strata' recorded by Moore and Cox (1943, p. 248), apparently from this locality, but red clay soil was observed at the level of the lower portion of this sequence on Twyn-y-Briddallt, ½ mile to the W.N.W.

A.W.W.

RHONDDA FACH AREA

Only the lower strata have been proved in shafts in the lower half of the valley and knowledge of the group is derived mainly from the mapping. The beds show a slight increase in thickness when traced southwards, from about 600 to 650 ft. There is also a general increase in the same direction in the proportion of sandstone, this being particularly noticeable in the middle and upper parts of the group.

The **No. 2 Rhondda** is split into two distinct seams in the south, the lower split retaining the seam name, and the upper one, though unnamed, being the correlative of the Fforest Fach of the Pontypridd area. The lower split was worked extensively in the Ynyshir–Wattstown area, mainly at Standard Colliery and Ynyshir Pit [02559262], where the section varied from 2½ to 3 ft. Northwards the seam thins and is absent, possibly due to washout, in certain of the south Ferndale shafts. North of Tylorstown the seam crops on either side of the valley immediately beneath the lowest of the massive sandstones. It was worked on a small scale from a level [00329700] above

Ferndale Nos. 1 and 5 pits and from numerous crop levels on either side of the valley. Throughout the Ferndale–Maerdy area the coal varies from 16 to 18 in, and is overlain by conglomeratic sandstones. At the western end of Cwar Isaf [98739808] the following section may be seen:

	Ft	In
Massive pennant sandstones	40	0+
Sandstone with irregular layers of conglomerate carrying numerous quartz and coal pebbles, and irregular rafts and streaks of coal ..	3	6
Hard sandstone with abundant rolled pellets of clay-ironstone and irregular streaks of coal	4	0
NO. 2 RHONDDA: **coal** 3½ **in**, parting ½ in, **coal 4 in**, parting ½ in, **coal 8½ in**	1	5
Mudstone-seatearth	5	0

The top split of the No. 2 Rhondda, 10 to 12 in thick, is developed at Lady Lewis and Ynyshir pits and at Standard Colliery, separated from the main seam by 20 to 40 ft of mudstones and seatearth. Traced northwards from Standard the coal disappears and the mudstones below are replaced by massive conglomeratic sandstones. The abundant rafts and pebbles of coal in the sandstone overlying the lower coal in the upper parts of the valley may have originated from the penecontemporaneous erosion of the Fforest Fach.

The massive sandstones above the No. 2 Rhondda crop extensively on the broad rolling moorlands around Maerdy. Upwards of 300 ft thick in this area they thin southwards to 150 ft in the neighbourhood of Ferndale. The thin coal mapped within the sandstone in the Cynon Valley is not developed in the Maerdy–Ferndale area; to the south, however, a seatearth and mudstone 'slack' develops about 30 ft from the top and in the Ynyshir area is associated with a thin coal, presumably the Daren Rhestyn. The No. 2 Rhondda Sandstone was formerly an important source of building stone and was quarried extensively at Maerdy and Ferndale. It forms the bold scarps of Cefn Craig Amos (east of Maerdy), which are accentuated by landslip scars, as well as the magnificent 120-ft bluffs of Craig Rhondda Fach, to the west of Ferndale. Where seen, as for example east of Ferndale, the top 20 ft or so of the sandstone carry intercalations of silty or sandy mudstone, while channel-lenses of silty mudstone are not uncommon among the main mass of sandstones.

On the east side of the valley at Blaenllechau the **No. 1 Rhondda** was proved in a trial level [00029773] close above the No. 2 Rhondda Sandstone. Said to be 18 in thick, the coal was evidently overlain by irony shale (blackband) containing abundant *Anthraconauta phillipsii* and *Carbonita sp.* The seam was worked from numerous strike levels on the west side of the valley (two concentrations being observed between 00589590 and 99389673). Abundant blackband spoil again yielded *A. phillipsii* and *C.* cf. *humilis* together with *Spirorbis sp.* At Tylorstown Railway Station the coal was exposed at platform level and the following section measured:

	Ft	In
Massive sandstones with 5 in of shale at base	15	5
COAL		6
Seatearth and sandstone	6	0
Silty mudstones	14	0
Dark shales and sandy ferruginous shales; *A. phillipsii, A.* cf. *tenuis* and *Spirorbis sp.*		6
NO. 1 RHONDDA: **coal, seen to 12 in**	1	0
Seatearth and mudstone	5	0

The coal was worked from several levels behind Furnace Road on the east side of the valley at Pont-y-gwaith, where the section was said to vary from 2½ to 3 ft. The seam dips beneath the valley bottom in the Wattstown area, emerging again at Ynyshir, about three-quarters of a mile south-east. In National Colliery shaft it is overlain by black shale and has the following section: top coal 10 in, clod 6 in, coal 8½ in. The horizon of the coal is next seen in a small gully [02629248], 170 yd N. of Lady Lewis Pit:

	Ft	In
Gravel overburden	—	—
Massive sandstones with silty mudstone lenses	14	0
Shales	2	0
NO. 1 RHONDDA	smut	
Seatearth passing into sandy mudstones and blue shales	13	2
Ganister-like siltstone	1	6
COAL	streak	
Seatearth and silty mudstone	13	6

The No. 1 Rhondda is closely followed by sandstone which thickens westwards and southwards from a few feet to 50 ft. At Blaenllechau the overlying **No. 1 Rhondda Rider** occurs 50 to 60 ft above the No. 1 Rhondda and has been worked from numerous strike levels over about a mile of crop north-north-west of Blaenllechau Farm [00609666]. Among the spoil from these workings blue-black paper shales with *A. phillipsii* and *A.* cf. *tenuis* have been recorded. The seam was also worked along the crop on the north-eastern flank of Mynydd Ty'n-tyle, west of Tylorstown. It was from this locality that Moore and Cox (1943, p. 245) recorded solid shells identified as '*Anthracomya pruvosti*' and '*A.* cf. *pringlei*' but the occurrence of these forms was not confirmed during the present survey.

North-east of Blaenllechau a thin sandstone follows a short distance above the Rider coal, but the bulk of the 50 ft of measures up to the foot of Craig y Gilwern crags is argillaceous. The massive sandstones of these crags are 50 to 70 ft thick above Blaenllechau, but they thicken rapidly when followed down the valley reaching at least 250 ft south of Ynyshir. At the base an impersistent dirty coal and rashings horizon, seen in places, as, for example, beneath the waterfall [00779682], 200 yd N.E. of Blaenllechau Farm, was worked from strike levels at several points above Tylorstown. The sandstone has been quarried extensively and in the Ynyshir area was the principal source of building stone.

On the west side of the valley above Ferndale and Tylorstown the No. 1 Rhondda Rider is overlain by 20 to 30 ft of sandstone, separated from the main sandstone development of Mynydd Ty'n-tyle by a 'slack' of argillaceous strata 50 ft thick which splits and thins when traced towards the north. The sandstone of Ty'n-tyle is equivalent to that forming the crags of Craig y Gilwern on the east of the valley, and is well exposed in the landslip scar behind Brynbedw [00769512]. It was also extensively quarried on the west side of the valley: opposite National Colliery [01959360], where 40 ft of massive pennant with irregular bands of grey-blue silty mudstone up to 6 ft thick may be seen; and opposite Standard Colliery [02159320], where a face 80 ft high in very massive false-bedded pennant with a few lenses of purplish-grey micaceous silty mudstone up to 10 in thick was measured. At the southern end of the face the base of the sandstone was seen to be conglomeratic with quartz pebbles up to ¾ in and rolled pellets of clay-ironstone. This overlies 4 ft of sandy mudstone resting on a dirty coal.

The sandstones of Craig y Gilwern are overlain by 80 to 100 ft of mudstones with a few thin sandstones, which form the top of the group. The top part of this sequence is presumably similar to that measured at Taren Pwllfa (p. 197). These mudstones thin southwards, but are nowhere exposed. A constant feature on the east side of the valley indicates a persistent sandstone up to 20 ft thick, lying 20 to 30 ft below the Brithdir, extending from the vicinity of Taren Pwllfa in the north to a point about ½ mile S.E. of Lady Lewis Pit in the south.

A.W.W.

NORTHERN RHONDDA FAWR AREA

The beds are fully preserved only to the west of Fernhill, where they are rather more than 600 ft thick, and at Gelli, where they reach about 650 ft.

The **No. 2 Rhondda** has been mined extensively at the head of the valley, and to a smaller extent to the south-east. The infrequent sections recorded show the seam to consist of a top coal, about 10 in thick, and a bottom coal, which on the west side of the valley thickens southwards from 22 to 34 in beneath Mynydd y Gelli. The two coals are separated by mudstone which thickens southwards from about 8 in to the north-west of Fernhill to between 3 and 5 ft west of Graig-fawr corrie. The lower coal includes the Wernddu and Wernddu Rider of the Avan Valley, and the upper leaf appears to equate with the Field Vein. On the east side of the valley, south of Foel-goch, only scattered trials mark the crop of the seam, and a general deterioration in this direction can be inferred. In several places a thin impersistent rider coal lies a few feet above the top coal; it is best seen north of Blaen-y-cwm [91969947], where it comprises two thin leaves, each a few inches thick, separated by about a foot of sandstone.

The sandstone which follows thickens from about 120 ft west of Blaenrhondda to about 225 ft in the south of the area, where several mudstone partings develop within it. As in the Rhondda Fach the base carries up to 3 ft of conglomeratic grit containing pebbles of quartz and coal and abundant rolled ironstone. The sandstone is best exposed in the crags around Blaen-y-cwm, in the corries west of Treorky and on Cefn y Rhondda from Nant Orky to Mynydd yr Eglwys. It has been quarried on a large scale near Abergorky Colliery, at Ty'n-y-bedw and north-east of Pentre. The best section is seen at Ty'n-y-bedw quarry [96909700] where 145 ft of massive and false-bedded sandstones with several bands of siltstone and silty mudstone up to 10 ft thick may be seen, the base lying 10 to 20 ft above the No. 2 Rhondda.

The measures overlying this sandstone are very variable, see Fig. 25. In the south-west at Graig-fawr [92319600] about 15 ft of mudstones are overlain by a thin coal, which has a roof of massive sandstone, locally carrying rolled ironstone pellets near the base. Southwards at Graig Fach [92929542] the mudstones thin, first to as little as 5 ft, though the coal above is still present, and finally both mudstones and coal die out; since there is no obvious sign of erosion at the base of the overlying sandstone, the mudstone and its accompanying coal appear to be cut out from below. From its general relations this coal appears to be the **No. 1 Rhondda**. Northwards from Graig-fawr towards Taren Saerbren [92509755] the 'slack' thickens rapidly and at the latter corrie the following section has been measured:

	Ft	In
Massive sandstone	—	—
NO. 1 RHONDDA: **coal about 12 in**	1	0
Seatearth passing into mudstone; rootlets at base .. .. 21 ft to 29	0
Silty sandstones with mudstone bands	5	0
Quartzitic sandstone .. .. 10 ft to 15	0
Mudstone with fragmentary 'mussels' .. .. 0 to 2	0
COAL .. .. streak	
Sandy seatearth passing into siltstone and mudstone	9	0
Quartzitic sandstone; pebbly at base	6	0
Shaly mudstone	1	0
?DAREN RHESTYN: **coal about 18 in**	1	6
Seatearth passing to mudstone and siltstone	21	0
Seatearth and mudstone with sandstone ribs	4	0
Massive sandstone (overlying No. 2 Rhondda)	—	—

The continuity of the massive sandstone above the No. 1 Rhondda between Graig-fawr and Taren Saerbren is not in doubt, and at no point is there any sign of it cutting down into the underlying beds; it is also unlikely that the No. 1 Rhondda of Graig-fawr splits to give the three coals of the above section. It appears more probable that the bottom 60 to 70 ft of mudstones and associated quartzites described above were not deposited in the Graig-fawr area, where they were replaced by the upper part of the underlying sandstones, which formed a sand-bank against which the overlying mudstones, sandstones and coals lapped. This interpretation is supported by the inverse relationship between the thickness of the 'slack' measures and that of the underlying sandstone. In the light of this interpretation the first coal above this sandstone on Taren Saerbren has been tentatively referred to as the Daren Rhestyn, for it occupies a similar position to, and has much the same relations as that seam in the lower parts of the valley.

With minor variations the above section is maintained northwards to the head of the valley, though there is an overall thinning to about 65 ft north of Pen-pych. South-west of Blaen-y-cwm the middle coal thickens to 12 in and the shelly mudstone overlying it is not present. Farther north at Pen-pych [92000000] the lower quartzite is absent and the Daren Rhestyn is overlain by 20 ft of mudstone, the lower part of which is ferruginous and yields abundant *Anthraconauta phillipsii* and *A*. aff. *tenuis*, associated with *Carbonita humilis* and *Spirorbis sp.* A similar fauna was obtained from spoil from a trial on the coal [91960062], west of Fernhill.

On the east of the valley exposures are poor, but the strata between the Daren Rhestyn and the underlying sandstone are well seen on the Treherbert–Rhigos road [93070050] and are generally similar to those on the west of the valley. The Daren Rhestyn has been worked (under the name of No. 1 Rhondda) at Ty'n-y-Waun Colliery [93300005] where it is reputed to be 18 to 20 in thick. The outcrop of the mudstones below the No. 1 Rhondda can be traced around Mynydd Tynewydd and northwards towards the Rhondda Fach. Southwards on Cefn y Rhondda about 265 ft of measures are preserved above the No. 2 Rhondda but no coals are noted. Three impersistent mudstone 'slacks' have been mapped on the west of Mynydd yr Eglwys, lying 150, 200 and 225 ft above the No. 2 Rhondda respectively, but all disappear rapidly eastwards. While it is possible that these all lie within the No. 2 Rhondda Sandstone, it appears more likely that they represent the Daren Rhestyn–No. 1 Rhondda group.

A massive false-bedded sandstone, 100 to 150 ft thick, overlies the No. 1 Rhondda. It is well exposed on the roadside east of Fernhill, and on the west of the valley it is best seen in the corries above Cwm-parc. On Mynydd Ty Isaf [91959781] it thins to 60 ft; and on Craig Blaenrhondda [91840065] and in Cwm Lluest [91440006] a poorly exposed shale, about 10 ft thick, occurs in the middle.

The **No. 1 Rhondda Rider** succeeds a thin mudstone above this sandstone. The mudstones, 30 ft thick on Mynydd Ystradffernol, thin westwards, and locally on Mynydd Blaenrhondda are cut out by the overlying sandstone. They are barely 5 ft thick on Mynydd Ty Isaf, but they thicken again southwards to 15 ft on the south side of Graig-fawr corrie. The Rider coal can be seen in a few places; it is unlikely that it exceeds 12 in. in thickness and no attempt has been made to work it in this area.

The sandstone above the Rider caps Mynydd Tynewydd and forms much of the plateau west of the valley. It has been quarried on the south side of Graig-fawr where two thin and impersistent mudstones are developed about 70 and 100 ft above the base. In the extreme north-west of the area a mudstone in the latter position forms a feature on the hillside north of Bryn Bach.

The mudstones at the top of the group are confined to an outlier on the Corrwg-Rhondda watershed, extending two miles northwards from Cefn Nant-y-gwair. Much of their crop is concealed beneath peat. On Cefn Nant-y-gwair they are 30 to 40 ft thick and lie little more than 100 ft above the No. 1 Rhondda Rider. The underlying sandstone here is, however, abnormally thin, and to the east and south-east upwards of 150 ft of sandstones are seen with no overlying mudstones.   W.B.E.

SOUTHERN RHONDDA AREA

The group is thinnest in the north and east of the area, being about 650 ft between Llwyn-y-pia and Gelli and 700 ft at Pontypridd and Llantwit Fardre. The measures thicken steadily south-westward and westwards from these localities to rather more than 800 ft to the south-west of Gilfach Goch and west of Tonyrefail. Much of the surface of the area is occupied by strata belonging to this group, the sandstones dominating all the high ground.

The **No. 2 Rhondda** is by far the most important coal of the group and in former days was worked extensively. It forms a single seam over most of the area, but splits into two coals (the Fforest Fach or Bedw Rider above, and the No. 2 Rhondda or Cymmer Vein below) to the north-east of a line extending from Mynydd Troed-y-rhiw southwards towards Trebanog, and thence taking a more easterly course along the southern limb of the Pontypridd Anticline towards Pen-y-coedcau, between Pen-y-rhiw and Cwm collieries.

The seam was worked beneath Mynydd y Gelli, the average section consisting of 34 in of coal. To the west, in the workings beneath Mynydd Bwllfa and Mynydd Ton, the seam thickens and develops a three-leaf structure: top coal 4 to 6 in, clod 7 to 10 in, coal 14 in, clod 12 to 14 in, coal 27 in. In Cwm Clydach, where the crop closely follows the upper limit of the boulder clay, the two top leaves unite to give a double-coal section: top coal 13 to 18 in, rashings 1 in, coal 22 in. At Llwyn-y-pia on the north-east side of the Cymmer Fault, the seam crops beneath drift towards the bottom of the valley. It was mined, 40 in thick, at Sherwood Level, the workings extending towards the Rhondda Fach. In the Porth area some of the oldest pits in the valley exploited the seam on the north side of the valley; the section here was top coal 18 in, holing 2 in, coal 24 in. At Cymmer Colliery, where the seam is split, workings on the lower coal extended southwards to the axis of the Pontypridd Anticline. The section here and farther down the valley towards Llwyncelyn and Hafod varied from 30 to 39 in. Limited areas were also worked in the Pontypridd area, where it was thinner than the upper split or Fforest Fach.

The full section was again worked in the area of the Ely Valley extending from Pen-y-graig to Penrhiw-fer, Collena and Glyn, and showed little variation: top coal 14 to 24 in, holing $\frac{1}{2}$ to 6 in, coal 24 to 27 in. In the Ogwr Fach the outcrop of the seam encircles Gilfach Goch and the coal has been worked extensively; compared with the Ely Valley a thickening of the bottom coal is noticed: top coal 16 to 17 in, holing 8 to 9 in, coal 31 to 34 in. Towards the south-west at Glynogwr [97728912] and Etna [98108910] the parting dies and 48 to 51 in of clean coal are recorded. In the extreme south of the area the coal has been worked only at Coedely Colliery where it was 48 to 51 in thick with plane partings 6 and 27 in from the top of the seam.

The **Fforest Fach** was thick enough to work only in the Trehafod–Pontypridd area. At Pen-y-rhiw Pit the section reads: top coal 16 in, clod 2 in, coal 20 in; the seam is separated from the No. 2 Rhondda by 10 ft of fireclay. North-westwards the seam deteriorates as shown by the following sections: Gyfeillon [05369097], top coal 8 in, dirt 7 in, coal $5\frac{1}{2}$ in, dirt $\frac{1}{2}$ in, coal $7\frac{1}{2}$ in; and Llwyncelyn [03419100], top coal 6 in, dirt 7 in, coal 12 in. The separation from the No. 2 Rhondda increases to a maximum of about 27 ft in the Trehafod area, but decreases north-west of Porth.

The roof of the No. 2 Rhondda seam is often massive sandstone, but where mudstones are developed plants are common. Thus at Ton Level [95759399] the following forms were collected: *Annularia sphenophylloides*, *Asolanus camptotaenia* Wood, *Calamites sp.*, *Neuropteris heterophylla*, *N. ovata*, *N. tenuifolia*, *N.* cf. *scheuchzeri*, *Lepidodendron simile* Kidston *sensu* Némejc, *Lepidostrobus lanceolatus* (Lindley and Hutton), *Sphenophyllum cuneifolium*, *S. saxifragaefolium* (Sternberg), *Trigonocarpus sp.*

The **Daren Rhestyn** and its associated mudstones are not developed south-west of a line drawn from Clydach Vale through the Ely Valley towards Castellau. In these areas massive pennant (varying in thickness from 200 ft on Mynydd Ton to 300 ft at Coedely) continue upwards from the No. 2 Rhondda to close beneath the No. 1 Rhondda. In the main valley and at Cwm Colliery the Daren Rhestyn is always present and the sandstones directly above the No. 2 Rhondda are reduced in thickness; they are 60 to 70 ft at Trehafod, but thicken to 90 or 100 ft to the north-east of Llwyn-y-pia, and to the south at Cwm. The sandstones have been extensively quarried for building stone throughout their outcrop. Though the base is seen in only a few places, the characteristic conglomerates seem to be widely developed.

In the upper reaches of Cwm Cae-Dafydd, separating Blaen Clydach from Clydach Vale, the following section between the No. 2 and No. 1 Rhondda seams was measured:

	Ft	In
NO. 1 RHONDDA: **coal about 18 in**	1	6
Mudstones, shales, etc.	20	0
Massive and flaggy sandstones with silty partings	9	0
Buff-weathered shaly mudstone	10	0
DAREN RHESTYN: **coal** (not seen)	—	—
Dark carbonaceous shale on irregularly bedded blue-grey silty mudstone	2	0
Massive and flaggy sandstones with thin mudstone band	7	6
Dark blue-grey mudstones and flaggy sandy mudstones; poorly exposed	25	0
Grey-blue silty mudstones interbedded with thinly flaggy sandstones ..	29	0
Massive false-bedded sandstones .. .. about	100	0
NO. 2 RHONDDA: **coal and parting about 48 in**	4	0

The Daren Rhestyn was worked from numerous bell-pits, while the No. 1 Rhondda crop is marked by a continuous line of strike-levels.

On the south side of Cwm Clydach these strata are again well exposed in the crags of Taren Ty-cneifio and in the gulleys between here and Craig Pwll-yr-hebog. In one gully [97449235], 550 yd S.E. of Cambrian Colliery, the following section is exposed:

	Ft	In
NO. 1 RHONDDA and associated mudstones	—	—
Massive false-bedded sandstones	50	0
Gap	7	0
Blue-grey shales; *Anthraconauta phillipsii*	4	0
DAREN RHESTYN	smut	
Brown-weathering mudstones and shales	28	0
Massive and false-bedded sandstones	80	0
NO. 2 RHONDDA	—	—

When the argillaceous beds associated with the Daren Rhestyn are traced westwards into the precipitous crags of Taren Ty-cneifio they are seen to die out, leaving only massive sandstones between the No. 2 Rhondda and the 'slack' of the No. 1 Rhondda. In a similar fashion the Daren Rhestyn horizon fails when traced westwards from the slopes above Gelli to Tarren-y-bwllfa, as does the feature associated with the No. 1 Rhondda.

Both the **Daren Rhestyn** and the **No. 1 Rhondda** have been worked from numerous crop levels on the hillside beneath Cefn Pen-Rhys and Mynydd Brith-weunydd, east of Llwyn-y-pia. Here the 90 ft or so of sandstones above the No. 2 Rhondda are succeeded by about 100 ft of almost wholly argillaceous strata, the Daren Rhestyn lying about 60 ft above the base and the No. 1 Rhondda at the top. The lower coal was worked commercially from Mountain Level [99979350], where the section was given as top coal 10 in, rashes ½ in, coal 2 in, rashes ¾ in, coal 11 in. The No. 1 Rhondda above is overlain by blackband containing *Spirorbis sp.*, *Anthraconauta phillipsii* and *Carbonita* cf. *humilis*. On the west side of the valley the lower coal was worked from levels [99209360], while the No. 1 Rhondda, 30 ft above, is separated from its overlying sandstone by only 9 in of shelly blackband.

In the Pen-y-graig and Williamstown areas the No. 1 Rhondda, about 12 in thick, has been proved in several levels on either side of Nant Ffrwd-amws, while the Daren Rhestyn below is overlain by sandstone which dies out south-eastwards.

On the north-west side of Mynydd y Cymmer in Nant Graig-ddu the following section was measured:

	Ft	In
Massive sandstones	100	0
COAL	thin	
Measures, mainly mudstone	20	0
NO. 1 RHONDDA: **coal 30 in**	2	6
Measures, mainly mudstone	18	0
DAREN RHESTYN: **coal 10 in**, clod 6½ in, **coal 3 in,** clod 1 in,		
coal 15½ in	3	0
Measures, mainly mudstone	30	0
COAL		6
Measures, mainly mudstone	20	0
Massive sandstones .. .. about	100	0
NO. 2 RHONDDA	—	—

In the Hafod area the Daren Rhestyn, a compound dirty coal, lies about 120 ft above the Fforest Fach and is separated from the No. 1 Rhondda by 23 ft of fireclay and shale. The No. 1 Rhondda was worked at several points along the crop in the area; its section in Gyfeillon Pit is as follows: top coal 4 in, carbonaceous shale 12 in, coal 18 in, clod 3 in, coal 15 in. Traced north-westwards above Cymmer the seam splits, both leaves being worked from levels on the crop. In spoil from Johnson's Level [02379022], 240 yd S.E. by S. of Ty'n-y-berllan the following flora and fauna was collected (presumably from the roof of the lower split): *Annularia sphenophylloides*, *Asterotheca miltoni, Mariopteris nervosa, Neuropteris flexuosa* Sternberg, *N. rarinervis, N. scheuchzeri, Sphenophyllum cuneifolium*; *Spirorbis sp.*; *Anthraconauta sp.* mainly *phillipsii*; *Carbonita* cf. *humilis*; '*Estheria*'?; fish debris.

The No. 1 Rhondda was worked at several localities in the Hopkinstown area, and on either side of Nant Gelli-wion above Maritime Colliery. Spoil from levels near Dan-y-Coedcae Road [07008930] yielded the typical shelly blackband.

Little is known of the No. 1 Rhondda horizon in the Ogwr Fach Valley. The 'slack' feature associated with the seam can be followed on Mynydd Maesteg and on the north-east slopes of Mynydd Maendy, to the west of the valley. On the north-east side, however, the feature fails locally, and the coal and its accompanying mudstones may be cut out by washout. In the Cilely area the seam is split in much the same manner as at Cymmer; the lower coal was worked on a small scale at Edmondstown and north of Cilely.

The No. 1 Rhondda has not been proved on the south limb of the anticline, though the associated feature can be seen; the Daren Rhestyn on the other hand appears to be absent. At Coedely Colliery the presumed equivalent of the No. 1 Rhondda (top coal 21 in, clod 8 in, coal 15 in) lies 316 ft above the No. 2 Rhondda, the intervening strata consisting almost entirely of pennant. At Cwm Colliery both the Daren Rhestyn and No. 1 Rhondda appear to be developed, the former 110 ft above the No. 2 Rhondda and the latter 40 ft higher.

The sequence above the No. 1 Rhondda was measured on the southern slopes of Mynydd Bwllfa [97109320], ½ mile W. of Cwm Cae-Dafydd:

	Ft	In
NO. 1 RHONDDA RIDER	—	—
Measures, mainly mudstones	about 50	0
Massive irregularly bedded sandstone	40	0
COAL	thin	
Buff-grey seatearth passing to black, rusty weathering mudstones ..	10	0
Massive false-bedded sandstones passing westwards into strong silty mudstones with thin bands of sandstone	30	0
Blackband shale; abundant *Anthraconauta phillipsii*		6
NO. 1 RHONDDA: **coal 18 in**	1	6

Traced westwards towards the head of Cwm Clydach, the sandstone immediately above the No. 1 dies out and the thin coal above probably joins with the main seam.

Throughout the Llwyn-y-pia–Trealaw area the No. 1 Rhondda is followed by massive sandstones thickening southwards from 40 to 70 ft. On Mynydd y Cymmer, where they reach about 100 ft, they form a gently sloping flat cap to the hill, surrounded by a striking, almost circular, scarp, accentuated by landslip on the northern side and by quarries on the east. Irregular bands of conglomerate with impersistent streaks and rafts of coal are common, and are particularly well seen in the quarries. In the Hafod area the sandstones, 50 to 80 ft thick, have been quarried at several points. A marked thickening evidently takes place south-eastwards for in the Cwm shafts 140 ft was proved, the thickest record in the east of the district.

The **No. 1 Rhondda Rider** is developed almost everywhere, but it has seldom been worked and little is known about it. It lies 100 to 150 ft above the No. 1 Rhondda and is underlain by well-developed mudstones, which make a readily traceable feature. One of the few places where the coal, and its associated strata, can be seen is in Cwm-George [04339140], on the north side of the valley at Hafod:

	Ft	In
Massive pennant sandstone	about 220	0
COAL	1	0
Seatearth and mudstone	15	0
Massive sandstones	12	0
Mudstones	4	6
NO. 1 RHONDDA RIDER: **coal 23 in**	1	11
Mudstones, etc.	20	0
Massive sandstones	40	0
Grey shales	2	0
NO. 1 RHONDDA: **coal 31 in**	2	7
Seatearth and mudstones	30	0+

A similar sequence evidently persists to the Ely Valley where on Mynydd Pen-y-graig, to the west of Williamstown, the No. 1 Rider was proved in levels 20 ft below a thin coal at the base of massive sandstones, the strata between the two coals being wholly argillaceous. Spoil from a level [99859048] on the upper coal, which is 12 in thick, yielded blue-black shales with *A. phillipsii*. When traced to the western parts of the area, and southwards to Coedely and Cwm collieries, the coals appear to coalesce for only one thin coal is developed more or less at the base of the main sandstone. At Cwm it lies 180 ft above the No. 1 Rhondda, but at Coedely this distance appears to be reduced to about 50 ft.

The No. 1 Rhondda Rider is overlain by one of the principal sandstones of the Pennant in this part of the Coalfield: 200 to 250 ft thick over most of the area, it thickens south-westwards to nearly 400 ft at Coedely. It caps the hills between the Gelli–Nant Ian area and Cwm Clydach and between the latter and the Ogmore and

Gilfach areas; it forms the bold slopes of Mynydd Brith-weunydd and the watershed area of Cefn Pen-Rhys; it caps Mynydd Pen-y-graig and Mynydd Dinas (the south-western scarp of which is one of the boldest features in the central part of the coalfield) and has extensive outcrops around the flanks of Mynydd y Glyn.

The uppermost strata of the group, which underlie the Brithdir coal horizon, are mainly argillaceous. They are preserved only on the highest ground of Twyn Bryn-y-beddau and Mynydd Troed-y-rhiw and on the southern limb of the anticline between Gilfach Goch and Pontypridd. They are poorly exposed. On Mynydd Troed-y-rhiw about 120 ft of strata crop out, and a thin coal occurs at about the middle; mudstones predominate below the coal, and flaggy sandy mudstones above. On the south limb of the anticline trial levels west of Tylawynder [04048877] and on the Graig, south of Pontypridd, indicate that the Brithdir lies at the top of the 'slack'. At Coedely 100 ft of argillaceous strata, including several fireclay horizons, occur at the top of the Downcast Shaft, but there is no associated coal. The shafts at Cwm show about 120 to 130 ft of mudstones, with much fireclay and several thin coals, including a dirty coal and rashings horizon at the top, which presumably represents the Brithdir. A.W.W.

OGMORE–GARW–LLYNFI AREA

The group is fully developed only in the southern part of the area on the south limb of the Maesteg Anticline. In general the beds thicken from north-east to south-west and 1100 ft are present to the south of Maesteg.

On the east side of the Ogmore Valley the **No. 2 Rhondda** comprises a single seam, worked from Aber [93479055] and Cwm y Fuwch [94909100] collieries with a section varying as follows: top coal 18 to 26 in, clod 27 to 33 in, coal 33 to 42 in. At Rhondda Main [93608900] the section was more complicated, the top coal splitting into two or three leaves, as beneath Craigcaedu [92808990]: top coal 22 in, 'rubbish' 5 in, coal 3 in, 'rubbish' 2 in, coal 3 in, 'rubbish' 33 in, coal 34 in. The seam splits west of Ogmore Vale, a separation of 5 ft being recorded at Tynewydd [93059064], and from here the line of split follows a north-westerly course to the head of the Garw Valley. To the south-west only the lower coal (still called the No. 2 Rhondda) has been worked to any degree. Throughout the area extending from Mynydd Llangeinor, west of Ogmore Vale, to Mynydd Moelgilau on the south side of the Garwfechan the section varied from 32 to 40 in of clean coal. At the northern end of the Garw a separation of 3 ft between the two coals was proved in a level [90209460] near the landslip on Werfa, while on the western side of the valley the lower coal was worked extensively from the Glenavon–Garw Colliery with openings above International Colliery [89089360] and from Blaencaerau Level in the Avan Valley [87469493].

On the north and north-west sides of the Llynfi Valley the lower split (or **Wernddu**) has been worked beneath Foel Fawr from Glyncymmer Level [85669595] in the Avan Valley, and on either side of Foel y Duffryn. On Mynydd Caerau and Foel Fawr the separation between the two splits is in places as little as 2 ft, but on Foel y Duffryn it appears to reach 50 ft with the development of a prominent sandstone between the coals. In this area the Wernddu is about $2\frac{1}{2}$ ft thick and the top split (or **Field Vein)** little more than 12 in. On the south side of the anticline the two coals crop on the western side of Moelcynnhordy, the Wernddu being worked from numerous openings on the crop. At Gelli-hir Colliery the section of the two coals is:

	Ft	In
FIELD VEIN: **coal 18 in**	1	6
Measures, mainly mudstone	30	0
NO. 2 RHONDDA or WERNDDU: **coal 10 in**, parting 2 in, **coal 34 in**,		
rashes 6 in	4	4

In the Garwfechan the No. 2 Rhondda lower split is overlain by blocky mudstones containing an abundant flora, and the following forms were collected from spoil 1300 yd S.S.W. of Pontycymmer Church [89829034]: cf. *Asterotheca miltoni*, *Lepidodendron sp.*, *Lepidophloios laricinus*, *Mariopteris nervosa*, *Neuropteris heterophylla*, *N. ovata* Hoffman, *N. ovata* forma *flexuosa* Sternberg, *N. rarinervis*, *N. scheuchzeri*, *Sigillaria ovata*, *Sphenophyllum cuneifolium*, *S. emarginatum* Brongniart, *S. majus* Bronn, *Sphenopteris neuropteroides* Boulay, *S. striata*.

In the Ogmore Valley the unsplit seam is overlain by 30 to 40 ft of blocky, spheroidally weathering mudstones. At the head of the Garw Valley this mudstone dies out and the coal is overlain immediately by massive pennant sandstone. Elsewhere in the Garw and in the Llynfi Valley where the seam is split the Field Vein lies at or near the top of its associated 'slack', overlain by massive pennant.

The massive sandstone above the No. 2 Rhondda varies from about 200 ft at the head of the Ogmore to about 450 ft in the south-west around Mynydd Moelgilau and Moelcynhordy. The base is usually marked by the development of irregular rafts and streaks of coal and layers of conglomerate with abundant rolled pellets of clay-ironstone.

On the north-eastern slopes of the Ogmore Valley the No. 1 Rhondda appears to be developed only intermittently, if at all, and the bold steep scarps of Craig-y-Geifr and the west side of Mynydd William Meyrick are formed by the combined massive sandstones between the No. 2 Rhondda and the No. 1 Rhondda Rider seams. The No. 1 Rhondda horizon and its associated argillaceous beds also appear to fail on the steep corrie slopes of Craig Ogwr at the head of the valley. Good sections in the massive pennant above the No. 2 Rhondda are seen here, particularly in the cuttings associated with the road over Bwlch-y-Clawdd, where the following section may be seen [93769458]:

	Ft	In
Massive, false-bedded pennant sandstone with irregular rafts and streaks of coal at base ..	—	—
Shales and silty mudstones ..	6	0
Irregular sandstone ..	2	9
Blocky silty mudstone with thin sandstone layers and impersistent coal streaks ..	4	6
Massive, false-bedded pennant sandstones .. about	120	0
NO. 2 RHONDDA position ..	—	—

It is possible that these irregular mudstones and streaky coal developments are associated with washout conditions in the No. 1 Rhondda.

The No. 2 Rhondda Sandstone is well exposed in the scars at the back of the large landslip on the hillside west and north-west of Wyndham, and it forms the broad mountain top of Mynydd Llangeinor in the crest of the Maesteg Anticline. At the head of the Garw Valley the sandstone is about 330 ft thick and contains several relatively persistent mudstone and shale bands. Good sections can be seen in landslip scars at Tyle'r Fedwen and Darren-goch. Mudstone 'slacks' can also be seen on the south limb of the Anticline on the Rhyl between the main valley and the Garwfechan and on the adjoining slopes to east and west.

In the Llynfi area the sandstone above the Field Vein is thickest at Moelcynhordy, where it is 350 to 400 ft. On this hillside it again contains several argillaceous bands, one being particularly prominent and lying about 200 ft above the base. On Foel y Duffryn the thickness of the sandstone is difficult to measure since the position of the No. 1 Rhondda is in doubt, but it appears to be about 325 ft and includes a thin mudstone associated with an impersistent coal about 60 ft above the base. On Mynydd Caerau the sandstone is nearly 300 ft thick but locally, as at the east side of Blaencaerau Farm, it thins to about 225 ft.

Little is known about the **No. 1 Rhondda** in the Ogmore Valley. Away from the north-eastern slopes its presumed position can usually be followed by features and scattered shale outcrops. A smut about 180 ft above the No. 2 Rhondda may be seen on a narrow ledge among the crags on the eastern slopes of Braich-yr-Hydd [92709450] at the head of the valley, where it is associated with a thin development of shale and silty mudstone. The crop of the seam is presumably marked by a line of springs and seepages on the south side of Cwm y Fuwch. The coal is evidently thin: in the Rhondda Main shafts it appears to be represented by 8 to 12 in of coal and rashings about 450 ft above the No. 2 Rhondda. The section improves in the Garw Valley. In a level [89339414] on the south-east slopes of Llyndwr Fawr, north-west of Blaengarw, it is reported to be: top coal 6 in, parting $\frac{1}{4}$ in, coal 19 in, holing 3 in, coal $6\frac{1}{2}$ in. On the south limb of the anticline the seam has been worked from numerous levels on the eastern slopes of Mynydd Moelgilau and on either side of the Cwmcedfyw Valley. Here the section varies as follows: sandstone roof, clod 12 to 21 in, rashings (sometimes recorded as inferior coal) 9 to 10 in, coal 19 to 22 in, fireclay 6 to 12 in, coal 3 in.

In the Llynfi Valley the coal lies at or near the top of a 20-ft mudstone 'slack', which appears to fail westwards. It has been worked from several levels near Cynhordy Farm [88458905] with sections similar to those in Cwmcedfyw. On Mynydd y Caerau the main coal lies 10 ft below the overlying sandstone, the intervening mudstone including a second thin smut.

The roof of the No. 1 Rhondda has not been seen during the present survey but spoil from a level, 1000 yd N.N.W. of International Colliery, includes blackband-shale containing *Anthraconauta phillipsii*, *A. sp.* intermediate between *phillipsii* and *tenuis* and *Carbonita sp.* The level-mouth showed massive pennant sandstone with a band containing abundant rolled clay-ironstone pellets and a few coal streaks towards the base.

The strata between the No. 1 Rhondda and the No. 1 Rhondda Rider seams vary considerably both in thickness and lithology. On the south side of Braich-yr-Hydd, about 1 mile N.W. of Nant-y-moel, the separation is about 100 ft, the bulk of which is massive pennant sandstone. Traced southwards in the Ogmore Valley this appears to thicken to about 250 ft in Cwm y Fuwch and Cwm-cyffog; sandstones again predominate, but an impersistent band of silty mudstone appears about 150 ft above the No. 1 Rhondda. In the Garw Valley the two seams are about 100 ft apart, the sandstone overlying the No. 1 Rhondda again predominating. Traced westwards over Mynydd Moelgilau the sandstone thins, while the argillaceous strata beneath the No. 1 Rider thicken, and in parts of Cwmcedfyw they make up the bulk of the measures between the two seams.

In the Llynfi Valley the No. 1 Rhondda is succeeded by pennant sandstone varying from about 70 ft on the east side of the valley to about 125 ft north-west of Caerau. On Moelcynhordy the base of the sandstone is conglomeratic. The mainly argillaceous strata at the top of the cyclothem varies from 25 to 50 ft to the west of Foel Fawr, where it contains a sandstone intercalation.

The **No. 1 Rhondda Rider** Seam is almost unknown within the area, though its position is traceable almost everywhere from the feature made by the underlying mudstones. A thin smut is seen on the eastern slopes of Llyndwr Fawr and trial levels are located on the seam on the south side of Cwm Fforch and in Cwmcedfyw. At this last locality the spoil includes blue-black paper-shale with *A. phillipsii*.

The succeeding massive sandstones are about 300 ft thick and cap the hilltops of Mynydd William Meyrick, Braich-yr-Hydd, Werfa, Llyndwr Fawr, and Mynydd Caerau, around the heads of the three valleys. Only to the south of the anticline (on the flanks of Mynydd y Gwair, Pen-y-foel, and Mynydd Moelgilau) is the full thickness preserved. Here the mudstones at the top of the group, which immediately underlie the presumed Brithdir horizon, are thin, seldom exceeding 25 ft. Though they make a readily identifiable 'slack' feature only rarely can exposures be seen.

A.W.W., W.B.E.

AVAN VALLEY AND BRYN AREA

These are the most variable beds of the Pennant sequence in this area, and they range in thickness from less than 650 ft in the Glyncorrwg area to over 950 ft at Cwmavon.

The **No. 2 Rhondda** is a single seam at Glyncorrwg: top coal 10 in, parting 10 in, coal 34 in. Traced southwards it is seen to split into three distinct seams, as shown by the following sections:

		Whitworth		Duffryn Rhondda		Glyncymmer		Ynyscorrwg	
		Ft	In	Ft	In	Ft	In	Ft	In
FIELD VEIN { coal	..	1	3	1	4	1	4		
parting	..		6		7	1	3	1	3
coal	..		8		10		10		
Measures ..	..	35	0	30	0	1	6		6
WERNDDU coal ..	..	2	10	2	9	3	0	2	10
Measures	..	10	0	2	4		2		8
WERNDDU RIDER coal..			7	1	0		3		4

Around Cymmer a further parting develops locally near the top of the main (Wernddu) leaf. The main split between the Field and Wernddu runs approximately north-west to south-east through Cymmer towards Blaen-gwynfi, where it swings north-eastwards. In the latter area the section resembles that at Glyncorrwg except that the parting between the two leaves reaches about 6 ft. Towards the south-west of the area the bottom leaf of the lower seam splits away and thickens locally into a workable coal termed the Wernddu Rider. The term 'Wernddu' should properly be restricted to this area, but for convenience it is also applied to those areas where only the Field Vein has split away. In the Neath Valley workings from Glyncastle, Ynisarwed and Glyn Merthyr collieries, in the Merthyr Tydfil district, extend into the present area. The section of the 'Ynisarwed' Seam at that colliery is: top coal 11 in, parting ½ in, coal 17 in, parting 2 in, coaly rashes 12 in; and of the 'No. 2 Rheola' Seam at Glyncastle: top coal 7 in, parting ½ in, coal 3 in, parting ½ in, coal 16½ in, parting ½ in, coal 11 in. Robertson (1933, p. 172) suggested that the 'No. 2 Rheola' represents only the upper part of the No. 2 Rhondda; it now seems, however, that it and the 'Ynisarwed' are both the equivalents of the whole of that seam.

The seam has been worked extensively along both flanks of the Cefnmawr Syncline and its continuity with the No. 2 Rhondda of the upper Rhondda Fawr has been clearly established by workings from Blaen-gwynfi and Glyncorrwg. Extensive workings underlie the watershed between Bryn and the Avan, from levels in Cwm-yr-Argoed, and from Cynon, Parc y Bryn, and Oakwood collieries.

The roof of the Wernddu, and locally that of the Field Vein, yield plants, and the following species have been identified in material from Glyncymmer [85679595] and North Rhondda [89000100] levels: *Annularia radiata*, *A. sphenophylloides*, *Asterotheca abbreviata* (Brongniart), *Lepidophloios laricinus*, *Neuropteris flexuosa*, *N. ovata*, *N. rarinervis*, *Sphenophyllum emarginatum*, *S.* cf. *majus*, *Sphenopteris neuropteroides*.

The roof of the Field carries several feet of planty mudstone in a few places, but more usually it consists of massive sandstone, commonly conglomeratic at the base. These sandstones continue upwards for about 200 ft in the north-east of the area and for 350 ft or more around Pont Rhyd-y-fen. Two impersistent mudstones are developed. The lower lies 10 to 40 ft above the Field Vein and is about 10 ft thick in Nant Gwyn, and in Cwm-yr-Argoed carries a thin coal. The upper one is developed only on Mynydd Pen-hydd, where it is up to 30 ft thick and lies 150 ft above the Field.

The sandstone is succeeded by a variable group of mudstones, sandstones and coals associated with the horizon of the No. 1 Rhondda. These are best developed in the Corrwg Valley, where the sequence is very similar to that in the northern parts of the Rhondda Fawr. On Moel Yorath [89300005] the following section was measured:

	Ft	In
Massive sandstone with quartz conglomerate at base	—	—
NO. 1 RHONDDA: **coal about 18 in**	1	6
Gap with seatearth	15	0
Quartzitic sandstone	8	0
Mudstone; fragmentary 'mussels' and ostracods near base	2	0
Striped beds; sporadic roots	4	0
Gap	8	0
COAL	2	3

The lower seam, called the 'Fforest Fach' at Blaencorrwg Colliery [89460067], is believed to be the equivalent of the Daren Rhestyn of the Rhondda. Farther down the valley the shafts at Ynyscorrwg and Nantewlaeth provide similar sections:

Ynyscorrwg	Ft	In	Nantewlaeth	Ft	In
Sandstone	—	—	Puddingstone	—	—
NO. 1 RHONDDA: **coal**			NO. 1 RHONDDA: **coal**		
21 in	1	9	**15 in,** holing 2 in, **coal**		
Fireclay and clift ..	47	3	**4 in**	1	9
COAL	1	0	Fireclay and clift ..	11	9
Fireclay and rashes ..	12	0	Rashes	1	6
COAL		10	Clift	18	0
Rashes and clift	17	2	COAL		5
DAREN RHESTYN: **coal**			Rashes and fireclay ..	3	6
16 in, stone 2 in, **coal**			COAL		9
10 in	2	4	Rashes and fireclay ..	5	6
Clift with rock bands ..	47	0	COAL	1	5
Sandstone	—	—	Rashes and fireclay ..	8	6
			DAREN RHESTYN: **coal**		
			26 in, rashes 4 in ..	2	6
			Fireclay and pouncing ..	3	6
			Sandstone	11	3
			Rashings, clift and fireclay	21	2
			Sandstone	—	—

Traced westwards towards the Vale of Neath the sequence changes. At Glyncastle the **No. 1 Rhondda** can be seen just above pit-top. It has been worked from the adjoining Ffalydre Colliery where a massive sandstone overlies the following section: bast 1 to 4 in, coal 20 to 24 in, parting 2 to 4 in, coal 6 in. In the Glyncastle shaft a coal 50 ft below the No. 1 Rhondda, and 30 ft above the underlying sandstone, appears to represent either the Daren Rhestyn or the composite coal horizon just above; its section is: top coal 2 in, parting 7 in, coal 12 in. These lower coals fail south-westwards, and on the west of the valley [80650172] near Ynisarwed Farm the entire group is represented by 20 to 30 ft of mudstones with the No. 1 Rhondda at the top. Similar conditions obtain in the workings from Glyn Merthyr Colliery, where sections of the No. 1 Rhondda show a top coal of 14 in and a bottom coal of 10 in; and at Clyne Merthyr [80430037], where, as the Penygraig Vein, it was proved in the Clyne Pit with the following section: top coal 18 in, clod 15 in, coal 9 in.

Southwards from Nantewlaeth similar changes occur. In Nant Gwynfi and also west of Aber-gwynfi [89139613] the No. 1 Rhondda 'slack' thins considerably, but still carries two coals. It is further reduced in Nant Gwyn and may die out altogether in Nantcynon. The No. 1 Rhondda has been worked from minor levels on the south side of Mynydd Blaengwynfi, and on the south side of Cwm Nant-ty. To the south of

Nantyfedw the seam again lies at the top of 10 to 15 ft of mudstone, but westwards from here to Cwm-yr-Argoed the horizon cannot be recognized with certainty. In the intervening ground two impersistent mudstones can be traced on Foel Trawsnau, but they cannot be positively correlated. From the slope between Tarren Forgan [81409450] and Mynydd Pen-hydd to the western margin of the district a thin 'slack' can be followed in which the associated coal is probably the No. 1 Rhondda.

The No. 1 Rhondda is usually succeeded by 60 to 120 ft of pennant sandstone, but these thin northwards from Nantewlaeth and are less than 10 ft thick at Ynyscorrwg before apparently thickening again north of Glyncorrwg.

The **No. 1 Rhondda Rider** is persistent throughout the area though neither the coal nor the associated mudstone is ever very thick. Around Glyncorrwg the 'slack' is barely 10 ft thick and at Ynyscorrwg the coal is only 6 in. At Nantewlaeth the mudstones total about 25 ft and the Rider at the top is 14 in thick with a second 5-in leaf 10 ft below. Exposures are poor around Cymmer but the coal has been proved by trials at several localities. Around Pont Rhyd-y-fen it has been worked in a series of small mines; here it is called the **Graig Isaf** and is reputed to be 22 to 24 in thick. The coal appears to have been proved at a depth of 858 ft in the Whitworth shafts [79929680] with the section: top coal 18 in, fireclay 12 in, coal 6 in. In the Vale of Neath it crops out west of the river near Ynisarwed Farm, but here appears to be little more than 12 in thick.

Around Cymmer the succeeding sandstones attain 400 ft and they form the thickest individual development of sandstone in the area. They thin to about 250 ft at Pont Rhyd-y-fen, 200 ft at Blaen-gwynfi, 350 ft at Ynyscorrwg and 250 ft at Glyncorrwg. In the north of the area mudstones develop in the middle. In the Vale of Neath these thicken to 30 or 40 ft and are separated from the Rider by 200 ft of sandstones, and from the mudstones beneath the Brithdir by about 120 ft.

The argillaceous beds at the top of the group are nowhere well exposed, but they appear to maintain a constant thickness of 30 to 40 ft. In Melin Court Brook [92540169] a thin coal is present near the base.

SOUTH CROP

Cwm Wernderi area: The **Wernddu** has been worked extensively under the name 'Rock Fawr' from Cwmgwyneu and Glenhafod collieries. At the former the section is recorded as: top coal 40 in, rashes 6 in, coal 6 in; to the north the seam deteriorates rapidly, probably as a result of washout. At Glenhafod the seam consists of: top coal 24 in, rashes 4 in, coal 18 in, inferior coal and rashes 12 in. Between 20 and 40 ft above, a thin coal has been proved and 9 ft above this *A. phillipsii* and ostracods are recorded. This horizon is followed upwards by 70 to 90 ft of sandy mudstones and thin-bedded sandstones to the Field Vein, which usually varies from 18 to 24 in with a thin parting about 8 in from the top.

About 400 ft of massive sandstones, broken only by a thin impersistent mudstone about 300 ft above the base, succeed the Field Vein. The **No. 1 Rhondda** can be traced on the south side of Cwm y Garn, but northwards it deteriorates and appears to die out south of Cwmgwyneu Colliery. The **No. 1 Rhondda Rider** occurs about 100 to 150 ft above; except at Bryn, where they attain 60 ft, mudstones at the top of this group are normally about 25 ft thick, The Rider is exposed in Cwm Wernderi [81459028] though no section is available. There are numerous exposures of the coal, about 12 in thick and overlain by sandstone, south of Bryn in Nant Farteg-fach and the stream a few hundred yards to the east. A second thin coal occurs 50 ft or so below near the base of the mudstones, but no details are known.

The sandstones above the Rider vary from 200 ft south of Bryn to 250 ft at Cwm Gwineu. A mudstone, 20 ft thick, lying just above the middle may be at the same horizon as that noted in the Vale of Neath. The sandstones have been quarried near

Maesteg and adits [83459133] have been driven in them to augment the water supply of that town.

The mudstones at the top of the group are usually about 30 ft thick and are exposed on Mynydd Bach. Near the head of Nant Sychbant, in excavations for a reservoir [830898], they are extremely thin, and it is possible that they and the Brithdir coal are both affected by washout associated with the overlying sandstone. In Cwm y Garn [809895] the 'slack' thickens to nearly 100 ft and includes a sandstone, locally up to 40 ft thick.

The full thickness of the Rhondda Beds in this area is about 900 ft.

West of the Ogmore Valley: The Rhondda Beds here reach their thickest in the district, ranging from about 1000 ft north of Aberkenfig to 1150 ft north of Margam. They are predominantly arenaceous and the mudstones are generally thin and confined to the top of each cyclothem.

The **No. 2 Rhondda** is everywhere split. The lower seam, the correlative of the Wernddu, was known over most of the area as the **Malthouse**, though at Brombil, in the extreme west, the term 'Rock Fawr' was used in mis-correlation with the No. 3 Rhondda (see p. 189). At this latter locality the worked section was about 4 ft. At Aberbaiden Colliery a borehole recorded top coal 41 in, dirt 48 in, coal 15 in, dirt 31 in, coal 6 in. The coal was worked opencast between Nant Craig yr Aber and Baiden Farm [87058505]; the section here proved to be very variable, being much affected by irregular cutting down of the massive sandstone roof. A small area was worked at Coytrahen Colliery with the following section: weak clift roof, coal 49 in, rashes 16 in, fireclay 34 in, coal 12 in, rashes 17 in, fireclay 52 in, rashes 12 in, coal 6 in, fireclay 42 in.

Little is known about the upper split or **Field Vein**: it appears to overlie thin mudstones, which are separated from the Malthouse by about 100 ft of sandstone. At Brombil it was known as the 'Rock Fach' but was not worked. Its smut may be seen on the roadside [79558727] 300 yd N.W. of Crugwyllt Farm, and the following section was proved in a trial pit nearby: sandstone roof, top coal 12 in, coal and clay 14 in, coal 16 in, inferior coal 5 in. The feature associated with the coal is not everywhere distinct, and the horizon may fail locally.

The Field Vein is succeeded by massive pennant sandstones, with only a few thin and impersistent silty mudstone layers. These sandstones vary in thickness from about 270 ft in the Ogmore Valley to about 500 ft in the Margam area, and they occupy a narrowing belt extending from Cefn Crugwyllt and Craig Cwm-Maelwg, along the southern slopes of Moel Ton-mawr, Mynydd Ty-talwyn and Mynydd Baiden to Coytrahen. They are overlain by mudstones associated with the **No. 1 Rhondda** or **Cildeudy** Seam (formerly referred to as the Cildy-du), the coal apparently lying at, or close to the top of a 'slack' feature, which can be readily traced.

The smut of the Cildeudy has been seen at several localities, but details of the seam are known only on the southern slopes of Mynydd Baiden. Here an opencast trial pit [87328523] showed: top coal 3 in, clay 16 in, coal 8 in, clay 45 in, coal 24 in, clay 2 in, coal 12 in. The seam was formerly worked on a small scale from levels near Cildeudy Farm [84438524], on both sides of the Llynfi Valley at Coytrahen [89028568; 89408572], and near Abernantclydwyn Farm [91408540] in the Ogmore Valley.

Nothing is known of the section of the No. 1 Rhondda Rider, which lies at or near the top of a well-developed 'slack' traceable about 100 to 200 yd north of the Cildeudy crop. The intervening strata consist predominantly of sandstone surmounted by about 20 ft or so of argillaceous beds. Small crop-levels have proved the coal at several localities: as for example 75 yd E.N.E. of Llan-ton-y-groes [79808824] on the northern slope of Cwm y Brombil, and immediately beneath the old tramway to the north-west of Coytrahen [88778577; 88898585], where the spoil yielded striped blackband with *Anthraconauta phillipsii*. In the Ogmore Valley several old levels worked what is thought to be this coal at the edge of the alluvium on the north bank of the river north-north-east of Ynyslas Isaf [92308590].

The highest beds of the group consist of sandstones, 500 to 550 ft thick over most of the area, but thinning to 380 ft on Mynydd Margam, Mynydd Baiden and Mynydd Ty-talwyn. A thin mudstone is traceable 80 to 100 ft above the Rider, while on Mynydd Ton-mawr and Mynydd Margam a second 'slack' in the upper part of the sandstones can be followed into the upper reaches of Cwm Gwineu where it widens and unites with the mudstone 'slack' beneath the Brithdir to form almost 100 ft of unbroken argillaceous strata.

East of the Ogmore Valley: The lower coal of the **No. 2 Rhondda** was worked at Ty'n-y-waun Colliery from a slant [93118475] near Heol-laethog, the seam being recorded as 3½ ft of clean coal. The crop was proved by opencast exploration between this point and Nant Crymlyn with the section: top coal 2 to 2½ in, dirt 1½ to 9½ in, coal 35 to 40 in. The upper split (**Field Vein**) is stated to be 18 in thick and crops at Heol-laethog about 100 ft higher in the sequence. Traced eastwards along the slope above Heol-y-Cyw the two coals appear to unite in the neighbourhood of Nant Cwm-llwyd, the full section being worked at Brynchwith No. 2 Drift of Raglan Colliery [95148545]; clift roof, coal 7 in, parting 1½ in, coal 16 in, clod 20 in, coal 38½ in. The section at Tai-hirion Colliery [97138484] east-north-east of Werntarw Colliery was given by Jordan (1908, p. 19) as: top coal 6 in, parting 3 in, coal 7 in, parting 3 in, coal 16 in, parting 1 in, coal 30 in.

The seam crops at pit top at South Rhondda Colliery, where it was formerly worked from adits [99158487]. It was also worked from numerous levels on Mynydd Coed Bychan and in the area above Meiros Colliery. Between the South Rhondda–Meiros area and Llanharan the seam was thrown southwards by the Meiros faults; in the Brynheulog area [99208390] the crop was proved by opencast exploration, while farther south it can be traced from Gelli-fedi through Hendre-wen and beneath the Triassic rocks between Dolau and Bedw-bach [02808256]. There was formerly much speculation concerning the correlation of the Hendre-wen seam. It was thought by Jordan (1876, p. 261) to be the White Seam of Maesteg, while in the previous editions of this memoir it was stated to be the Wern-tarw seam of Raglan Colliery. There is now little doubt that it is the No. 2 Rhondda, and its position so far south can be fully explained by the intervention of the Meiros Fault and its associated anticline. The following sections in this area show close similarity:

	1		2		3		4		5	
	Ft	In	Ft	In	Ft	In	Ft	In	Ft	In
Coal ..				4		6 ⎫		⎧		9
Dirt ..				2		1½ ⎬	1	4 ⎨		3
Coal ..				5		4 ⎭		⎩		9
Dirt ..				1		4		6		2
Coal ..	1	8	1	6	1	3		11	1	9
Dirt ..				3		1		2		3
Coal ..	2	8	2	6	2	9	2	9	3	0

1. Worked section, South Rhondda Colliery; 2. Meiros Colliery; 3. Opencast trial pit [99428378], 90 yd E. of Brynheulog Farm; 4. Opencast trial pit [00038402], 180 yd E. of Rhiw-perra Cottage; 5. Hendre-wen Colliery [99428328].

At Llantrisant Colliery, where the seam was worked on a small scale, the section in No. 3 Shaft [03338394] reads: top coal (with soft parting) 23 in, bottom coal (with parting) 54 in. In the area south of Llantrisant the seam crops, not at the foot of the pennant scarp as might be expected, but beneath the low-lying drift-covered area between the Llantrisant–Lanelay road and the Mwyndy Branch of the railway. In the area immediately east of Newpark [04308280] it was worked from shallow pits, the disposition of which indicates a gentle northward dip marking the dying out eastwards of the Meiros Anticline. Spoil from these workings includes quartz-conglomerate with rolled clay-ironstone pellets, characteristic of the roof of the seam over much of the district to the north. In the Rhiwsaeson area, the seam crops at the foot of the steep scarp of the Caerau, where it was worked under the name Rock Fawr at Tor-y-coed Colliery [06768270]. The section is given as 50 in of coal with four beds of shale each 1 in thick.

The sandstones above the No. 2 Rhondda thin from about 400 ft in the western part of the area to about 100 ft east of Llantrisant. Little is known concerning the **No. 1 Rhondda** and **No. 1 Rhondda Rider** horizons. Feature-mapping, supported by occasional outcrops suggests that they are separated by 60 to 100 ft of variable strata. In the South Rhondda area the sandstone usually present above the No. 1 Rhondda appears to fail, and the two coals are assumed to crop within a single mudstone-'slack'. In several areas, as for example, on Mynydd y Gaer, no feature attributable to the No. 1 seam can be seen, and the coal and its associated mudstones may be absent. The seam was proved in a trial pit [99648356], 430 yd W. by S. of Pentwyn Farm, north-west of Llanharan, the section being given as: sandstone roof, clay 12 in, coal 23 in, clay 2 in, coal 46 in, clay 15 in, coal 6 in, coal and shale 30 in, fireclay floor. The seam was formerly worked from shallow drifts and bell-pits in the wood about 600 yd N.W. of Llanharan Church. The three shafts at Llantrisant Colliery showed sections varying in detail; in No. 3 Pit the section was closely similar to that from the opencast pit given above, while No. 2 Pit also shows a thick dirty coal: top coal 47 in, fireclay 7 in, coal 32 in, fireclay 18 in, coal 6 in.

The only sections of the No. 1 Rhondda Rider coal known from the area are in the shafts at Llantrisant Colliery: 15 to 20 in of coal overlain by thick massive pennant occur about 60 ft above the No. 1 Rhondda with 4 to 6 in of coal lying at the base of the mudstone about 30 ft below. Spoil from a trial level on a coal, presumably the Rider, lying near the top of a composite 'slack', 400 yd N. by W. of South Rhondda Colliery, yielded blue-black paper shale and blackband with *Anthraconauta phillipsii, Pruvostina?* [*Carbonita* pars], and *Rhabdoderma sp.* Massive pennant sandstones with a few impersistent mudstone and siltstone developments, upwards of 400 ft thick continue the sequence upwards, the group ending with a thin mudstone development lying immediately below the Brithdir horizon. Sandstones belonging to this part of the sequence have been quarried near Llanharan House [01058326], and 250 yd N.N.E. of Rhyd-y-melinydd [01758302], where red Triassic staining was observed, especially on the joints; on the Graig at Llantrisant [04168352]; at Erw Hir [05300328, see section p. 221]; and on the Caerau near Rhiwsaeson. A quarry [07858300], 350 yd N. by E. of Caesar's Arms at Rhiwsaeson shows a 110-ft section in variable pennant.   A.W.W.

C. BRITHDIR BEDS

TAFF–CYNON AREA

This group, almost entirely sandstone, varies from about 500 ft in the north of the area to about 600 ft in the south. The **Brithdir** Seam at the base was worked extensively on the east side of the Taff Valley at Merthyr Tydfil, just to the north of the district, but it deteriorates rapidly when traced into the present area. It was mined on a small scale in the Merthyr Vale area from old levels at Hafod-Tanglwys [06850015] and near the river below Perthigleision Crescent [07519950], and from Perthigleision Pit [07289958], where the worked seam consisted of 21 in of clean coal. Numerous small levels have also been driven on the seam on the north-eastern flanks of Cnwc, ¾ mile S.W. of Troed-y-rhiw; the section here consisted of 'two thin coals separated by a thick clod'. The clod yielded: *Annularia sphenophylloides, Lepidophloios laricinus, Lepidophyllum sp., Neuropteris ovata* var. *flexuosa, N. rarinervis, N. scheuchzeri* and *Sphenophyllum emarginatum.* The spoil from the workings at Hafod-Tanglwys also yielded abundant plants and the following species have been identified from material collected during the present survey: *Alethopteris sp.*, *Annularia radiata*, *A. spheno-phylloides*, *Asterotheca cyathea* (Schlotheim), *Cordaites borassifolius* (Sternberg), *Diplotmema sp.*, *Eremopteris artemisaefolia* (Sternberg), *Lepidophloios laricinus*, *Lepidophyllum sp.*, *Lepidostrobus lanceolatus* (Lindley and Hutton), *Mariopteris sp.*, *Neuropteris flexuosa*, *N. rarinervis*, *Pecopteris abbreviata* Brongniart, *?Ptychopteris unitus* Brongniart, *Sigillariophyllum bicarinatum* (Lindley and Hutton), *Sphenophyllum emarginatum* and *?Sphenopteris pecopteroides* Kidston.

On the east side of the Cynon Valley the crop of the seam rises from beneath the alluvium at Cwm-cynon and can be traced along the hillside to the old George Pit, north of Cefnpennar. Several old levels are located beneath the crags east of Newtown and the following section was noted in one of them: sandstone roof, coal 2 in, shale 8 in, coal 2 in, rashings 8 in, coal 14 in, fireclay and shale 6 ft. On the west of the valley the seams lie at the base of the precipitous crags above Graig Daren-las, Graig Isaf and Graig-hwnt and several old levels are located on Craig Abercwmboi. Old workings also occur at Llesty Colliery [01059875] on the western slopes of Coetgae Aberaman, where reddish mudstone and fireclay occur in the spoil. No sections are known in these areas. At Taren Pwllfa (see p. 197) the weathered seam can be seen at the top of the landslip scar, where the following section was measured (all coals occurring as smuts): sandstone roof, carbonaceous shale 10 in, coal $\frac{1}{2}$ in, carbonaceous shale 5 in, coal 4 in, fireclay 3 in, coal $2\frac{1}{2}$ in, carbonaceous shaly seatearth 24 in, coal 2 in, dark shale 4 in, coal 3 in, shale 2 in, coal 1 in, shale 4 in, coal 7 in, fireclay floor.

To the south at Penrikyber Colliery, the seam was formerly mis-correlated with the No. 2 Rhondda; the shafts proved: top coal 6 to 9 in, clod 44 to 48 in, coal 13 to 20 in. The seam was very dirty at Lady Windsor Colliery: top coal 4 in, rashings and fireclay 6 ft 1 in, coal $16\frac{1}{2}$ in, rashings and fireclay 22 in, coal $6\frac{1}{2}$ in, rashings 19 in; but in the Mynachdy Borehole (see p. 348) a surprisingly good section is recorded: top coal 3 in, black stone and 'seggar' 4 in, coal 5 in, 'seggar' and black stone with coal partings 11 in, coal 34 in.

The sandstones overlying the Brithdir are very massive and stand out as bold precipitous crags. About 150 ft thick they make up most of the strata between the Brithdir and Brithdir Rider seams. They have been quarried for building stone at Aber-fan, near Newtown, and at Miskin. The quarry at Newtown [05809750] shows, immediately overlying the Brithdir, a massive bed of pennant 25 ft thick and almost devoid of bedding, overlain by 20 ft of sandstone in beds up to 2 ft thick, followed upwards by further massive sandstones.

The **Brithdir Rider** Seam has not been proved with certainty in the Taff Valley, though it evidently crops near the summit of Cnwc, and a trial [06869978], 200 yd N.E. of Perthigleision, was probably driven on the coal. In the Cynon Valley several trials occur along the crop on the east side of the valley (e.g. 05559965, 500 yd N. by E. of Fforest-uchaf, and on the slopes of Mynydd Merthyr above Gelli-ddu); while on the west it was worked from numerous adits above Graig Daren-las, Miskin and Penrhiwceiber. Above Daren-las the seam consisted of 16 in of clean coal and was overlain by blue-black shales with *Anthraconauta* cf. *phillipsii* and *A. tenuis*. In the No. 3 Pit at Penrikyber Colliery the seam is 15 in thick, and 9 in was proved in the Mynachdy Borehole.

The overlying sandstone is one of the thickest in the north-eastern part of the coalfield; it varies from 200 to 400 ft, thickening southwards. It has been quarried extensively, particularly where it crops on the lower slopes. A 90-ft face of massive pennant with a few lenses of silty mudstone is seen at Ynys-Owen [07859925], to the south of Merthyr Vale. Quarrying on a considerable scale has also been carried out above Graig Daren-las, at Penrhiwceiber, and at Ynys-y-bwl. The lowest 100 ft of the sandstone is commonly characterized by regular thin bedding; such 'tilestones' have been quarried on Cnwc and on Mynydd Merthyr above Cwmpennar.

A.W.W.

RHONDDA FACH AREA

A small patch of the basal beds caps Mynydd Troed-y-rhiw, between Ynyshir and Dinas; elsewhere the outcrops are confined to the east side of the valley where they form the highest ground on Cefn Gwyngul to the west of the Llanwonno Fault. In total thickness they range up to about 600 ft.

The crops of both the Brithdir and Brithdir Rider seams can be readily traced, since their associated mudstones make well-developed 'slack'-features. The **Brithdir** was worked from numerous small levels on the crop above Tylorstown, extending about 1 mile northwards from Penyrheol [01659505]. At one of these [01459615] the following section was proved: sandstone roof, fireclay with coal streaks 38 in, coal 4 in, fireclay 3 in, cannelly rashings 5 in, coal 4 in, rashy fireclay 2 in, coal 22 in, fireclay $5\frac{1}{2}$ in, coal $3\frac{1}{2}$ in, fireclay 7 in, rashings 2 in, coal 7 in, fireclay floor. Sporadic levels are also located on the seam between Pont-y-gwaith and Ynyshir, on the sides of Twyn Llechau and on either side of the National Colliery mountain tip.

The **Brithdir Rider** in general lies about 150 ft higher in the sequence, its position being marked by a well-developed line of springs or seepages. The coal rests on about 20 ft of argillaceous strata, and it is usually succeeded by mudstones, which may be as much as 20 ft thick on the northern flanks of Cefn Gwyngul. The immediate roof of the coal consists of blue-black paper shales with *Anthraconauta phillipsii* and *A. tenuis*. These may be seen [00929744] to the east of Ferndale, where the coal was proved by augering to be 18 in thick. Trial levels occur above National Colliery but no details are known.

Upwards of 200 ft of sandstones are developed above the Rider on Cefn Gwyngul. In general they are here too far removed from the valley floor to have been extensively quarried, but they have been dug for walling-stone at numerous points.   A.W.W.

NORTHERN RHONDDA FAWR AREA

An outlier, two miles long by half a mile wide, composed only of the lower beds of the group, caps the watershed between the Rhondda and Corrwg valleys, between Cefn Nant-y-gwair, west of Blaen-y-cwm, and Cefntyle-brych, west-north-west of Blaenrhondda. The smut of the **Brithdir** coal has been seen at the southern end [90549853] but no section was measurable. Massive sandstones occur in the roof and reach about 100 ft in thickness over the outlier as a whole. The area is extensively covered by peat and few exposures are seen.   W.B.E.

SOUTHERN RHONDDA AREA

The group is fully developed only on the southern limb of the Pontypridd Anticline, in the Llantwit Syncline, and in the trough between the Llanwonno and Daren-ddu faults at Pontypridd. In these areas a total thickness of about 600 ft of beds is present.

Nothing is known of the **Brithdir** in its small area of outcrop on Mynydd Troed-y-rhiw, north of Dinas. On the northern slopes of the main valley above Hafod and Hopkinstown, several old levels were driven on the seam in Coed yr Hafod Fawr-uchaf [05109140; 05509130] but no details are available. The feature associated with the seam can be traced around the flanks of Mynydd y Glyn and Mynydd Gelliwion, south of Hafod, the coal lying about 150 to 200 ft below the crest of the ridge. The coal, which appears to be thin and dirty, is immediately overlain by massive sandstones. On the south limb of the anticline several trials were driven on the coal on the Graig, above Pen-y-rhiw and Maritime collieries, and a further level [04048877], 750 yd W. by S. of Tylawinder, showed dirty coal and rashings lying close beneath crags of massive pennant. In the Mildred shaft at Cwm Colliery the seam is represented by a series of thin and dirty coals associated with rashings, spread through 13 ft 2 in of strata.

The **Brithdir Rider** lies 150 to 200 ft above the Brithdir, the intervening measures, apart from about 20 ft or so of mudstones at the top, consisting of sandstones, which thicken when traced southwards. The seam was apparently proved in a level adjacent to the Llanwonno Fault in Nant Blaenhenwysg [05369216]. The highest points of both Mynydd y Glyn and Mynydd Gelliwion are formed of several small outliers of the coal and its overlying sandstone, and the smut can be seen in subsidence chasms [04809017], 400 yd N.E. of the summit of Gelliwion. Near Pontypridd the coal (locally termed the Pont Shon Norton Seam) was formerly worked from a level [07889124] at the foot of Craig-yr-hesg. On the south side of the Anticline the seam has been proved only on the slopes above Treforest: an old level is located alongside the stream [07408827], 250 yd S.S.W. of Fforest Uchaf Farm, but no details are known. At Cwm Colliery two thin coals, 4 in and 5 in thick and separated by 6 ft 8 in of rashings and fireclay, lie about 180 ft above the Brithdir.

A third thin coal (possibly the Glyngwilym), occurs about 100 ft above the Rider on the spur between Cwm Hafod and Cwm Blaenhenwysg, and was proved in trials [05239216], about ¼ mile N.W. by N. of Blaenhenwysg Farm. The coal and its underlying thin mudstone make a strong 'slack'-feature immediately beneath bold crags on the south side of the spur, but when traced north-westwards on either side of the ridge they die out within a mile.

The sandstones overlying the Brithdir Rider are elsewhere particularly thick and massive, and the magnificent bluff of Craig-yr-hesg is formed of 300 ft of unbroken sandstone. Craig-yr-hesg Quarries [07909180], with faces of 130 ft, are by far the largest working quarries in the district; the bottom 40 ft of strata are particularly massive, with individual beds reaching 12 ft or more in thickness. Only the highest beds are exposed in the trough between the Llanwonno and Daren-ddu faults at Pontypridd. The sandstones were quarried on Lan Wood [07409090], on Coed-Pen-Maen Common [07759035], and on the north-eastern slopes of the Graig [07758945]. The main Cardiff to Merthyr road at Coed-Pen-Maen [07679032] shows a 40-ft section in these massive false-bedded sandstones, in which irregular lenses of silty mudstone fill erosion channels. The uppermost sandstones of the group are regularly and thinly bedded and have been quarried for tiles and paving stones near Maes-y-grug [06408820; 06808820] and 500 yd N.W. of Maendy Farm [07788732]. At the first-mentioned locality certain high-quality beds were worked from underground chambers.

A.W.W.

OGMORE–GARW–LLYNFI AREA

The Brithdir Beds are here confined to the southern limb of the Maesteg Anticline on the north side of the Moel Gilau and Ty'n-y-nant faults. The full development of about 600 ft occurs only on the eastern slopes of the Dimbath Valley. To the west the higher strata have been lost by erosion, but the lower beds have been preserved on Mynydd y Gwair, Craig Pentre-Beili, Pen-y-foel, Mynydd Moelgilau and Craig-yr-hudol.

The two main coals of the group are about 200 ft apart and are, as usual, poorly developed. In the Dimbath Valley a trial [94948958], 400 yd S.E. of Hendre-fased, yielded spoil of shale, fireclay and coal fragments and is probably on the horizon of the **Brithdir.** A strong feature running around the northern side of Pen-y-foel presumably marks the crop of the seam and of the associated mudstone, and a similar feature may be traced around Craig-yr-hudol. Several old levels appear to have worked the coal on a small scale on the slope above Moelgilau-fawr [90108830], but no details are known.

Several trials exist on what is taken to be the **Brithdir Rider** Seam in the Dimbath Valley. At a point in the valley bottom [95088914], 650 yd N.E. by N. of Penllwyn-gwent Farm, 8½ in of coal were observed, overlain by 20 ft of flaggy sandstone and underlain by 3 ft of fireclay with coal streaks; and in an old level mouth [94508993], 270 yd W. of Hendre-fased, the following section of the seam was measured: massive sandstone roof, coal 7½ in, grey fireclay 11 in, coal 1½ in, grey fireclay 13 in, coal 2½ in, hard dark clod 2½ in, coal 1 in, clay with coal streaks 8½ in, coal 6½ in, fireclay floor. The same coal was worked on the west side of Mynydd y Gwair [93908958], 630 yd N.N.E. of Catherine Pit, Rhondda Main Colliery. The coal was stated by Jordan (1915, p. 96) to be inferior in quality with a high ash and sulphur content, and he gives the following section: clift roof, coal 12 in, clod 10 in, coal 1½ in, clod 12 in, coal and rashes 9 in, coal 20 in.

A further thin coal was noted on Mynydd y Gwair, and in the Dimbath Valley it lies about 200 ft above the Brithdir Rider and has been worked on a small scale at two localities. According to the abandonment plan, a level [94868792], 200 yd S.S.W. of Pant-y-fid, was driven on a 7-in coal, while the earlier editions of this memoir record that near Rhiw-glyn [93908857] the coal was worked with the following section: rock roof, coal 5 in, holing 4 in, coal 20 in, fireclay 24 in. This coal is thought to be at about the horizon of the Glyngwilym.

The argillaceous strata associated with these seams are thin and pennant sandstones make up by far the greater part of the group. Quarrying on a commercial scale has been carried out on Craig Pentre-Beili and near Llangeinor, where faces up to 45 ft in massive false-bedded sandstones may be seen. Up to 90 ft of sandstone above the Brithdir Rider are particularly well-exposed in the chasms of Daren y Dimbath.

A.W.W.

AVAN VALLEY

The Brithdir Beds underlie much of the area between the Neath and Avan valleys, west of the Glyncorrwg Fault. They show little overall variation in thickness ranging from 625 ft in the north-east to about 750 ft in the south-west.

Between Duffryn and Blaen-gwynfi a few trials have been driven on the **Brithdir** Seam, but it seems unlikely to exceed 12 in. in thickness. In the Corrwg Valley it has been proved [88869952] south of Glyncorrwg Colliery between the two branches of the Glyncorrwg Fault, and its seatearth can be seen in a stream [86539803] near Nantewlaeth Colliery. Near Pont Rhyd-y-fen the coal thickens and it has been worked under the name **Graig Uchaf** on the southern slopes of Mynydd Pen-Rhys to the north of the village, where it varies from 18 to 26 in. To the north in the Whitworth shafts its probable equivalent is recorded as: top coal 8 in, bast 1 in, coal 12 in. It has been proved on either side of the Neath Valley; on the east side at Melin Court [81670140], and on the west beneath Craig-Lletty'r-afel [79500080], but no sections have been preserved.

The sandstones above the Brithdir vary in thickness from about 175 to 300 ft between Cymmer and Pont Rhyd-y-fen, and are overlain by the mudstones associated with the Brithdir Rider. Farther north, mudstones with an associated coal develop about 60 to 120 ft below the top of the sandstones. These can be seen on Pencraig-isaf, north of Cymmer, where old trials [85879711] have proved a thin coal overlying 10 to 15 ft of mudstone. Farther east it lies near the top of Coetgae Isaf [881962], and it has also been worked on the north side of Nant Boeth [88229720] and the west side of Nant Gwynfi [89009740]. Still farther north, it is exposed on Mynydd Corrwg-Fechan about 150 ft above the Brithdir. Here the coal is overlain by about 15 ft of mudstones, those below the seam having thinned markedly. Near the axis of the Cefnmawr Syncline the Whitworth shafts struck the seam 71 ft below the Brithdir Rider, the section being recorded as: top coal 6 in, rashes 7 in, coal 6 in. In the Vale of Neath the coal was called the **Graig** (locally also Graig Isaf); numerous small levels were located on the hillside between the Twrch Brook at Clyne [80800025] and Cwm-syfirig [82800160], east of Melin Court, and the section is reputed to vary from 18 to 24 in.

Spoil from one of these levels [82180150] yielded abundant *Anthraconauta tenuis*, *A. phillipsii* and ostracods. The coal was also worked on the west side of the valley at Clyne; here a sandstone is developed immediately above the coal, which in consequence lies in the lower of two closely associated 'slacks'.

The **Brithdir Rider** lies at the top of 10 to 35 ft of mudstones; it has been worked only to the north-west of Pont Rhyd-y-fen, where it was called the **Dirty**. Locally, as near Blaen-gwynfi and in the Neath Valley west of Clyne, it appears to be absent, possibly due to washout. The Whitworth shafts record it as: top coal 3 in, rashes 31 in, coal 2 in.

The succeeding sandstones are usually about 250 ft thick, but in the north and north-east they thin to little more than 150 ft. Towards the base beds with coal 'pebbles' and clay-ironstone pellets are common. The sandstone is overlain by 20 to 30 ft of mudstones at the top of which lies the Glyngwilym Seam. These mudstones are exposed in the Melin Court Brook [83360090], south-westwards of which they have a relatively wide outcrop.

The **Glyngwilym** Seam takes its name from the farm [83000117] in the Neath Valley, 1 mile E.S.E. of Melin Court, and it is in this area that the coal is best developed. It has been worked down-dip for almost a mile from Garth Merthyr Colliery [82800102], and on a smaller scale from Premier Merthyr Colliery [81810077], the coal averaging 22 to 26 in. in thickness. Elsewhere only scattered trials have proved the seam. In Cwmgwenffrwd off the Pelena Valley [79829688] it is reputed to be 18 in thick, but it deteriorates eastwards, for only 8 in are recorded near Gyfylchi Farm [81109567]. Fossils were found only in a stream [82969614] on the southern slopes of Mynydd Nant-y-bar, north-west of Duffryn, where dark shaly mudstone yielded 'mussel' fragments and ostracods.

The cyclothem above the Glyngwilym is very variable both in thickness and lithology. In the Corrwg and Nant Gwynfi areas it is 30 to 40 ft thick and almost wholly argillaceous; on Moel yr Hyrddod and in the Melin Court Brook area it is 65 to 75 ft thick and silty mudstones predominate. To the south-west of these areas massive sandstones develop a few feet above the seam, eventually descending to form the immediate roof. This sandstone thickens south-westwards, and to the west of the Cregan is 120 to 150 ft thick. It has been quarried extensively on either side of Cwm Cregan, on Penmoelgrochlef and to the west of Ton-mawr. Where the sandstone is present it is overlain by 10 to 20 ft of mudstones which complete the sequence of the Brithdir Beds.   W.B.E.

SOUTH CROP

Cwm Wernderi–Mynydd Bach–Waun Lluest-wen Area: The Brithdir Beds here occupy a considerable part of the surface, though they are fully developed only in Cwm Cerdin immediately to the north-west of Llangynwyd, where about 700 to 800 ft of strata are present.

The **Brithdir** Seam has been proved by numerous trials. It crops out on the steep slopes to the west of Duffryn Brook, where a level [79209023] shows it to lie immediately beneath massive sandstones. On the opposite hillside the coal was proved to be 10 in thick at Gallt-y-cwm Farm [80349115]. Other trials were located to the east of Moel Gallt-y-cwm [81609084], near Ty-draw Farm [83659078] where the coal was reputed to be 18 in thick, to the south-east of Cwmcerwyn Cottages [84429039], and behind Vicarage Terrace on the south-west side of Maesteg [83669120].

The sandstone above the Brithdir varies from about 150 to 250 ft, being thickest to the west. A thin seatearth is developed locally in the middle of this sandstone and can be seen in the bank of the Sychbant near Moelsychbant [84209010].

Near the summit of Moel Gallt-y-cwm the **Brithdir Rider** appears to consist of two coals 6 to 9 in thick, separated by 3 ft of seatearth. Both here and to the south-east of Hafod Farm [80958952], it lies within a foot of the overlying sandstone. Similar conditions obtain south of Maesteg [84629084] and west of Llwydarth [85448986] where the seam has been proved by small trials; at the latter locality the coal is said to be only 8 in thick.

The sandstone succeeding the Brithdir Rider forms the summits of Moel Gallt-y-cwm and Mynydd Margam, and is probably the thickest sandstone in the sequence in this area. To the south of Maesteg several impersistent bands of mudstone can be traced within it, the lowest lying barely 80 ft above the Rider. The most persistent of these bands forms a 'slack'-feature traceable around Moel Troed-y-rhiw [858892] towards Pont Rhyd-y-cyff. On the east side of the hill trial levels have shown the presence of a coal within this 'slack' and its position, about 350 ft above the Brithdir Rider and 150 to 200 ft below the Bettws Four-Feet, suggests that it may represent the Glyngwilym horizon.

The highest beds of the group consist of about 40 ft of mudstones and seatearth underlying the Bettws Four-Feet Seam, and these are seen only in the vicinity of Llangynwyd.   W.B.E.

West of the Ogmore Valley: In this area the Brithdir Beds vary from about 650 to 800 ft, the thickening taking place in a general westerly direction. Made up almost entirely of pennant sandstones the group has a broad outcrop extending from Mynydd Margam and Moel Ton-mawr to the northern flanks of Mynydd Ty-talwyn and Mynydd Baiden, the higher sandstones making up the long dip slopes from these hills towards Llangynwyd and the Gadlys Valley. Little is known about the two main coals, and the features made by their associated mudstones are not always easy to follow.

The **Brithdir** Seam is here correlated with the locally-named **Pentwyn,** formerly worked on a small scale from a level [89108612] on the crop, 200 yd E. of Pentwyn Farm at Coytrahen. The section was recorded by the former surveyors, as 57 in of coal, but this almost certainly includes several dirt bands.

A thin and apparently impersistent coal appears to have been proved in an old trial level [86528704], 150 yd S.W. of Maes-cadlawr. While this may represent the horizon of the Glyngwilym, it certainly appears to be the same seam, whose smut may be seen at the site of a spring on Bryn-y-Wrach, south of Llangeinor, at a point [91828640], 250 yd W.S.W. of Blaenclydwyn Farm.

East of the Ogmore Valley: The two main coals are difficult to trace and correlate throughout this area. North of Waun-wen and Mynydd y Gaer the beds crop out on the dip slopes towards the Ogwr Fach Valley and in consequence the features are poorly developed. However, to the east of Mynydd y Gaer 'slacks' attributable to the Brithdir, Brithdir Rider, and ?Glyngwilym horizons may be traced almost continuously; but coals within them have been proved only rarely. An old trial on the west bank of the stream [00258516], 1250 yd N. by W. of Meiros Colliery, is believed to be on the Brithdir Rider, and the same seam appears to have been proved on the hillside [03018402] about 300 yd W. by N. of Ynys-maerdy Farm. The exact horizon of old workings near the south-west bank of the Ely River [02858460] just south-west of Dyffryn-isaf is in doubt because of faulting in the adjacent ground, but it is thought that these too were in the Rider coal. Similar doubts attend the identification of the seam worked from numerous old bell-pits and levels along the banks of the stream ½ mile W.N.W. of Graig Fatho [01258540]; and though on the map this has been interpreted as the Brithdir, there are strong reasons for believing that it may equally well be the Rider. The correlation of the coal proved in an old trial [02008594], 150 yd W.N.W. of Garth Hall on the opposite side of the Ely Valley, with the Brithdir Rider depends to a large extent on this interpretation, but it is possible that this seam represents the impersistent Glyngwilym horizon.

The presumed Brithdir and its associated mudstones make a slight feature along the higher part of the scarp at Llantrisant, but the horizon is traversed by quarries at Erw Hir to the east of the village. The following composite section was measured in these quarries.

	Ft	In
Massive, false-bedded pennant	60	0
Irregularly bedded sandstone with much 'timber', fugitive coal streaks and coal 'pebbles'	8	0
COAL	streak	
Seatearth and shales .. .. seen to	2	0
COAL		6
Blue-grey seatearth with ironstone nodules passing down into carbonaceous shales with a few coal streaks	12	0
Grey-blue shaly mudstone	6	0
Very massive, false-bedded pennant	40	0

Similar features are developed near the top of the Caerau scarp at Rhiwsaeson, but no coals have been proved.   A.W.W.

D. HUGHES BEDS

These strata occur in three main areas: (a) in the north-east, between the Taff and Cynon valleys, within the Abercynon Syncline and in the extensive fault-troughs between Llanwonno and Pontypridd; (b) in the north-west, on the high ground between the Avan and Corrwg valleys and the Vale of Neath; and (c) along the syncline between the South Crop and the Pontypridd–Maesteg Anticline; west of Llandyfodwg the outcrops lie entirely south of the Moel Gilau Fault and so form part of the South Crop area, but to the east they extend into the Southern Rhondda area.

TAFF–CYNON AREA

The group is nowhere completely preserved, although on Cefn y Fan (between Mountain Ash and Merthyr Vale), on Pen-y-foel (between Abercynon and Ynys-y-bwl), and in the Llanwonno–Pontypridd trough only the highest beds are missing.

The **Cefn Glas** has been worked at numerous localities though nowhere to any great extent, its high sulphur content greatly reducing its value. The seam was named after Cefn-glas Colliery [07939688], situated on the west side of the Taff Valley, 150 yd N.W. of the eastern entrance to the railway tunnel at Quaker's Yard. The seam crops in the valley bottom at Buarth-glas [08009750] and was worked from levels adjacent to the Kilkenny Fault, some 600 yd to the north-west. Northwards the section deteriorates and little working has taken place, though trials are located on the east side of the valley at Mount Pleasant, Ynys-Owen and to the north-east of Dan-y-deri, and on the west side immediately to the north-west of Taren-y-gigfran landslip. The seam is also involved in the landslip, where it may be seen [07179890] carrying a roof of blue-black shale with *Anthraconauta phillipsii* and *A. tenuis*.

On the east side of the Cynon Valley the broad shelf associated with the mudstones below the seam may be traced below the summit of Twyn Brynbychan, descending into the valley bottom east of Ynys-boeth, where the coal was worked from numerous levels [07539650]. On the west side of the valley the seam was worked in recent years from a road-side level at Nant-y-fedw [07199606] and its crop can be followed by levels rising north-westwards towards the top of Penrikyber Colliery tip. Throughout this area the roof consists of several feet of blocky blue-black shales carrying an abundant fauna of variants of *A. phillipsii* and *A. tenuis* (many up to $1\frac{1}{2}$ in long), associated with the ostracods *Carbonita humilis* and *C.* cf. *wardiana*; a few plants including *Lepidodendron obovatum* and *Neuropteris scheuchzeri* were also found. At Nant-y-fedw Colliery the section of the seam was: top coal 8 in, clod 10 in, coal 22 in, on very hard fireclay.

South of Carnetown the crop reappears from beneath the drift and ascends the hillside of Coed-Pen-y-parc, where it has been worked on a moderate scale from several levels. On the east side of the Clydach Valley a well-developed feature follows the coal from Gellifendigaid [07109340] to Ffynnon-dwym [05489597] and Gelli-Wrgan [04159695] and beyond. There are numerous indications of coal and seatearth at crop, but working has only taken place at one point [05979511], where the spoil again yielded *A. phillipsii*, *A. tenuis* and ostracods.

On the west side of the Clydach Valley the Cefn Glas was proved in a trial [05919386] above Crawshay Street at Ynys-y-bwl, and the shelly roof shales may be seen in the stream 300 yd to the north-north-west [05779409]. The crop has also been proved in the cemetery [05639465] north of Ynys-y-bwl. Farther north, in the area between Old Ynys-y-bwl and Llanwonno, the seam has been worked at several localities: at Mynachdy [05009507], near Pistyll-goleu [03509633], and in the forest above [03469615]. Throughout this area the spoil is again characterized by an abundance of large specimens of *Anthraconauta* associated with ostracods. The seam also crops out in the trough between the Daren-ddu and Llanwonno faults on the middle slopes of Lan Wood and Pantygraigwen to the north-west of Pontypridd. Several small levels in what was known as the Stinking Vein proved the coal but details are lacking.

The **Daren-ddu** lies about 150 ft higher in the sequence, all but about 20 ft of this thickness consisting of pennant sandstones. The seam is best developed within the structural trough between Llanwonno and Pontypridd, and throughout this area the coal has been virtually exhausted. A continuous line of small levels marks the crop immediately beneath the bold scarp above Lan Wood and Pantygraigwen, and collieries were located at the following points: in Lan Wood [07279150]; Daniel's Level [06169147] above Troed-rhiw-trwyn; Daren-ddu [06509280], south of Ynys-y-bwl; Mynachdy [04759453]; Screens Level [05339527]; and Black Grove [04659596], to the east of Dduallt. Sporadic levels follow the crop around the eastern and northern slopes of Dduallt to a point near St. Gwynno's Church [03009560], where it is cut off by the Llanwonno Fault. The section of the seam near Pontypridd is given as: top coal (soft with shale) 4 in, middle coal 22 in, holing 6 to 8 in, bottom coal 8 in. At the northern end of the workings towards Llanwonno the following general section is recorded on the abandonment plan: rock roof, clod and chippings 13 in, coal 18 in, rashings 14 in, coal 7 in, rashings and fireclay 48 in.

North-eastwards from the trough the seam evidently deteriorates and nowhere has it been worked to any degree. Its crop encircles the hill between the Clydach and Cynon valleys south of Tyntetown, lying everywhere at the foot of a bold feature. A few levels mark past efforts to prove the seam: near Ysgubor [06199523]; above Lady Windsor Colliery [06679455]; at Carnetown [07679475]; near Ty-dan-daren [06959598]; and on the slopes adjacent to Penrikyber Colliery tip [06159635; 06649628].

Evidence of the Daren-ddu horizon was not fully established on the east side of the Cynon Valley or in the Taff Valley, and it may well be that the coal fails in these areas. A weak feature encircles Twyn Brynbychen 100 ft or so above the Cefn Glas, but no coal appears to be present. On the east side of the Taff at Merthyr Vale the hillside rises 300 to 400 ft above the Cefn Glas but no significant features are developed. A 6-in coal was exposed in a quarry [08109951] at Ynys-Owen about 90 ft above the Cefn Glas but its correlation with the Daren-ddu is uncertain.

In the Ynys-y-bwl area a persistent, though thin, mudstone horizon is developed about 150 ft above the Daren-ddu. This produces a well-marked bench just beneath the summit of Pen-y-Foel with outliers near Pen-twyn-uchaf [06179617] to the north-west, and on Twyn y Glog and Pen y Lan to the west of Ynys-y-bwl. On the west side of Twyn y Glog a thin smut and fireclay were observed associated with the mudstone.

The sandstones of the group are well developed and make prominent features. Those above the Daren-ddu are particularly massive and produce bold lines of crags as on Lan Wood at Pontypridd, Daren y Foel above Carnetown, on the Dduallt at Llanwonno, on either side of the Cynon Valley between Penrhiwceiber and Abercynon, on Twyn Brynbychan east of Mountain Ash and on Cefn Merthyr east of Merthyr Vale. On the south-west slopes of Twyn Brynbychan the crags are associated with spectacular pillars. Quarrying has taken place at numerous localities: from above the Cefn Glas at Ynys-boeth and Ynys-y-bwl; from above the Daren-ddu at Lan Wood and Troed-rhiw-trwyn, north-west of Pontypridd, at Carnetown and Abercynon, at Daren y Celyn, east of Ynys-boeth and at Mount Pleasant. The rockfaces of Taren-y-gigfran, which mark the scar of the great landslip to the south-west of Merthyr Vale, are partly in the sandstones above the Cefn Glas and partly in those above the Daren-ddu, the Kilkenny Fault, which traverses the face, bringing the two groups of sandstones together.   A.W.W.

AVAN AND NEATH VALLEYS

The group is complete only under the western spur of Cefnmawr [842986], where it is rather more than 800 ft thick.

At the base the **Wenallt** and its rider coals form a single composite seam at Garth Merthyr Colliery [82740093], on the northern slopes of Carn Caca: sandstone roof; **Top Rider**—coal 8 in, coal and rashes 5 in, rashes 5 in; parting 22 in; **Wenallt Rider**—coal 7 in, coal (locally inferior) 12 in, rashes 4 in, coal 9 in, rashes 1 in; parting 22 in; **Small Rider**—coal 8 in; parting 48 in; **Wenallt**—inferior coal and rashes 27 in, stone 9 in, coal 12 in, coal and rashes 21 in. The Wenallt has been extensively worked in Cwm Pelena, but the Wenallt Rider is unimportant. In the east of the Garth Ton-mawr [81639720] colliery-taking the section is similar to that previously described, though, locally, the overlying sandstone has an irregular base and cuts out the Top Rider and even part of the Wenallt Rider. A typical section of the Wenallt is: inferior top coal 10 in, parting 1 in, inferior coal 6 in, mudstone 8 in, coal 1 in, stone 8 in, coal 21 in; only the bottom coal is extracted.

To the east the two main seams diverge and in Cwm Cregan [843970] and beneath Cefnmawr are up to 30 ft apart. The Wenallt maintains a workable section of about 2 ft and has been mined at West End Colliery [84309687]. The Wenallt Rider is the better seam in this area; at Avan Vale [84429735], Blaen-y-cwm [84659977] and Llwyn-coedwr [84879971] collieries its section is: top coal 4 to 6 in, clod with coal 13 to 30 in, coal 18 to 20 in, holing 2 to 5 in, coal 12 to 17 in. It is likely that the Top Rider is included in this section. At Corrwg Vale [84589757] and West End collieries, the overlying sandstone again appears to have an irregular base and to rest directly upon the middle coal of the above section.

Still farther east the outcrops of the Wenallt coals are marked by many levels, now all abandoned; these are particularly numerous around Glyncorrwg [e.g. 86419879; 86629927; 87789984], but nowhere has there been major exploitation. The only known section from these workings is given by Plummer (1878, pl. 30) for a seam stated to be the Wenallt Riders, which crops near Llwyn-y-ffynnon Farm [87509980]; it is said to comprise 14 in of coal with a 10-in parting.

In the extreme west the Wenallt and Wenallt Rider are about 15 ft apart. At Bush Colliery [79929955] the Wenallt maintains its characteristic section: bastard coal 6 in, coal 19 in, parting 1 in, coal 4 in, stone 10 in, coal 20 in. Similar sections were proved at Clyne Merthyr Colliery [80969978], and at East End Colliery [79549645], $\frac{1}{2}$ mile W. of Ton-mawr. The Wenallt Rider has been worked only at New Forest Colliery [97569658] and nearby openings, but no reliable sections are known.

Westwards of a line from near Pen-rhiw-Angharad Uchaf Farm [80039944] to Cwmgwenffrwyd [797975] up to 75 ft of mudstones are developed between the Wenallt Rider and the overlying sandstone. Immediately beneath the sandstone lies a thin coal, the **Garlant,** which, on the southern slopes of Cefnmorfudd [79279630] has the following section: sandstone roof, coal 6 in, mudstone with coal streaks 4 ft, coal 12 in. This seam may be the Top Rider of the Garth Merthyr section, or, together with the underlying mudstones, it may be affected by regional washout to the east.

Workings on the Wenallt Rider between Mynydd Nant-y-bar [83659660] and Blaen Cregan [85439875] yield spoil of purple-grey, ferruginous mudstone containing abundant variants of *Anthraconauta phillipsii* and *A. tenuis* associated with *Carbonita* cf. *humilis*; the precise position of this fossiliferous horizon within the complex Rider section is not known.

The sandstone overlying the Wenallt 'slack' is about 200 ft thick in the north-east and this increases westwards to about 250 ft. West of Glyncorrwg its base is conglomeratic and locally transgressive. In the Pelena Valley impersistent shale or siltstone partings occur in the upper parts, and on Mynydd Nant-y-bar [82309665] about 4 ft of muddy seatearth are exposed some 15 ft from the top. Mudstones above the sandstone are everywhere 20 to 40 ft thick, and are overlain by the Mountain Seam.

The **Mountain** is best developed in Cwmblaenpelena and Cwmgwenffrwd, where it has been worked on a small scale at several localities. At Craigavan Colliery [81479765] the section is: sandstone roof, top coal 12 in, clod 18 in, coal 7 in, dirt 2 in, coal 8 in, stone 1 in, coal 6 in. The main parting varies considerably in thickness and when excessive the seam is of little value. Locally the overlying sandstone cuts out the top leaf of the seam. Away from this area only sporadic trial-levels mark the crop; from one of these [possibly 87510011] to the north of Glyncorrwg, Plummer (1878, pl. 30) records 15 in of coal with 3 in of dirt. In the Corrwg-fechan [87960128] the smut is visible, with 6 in of seatearth between it and the overlying sandstone. Near Corrwg Vale Colliery [84129751] spoil from workings on the seam yielded *Asterotheca abbreviata, Carpolithus sp., Lepidodendron sp., Neuropteris scheuchzeri, Sphenophyllum emarginatum, Sphenopteris neuropteroides.*

Above the Mountain Seam another thick sandstone forms much of the plateau between the Neath and Avan valleys. On Cefnmawr it is 250 ft thick but it thins to 200 ft or less on Bryn Llydan [874008]. A silty development occurs about the middle on Mynydd Blaenafon [808968].

Apart from three outliers of mudstone on the summit of Twyn-gwyn [812981], the highest measures of the group are preserved at only two localities. The simplest sequence is west of Cefnmawr, where about 230 ft of mainly argillaceous beds continue the sequence to the Graigola. A few feet above the base of these mudstones an unnamed coal has been proved by two old levels [84219804; 83529875] and by opencast exploration. A trial pit proved the following section: top coal 15 in, parting 17 in, coal 11 in. This coal is evidently the **Westernmoor** of Neath, though the parting between the two leaves is here thicker than at Neath. About 120 ft above, another trial has been driven on a thin seam which lies in the position of the **Shenkin** Vein of Neath. The sequence is continued by 90 ft of mudstones to the Graigola, but much of these is concealed by peat and landslip.

While the above sequence is comparable with that proved around Neath, the following section on Bryn Llydan shows several anomalies:

							Ft	In
Silty sandstone	..	..	..	..	..	..	50	0
Seatearth	..	..	..	..	..	..	1 ft to 2	6
HENDRE-GAREG: **coal 14 to 16 in,** rashings 12 to 22 in, stone 0 to								
14 in, **coal 24 to 32 in**	..	..	..	..	..	..	5	7
Mudstones ..	..	..	..	..	..	..	15 ft to 25	0
Sandstone ..	..	..	..	..	..	..	0 ft to 15	0
COAL, inferior	..	..	..	..	..	..	thin	
Mudstones ..	..	..	..	..	..	..	0 ft to 15	0
Massive sandstones	..	..	..	..	..	..	160	0
MOUNTAIN	..	..	..	..	..	..	—	—

The previous edition of this memoir (p. 71) mentions a further coal capping Bryn Llydan, and overlying the sandstone above the Hendre-Gareg, but no trace of this was seen during the present survey. It is possible that the anomalies are a result of the Hendre-Gareg lying above an unconformity, for its section is unlike that of either the Westernmoor or Graigola. W.B.E.

SOUTH CROP AND SOUTHERN RHONDDA AREAS

Llantwit Fardre to Tonyrefail: On the north side of the Llantwit Syncline the beds have a broad outcrop extending from the neighbourhood of Cwm Colliery through Pen-y-coedcae to the Castellau and Mychydd valleys. The **Cefn Glas** horizon was intersected in the shafts at Cwm at a depth of 250 to 300 ft, but the details in the two sections differ. In the Margaret Shaft the following is recorded:

	Ft	In
Pennant rock	—	—
COAL		8
Fireclay and rashings	11	5
COAL		9
Fireclay, clift, rock and shale	40	0
COAL	1	5
Fireclay	3	0
COAL	1	1

The Mildred Shaft shows only one coal 21 in thick and its position in the above sequence cannot be determined. The seam crops out to the north of the colliery in Nant-ty'r-arlwydd, where it has been worked from numerous levels [067869]. From here the crop follows a northerly course to Pen-y-coedcae, where it encircles the north side of the hill. It was worked to the north-west of the hamlet [05728787] as the **Pen-y-coedcae** Seam with a section said to be: fireclay roof, coal (inferior) 15 in, clod and rashings up to 24 in, coal 15 in. Opencast prospecting in the area immediately to the south-west proved a thin and dirty coal very variable in section. A trial pit [05568775], 170 yd S.E. of Penbwch Uchaf, proved:

	Ft	In
Soil, drift and broken sandstone ..	7	0
Sandstone and broken shale	3	0
COAL (dirty)		6
Sandy shale and sandstone	4	6
Grey shales with thin irregular coal	2	6
Mixture of coal and clay ..	2	6
COAL	1	1
Fireclay		11

The coal crops or lies near the surface over an extensive area near the bottom of the Castellau Valley, and it was formerly worked from numerous bell-pits and levels as far west as the eastern margin of Wauncastellau. Between here and Tonyrefail the crop is much obscured by glacial drift, but levels in Rackett Wood [03858773] and north-west of Treferig House [03108784] are believed to have been in the seam. At Rackett the seam-section is given by Jordan (1903, pl. IV) as follows: coal 6 in, rashes 3 in, clod 9 in, coal 6 in, rashes 2 in, coal 6 in, clod 6 in, coal 6 in.

In the Tonyrefail area the seam was worked extensively from levels under the name of **Ty-du.** The principal openings were: Tylcha-fach Level [01328694], Ty-du Level

[00828742] and Duke Level [00828765], all on the east side of the valley at Thomastown; and Ely-Llantwit Level [00448742] to the west of the river. Although the coal was sulphury the section throughout these workings was a good one: top coal 20 to 22 in, plane parting, coal 12 in, stone ½ to 2 in, coal 16 in. The seam also appears to have been worked, though to no great extent, in Nant Llan just to the west of Llanilid Farm [99858675], as the eastward continuation of the crop of the Lower Glynogwr Seam (see p. 227).

At none of the above localities was any shelly roof material noticed in the spoil. However, an old pit [04778547], 400 yd S.W. by W. of Capel Castellau, sunk 70 yd from a horizon just below the Daren-ddu, yielded spoil containing a fauna identical with that from the Cefn Glas in the Taff–Cynon area, but there are no details of the coal worked.

The **Daren-ddu** was encountered in the Margaret Shaft at Cwm Colliery immediately beneath the drift and about 220 ft above the Cefn Glas; the section being: top coal 6 in, fireclay 2 in, coal 5 in, fireclay 2 ft, coal 23 in. The seam has not elsewhere been proved in the Cwm–Castellau area. A seam worked from several shallow pits and a level [04018658] to the west and south-west of Treferig-isaf in Nantmychydd is presumed to be the Daren-ddu; no details are available but it seems to lie close beneath massive sandstone.

Near Tonyrefail two outliers of the coal, here termed the **Dirty** Seam, occur on either side of the Ely Valley. The seam and its associated mudstones make a well-developed 'slack'-feature where not obscured by drift, and numerous old workings are located on the crop. On the east side of the valley the seam may be traced north-eastwards from Tylcha Fawr Farm [01158765], marked by old levels, for nearly half a mile, after which it plunges beneath thick drift. The crop courses southwards and eastwards around the hill past Tylcha Ganol before again being lost beneath drift. The section of the coal at Tylcha Fawr is recorded by Jordan (1903, p. 149): coal 5 in, shale 5 in, coal 4 in, shale 6 in, coal with plane parting 14 in; a section very similar to that in the Margaret Shaft.

On the west side of the valley the coal encircles the hill which lies between Gelli-seren [00358717] and Gelli'r-haidd-uchaf [99258700]. Numerous levels and bell-pits occur on the western and northern sides of the hill, but few details are available. Jordan gives the section of the Dirty Seam at Gelli-seren as follows: coal 8 in, fireclay 26 in, coal 12 in, parting 1 in, coal 12 in, parting 1 in, coal 6 in. Whereas on the Tylcha side of the valley the seam is overlain by massive sandstones, at Gelli'r-haidd these appear to lie about 20 ft or so above the coal.

A thin and seemingly impersistent coal occurs 90 ft above the Dirty Seam on the hillside east of Tylcha Fawr. A single trial [01538774] proves the coal but it has not been worked to any extent. In the area south-east of Cwm Colliery about 150 ft of measures, largely obscured by drift but presumably mainly sandstone, occur between the Daren-ddu and the No. 3 Llantwit seams.

Massive pennant sandstones again make up most of the group's total thickness and they have been quarried at several places. A quarry showing 60 ft of massive, false-bedded sandstones is located [06558665] mid-way between the Cefn Glas and Daren-ddu, just north of Cwm Colliery. At Pen-y-coedcae, extensive quarries show the sandstones near the base of the group to be thinly flaggy and false-bedded. The sandstones above the Ty-du Seam were also quarried on the east side of the Ely Valley at Thomastown; and there are 50-ft faces in almost horizontal beds near Ty-du [00958765] and Pen-y-gareg [01108710].

On the south side of the Llantwit Syncline, the outcrop of the group continues on the east side of the Ely Valley from the area south of Tonyrefail across Llantrisant Common and the northern slopes of the Caerau, obscured almost everywhere by thick drift. On Llantrisant Common two of a series of old trial pits appear to have proved the Daren-ddu Seam. In one of these [04188447], 400 yd S. 26° E. of Glanmychydd-fawr Farm, the seam had the following section: coal 8 in, clod 7 in, coal 7 in, soft clod with fireclay 4 ft 2 in, coal 18 in; a section not dissimilar to that of the Dirty Seam at Tylcha Fawr. The feature associated with the Cefn Glas is well developed on the dip slope of the Caerau running east-south-eastwards at the back of Rhiw-brwdwal, but the coal has not been proved. No details are available concerning the 150 ft or so of measures intervening between the Daren-ddu and No. 3 Llantwit in this area, except that no prominent sandstone feature is developed above the former.

Ogwr Fach and Dimbath valleys: In this much-faulted area surrounding the hamlet of Llandyfodwg the Cefn Glas or Ty-du Seam was usually called the **Lower Glynogwr.** It was worked on the south limb of the Pontypridd Anticline at a point [97758830] on the west bank of the river near Gilfach Garden Village; the seam was 31 in thick with 2 in of rashings in the middle. On the other side of the synclinal axis the section at Jenkins Merthyr Colliery [97888750] was: top coal 16 in, clod 6 in, coal 11 in. The Ty'n-y-nant Fault crosses the Ogwr Fach Valley in the vicinity of Tynewydd Farm [97658715], and on the south side and to the east of the Aber Fault the crop of the Ty-du follows a well-defined 'slack'-feature from just west of Tynewydd along the Caradoc Valley into the upper reaches of Nant Llanilid where it is again cut off by the Ty'n-y-nant Fault. Throughout this area the seam has been worked from a number of levels and from numerous bell-pits which line the crop for three-quarters of a mile. At Caradoc Vale Colliery [97558715] the section of the seam was: sandstone roof, coal streak, rashings and stone $1\frac{1}{2}$ in, coal 6 in, soft blue clay 12 in, coal 22 in, rashings 2 in, coal 12 in. Traced south-eastwards along the crop in this area the massive sandstone roof rises from the coal and up to 20 ft of strong silty mudstone may intervene.

On the west side of the Aber Fault the seam is thrown southwards nearly half a mile, and the crop-feature is developed, similar to that of Caradoc Valley, running south-eastwards from Pen-yr-heol [96958645]. Here the coal has been worked from numerous levels and bell-pits.

A second north-south fault intervenes in the vicinity of Pen-yr-heol, which displaces the outcrop of the seam northwards some 250 yd. From just north of the farm the crop follows a west-north-westerly course to the Ogwr Fach Valley near Llwyn-helyg, and along the lower slopes of the hillside below Llandyfodwg into the Dimbath Valley north-east of Glynllan [94508715]. At Glynogwr Colliery [95808693], in the valley bottom, the following section was recorded by Jordan (1903, pl. IV): coal 3 in, fireclay 18 in, coal 4 in, fireclay 10 in, coal (sulphurous) 8 in, tough fireclay 9 in, coal 28 in, black shale 9 in, coal 15 in, fireclay 24 in, coal 15 in. Jordan referred to this seam as the No. 3 Llantwit, but this correlation is here disputed. Several levels were driven on the crop of the seam on the south-east bank of the Iechyd stream in Cwm Dimbath about 250 to 500 yd north-east of Glynllan. Throughout this area the seam is again overlain by about 20 ft of mudstones which include a thin coal near the top.

Thrown upwards by the successive Moel Gilau and Ty'n-y-nant faults the Lower Glynogwr again crops on the eastern slopes of the Dimbath Valley about 300 ft above the stream. The mudstones below the coal produce a well-developed bench feature, and the coal has been worked at several points [95328812; 95528847; 95238882], 200 yd N. 16° E., 600 yd N. 26° E. and 950 yd N. 5° W. of Gadlys Farm respectively. At the last mentioned an Opencast trial pit [95188862] showed the following section immediately beneath drift: coal and clay 6 in, clay 1 in, coal 6 in, clay 20 in, coal 18 in, coal and shale 14 in, coal 15 in.

The Lower Glynogwr again crops on the hill immediately behind Dolau-Ifan-ddu [93608730], between the Dimbath and Garw valleys to the north of Blackmill. This outlier, cut off on the north side of the Moel Gilau Fault, was worked from levels overlooking the Garw Valley at Frithwaun. An Opencast trial pit [93478736] proved: sandstone roof 12 ft 9 in, coal 8 in, shale and clay 16 in, soft coal $4\frac{1}{2}$ in, coal and rashes $3\frac{1}{2}$ in, coal 22 in, on fireclay. A thin coal lies 20 to 30 ft above the seam, sandstone between the two horizons dying away eastwards.

The sandstones lying immediately or close above the Lower Glynogwr Seam vary from about 100 to 120 ft in thickness and are overlain by 20 to 30 ft of mudstones. The succeeding **Daren-ddu** or **Upper Glynogwr** Seam is developed in three areas. In the 700 yd wide trough on the west side of the Aber Fault the coal and associated argillaceous strata make a strong 'slack'-feature following the course of a small stream on the north-west flanks of Mynydd Maendy about 300 yd north of the Lower Glynogwr outcrop. The coal has been worked from a number of old adits near the crop [974864], about ¼ mile E. of Pen-yr-heol. Craig-las Colliery [97218686] presumably worked this coal, but the section given on the abandonment plan is surprisingly thick: rock roof, coal 27 in, rashings 6 in, coal 12 in, rashings 6 in, coal 9 in, fireclay 12 in, coal 12 in. The seam crops around the hill on which stands Llandyfodwg, and it was worked from Cwmogwr Colliery [96018713], 400 yd E. 16° S. of the church; the section is given as: strong clift roof, clod 3 in, coal 24 in, holing 2 in, fireclay 1½ to 4 ft, coal with bands of clod 36 in, strong clift 6 in, coal 6 in, on fireclay floor.

On the north side of the Ty'n-y-nant Fault the Upper Glynogwr crops along the slopes above Cwm Gadlys. Workings from levels [95908890] are located alongside the mountain road running north from Llandyfodwg about half a mile north-north-east of Llwyn-yr-ysgol. This is the locality of the Gadlais Seam of Jordan (1876, p. 263), who thought that it occupied a position between the Upper and Lower Glynogwr seams. An Opencast trial pit [95878906] in this area proved the following section: coal and clay 6 in, clay 5 in, coal 10 in, clay 5 in, coal and clay 3 in, clay 15 in, coal 18 in, clay 3 in, coal 12 in, on fireclay floor. Evidently only the two lowest coals are included in the section given by Jordan and quoted in the previous edition of this memoir (p. 87).

The Hughes Beds are nowhere fully developed in this area, but at Llandyfodwg and to the north about 150 ft of beds, mainly pennant sandstone, are developed above the Upper Glynogwr. On the hillside below Llandyfodwg a 'slack', indicative of about 10 to 20 ft of argillaceous strata, occurs 30 to 50 ft above the coal, while to the north of the Ty'n-y-nant Fault a similar feature, associated with a thin coal near the top, lies 50 to 100 ft above the same seam.

Llangeinor, Bettws and Llangynwyd: The Hughes Beds occur as a number of more or less disconnected outliers lying astride the axis of the syncline between the South Crop and the Moel Gilau Fault. The seam at the base of the group is known as the **Bettws Four-Feet** and it has been worked extensively, both underground and by opencast methods. It crops out on the hill between the Ogmore and Garw valleys immediately south-east of Llangeinor. Old levels are located north and west of Glyn-y-glowr Farm [91958735] and at Frithwaun [92658755] and numerous bell-pits and crop-works are seen on the northern part of Bryn-y-Wrach. The coal was proved by Opencast exploration to be very variable: about 4½ ft of coal with several partings appears to be average. At Glyn-y-glowr Colliery [91978715], a small working 250 yd S. 6° E. of the farm, the section on the abandonment plan is: weak clift roof, coal 30 in, clod and rashes 18 in, coal 27 in, on fireclay. In this area two thin mudstone bands occur about 30 to 80 ft above the coal.

The main workings in the seam occur on the south side of the Felin-arw Fault, between the Garw and Llynfi valleys. Around Bettws village the seam is much faulted and in recent years it has been worked by opencast methods in eight separate sites. In former years numerous pits, drifts and levels were located on either side of Nant Cedfyw. Alder Llantwit Colliery [89448674] recorded the following section: clift roof, lantern coal 4 to 6 in, holing, coal 28 to 36 in, fireclay 9 to 12 in, coal 16 to 18 in. A consistent section was proved by opencast works along the crop north of Ty'n-y-waun [89528742]: top coal 14 to 18 in, parting 12 in, coal 35 in, parting 3 to 14 in, coal 12 to 15 in.

In the Llynfi Valley the coal was worked from several levels on the crop immediately west of Shwt [890868]; from a pit [87958790] near the river, 300 yd W.S.W.

of Llwyn-y-brain; from Gelli-siriol Level [87538842]; and Maes-y-bettws [88008815]. On the western portion of this outlier the coal was also worked from levels [86708835] in the old tramway, east of Ty'n-y-waun, and just south of Brynllywarch-fach [87508725]. At the latter the following section is recorded: clift roof, coal 39 in, soft clift 10 in, coal 12 in. The coal was worked opencast at Maes-cadlawr Site [875875], which lies astride the old tramway to the west of Brynllywarch-fach. Sections here showed that both the coal and the measures above were very variable with washout conditions associated with both sandstone and shale over-burden. Shales up to 18 ft thick were observed to pass laterally into thick-bedded sandstone, the change being complete within 100 yd. The variable nature of the strata is shown by the following two sections measured only 30 yd apart:

	Ft	In			Ft	In
Shale	—	—	Shale		—	—
Sandstone		6	Sandstone	2 in to		6
Shale	10	0	Shale			10
COAL		10	Shaly sandstone		1	6
Soft clay		1	Shale		1	0
COAL and rashing		2¾	Sandstone			4
Clay		0½	Shale			1
COAL		1½	Sandstone			3
Fireclay and pyrite band		8¼	Shale		4	6
COAL		1½	COAL		1	0
Clay		0¼	Fireclay and rashings			5
COAL		2	COAL			4
Clay		3½	Fireclay with coal streaks		1	6+

Two further outliers appear to the west of the Pen-y-castell Fault in the area around and to the south of Llangynwyd. Opencast working has taken place on the slopes above the east bank of the Nant-y-castell stream immediately west and south-west of the village, and also on the small outlier of Waun y Gilfach to the west of the same stream. On the former site a two-coal section was proved in places: coal 12 to 16 in, dirt 1½ to 3 ft, coal 37 to 41 in. Locally the bottom coal appears to split and a three-coal section is recorded: coal 14 to 22 in, dirt 10 to 22 in, coal 28 to 33 in, dirt 3 to 12 in, coal 15 in. The coal has also been proved in a trial pit [85778766], on the north bank of the Nant Bryn-Cynon, 400 yd S. 41° W. of Gadlys Farm: shale roof, coal 15½ in, clay 3 in, coal 2½ in, clay 11 in, coal 5 in, clay 4½ in, coal 28 in.

A shallow pit [91078737] to the Bettws Four-Feet, 400 yd S. 39° W. of Llangeinor Station, yielded spoil including blue shale with *Anthraconauta tenuis*, *A. phillipsii* and ostracods, while at the opencast sites west of Bettws village *Anthraconauta* as large as those from the Cefn Glas in the Cynon Valley were associated with *Spirorbis sp.*, *Carbonita* cf. *humilis* and *C. wardiana* (Jones and Kirkby).

In the Bettws area the seam is overlain by 30 to 40 ft of sandstones, above which lies a thin mudstone associated with an insignificant coal. About 150 ft higher in the sequence a bold feature running around the back of the old village indicates the presence of a second mudstone horizon which also has a thin coal at the top. No details are known concerning this coal, and its correlation with the Upper Glynogwr is doubtful. This coal also crops out on the high ground west of Nant Cedfyw to the south-west of Tyle-coch [89058755] and around Celfydd Ifan [88708790]. About 100 ft of pennant sandstone overlying the coal mark the upward limit of the group in this area.

In the Llynfi Valley and the areas to the west the Bettws Four-Feet is overlain immediately by some 30 to 40 ft of mainly argillaceous measures, followed by pennant sandstones; no further coals have been reported. A.W.W.

E. Swansea Beds

Few details are available to supplement the general description of these beds given on p. 72. No sections are recorded of the **Graigola** or **Six-Feet Bituminous** workings on Cefnmawr, though the coal was said to be about 6 ft thick. It is directly overlain by massive sandstone. A seam which may equate with the Graigola was worked opencast over a small area nearly ¾ mile west of Llangeinor Station. A trial pit [90358765], 550 yd S.S.W. of Efail Moelgilau, proved: coal 19 in, parting 4 in, coal 14½ in, parting 13½ in, coal 61 in. Old levels nearby showed more than 20 ft of massive sandstones closely overlying the coal.

The presumed **Bettws Nine-Feet** was worked from numerous bell-pits in the area between Bedw-bach [89838803] and the railway, and Glannant Llantwit Colliery [89668796] was sunk 85 yd to the coal. Sections on the abandonment plan of the latter workings showed the following variation of section: top coal 36 to 54 in, clod 2 to 12 in, coal 36 to 48 in; locally a plane parting was developed about 12 in from the roof of the top coal. What appears to be the same seam is exposed on the roadside [88778842], 400 yd S.S.W. of Cefn Cedfyw, where the following section was measured: shales 20 ft, rashings 4 in, coal 2 in, clay 6 in, shale 1½ in, coal 3 in, rashings 1 in, fireclay 6 in, coal 2½ in, clay and rashings 4 in, coal 4 in, clay and rashings 2 in, coal 28 in; this may represent disturbed coal. The seam was worked at Capel-bach, presumably from the old shaft [88438851] on the south side of the road; according to the previous editions of this memoir the seam here was 'quite 9 ft thick'.

A.W.W.

F. Grovesend Beds

The outcrops of these beds within the Llantwit Syncline are extensively covered by glacial drift and knowledge of the sequence is derived almost entirely from mining activity which was, to a large extent, completed before the original six-inch survey. Few details can be added, therefore, to those already published in the earlier editions of this memoir. Little variation either of coals or sequence takes place within the area and the section encountered in Ystrad-barwig Shaft [07468462], on the axis of the syncline just south of Newtown Village, furnishes a virtually complete account of the stratigraphy of the group:

	Thickness		Depth	
	Ft	In	Ft	In
Gravel	42	0	42	0
Cliff with rock beds	54	0	96	0
Rock	54	0	150	0
Cliff	4	0	154	0
Rock	2	0	156	0
NO. 1 LLANTWIT: **coal 12 in,** clay 3 in, **coal 45 in,**				
rock in thin beds 9 ft, cliff 6 ft, **coal 36 in**	23	9	179	9
Cliff	36	0	215	9
Rock	48	3	264	0
Cliff	26	0	290	0
Red fireclay	61	0	351	0
COAL	1	0	352	0
Rock and cliff	21	0	373	0
Cliff	40	0	413	0
Rock	16	0	429	0
Cliff	17	0	446	0
NO. 2 LLANTWIT: **coal 33 in,** fireclay 6 in, **coal 12 in,**				
shale 12 in, **coal 24 in**	7	3	453	3

								Thickness		Depth	
								Ft	In	Ft	In
Cliff ..	..	..	..	..	..	..	..	31	3	484	6
COAL	..	..	..	..	..	..	..	1	0	485	6
Cliff ..	..	..	..	..	..	..	..	29	0	514	6
Rock	..	..	..	..	..	..	..	12	0	526	6
Cliff ..	..	..	..	..	..	..	..	39	0	565	6
NO. 3 LLANTWIT: coal 48 in ..	..	..	..	..				4	0	569	6
Fireclay	..	..	..	..	..	..	..	2	0	571	6

The **No. 3 Llantwit** crops on the north side of the syncline in the area to the south-east of Cwm Colliery. Drifts on or near the crop [07138624; 07388634, etc.] are located on either side of the road running northwards from The Cottage. West of the Myddlyn the seam is cut off northwards by the Ty'n-y-nant Fault and it does not crop again until the western closure of the syncline at Llantrisant Common. On the south side of the syncline the crop is marked by numerous bell-pits and by major drifts at the old Llantwit Main Colliery [06168382], Llantwit Wallsend Colliery [06968362] and Park House Colliery [07678400]. The seam is usually recorded as consisting of clean coal 48 to 58 in thick, though at Park House Colliery a 5½-in stone band is recorded 6 in from the base of the seam. Old pits [05088498] on the north side of Llantrisant Common yielded spoil with many plants: *?Asterotheca abbreviata*, *Lepidophyllum sp.*, *Lepidostrobus anthemis* (Koenig), *L. lanceolatus*, *Neuropteris rarinervis*, *N. scheuchzeri*, *Sigillariophyllum bicarinatum*, *Sphenophyllum emarginatum*, *Sphenopteris neuropteroides*.

The **No. 2 Llantwit** crops to the east of Nantmyddlyn, on the north side of the Ty'n-y-nant Fault about 150 yd to the south-east of the No. 3 crop. Levels occur [07108610] on either side of the road just north-west of The Cottage. The seam also crops on the eastern side of the Castellau Valley in the angle between the Ty'n-y-nant Fault and the easterly throwing fault east of Capel Castellau. On the south side of the syncline the crop (again marked by many old shallow pits) runs parallel to that of the No. 3 Llantwit about 70 to 140 yd to the north; a drift [07498405] associated with Duffryn Llantwit Colliery is on the crop, while the shaft [07448423] struck the seam at a depth of 197 ft. The seam section shows very little variation:

			Dehewyd Colliery [07848532]			Ystrad-barwig Colliery			Duffryn Llantwit Colliery	
			Ft	In		Ft	In		Ft	In
COAL	..	..		9	⎫			⎧	1	6
Parting	..	..	—	—	⎬ 2 9			⎨	—	—
COAL	..	..	1	6	⎭			⎩	1	6
Clod	..	..	1	0			6		1	0
COAL	..	..	1	0		1	0			6
Parting	..	..		3		1	0			4
COAL	..	..		6	⎫			⎧		6
Parting	..	..	—	—	⎬ 2 0			⎨	—	—
COAL	..	..	1	8	⎭			⎩	2	0

Spoil from an old drift [07628593], 300 yd N.N.W. of Dehewyd Farm, stated to be to the No. 2 Llantwit Seam, included minutely crumpled shale, which yielded *Spirorbis sp.*, *Anthraconauta* cf. *tenuis* and *Carbonita humilis*. Similar spoil was noted in the tip of Gelynog Colliery [05568530], to the west of Beddau, which worked both the No. 2 and No. 3 Llantwit seams.

The **No. 1 Llantwit** everywhere consists of two thick coals, between which the separation varies considerably. In Llest Llantwit Colliery shaft [07418521] the floor of the seam was struck at 103½ ft and the two coals were only 2 ft apart: top coal 4 ft, fireclay 2 ft, bottom coal 5 ft. At Ystrad-barwig Colliery, 650 yd to the south, the separation increases to 15 ft. The coal forms a small outlier, faulted on the south-west side, on the Foel [074857] to the north-west of Llantwit Village, and numerous old levels mark its crop. On the south side of the Ty'n-y-nant Fault, in the same general area, the crop of the seam is indicated by a level and several bell-pits to the west and north-west of Croes-ged House [07128525] and again about 100 yd N.W. of the Bush Inn [07708530]. Still on the north side of the syncline it crops on either side of the main road at Newtown where it was worked from a level [07128500] 100 yd W. by S. of The Gables. On the south limb of the syncline the crop is completely obscured by drift; on the south side of Gwaun Miskin a number of old bell-pits mark the coal but no details are known.

Pennant sandstone makes up the bulk of the measures above the No. 1 Llantwit. These cap the Foel and have been quarried [07608540] about 200 yd N.W. of the Bush Inn. They were also worked on a small scale just south of Ystrad-barwig Uchaf and a 20-ft section of massive sandstone may be seen in the railway cutting nearby [07208470]. They also crop through the drift on the ridge to the south of Beddau and 15 ft of massive beds are exposed in an old quarry [05648447] on the west side of the road, 250 yd E. 42° S. of Llwyncrwn Isaf.   A.W.W.

REFERENCES

JORDAN, H. K. 1876. The Pencoed, Mynydd-y-Gaer, and Gilfach Goch Mineral Districts. *Proc. S. Wales Inst. Eng.*, **9**, 250–70.
——1903. Notes on the South Trough of the Coalfield, East Glamorgan. *Proc. S. Wales Inst. Eng.*, **23**, 131–56.
——1908. The South Wales Coal-field: Sections and Notes. *Proc. S. Wales Inst. Eng.*, **26**, 1–84.
——1915. The South Wales Coalfield. Part III. *Proc. S. Wales Inst. Eng.*, **31**, 49–135.

MOORE, L. R. and COX, A. H. 1943. The Coal Measures sequence in the Taff Valley, Glamorgan, and its correlation with the Rhondda Valley sequence. *Proc. S. Wales Inst. Eng.*, **59**, 189–265.

PLUMMER, E. 1878. Observations on the Glyncorrwg Mineral District from the Avan Valley to the Neath Trough. *Proc. S. Wales Inst. Eng.*, **10**, 335–41.

ROBERTSON, T. 1933. Geology of the South Wales Coalfield. Part V. The Country around Merthyr Tydfil. 2nd edit. *Mem. Geol. Surv.*

G

FEET
- 2000

ML

- 1500

- 1000

- 500

- 0

M. F. P.

CHAPTER VII
MESOZOIC

UPWARDS OF several hundreds of feet of strata belonging to the Keuper, Rhaetic, and the lowest part of the Jurassic occupy four small areas along the southern margins of the district. They are parts of the northern edge of larger outcrops in the Vale of Glamorgan to the south, described fully in the memoir on the Bridgend district (Strahan and Cantrill 1904). All three formations are represented in the west, between Kenfig Hill and Pyle, but only the Keuper occurs in the Parc-gwyllt area (to the south of Bryncethin), between Dolau and Lanelay, and on the south side of the Clun Valley between Llantrisant and Rhiwsaeson. Although local unconformity and non-sequence can be proved when the rocks are traced regionally through the Vale of Glamorgan, they appear to be broadly conformable within the present district. The basal Keuper rocks, however, rest with great angular discordance on the folded and eroded Carboniferous rocks, which from south to north they overstep completely up to the Lower Pennant Measures.

KEUPER

The Keuper strata of the district fall into two distinct facies: normal red marls and 'littoral' deposits consisting predominantly of coarse angular limestone breccias and conglomerates, the so-called 'Dolomitic Conglomerate'. Only in the west, where overlain by Rhaetic, is the succession complete, and here the mapped junction with the Carboniferous is faulted. The full thickness is difficult to determine, but several hundred feet of beds are probably present. They consist mainly of red marls with green mottling; bands of Dolomitic Conglomerate are developed in the lower part, and near the top a 20-ft bed can be traced. Above this latter the strata consist largely of green, cream and buff marls with thin calcareous nodular bands, the equivalent of the widespread 'Tea-green Marls'; these are probably not more than 15 or 20 ft thick.

Traced eastwards, such thick deposits are not usually present, and the Dolomitic Conglomerate facies becomes relatively more important. The Conglomerate more usually approaches a breccia in character, consisting typically of angular or sub-angular fragments of Carboniferous Limestone embedded in an inorganic, ferruginous limestone matrix. This character persists even when the beds completely overstep the Lower and Middle Coal Measures, and rest against almost vertical 'cliff' faces of pennant sandstone, as they do in the area east of Llanharan. Where well developed the breccias are interbanded with red, buff, cream and whitish porcellanous limestones and red marly micaceous sandstone. In the area between Llanharan and Lanelay such a sequence, associated with red marls, occupies a hollow several hundred feet deep in the surface of the Coal Measures. This hollow appears to be completely enclosed, for although the area is heavily drift-covered, it seems certain that there can be no connexion with any of the neighbouring Keuper areas to the south and east. In the most easterly outcrops, south of Llantrisant, only the basal beds of the formation are present: they consist entirely of breccia, much veined by calcite containing traces of galena.

233

Although the area has clearly undergone much erosion since Jurassic times it seems unlikely that the Keuper rocks extended much farther to the north than they do at present. The Pennant scarp is evidently in much the same position now as it was in Triassic times, and the shallow water deposits probably lapped against it over much of its length. In the area between Llanharan and Rhiwsaeson the scarp, which presumably rose originally from the base of the lowest massive sandstones above the No. 2 Rhondda, was eroded back to about the level of the No. 1 Rhondda. The flat area to the south of Talbot Green, now occupied by the lowermost pennant sandstones, presumably represents a local sub-Keuper planation.

DETAILS

Water Street to Kenfig Hill: West of Pyle the rocks are almost entirely covered by drift, and they can be seen only in sporadic exposures in the banks of the River Kenfig. Here red marls with greenish intercalations, especially abundant close beneath the Rhaetic, are seen dipping southwards at about 15°. A single band of Dolomitic Conglomerate is developed, and this was formerly quarried on either side of the Port Talbot railway [83088290] about 300 yd S. of Waterhall Junction. This bed is probably truncated by the coalfield boundary fault beneath Kenfig Hill, but it reappears again as a small inlier against the fault about a quarter of a mile south-south-east of St. Theodore's Church. Here it was quarried [84158255] near Pen-y-castell Farm in beds dipping southwards off the fault at 20° to 36°. Massive coarse limestone-breccia passes upwards into fine-grained detrital limestone, in beds up to 2 ft thick, which is interbanded with deep red marl, calcareous marl with limestone fragments and a few layers of porcellanous limestone.

Farther west the British Industrial Solvents' borehole [79968316] at Water Street proved, beneath 65 ft of glacial drift, 161 ft of red marls, with Dolomitic Conglomerate layers in the lower part.

The highest beds are seen in a stream [84808243], 350 yd S. of the Methodist Chapel on Cefn Cribwr. Here the following section, faulted against Lower Coal Measures, was measured:

			Ft	In
Rhaetic	..	White sandstone	2	0
		Soft dark grey clay, disturbed..		6
Keuper	..	Red marls with subordinate grey and green streaks ..	3	0
		Grey, grey-green and yellowish marl with subordinate reddish streaks	8	0
		Dolomitic Conglomerate with much calcite ..	1	0

Parc-gwyllt to Heol-las [93148273]: In the vicinity of the Parc-gwyllt Mental Hospital and extending eastwards for about a mile, gently dipping Keuper rests with marked unconformity on Millstone Grit strata. Around the hospital itself only red marls are seen, but east of the Bryncethin–Coity road interbedded Dolomitic Conglomerate is widely developed. Throughout this area the gently falling surface of red clay soils is interrupted by numerous sink-holes indicating collapse into cavernous Dolomitic Conglomerate beneath. The Conglomerate is only rarely seen, but it is quarried at surface [92308227] ¼ mile S.S.E. of Pen-yr-heol, where 10 to 12 ft of massive breccia with rolling dips may be seen. Between Heol-las and Giblet [93358240] red marls dipping southwards at angles of 15° to 35° indicate that the junction with the Millstone Grit may here be faulted.

Llanharan to Lanelay: The western end of this largely drift-covered outlier forms the higher ground of the hill immediately east of Dolau. Several old quarries, now largely overgrown, show red marls with green mottling dipping in general south-south-eastwards at angles up to 15°. Locally the breccia is associated with beds of porcellanous limestone. Eastwards the strata plunge beneath the drift of the low-lying Melyn Valley, where they occupy a 'gulf' several hundred feet deep. An old pit [01658278] near Rhyd-y-melinydd, reputed to be in search of ironstone, appears from the size of the spoil to have reached a considerable depth, and to have passed through nothing but Keuper Marl. A borehole [02698257] near Garth Villas showed the following section:

	Thickness Ft	In	Depth Ft	In
Drift	16	0	16	0
Red marl	37	0	53	0
Conglomerate	19	0	72	0
Red sandstone with green bands near top and white bands in bottom half	42	0	114	0
Conglomerate	3	0	117	0
Red sandstone and conglomerate interbanded	27	0	144	0
Red marl	8	0	152	0
Conglomerate with a few limestone bands	93	0	245	0
Red marl	2	0	247	0
Red conglomerate	3	0	250	0
Red and mottled marl	12	0	262	0
Pennant sandstone	42	8	304	8
?NO. 2 RHONDDA: coal 54 in..	4	6	309	2
Red fireclay and grey shale	29	10	339	0

A second borehole [02478287] 400 yd to the north-west penetrated only 74½ ft of Triassic red rocks to Coal Measures, while in the stream, less than 100 yd farther north-west, pennant sandstones are exposed at surface. In the Melyn stream [01768297], about 700 yd to the east, a nearly vertical contact between the red marls and pennant sandstones was observed, and both here and near Garth Uchaf [02408295] the pennant shows red staining, especially on the joints.

Red marls with breccia bands are exposed in the railway cutting immediately west of Rhyd-y-melinydd as well as in the stream [01978224] 200 yd S.E. of Hendre Owen. In the latter area shallow pits to coal also penetrated a few feet of red marls. The eastern limit of this outlier is not known with certainty, but the Cardiff Navigation Colliery shafts [03058203], south of Lanelay, were sunk through drift directly into Coal Measures, and there is no evidence of red rocks on the low ground south of Llantrisant, where to the east of Newpark [04308280] numerous shallow bell-pits worked coal.

Pont-y-parc [05078230] *to Llwynmilwas* [06908225]: Boulder clay again largely obscures the rocks of this tract, though they are exposed in several small 'windows'. The Triassic rocks here appear to consist wholly of Dolomitic Conglomerate. No great thickness is present for several old bell-pits within its outcrop reached Coal Measures. Near Gwern-Efa [06328232] several shallow quarries were worked over for lead, and between there and Llwynmilwas, 600 yd to the east, numerous shallow pits were formerly sunk for the same purpose. Little can now be seen to give any clear idea of the mineral occurrence, but the lead (as galena) appears to have been associated with pyrite and chalcopyrite in small discontinuous veins, stringers and pockets throughout the rock. Dolomitic Conglomerate was also quarried [05248206] just north-east of Cefn-parc. To the west of Pont-y-parc the sites of numerous old pits for lead can be seen, and there are signs that the Conglomerate was also worked opencast. At one point a 12-ft section of breccia with calcite veins containing galena is still visible.

RHAETIC

The limited exposures in the Pyle and Kenfig Hill areas show that the Rhaetic here succeeds the 'Tea-green Marls' of the Keuper. The strata consist largely of pale sandstones with subordinate interbedded shales, clays, marls and limestones, the whole making up the so-called 'littoral facies' which characterizes much of the west of the Vale of Glamorgan. Francis (1959, pp. 160–1) recognized a four-fold sub-division, which may be adapted for the present district as follows:

Upper Rhaetic   ..   Cotham Beds: cream-coloured marls with bands of soft marly and harder concretionary limestones and pockets of red and green sandstone; 20 ft

Lower Rhaetic   ..   Upper Sandstone: largely massive, fine- to medium-grained white, yellow or drab sandstones with intercalations of brown, blue or green shale or clay at base; this is the 'Quarella Stone' so-called from the Quarella Quarry of Bridgend; up to 35 ft

Black Shales: black and dark green shales with nodular limestones; about 7 ft

Lower Sandstone: white or yellowish medium-grained sandstone, weathering grey; up to 12 ft

The three divisions of the Lower Rhaetic form the Quarella Sandstone of the one-inch map.

DETAILS

The quarry near Pyle Inn [82708280] is now largely overgrown but the section was seen by Tiddeman and described in the second edition of this memoir (p. 103) thus:

	Ft	In
" Soft drab and green sandstone with casts of *Natica pylensis*, *Cylindrites*		
oviformis, *Myophoria sp.*	12	0
Ochreous bed		4
Fine sand	1	3
Green and grey clay	3	6
Calcareous nodular bed, *Acrodus minimus?* Ag.		3
Shales	1	3
Calcareous nodular bed, with fish remains		3
Shales	1	8
Fine whitish sandstone with plants, and at top a seam of quartz-pebbles		
with remains of *Hybodus* and *Acrodus*; seen to	12	0 "

In addition to the above mentioned fossils Francis (op. cit., p. 162) found *Chlamys valoniensis* (Defrance), *Protocardia rhaetica* (Merian) and *Rhaetavicula* [*Pteria*] *contorta* (Portlock) in the lower shales, and *Cercomya praecursor* Quenstedt and *R.* [*P.*] *contorta* in the upper sandstone.

A quarry [82408260] behind St. James's Church is also largely obscured, but it was worked extensively in the past for building stone. Here 21 ft of pale sandstones with green marly partings have yielded '*Natica*' *oppelii* Moore, *Cercomya praecursor*, *?Eotrapezium depressum* (Moore), *Lyriomyophoria sp.* and '*Pullastra arenicola* Strickland'. These sandstones are overlain by about 6 ft of cream limestone with bands of green and red marls succeeded in turn by a foot or two of red and green marl, which represent the lower part of the Cotham Beds of Francis. The sandstones presumably represent the uppermost part of the Upper Sandstone.

Lias

The small area south of Kenfig Hill occupied by the basal beds of the Lias is completely drift-covered. The sequence, is presumed to consist of alternations of thin earthy limestones and grey marls as in the Bridgend district. Traces of such beds are to be seen in the cutting of the Port Talbot railway at Pyle [82928258], 400 yd E. by S. of St. James's Church.

REFERENCES

STRAHAN, A. and CANTRILL, T. C. 1904. The Geology of the South Wales Coalfield. Part VI. The Country around Bridgend. *Mem. Geol. Surv.*

FRANCIS, E. H. 1959. The Rhaetic of the Bridgend District, Glamorganshire. *Proc. Geol. Assoc.*, **70**, 158–178.

CHAPTER VIII

STRUCTURE

APART FROM differential down-warping of the crust which accompanies large-scale and prolonged sedimentation, and which can be measured in terms of the varying thicknesses of the strata, earth-movements belonging to at least four different periods can be recognized in the Carboniferous and later rocks of the Pontypridd district. The first two of these movements were of a relatively small-scale positive or elevational nature; the later two were orogenic in character. The main structural elements of the Coalfield were determined during the Armorican (or Hercynian) orogeny in post-Pennant, pre-Keuper times; and folding and faulting on a less severe scale also took place in post-Liassic (? Tertiary) times.

INTRA-CARBONIFEROUS MOVEMENTS

Sudetic movements are recognizable along the southern margins of the district by the unconformity developed between the Millstone Grit Series and the Carboniferous Limestone. The stratigraphical break increases when traced from west to east across the Coalfield, both by progressive erosion of the Lower Carboniferous strata and by successive overstep of the overlying Millstone Grit and Coal Measures, indicative of greater and more prolonged movement towards the Malvern Axis (Welch and Trotter 1961, p. 10). As far as the South Crop of the Coalfield is concerned the break is least in North Gower where, according to Mr. J. V. Stephens (personal communication), beds of Upper *Eumorphoceras* Age (E_2) succeed shales and mudstones of Upper *Posidonia* Age (P_2). In the area south of Aberkenfig (George 1933, p. 256; and p. 9 above) the basal beds of Lower *Reticuloceras* Age (R_1) lie close above 'Black Lias' of Lower *Posidonia* Age (P_1); at the Llanharry Haematite Mine, $1\frac{1}{2}$ miles south of Llanharan, black shales of Upper *Reticuloceras* Age (R_2) rest directly on standard type Carboniferous Limestone probably belonging to the Lower *Dibunophyllum* Zone (D_1); while in the Rudry area, 7 miles east of Llantwit Fardre, the *Gastrioceras cancellatum* horizon occurs only 26 ft above limestones of the Upper *Seminula* Zone (S_2) (Eyles 1956, p. 35; Dixey and Sibly 1918, pl. xvi). Similar conditions of unconformity have been described on the North Crop of the Coalfield, east of the river Neath (see, for example, Robertson and George 1929, p. 24; Robertson 1933, p. 50; Evans and Jones 1929, p. 172). The strike of the Carboniferous Limestone zones against the plane of the unconformity runs approximately north to south, and this is closely paralleled by the overstep in the Millstone Grit. It is evident, therefore, that the movements which produced this unconformity were Malvernoid in direction throughout the central and eastern parts of the Coalfield.

During the present re-survey no evidence has been found within the Pontypridd district of the unconformity described by Moore (1945) on the South Crop east of the Taff Valley. However, a break of some magnitude is thought to occur in the Upper Pennant Measures at a level just below the No. 3 Llantwit Seam

Geology of Pontypridd and Maesteg (*Mem. Geol. Surv.*) PLATE VI

A 7813

HILLS ENCLOSING THE HEAD OF GARW VALLEY, SHOWING COURSE OF TABLELAND FAULT

A. 9065

B. THRUST-FAULTS IN MIDDLE COAL MEASURES, EAGLE BRICK-PIT, CWMAVON.

[To face page 238

(= Mynyddislwyn of Monmouthshire). This is inferred from the correlation of the *Leaia–Anthraconauta tenuis* bands in the strata above the Mynyddislwyn with those above the Swansea Four-Feet or Wernffraith in the areas west of the Neath Valley (Woodland and others 1957, p. 9). The Pennant sequence is evidently complete in the latter area, but when traced eastwards, the Swansea Beds are presumably overstepped progressively, until, in the Llantwit Syncline and in Monmouthshire, the Grovesend Beds rest directly upon the Hughes Beds. Although upwards of 1500 ft of strata may be considered to have been cut out by this unconformity, the movements responsible must nevertheless be regarded as gentle since the gradient of uplift is less than 1°. This movement may be considered as belonging to the Malvernian phase of Trueman (1947, p. xcix).

POST-CARBONIFEROUS MOVEMENTS

The movements which culminated in the building of the Armorican–Variscan mountain chain across Europe, produced in South Wales most of the present-day structures of the Coalfield. The most obvious result of the general south to north compression was the formation of the broad structural basin, which extends from Monmouthshire in the east across Glamorgan to Carmarthenshire in the west, and even in the midst of more complex structures is still recognizable farther to the west in Pembrokeshire. In a south–north direction it can be traced from the Cowbridge Anticline (in the Bridgend district to the south), far into the Old Red Sandstone outcrops of Brecknockshire to the north.

Fig. 30 shows the main structural elements of the district including the surface positions of the axes of folding and the principal faults. In detail the folding is far from simple, and a number of subsidiary anticlines and synclines, some gentle, some sharp, interrupt the broad downfold; while faults of varying types and complexity are profuse. The following four groups of faults can be related directly to the folding and like it can be explained in terms of a general south to north compression:

1. Strike-thrusts, which are confined to the South Crop.
2. Normally inclined strike-faults, closely associated with subsidiary folds; they include the Cilely fault system on the north limb of the Pontypridd Anticline (Y–Y of Fig. 30), and the Meiros faults.
3. Dip- or cross-faults; these are aligned sub-parallel to the regional dip and so cut across the axes of main folding.
4. Incompetence-faults; these are very widespread and consist of multitudinous minor compressional dislocations developed particularly among the main coals of the Lower and Middle Coal Measures, and which have resulted from the incompetent nature of these strata; they are not shown on Fig. 30.

These structures, together with the folding, are the main manifestations of the Armorican orogeny, and for them a time-sequence can be inferred. On the other hand there are certain other large and important structures which do not fit into this sequence and which cannot be explained in terms of a simple south to north compression. The principal of these are:

1. The Jubilee Slide (JS of Fig. 30)—a complex fault-system extending from the Gilfach Goch area westwards towards Pont Rhyd-y-fen, a distance of more than 12 miles.

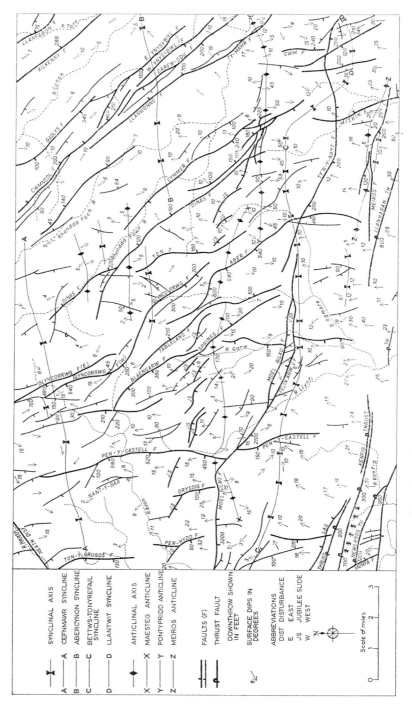

FIG. 30.   *The main structural elements of the Pontypridd District*

2. The Ty'n-y-nant–Moel Gilau Fault-belt, which can be traced throughout the length of the district from Llantwit Fardre in the east through Llangeinor and the area south of Maesteg westwards to the shores of Swansea Bay at Baglan.
3. The Neath Disturbance, which crosses only the extreme north-west corner of the district, and which is of much greater significance in the Merthyr Tydfil district (Sheet 231) to the north.
4. Minor thrusts and slides, developed among the Lower Pennant strata of the upper Rhondda Fawr Valley.

FOLDING

This is best illustrated by Plate VII, which shows the structural contours in the Six-Feet Seam. This diagram has been based directly on mine plans of those areas where the Six-Feet has been worked; in other areas contours have been constructed by extrapolation from those of other worked seams of the main group of coals. Throughout these areas, the pattern to some extent reflects the incompetence of the associated strata, but the swarms of minor structures that are present in many localities are not shown on the diagram. Where no mining in the main seams has taken place, as for example, beneath Pennant cover between the Moel Gilau Fault and the South Crop, and in the area between the Avan and Neath valleys, the contours have been projected downwards to the Six-Feet Seam from workings in Pennant seams or from surface outcrops: they are thus necessarily generalized since there is no accurate information concerning the varying intervals between Pennant horizons and the Six-Feet; nor has any allowance been made for the differences in structure between the Pennant and the Middle Coal Measures due to incompetency.

Broadly speaking, the folding of the Pontypridd district can be summarized as follows: a sharply synclinal tract, in the southern third of the district, is separated from a gently-rolling, broadly synclinal area to the north by a pronounced anticlinal belt, extending from Pontypridd westwards to beyond Maesteg.

Throughout the southernmost outcrops the Millstone Grit and Lower Coal Measures dip northwards at angles ranging in general from 25° to 35°. East of the Ely river and its tributary the Nantmychydd these dips continue to the axis of the Llantwit Syncline (D–D of Fig. 30), which can be traced from Llantrisant Common through Beddau to Llantwit Fardre, and eastwards into the Caerphilly basin in the Newport district. West of Nantmychydd the axis of this fold swings sharply north-westwards to Tonyrefail. From here the fold continues as the Bettws–Tonyrefail Syncline (C–C of Fig. 30); it follows a steady east to west direction about $2\frac{1}{2}$ to $3\frac{1}{2}$ miles north of the edge of the Coalfield, and between Tonyrefail and the Gilfach Garden Village in Cwm Ogwr Fach, lies at the foot of the steep southern limb of the Pontypridd Anticline.

In the area between Meiros and Talbot Green a sharp subsidiary anticline interrupts the general northerly dip towards Tonyrefail. The southern limb of this Meiros Anticline (Z–Z of Fig. 30), marked by dips of up to 25°, is partly replaced by the parallel Meiros Fault (see p. 247). From the crest the strata dip steadily to the Bettws–Tonyrefail synclinal axis (C–C of Fig. 30) at angles of 10° to 16°. In the area between Wern-tarw and Cefncarfan (1 mile north-east of Bryncethin) a second minor anticlinal flexure lies midway between the southern rim of the Coalfield and the main syncline: its southern limb is only feebly developed, dips of 2° to 5° being the greatest recorded, while horizontal sandstones are seen near Brynchwith (½ mile north-east of Heol-y-Cyw).

West of Coytrahen and Aberkenfig (in the Llynfi–Ogmore Valley) the south limb of the Bettws Syncline dips northwards to north-north-eastwards at angles which gradually decrease from about 30° on the Lower Coal Measures outcrop to an average of about 10° beneath the Pennant of Bryn Cynon, south-west of Llangynwyd. The area near the confluence of Nant y Gadlys and the Llynfi marks the deepest part of the syncline: here the lowest of the main coals, the Gellideg, is inferred to lie at a depth exceeding 4000 ft below O.D. To the west, beneath Mynydd Margam, the axis of the syncline turns west-north-westwards towards the Moel Gilau Fault, and the fold dies out in the vicinity of Cwm Duffryn.

The axis of the Pontypridd Anticline (Y–Y of Fig. 30)—known to many an old miner as the 'Atlantic Liner'—can be closely followed at surface from the northern slopes of the Graig, just south of Pontypridd, to the Ely Valley between Cilely and Collena, and westwards to Mynydd Maendy, south-west of Gilfach Goch, where it abuts against the Aber Fault (see p. 253). East of the Llanwonno or Ty-mawr Fault, the anticline is comparatively gentle and symmetrical, the dips seldom exceeding 10°. Westwards of the fault, however, the southern limb steepens sharply and the fold becomes monoclinal. Between Pen-y-rhiw and the Ogwr Fach Valley just south of Gilfach Goch, dips of 40° to 50° can frequently be observed, and throughout most of this area only three-quarters of a mile separates the axis from that of the Tonyrefail Syncline to the south. On the broad northern limb of the fold dips rarely exceed 10°. Between the head-waters of Nant Gelli-wion, immediately south of Mynydd y Glyn, and the Penrhiw-fer area this northern limb is interrupted by a group of northwards-throwing 'normal' faults, trending west-north-west to east-south-east, which are believed to form part of the anticlinal structure: these will be described later (see p. 250).

The crest of the anticline can be observed in Nantmychydd [02318896], just north-west of Glyn, where dips of 16° N. and 42° S. can be measured. The axis has been proved by mining in the No. 3 and No. 2 Rhondda seams in the Collena, Glyn, south Cymmer and Pen-y-rhiw areas, where there was little interruption of the workings. Only in Maritime Colliery has it been proved at greater depths in the main coals. Here, where the Four-Feet and Six-Feet seams are subject to incompetence-thrusting, the exact position of the crest is hard to determine. Still deeper, in the general horizon of the Yard Seam, intense compressional disruption was encountered in the core of the anticline.

The steep dips on the southern limb of the Pontypridd Anticline end abruptly against the Aber Fault. In the Mynydd Maendy area between the Ogwr Fach Valley and the Aber Fault a second anticlinal axis is developed on the north side of the main axis, and this can be traced north-westwards to Mynydd yr Aber, where it forms the eastern end of the Maesteg Anticline, which may, therefore, be regarded as a continuation *en echelon* of the Pontypridd Anticline.

The Maesteg Anticline (X–X of Fig. 30) can be followed almost due west through Ogmore Vale, Pontycymmer and Maesteg to the area south of Bryn where it dies out in much the same longitude as the Bettws Syncline. In the Ogmore and Garw valleys the southward dips vary generally from 15° to 20°, flattening towards the axis of the Bettws Syncline; on the north limb dips of 7° to 10° are fairly steady for several miles. In the Llynfi Valley and the area to the west the south limb is partly truncated by the Moel Gilau Fault and by convergence with the Bettws Syncline; dips are seldom greater than 10°. The dips on the north limb, however, are steeper, increasing steadily westwards as the anticline dies out, until they reach 20° or 30° in the Avan Valley south of Pont Rhyd-y-fen, where they merge into the general northerly dips of the South Crop. The form of the Maesteg Anticline is brought out by the scarps of the successive Pennant sandstone features on its opposing limbs, and this can be especially well seen by viewing the sky-line of the Garw–Lynfi watershed from the hillside east of Pontycymmer or west of Maesteg; the eroded crest of the fold can be seen in the col of Croes-y-bwlchgwyn.

To the north of the anticlinal belt lies a broad rolling synclinal tract in which dips in general vary from about 5° to 10°. To the east the shallow Abercynon Syncline (B–B of Fig. 30) reaches its deepest along an axis which extends in a west by south direction from the town of that name, dying out in the Llwyn-y-pia area. This fold forms the western extension of the Gelligaer Syncline of Monmouthshire. About 5 miles farther north-west the Cefnmawr Syncline (A–A of Fig. 30) develops *en echelon*, and this can be traced in a west-south-west direction from the Maerdy area through Glyncorrwg, and along the general line of the high ground between the Avan and Neath valleys, to become the main syncline of the Coalfield in the Neath Valley.

The Rhondda Fawr Valley between Llwyn-y-pia and Blaenrhondda, through which the *en echelon* stepping of the Cefnmawr and Abercynon synclines takes place, is an area of rolling dips in which at least two gentle minor anticlines and associated synclines can be recognized. An anticlinal axis through Pen-yr-englyn is succeeded in turn southwards by a syncline at Treorky, an anticline extending from Cwm-parc to Pentre, and a second syncline in the neighbourhood of the watershed between Cwm-parc and the Ogmore Valley. The axes of these gentle folds are broadly parallel to those of the flanking main synclines.

The general west-south-west trend of the folds north of the Pontypridd–Maesteg Anticline contrasts with the axial directions of the structures to the south; these are generally east-west but locally approach more nearly west by north to east by south. When viewed in relation to the Coalfield as a whole, however, it is seen that all of the main fold axes have similar shapes, curving through varying directions from east-north-east to west-south-west in the eastern parts of each fold, then through east to west, and finally turning slightly west by north in the western parts. These curved fold-axes are not fully concentric but replace each other *en echelon*; they follow a trend which is continued beyond the Coalfield into the Lower Palaeozoic areas of central and south-west Wales. The curvature of the axes is believed to be due to the impingement of the rising Armorican structures against a more rigid basement to the north-west, which had, in this central district, a Caledonoid trend. Each Armorican fold was thus anchored relatively towards its western end, whilst to the east it was free to move northwards, resulting in greater severity and closer spacing of the folds and associated thrusts in Pembrokeshire and Gower, when compared with the relative simplicity of structure in Monmouthshire.

STRIKE-THRUSTS

These are confined to the southern margin of the district, mainly to the outcrop of the Lower and Middle Coal Measures. Two distinct and separate developments are recognizable: (*a*) the Margam Thrust Belt, extending from

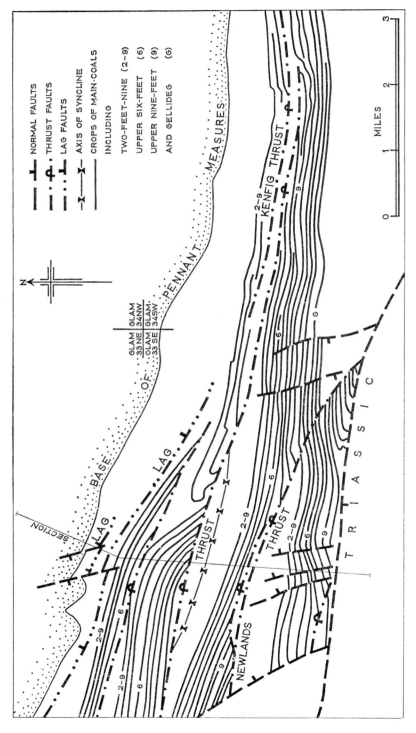

FIG. 31. *The South Crop area east of Margam showing repetition of the outcrops of the main coals caused by thrusting*

Margam and Kenfig Hill to within about a mile of Aberkenfig; and (*b*) the Llanharan Thrust, which can be traced from Heol-y-Cyw to Talbot Green.

In the Margam area a series of four major thrusts with at least two associated lag-faults has been proved by, or inferred from, mining operations and borehole exploration. Of these faults the **Newlands Thrust** has the greatest individual throw. It set a northward limit to the workings of the now abandoned Cribbwr Fawr Colliery, as it did at Newlands Colliery until penetrated by the recent North Drift roadway. This latter drivage proved a stratigraphical throw of some 750 ft, the measures just above the Yard (South Crop Six-Feet) being found in juxtaposition on the north side of the thrust with the Six-Feet (Esgyrn) on the south. The structure includes two main smash-belts about 80 yd apart, but intense associated crushing and slickensiding conceals the exact planes of major dislocation. The individual planes of fracture are all more or less parallel and dip northwards at about 50°, thus making an angle of about 30° with the regional dip of the strata. In direction the thrust maintains a roughly east-west course across the north of the Newlands workings, but eastwards it turns east-south-eastwards, and passes beneath the Trias near Waterhall Junction, north of Pyle. Here the Five-Feet and Gellideg seams are in approximate line of strike with the Cefn Cribbwr Rock and the Crows Foot Coal, indicating a stratigraphical throw of about 1000 ft which continues into the Millstone Grit. The repetition of the main coals is also proved by the relationships of the workings of Cribbwr Fawr Colliery with those of Bryndu Colliery, three-quarters of a mile north-east.

A second thrust about half a mile to the north is inferred from the position of the main coal crops in the Bryndu area, and those of the Four-Feet and Two-Feet-Nine in the opencast workings extending from East Lodge [81708490] towards Pen-y-bryn [83408455]. Its upper limb was exposed in the Two-Feet-Nine cut, at the western end of which a gentle anticlinal drag fold against the thrust was seen. Closely associated with this structure, and barely 150 to 200 yds to the north of it, a further thrust was proved in the Kenfig Opencast workings [848841], where it repeated the Two-Feet-Nine and the Upper and Lower Four-Feet coals. Both thrusts die out eastwards in the area north of the old Park Slip Colliery. The section proved in the Margam Park No. 2 Borehole implies the existence of yet a fourth thrust, succeeded upwards by two lag-faults in the area of Margam Park. The total effect of these structures on the crops of the main coals is shown in Fig. 31.

The thrusts seem generally to be associated with synclinal drag-folds developed beneath the major planes of movement. Such folds were recorded in the Newlands workings beneath the Newlands Thrust, and in the North Drift beneath the succeeding thrust to the north. A sharp synclinal drag was also seen beneath the thrust exposed in the Kenfig Opencast site. Fig. 32, which is a section across the whole of the Margam Thrust-Belt, shows the positions of all the main dislocations as well as the drag-folds where proved. In this section the complexity shown on the under limb of the Newlands Thrust contrasts sharply with the simple pattern shown in the areas to the north, but this apparent simplicity is due to lack of detailed evidence in these northern areas, where in fact a similar degree of structural complication almost certainly exists.

The **Llanharan Thrust** is illustrated by Figs. 33 and 34. Its surface position extends from the northern parts of Hirwaun Common, just south of Heol-y-Cyw, in a general east-south-east direction to a point about 300 yd south of Llanharan South Pit, thence turning eastwards to die out just east of Mwyndy Junction.

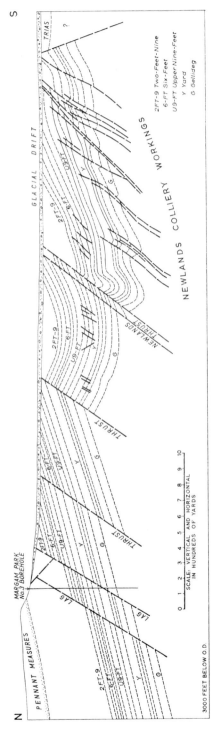

FIG. 32.   Section across the Margam Thrust-belt; the line of section is shown on Fig. 31

The Middle Coal Measures are widely repeated over most of this tract and, near Llanharan Colliery, Llynfi Beds are thrust against the higher beds of the Lower Coal Measures. On the underside of the thrust a major synclinal drag-fold has been proved in the colliery (see Fig. 34). In one part of the workings the northern limb of the fold was vertical for about 500 ft; in another some 640 ft of vertical or near vertical strata were present between the fold-axis and the thrust-plane. Despite its considerable size at depth, the fold dies out quickly upwards, its axial plane converging with the thrust at a point not far north of the North Pit, and in the sinkings themselves no abnormal dips were encountered.

The Llanharan and Newlands thrusts show close similarity. The presence of large drag-folds beneath the individual thrust-planes suggests that the thrusts developed by the under-limbs being pressed northwards beneath the upper-limbs; they are accordingly believed to be *underthrusts*. Continuing in the one case into the Pennant, and in the other into the Millstone Grit, both thrusts are clearly different in origin from the incompetence-structures discussed below, which are largely confined to the relatively narrow belt of strata containing the main coals. The difference is illustrated in Fig. 34, which shows incompetence-thrusts, with stratigraphical throws of up to 300 ft, dying out upwards before the Pentre Seam is reached, in contrast to the Llanharan Thrust, which continues into the Llynfi Beds at the surface. Many minor thrusts and lag-faults associated with the major underthrusts may be genetically linked to them rather than to the incompetence-structures, but such is the general complexity along the South Crop that it is impossible to separate the two sets.

Both the Llanharan and Newlands thrusts are cut by, and so pre-date, the cross-faults. In the case of the former, seam-plans show that it is displaced by the Pit Fault; and comparison of the plans of workings at Newlands and Morfa collieries shows that the Newlands Thrust is affected in like manner by the Morfa Fault. Evidence at Cribbwr Fawr Colliery suggests that the thrusts may belong to the early stages of the compression. Here two sets of shears are developed, one parallel and apparently related to the Newlands Thrust, dipping north at 50° to 60°, and the other, subordinate in incidence, of parallel strike but dipping only about 6° to the north (see Fig. 35). Anderson (1951) claims that thrusts should develop parallel to two planes having the same direction of strike, namely at right angles to the direction of the main tangential pressure, and equally inclined to the plane of that pressure. The two fault-sets at Cribbwr Fawr meet these conditions, except that the bisectrix between them, instead of being horizontal, is inclined to the north at about 30°, which is approximately the angle of regional dip. This suggests that the planes of thrusting were determined when the strata were horizontal and have later been tilted during the folding. It is interesting to note that in the Rühr Coalfield apparently analogous thrusts were formed at an early date since many are folded to the same extent as the strata (see, for example, Kukuk 1938, pp. 319 ff. and fig. 367).

NORMAL STRIKE-FAULTS

Both near Meiros and Cilely, strike-faults with a normal attitude are developed on anticlinal folds. The association is too close to be coincidental and there seems little doubt that the faults formed during the growth of the folds.

The **Meiros Fault** extends from about Wern Fawr, half a mile south of Werntarw Colliery, in an easterly direction to the north of Brynna and Llanharan, and thence across the lower slopes of Mynydd Garth Maelwg to Talbot

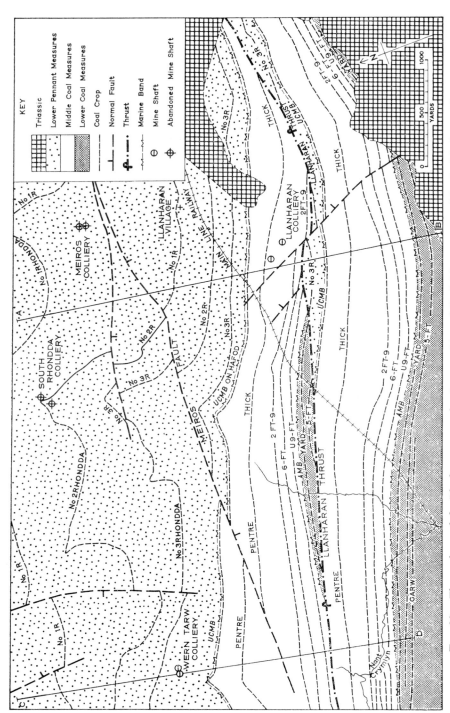

FIG. 33.   The geology of the Llanharan area, showing repetition of Coal Measures caused by the Llanharan Thrust.

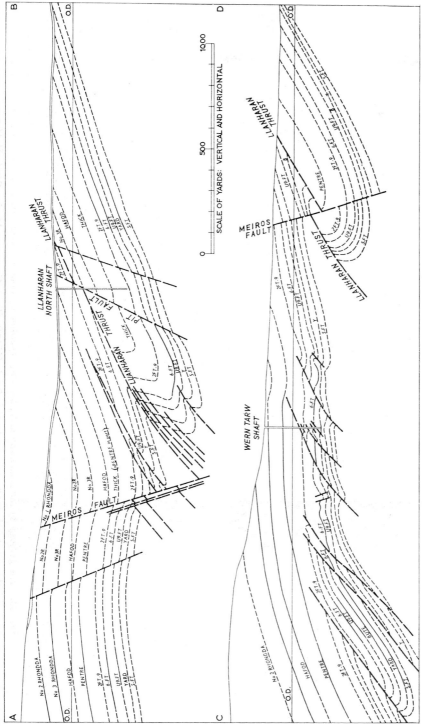

FIG. 34. *Sections illustrating the development of the Llanharan Thrust; the lines of section are shown on Fig. 33*

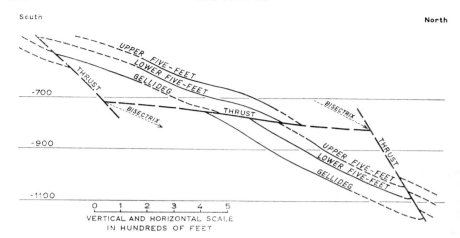

FIG. 35. *Section showing planes of thrusting at Cribbwr Fawr Colliery. Levels are shown in feet below Ordnance Datum (O.D.)*

Green. It is co-extensive with, and lies just south of, the crest of the Meiros Anticline; it has a maximum throw to the south of about 350 ft in the area north of Llanharan. A second fault, throwing south a little more than 100 ft, is located about 60 yd to the north in the Meiros Valley, coinciding with the axis of the anticline. Traced westwards it bends away from the Meiros Fault on to the northern limb of the fold, and dies out about half a mile south-west of the South Rhondda shafts.

The **Cilely group of faults** is limited to that part of the gently dipping northern limb of the Pontypridd Anticline which lies directly opposite the area of steepest dips on the southern limb. They can be traced at surface from Mynydd Pen-y-graig through the general area of the Cilely shafts to the south side of Mynydd y Glyn. Several distinct dislocations, in *en echelon* arrangement, can be recognized, in which individual faults are not exactly parallel to the fold axis, but diverge from it towards the west-north-west; the extension of the fault swarm, however, is more nearly aligned east-west. Fig. 36 shows surface positions of these faults in the area between Penrhiw-fer and the Glyn area of Nantmychydd. They have been proved underground at Cilely in workings ranging from Lower Pennant Measures to Lower Coal Measures, and Fig. 37 shows their disposition in section. The throws range upwards to about 300 ft and the fault planes dip northwards at relatively low angles between 45° and 55°: immediately behind Cilely No. 1 Pit Winding House one of the faults may be clearly seen with a dip of 45°. A similar dip can be calculated for the faults on the south side of Mynydd y Glyn from their positions in the No. 3 Rhondda and Hafod workings; these faults, however, were absent in the workings of the main seams on the south side of Lewis Merthyr Colliery. It appears, therefore, that these easternmost faults die out in depth within the upper part of the Middle Coal Measures. Interpretation of the plans of the Cilely workings suggests that these east-west faults are cut by, and so pre-date, north-north-west cross-faults.

Although not proved underground, an east-west fault mapped at the surface about 300 yd south of Cilely No. 3 Pit is probably a thrust related to the steep south limb of the fold.

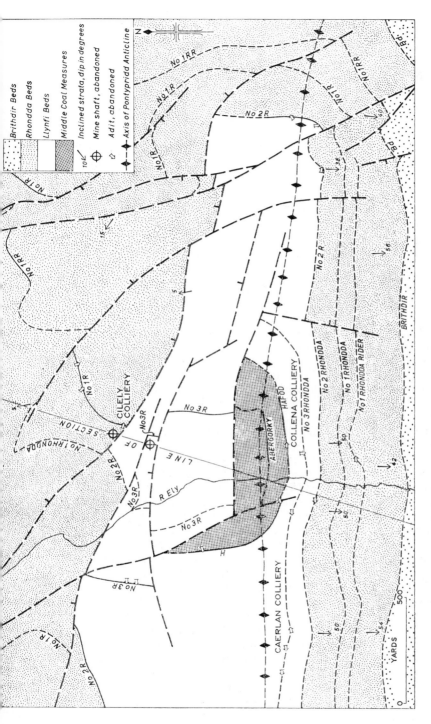

FIG. 36. *Sketch-map of the Cilely area, showing coal outcrops and distribution of faults at the surface*

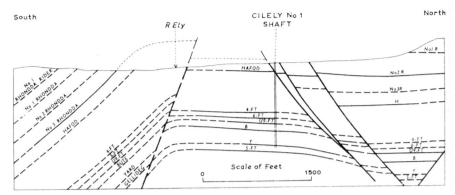

F̲ɪɢ̲. 37. *Section through Cilely Colliery shafts, showing the development of strike-faults. For line of section see Fig. 36*

CROSS-FAULTS

The term cross-fault is here used to describe faults of a relatively high normal dip, which are aligned more nearly parallel to the regional dip than to the strike and to the direction of the fold axes. They include most of the faults shown in Plate VII. Their vertical throws are not large, in only four cases (the Aber, Glyncorrwg, Llanwonno, and Pen-y-castell) exceeding 500 ft. In direction they mostly range from north to south or north by west to south by east in the western and southern areas, to north-west to south-east in the north-easterly areas. This gradual change in direction is related to the swing in direction of the fold axes, for the preferred orientation of the faults almost everywhere shows an anti-clockwise shift of 10° to 20° from the normal to the fold directions. A second series of cross-faults, numerically less and with small throws, shows a clockwise shift of the same order. These faults are mostly too small to detect at surface, and are thus not conspicuous on Fig. 30, but they have been proved underground, being best known in the Rhondda Fawr. Both sets are shown in Fig. 38, in the Lower Six-Feet workings of Cambrian and Eastern collieries.

It is seldom possible to measure or calculate the dip of the fault-planes with accuracy. Surface evidence is rarely conclusive, and underground workings hardly ever strip the precise fault-plane. However, there is some evidence that the dip of fault-planes may vary with the competence of the intersected strata, being appreciably steeper in the Pennant sandstones than in the underlying more argillaceous beds. On average, however, most of the faults are relatively steep, dips of about 70° being the general rule.

The cross-faults clearly post-date the development of the South Crop thrusts. Many show features which suggest that they were formed concurrently with the folding. In many cases there is an obvious relationship between their linear extent and the limits of the major synclines; the larger faults and fault-belts seem related to the limits of the coalfield synclinorium, dying out near the southern margin of the basin—the only exception of any size within the district being the Miskin Fault. The smaller faults are often confined to the subsidiary synclines: many of the faults in the northern syncline fail to cross the crest of the Maesteg–Pontypridd anticline, and within the Llantwit Syncline there is a particularly

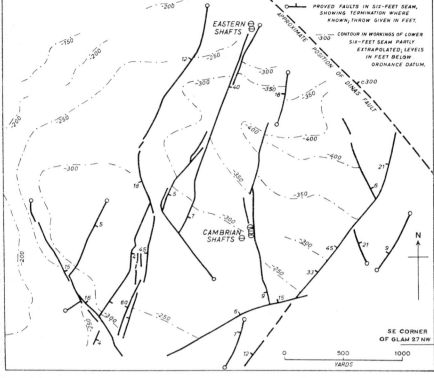

FIG. 38. *Cross-fault directions in Eastern and Cambrian colliery areas*

striking series of faults, each of which develops its maximum throw along the axis of the fold (see Fig. 39).

Such behaviour implies differential folding on either side of the fractures, and this is best illustrated by the **Aber Fault.** The steep southern limb of the Ponty-pridd Anticline ends abruptly against the eastern side of this fault near Mynydd Maendy. On the west side, the fold is much gentler and its axis has suffered an apparent northward, or dextral, displacement of about $1\frac{1}{2}$ miles. This is not a case of a tear-fault displacing the axis of an existing fold; but is due to the different behaviour of the rocks under compression on either side of a fault which was developing at the same time as the folding. In consequence the vertical throw of the fault changes rapidly along its length. From its southern hinge-point near Zoar Chapel (nearly $\frac{1}{2}$ mile east-south-east of Werntarw Colliery) the stratigraphical throw of the fault grows to about 600 ft on Mynydd Maendy, about $1\frac{1}{2}$ miles east-south-east of Llandyfodwg; along the axis of the Bettws–Tonyrefail Syncline its throw is reduced to about 250 ft near Hendre-Ifan-goch; it increases again to 600 ft on the south limb of the Maesteg Anticline in the Dimbath Valley, before dying-out on the northern limb of the anticline, north-west of Nant-y-moel.

Another example of the inter-relationship of folding and cross-faults is provided by the workings at Glyncorrwg Colliery. About a mile north of the shafts the Nine-Feet Seam is folded into a gentle syncline, the axis of which is

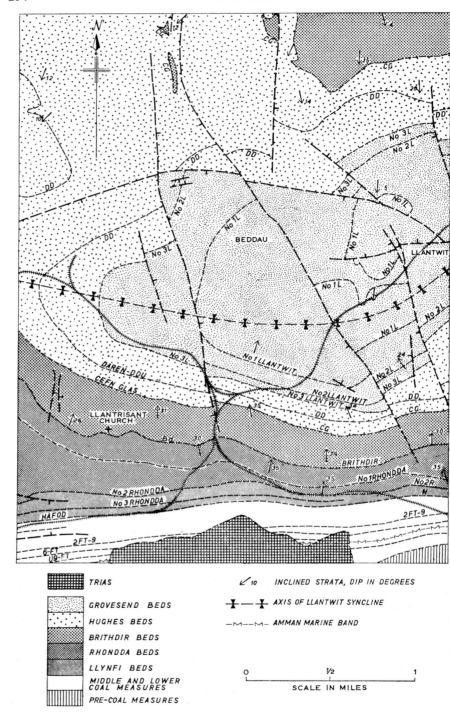

TRIAS

GROVESEND BEDS

HUGHES BEDS

BRITHDIR BEDS

RHONDDA BEDS

LLYNFI BEDS

MIDDLE AND LOWER
COAL MEASURES

PRE-COAL MEASURES

↙10   INCLINED STRATA, DIP IN DEGREES

✛—✛   AXIS OF LLANTWIT SYNCLINE

—ᴍ—ᴍ— AMMAN MARINE BAND

0   ½   1
                SCALE IN MILES

FIG. 39.   *Sketch-map showing the cross-faults of the Llantwit Syncline*

apparently displaced about 300 yd in a sinistral direction by a cross-fault, throwing about 85 ft down to the east. Since the fault dies out southwards within half a mile, it seems impossible to eliminate so considerable a horizontal movement in so short a distance, particularly as the workings have shown that compensatory thrusts and folds oblique to the fault are absent. Differential folding on either side of the fracture plane appears to offer the only reasonable explanation.

The nature of minor compressional structures may vary on either side of a cross-fault. Thus in the case of the **Cwm Fault,** intense minor crumpling on its western side appears to be matched by minor step-faulting and thrusting several hundred yards to the north on its eastern (and downthrow) side. Many instances are known where corresponding structures, even though different in detail, show apparent displacement on either side of a cross-fault. In all cases those on the downthrow side of the fault are situated farther to the north than those on the upthrow side. In other words eastwards throwing faults appear to be sinistral and westwards throwing faults dextral.

At depth the appearance of cross-faults may be quite different from that seen at the surface, and several examples have been chosen to illustrate this. The **Cwmneol Fault** which appears to be a relatively simple plane of fracture in the Pennant, in the incompetent lower measures divides into a number of separate fractures arranged more or less *en echelon*; between the individual fault-planes a series of hinged strips is thus developed in which the Four-Feet Coal was worked on a scale not usually practicable. Fig. 40 shows that it was possible to penetrate from one side of the fault zone to the other by threading a way between the individual fractures without leaving the horizon of the seam. Despite the great and rapid variations in the throw of the individual faults, their combined throw across the fault-zone remains at about 400 ft.

Different behaviour at depth is shown by the **Ton Fault** in the Cambrian–Maindy area. Fig. 41A shows the surface geological features and the structure-contours in the No. 2 Rhondda Seam. A fault mapped at the surface with a vertical displacement of 180 ft on Mynydd Ton is confirmed by the workings in the No. 2 Rhondda, but no sign of this fault is seen in the workings of the Pentre and underlying seams; Fig. 41B shows the structure-contours in the Two-Feet-Nine, which was worked out completely from beneath this area. The fault must have been formed by lateral movement which affected only the upper measures, and it presumably terminated downwards by gliding more or less along the bedding at some horizon between the No. 3 Rhondda and the Pentre. In Cwm Ian [95089480] the fault plane is vertical, and the strata adjacent to the fault show horizontal drag.

Rarely does the point of termination of a fault at depth lie vertically beneath its point of termination on the surface, and in extreme cases the distance between these two points may be considerable. The **Caerau Pit Fault,** with a downthrow eastwards of about 200 ft, dies out southwards on the northern slopes of Mynydd Bach. It extends progressively farther south in successively deeper seams, its termination in the Upper Five-Feet being at least a mile farther south than at surface some 1500 ft above. A similar southwards extension at depth is seen in the westerly throwing **Nantybar Fault,** proved in workings at Duffryn Rhondda and Bryn collieries. This fault continues at least 1600 yd farther south in the Upper Five-Feet than it does at the surface. The average northward rise of the lines along which both these faults are initiated is of the order of 15° to

the horizontal. In an analogous fashion the points of greatest throw of many of
the cross-faults show a similar displacement in succeeding seams.

All these features suggest the close association of the cross-faults with the
folding and they cannot be considered as being tensional phenomena subsequent
to the compressive phase (2nd Edition, p. 105). Faults with directions similar
to those discussed above are characteristic of the Coalfield from Monmouthshire
to Pembrokeshire. In Gower, George (1940) concluded that 'they clearly came
into operation very shortly after the initiation of the folding, and long before
the folds ... reached their final form' (p. 187). Both he and Dixon (1916,
pp. 149–50; 1921, pp. 181–3), the latter discussing the tectonics of the Pembrokeshire Carboniferous, concluded that the cross-faults were tear (or wrench) faults.
Anderson (1951, pp. 15, 64–6) agreed, and claimed that such faults should form
in two complementary directions, each making an angle of up to about 30° with
the direction of maximum stress; each should be essentially vertical and show

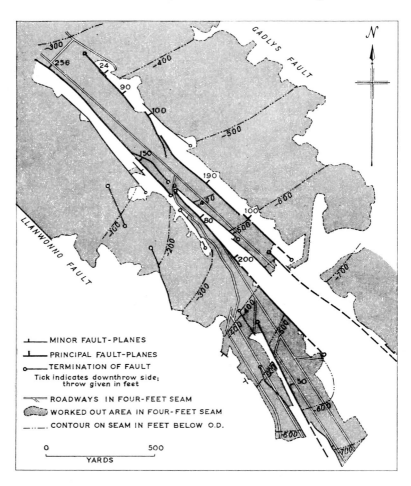

FIG. 40.   *Workings in the Four-Feet Seam at Cwmneol Colliery, showing the development
of the Cwmneol Fault*

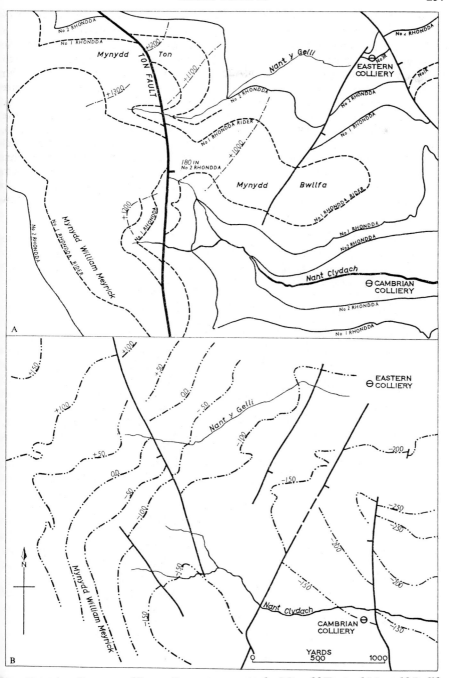

FIG. 41. A. Outcrops of Lower Pennant seams in the Mynydd Ton and Mynydd Bwllfa area, together with the surface position of the Ton Fault, and underground contours in No. 2 Rhondda Seam

B. Contours and fault positions in Two-Feet-Nine workings in the same area. Levels are shown in feet above and below Ordnance Datum (O.D.)

horizontal displacement. Such fault-swarms, all thought to be compressional in origin, and diverging about 10° to 30° from normal to the main folds, have been described from orogenic belts in many parts of the world (de Sitter 1959, pp. 169–72). In direction the cross-faults of this district all closely conform to Anderson's theory. The fault-planes are, however, not vertical (though Anderson claimed that they are steeper than should be the case for tension-faults), and the vertical displacement usually appears to be more important than the horizontal. It seems possible that they are incipient tear-faults arrested at an early stage in their development, and that the faults of Gower and Pembrokeshire represent a more advanced stage. There is no evidence within the district for two parallel systems of cross-faults, one of Armorican wrench origin and the other Tertiary and tensional, as suggested by Trotter (1947) in the Ammanford district.

STRUCTURES OF INCOMPETENT STRATA

Throughout the district swarms of thrusts and lag-faults affect much of the measures between the Gellideg and the Two-Feet-Nine. The thrusts are much the more numerous, and are known locally as 'laps' since they produce doubling or overlapping of coals. They range in size from little more than bedding-plane slips to ones with vertical throws of 250 to 300 ft. Their size and frequency give an unenviable reputation to the South Wales Coalfield, since they are particularly concentrated in that part of the measures which contains the most important of the coals.

It has long been recognized that these structures result from the compression of incompetent measures sandwiched between competent formations, for they die out rapidly both upwards and downwards (1st Edition, pp. 22–3). The group chiefly affected lies about midway between the Carboniferous Limestone and the Pennant Measures, and is more exactly bounded by the moderately competent basal Coal Measures and the Cockshot Rocks. Similar structures are, however, liable to develop wherever incompetent mudstones are closely bounded by thick sandstones, as in the Wenallt around Ton Mawr.

Only the larger 'laps' and lags can be traced with certainty from seam to seam, and the fault-pattern of any one seam may be very different from that of adjacent coals. Fortunately all the main coals are not equally affected: in most collieries certain seams are relatively undisturbed, while others are affected to greater or less degree. The belts of strata principally affected by these structures vary in different parts of the district, and this is illustrated by the sections in Fig. 42. In the Cynon and Rhondda Fach valleys they are largely the measures below the Nine-Feet: higher seams such as the Six-Feet and Four-Feet are almost free from such disturbances; even thrusts with throws of up to 150 ft are absorbed in, or immediately above, the intensely disturbed Nine-Feet. In the southern part of the Rhondda Fawr, beds between the Five-Feet and the Bute are principally affected; in the Garw Valley those between the Yard and the Caerau. Around Maesteg most of the measures below the Six-Feet are disturbed in some degree, and in the Avan Valley several comparatively large thrusts extend upwards through the Six-Feet. Along the South Crop the whole of the main coal sequence is affected, though the degree of shattering varies in its intensity from seam to seam and from place to place; west of the Ogmore the strata from the Six-Feet to the Two-Feet-Nine are almost completely shattered, but at Llanharan these seams have been those most extensively worked.

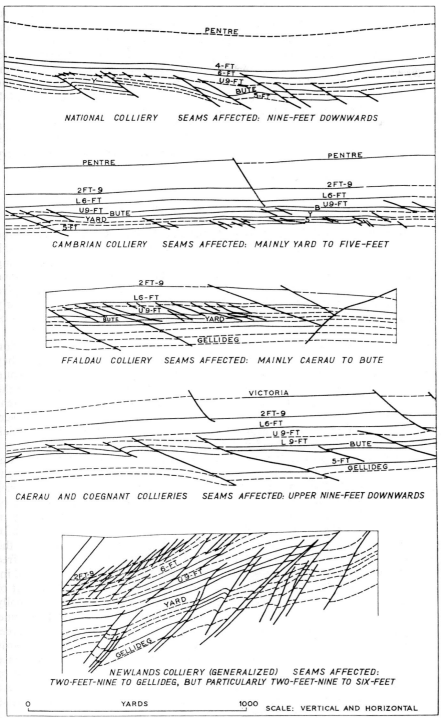

NATIONAL COLLIERY   SEAMS AFFECTED: NINE-FEET DOWNWARDS

CAMBRIAN COLLIERY   SEAMS AFFECTED: MAINLY YARD TO FIVE-FEET

FFALDAU COLLIERY   SEAMS AFFECTED: MAINLY CAERAU TO BUTE

CAERAU AND COEGNANT COLLIERIES   SEAMS AFFECTED: UPPER NINE-FEET DOWNWARDS

NEWLANDS COLLIERY (GENERALIZED)   SEAMS AFFECTED:
TWO-FEET-NINE TO GELLIDEG, BUT PARTICULARLY TWO-FEET-NINE TO SIX-FEET

0   YARDS   1000   SCALE: VERTICAL AND HORIZONTAL

FIG. 42.   *Sections illustrating incompetence-structures; all sections are aligned
approximately north-south, north being on the left side*

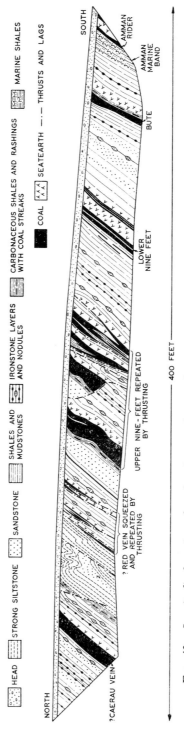

Fig. 43.   *Longitudinal section of exploration trench at Cefn Opencast Site, showing disturbances in the measures between the Amman Rider and the Caerau Vein*

Trenches on the Cefn Opencast Site, just over a mile west of Aberkenfig, gave opportunities to study these structures in a way rarely possible underground. In Fig. 43 an exploratory trench, extending from the Amman Rider to the Caerau, showed that while the detrital sediments are in general free from disturbance, both the Upper Nine-Feet and the Red Vein are intensely disturbed and show many local repetitions. The Two-Feet-Nine, worked almost continuously for a distance of 4 miles from Margam Park to Cefn, was so disturbed that nowhere could a true section be determined: its thickness varied from nothing to over 25 ft as a result of multiple repetitions (Fig. 44). Other sections showing schuppen-structure on a larger scale in the measures associated with the Amman Rider, can be seen in Tondu Brick-pit (Fig. 45), where multiple repetition is largely cancelled out by lag-faulting in the overlying strata.

This failure of incompetent horizons is not limited to the South Crop—though, in general, it is less easily studied in colliery workings. The Nine-Feet through-out much of the Cynon and Rhondda Fach valleys is so intensely disturbed that it is practically unworkable, despite its great thickness and high quality. In adjacent stalls the seam can vary in thickness from nothing to 30 ft; and where thin, considerable amounts of structural rashings are developed in its roof. The

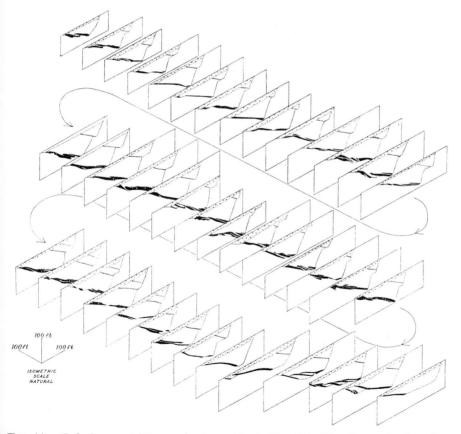

FIG. 44.   *Cefn Opencast Site: production-cut in the Two-Feet-Nine Seam; the three lines of section form part of one continuous excavation*

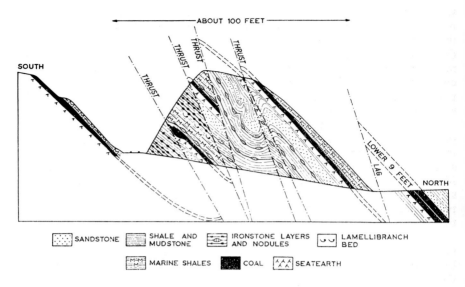

FIG. 45.   *Thrusts and lag fault between the Amman Rider and the Lower Nine-Feet seams in Tondu Brick-pit*

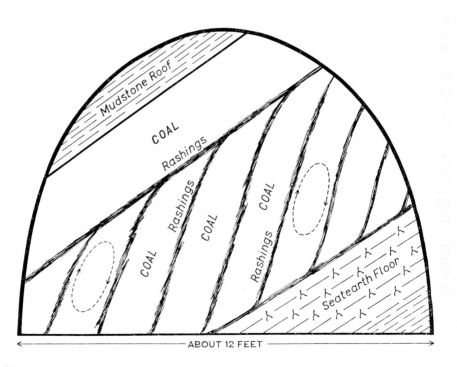

FIG. 46.   *Face of strike-heading in the Six-Feet Seam on the South Crop, showing rotation between sigmoidal shears in the bottom coal*

smaller incompetence-structures may be confined to the immediate horizons of particular seams; separate, but similar, fractures being present in other seams in the same area. The Caerau in the Garw Valley is particularly subject to these disturbances, though the Six-Feet, only some 30 ft above, is little affected. In a few instances individual leaves of compound seams react differently to the compressive forces. Over wide areas of Cwm Colliery, the Yard and the upper part of the Seven-Feet form a single seam, and, while the lower part of the section is undisturbed, the overlying Yard is completely shattered. Similarly along part of the South Crop the upper leaf of the Six-Feet is undisturbed, though the lower part of the seam has suffered complete internal rotation as a result of gliding movements parallel to the bedding; layers of structural rashings extending in sigmoidal curves from the floor to the roof of the lower leaf (see Fig. 46). Many other seams, although not so greatly disrupted, show gently undulating structural variations in thickness from this cause; bedding-plane gliding, producing slickensides with striae arranged normal to the local strike of the thrusts, is common in the finer mudstones.

Mine-plans show that these structures, large and small, are arranged in two swarms; one striking west-north-west to east-south-east, the other between west by south to east by north and west-south-west to east-north-east (see Fig. 47). One group is usually dominant, but rarely to the complete exclusion of the other. The bisectrix of the obtuse angle between the two directions is usually parallel to the direction of principal stress as determined by the major fold axes. The individual fault-planes are generally curved, decreasing in dip both upwards and downwards as they approach the confining competent beds. Along the South Crop they dip northwards on average at about 60°, i.e. at about 30° more than the dip of the beds, and the thrusts all throw down south, in both respects simulating the major underthrusts. North of the main syncline most dip southwards at about 20°–40°, and throw down north. Locally, as near Cwmavon, both sets of thrusts may be developed (see Plate VIB). Several of the larger thrusts, particularly in the northern Rhondda valley, have sharp synclinal drag-folds beneath them; above the thrusts the strata are relatively undeformed, whilst the dips below the shear-planes produce the repetition and stratigraphical throw.

Some of the larger thrusts pre-date the main movement along the cross-faults, and by inference the major part of the folding. At Park and Dare collieries, a thrust, proved about 1150 yd north of the Park shafts to have a downthrow of about 150 ft to the north, is displaced by a down-west cross-fault of similar size. The interaction of these two structures has resulted in workings in the Yard lying at about the same level on either side of the cross-fault, those on the east of that fault being beneath the thrust and those to the west above it.

The symmetrical arrangement of the dip of the shear-planes about the principal synclinal axis of this part of the coalfield, suggests that many of these planes originated early in the folding before the subsidiary flexures became well-established. Nevertheless it appears likely that minor incompetence structures continued to form throughout the entire period of earth-movements.

THE JUBILEE SLIDE

A system of at least sixteen somewhat arcuate normal faults extends for nearly 12 miles from the east side of the Ogwr Fach Valley above Gilfach Goch westwards to the neighbourhood of Pont Rhyd-y-fen (see map, Plate VIII). In general these faults have a downthrow south of about 150 to 200 ft, though on Foel

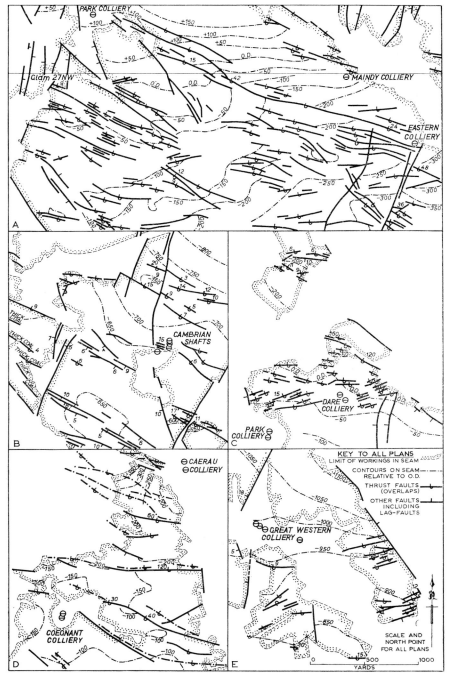

FIG. 47.   *Sketch-plans showing directions of incompetence-structures in different parts of the Pontypridd District: A, Lower Six-Feet Seam in the Eastern–Maindy area; B, Five-Feet Seam in Cambrian Colliery area;   C, Yard Seam in Dare Colliery area; D, Bute Seam in Caerau–Coegnant area; additional positions in Lower Nine-Feet are shown as broken lines;   E, Bute Seam in Great Western Colliery area. Contours are shown in feet above and below Ordnance Datum (O.D.)*

Gwilym Hywel (2 miles N.E. of Maesteg) and Twyn Disgwylfa (1 mile N. of Gilfach Goch) throws of 250 and 300 ft respectively have been proved. The individual faults trend roughly west-north-west to east-south-east and are arranged *en echelon*. Anomalous northward dips into the fault are general on the downthrow side. Mining has shown that these faults are all branches of a continuous dislocation which is flat-lying in the middle portion of the main group of seams (see sections 1–4, Plate VIII). The whole system is referred to as the Jubilee Slide, from the local name given to the fault in the Two-Feet-Nine workings at Wyndham Colliery.

The hade of the faults at surface can be measured at two points: in a landslip scar on Mynydd Llangeinor [92149225], 1 mile W.S.W. of Nantymoel; and on Pant y Ffald [87109260], 1¼ miles N.E. of Maesteg. At each locality the fault plane dips to the south-south-west at 45° with slickensides showing evidence of movement in a vertical plane. Mapping of these faults suggests that such dips are prevalent at and near the surface, and mining has shown that they continue downwards to about the level of the Two-Feet-Nine, below which the fault plane flattens sharply. At Penllwyngwent Colliery a fault belonging to this system courses west-north-west to east-south-east in the Two-Feet-Nine; it has a southwards throw of about 150 ft and a dip of 40°–45° (see Fig. 48). This dip cannot continue downwards for there is no fault in the appropriate position in the Bute workings beneath. To the north of this fault the Two-Feet-Nine and Bute seams in normal sequence are about 400 ft apart; to the south, however, workings in the two coals are separated vertically by only 160 to 230 ft, and boreholes drilled between them show that all the seams from Six-Feet to Lower Nine-Feet inclusive are missing over an east-to-west belt at least half a mile wide. It would thus appear that the fault proved in the Two-Feet-Nine flattens southwards and becomes almost parallel to the bedding in the ground between the abnormally close Two-Feet-Nine and Bute workings (see Plate VIII, section 2). Despite a horizontal movement of at least half a mile both coals are remarkably free from associated disturbances.

A similar but more complex example occurs in the St. John's–Coegnant area, to the north-east of Maesteg. A west-north-west fault with a southward throw of about 250 ft traverses the surface of Bryn Siwrnai and Foel Gwilym Hywel. In the Victoria workings it consisted of two separate fractures, each dipping to the south at 45° and having a combined throw of 240 ft. Between the

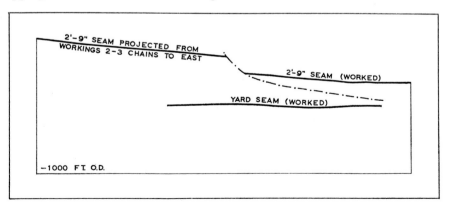

FIG. 48.   *The Jubilee Slide at Penllwyngwent Colliery*

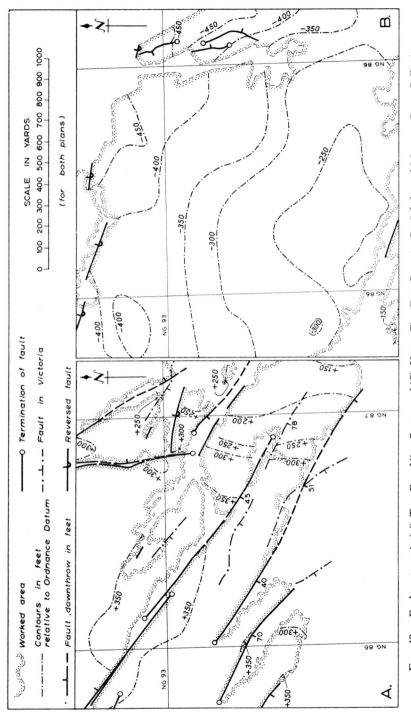

FIG. 49. *Fault-pattern in (A) Two-Feet-Nine Seam, and (B) Upper Five-Feet Seam at St. John's and Maesteg Deep Collieries, showing the positions of branches of the Jubilee Slide in the Two-Feet-Nine workings; and their absence from those of the Upper Five-Feet*

Victoria and the Two-Feet-Nine the fracture flattens, and in the latter seam it is nearly horizontal at a point about 600 yd N.E. of the shafts. To the south the Two-Feet-Nine workings are little more than 120 ft above those of the Lower Nine-Feet, and 250 ft of strata, including the Upper and Lower Six-Feet, Caerau, Red Vein and Upper Nine-Feet are cut out over an extensive area. Farther to the west a succession of steeply dipping faults arranged *en echelon* rise to the surface from the flat part of the Slide (sée Plate VIII, section 3). These faults, all proved in the Two-Feet-Nine and Victoria seams, have been located at the surface. In Fig. 49 the fault-pattern in the Two-Feet-Nine and Victoria is contrasted with that in the underlying Upper Five-Feet.

A similar behaviour at depth has been observed in the other collieries affected by the structure. At Britannic Colliery the Five-Feet has been worked extensively below the Slide. A cross-measures drift rising northwards at about 1 in 9 followed the fault-smash for about 800 yd, emerging in undisturbed strata only where the Slide started to rise sharply towards the surface. At Wyndham Colliery there has been no large-scale development in the southern districts affected by the flat portion of the Slide; boreholes, however, have proved its presence, while a branch fault dipping southwards at 45°, has been proved in the Two-Feet-Nine. At Ffaldau the effects are less than elsewhere, only the Upper Nine-Feet being cut out in the shafts. At Bryn two branches are known at the surface, and each flattens at depth cutting out 200 to 250 ft of strata close below the Six-Feet.

The structure ends in the east on Carn-y-celyn, on the watershed between Gilfach Goch and Pen-y-graig, where the arcuate surface fault bends sharply towards the south-south-east. The presence of the Slide in the west of the district is shown by a surface fault branching off the Pen-hydd Fault towards Pont Rhyd-y-fen. The southern limits at depth have nowhere been fully explored. At St. John's, where workings have extended farthest south, the Six-Feet above the Slide is affected by a multiplicity of small folds and thrusts which give the seam a marked northerly dip, although the underlying Upper Five-Feet beneath the Slide dips southwards. This suggests that the structure terminates in a compressional zone parallel to the strike of the fault.

As shown in the sections of Plate VIII, antithetic faults of limited northwards throw are developed above the Slide, and are associated with somewhat exaggerated northerly dips of the beds towards the main fault. There is clear evidence at Britannic, Maesteg Deep and Bryn collieries that these faults terminate along the main plane of movement and do not extend below it. In direction they vary from west-north-west to west-south-west.

The age of the Slide is best considered in relationship to that of the cross-faults and folds. At Penllwyngwent the Aber Fault clearly displaces the flat portion of the Slide, following, along part of its course, one of the steep branches. In a similar way the Pen-hydd Fault in Cwm Ifan-bach appears to be deflected along another steep branch. There is also evidence, particularly at Bryn Colliery, that the flat portion of the structure is itself folded to much the same extent as the beds. The Slide, therefore, appears to pre-date the main Armorican folding.

The Jubilee Slide is not a lag-fault in the accepted sense; such structures are characteristic of compressional zones and are accompanied by thrusting of at least the same magnitude. In many respects the Slide resembles a rotational landslip, and it is noteworthy that the movement was directed towards the area of thickest deposition. It is possible, therefore, that sliding under gravity in conditions of instability took place during late Coal Measures times.

THE TY'N-Y-NANT–MOEL GILAU FAULT SYSTEM

The Ty'n-y-nant Fault extends westwards from Llantwit Fardre on the eastern margin of the district for 8½ miles to the Dimbath Valley, about a mile north-west of Llandyfodwg. In this area it is paralleled about a quarter of a mile to the south by the Moel Gilau Fault, which can be traced continuously for a further 13 miles through the Maesteg area to the eastern shores of Swansea Bay at Baglan. It is probable that these southward throwing faults are elements of the same system arranged *en echelon*. Although no comparable faulting has been identified on the west side of Swansea Bay, the Moel Gilau fracture clearly extends well beyond Baglan for its throw there is still very substantial.

The Moel Gilau Fault is by far the largest individual fracture in the district. Paradoxically it is one of the least well-known structures, for it has remained unproved by underground workings. From its eastern limit near Llandyfodwg to the neighbourhood of Pont Rhyd-y-cyff (2 miles S.E. of Maesteg) its course is bounded on both sides by Pennant Measures; westwards to Swansea Bay it separates Pennant on the southern and downthrow side from Middle or Lower Coal Measures.

Between Pant-yr-Awel in Cwm Ogwr Fawr (north-north-west of Black Mill) and Pont Rhyd-y-cyff the main Moel Gilau Fault is flanked about a quarter of a mile to the south by a parallel structure with opposed throw, here called the Felin Arw Fault. Within this 3½-mile long trough, strata belonging to the Hughes, and probably also the Swansea, Beds are brought against Brithdir Beds to the north of the Moel Gilau Fault, and against Brithdir or Hughes Beds to the south of the Felin Arw Fault. In this sector the throw of the Moel Gilau Fault grows rapidly from 200 or 300 ft in the Ogwr Fawr to about 2000 ft or more in Cwmcedfyw, while that of the Felin Arw Fault ranges up to about 500 or 600 ft. These figures are difficult to determine with accuracy since the stratigraphical position of the strata within the trough is not known with certainty. However, it seems likely that both the Moel Gilau and Felin Arw faults are closely related and must be considered as parts of the same fault-complex. The resultant downthrow of the structure therefore ranges from about 200 ft at Pant-yr-Awel to 1500 ft at Cwmcedfyw. Within the trough the strike of the strata is generally north-west to south-east, contrasting sharply with the nearly east-west strike of the beds to the north and south. This change in strike is associated with several north-westerly trending faults, probably of relatively small throw, which cannot be traced beyond the main bounding faults.

At Pont Rhyd-y-cyff, where only the single Moel Gilau Fault is recognized with certainty, the throw is about 2000 ft. At Maesteg and Bryn it is about 2300 ft, but to the west it increases in a short distance to rather more than 3000 ft at Cilygofid [80009156], and reaches a maximum of about 3750 ft in the Avan Valley near Cwmavon. Thereafter the throw diminishes but is still of the order of 2700 ft when last measureable at Baglan.

Westwards from Pont Rhyd-y-cyff the Moel Gilau structure lies at the foot of a north-facing scarp, which marks the juxtaposition of the sandstones of the Pennant on the south side and the softer mainly argillaceous strata of the Middle and Lower Coal Measures on the north, which, from Maesteg westwards, are intensely disturbed. There is, however, no evidence that these structures are primarily due to the effects of the fault, for similar incompetence-disturbances are common elsewhere in these strata. On the south side of the fault Pennant dips of about 20° towards the fault (in a direction opposite to that of downthrow drag) can be seen in quarries east of Mount Pleasant. In a railway cutting [809916] south-west of Farteg-fawr intensive disturbances in massive Pennant extend for a distance of up to 150 ft south of the fault, and a similar belt of dislocation can be seen near Cwmavon.

The only evidence bearing on the dip of the Moel Gilau is that provided by the course of the fault-trace. This suggests that the fault-plane has a low inclination in the west, but steepens rapidly towards the east. The Second Edition of this memoir (p. 110) inferred that the dip of the fault-plane might be as low as 18° between Moel Gallt-y-cwm and Baglan: the present mapping confirms a low dip, but suggests a figure of 20°–30°. Eastwards, towards Maesteg, thick Head and Boulder Clay along the foot of the Pennant scarp conceal the exact line of fault, but it is probable that the dip of the fault-plane increases steadily eastwards. The Second Edition of this memoir (p. 108) recorded a northward dip of 55°–65° for the fault in the Garw Valley west of Llangeinor but the relevant exposures are now walled up. The trace of the fault on the hill side between the Garw and Ogwr Fawr valleys, however, implies a southerly dip of about 70°. The change in amount of dip of the fault-plane along its length partly explains the sharp change in direction of the fault-trace near Maesteg. This also suggests that the fault-plane flattens with increasing depth. If this be so, the portion of the fault west of Pont Rhyd-y-cyff represents increasingly lower tectonic levels in the structure, while to the east the steeply dipping portion represents high levels.

In the western part of the district several faults occur in the Pennant to the south of the Moel-Gilau Fault, and are presumably genetically linked and antithetic to it. In the Wernddu workings at Cwmgwyneu and Glenhafod collieries, the larger ones have throws of 100 to 300 ft northwards and dips towards the Moel Gilau Fault of between 50° and 60°. Their planes would thus appear to lie at right angles to the plane of the main fault.

The Ty'n-y-nant Fault is known with certainty only from surface mapping, except around Beddau and Llantwit Fardre, where its eastern extension was proved by old workings in the Llantwit seams. Through much of its length its course is concealed beneath drift. The throw (measured by the stratigraphical gap between the strata on either side of the fault-trace) is small by comparison with that of Moel Gilau Fault, and over most of its extent, is little more than 300 ft. The surface course of the fault implies a steep dip towards the south. Recent exploration at Llanharan Colliery points to a possible flattening of the fault-plane at depth. About 1 mile north of the shafts the Gellideg has been proved to lie close beneath the Upper Nine-Feet, the two seams being separated by a flat-lying fault producing a loss of ground of about 200 ft. This fault may be an incompetence-lag-fault, but it is possible that it is the Ty'n-y-nant Fault which, at surface, lies nearly a mile to the north. If so, this structure pre-dates the folding, for otherwise it would affect other districts of the colliery.

The Moel Gilau–Ty'n-y-nant Fault system was formerly believed to have developed contemporaneously with the folding, and to have replaced the Maesteg anticline to the west. The present resurvey has, however, shown that its relationship to the folding is by no means close. To the east of the Llynfi Valley the faulting is more closely related to the Bettws-Llantwit Syncline than to the Maesteg–Pontypridd anticlines. To the west of Pont Rhyd-y-cyff the trace of the Moel Gilau Fault makes an angle of almost 45° with the fold axes; and it is only between Maesteg and Bryn that the fault and the anticline are more or less coincident. West of Bryn the fold axis crosses the fault and can be traced in a south-westward direction towards Mynydd Bach and Cwm Wernderi, finally dying out near Cwmgwyneu Colliery only a few hundred yards north of the western extremity of the Bettws Syncline: the fault on the other hand, continues as a major structure as far west as Swansea Bay.

The limited available evidence for dating the fault system is contradictory. A comparison of the system with the Jubilee Slide shows a number of common features. These are their direction; the inferred decrease in the dip of the fault-planes at stratigraphically lower horizons; and the associated parallel antithetic faults to the south. Since the Jubilee Slide is demonstrably pre-folding, these similarities suggest that the Moel Gilau–Ty'n-y-nant is of a similar age. On the other hand no cross-faults can be shown to displace the Moel Gilau system[1]—indeed, several appear to have their continuations displaced sinistrally by 250–900 yd along the Moel Gilau. After allowance has been made for the vertical displacement of the Moel Gilau a net sinistral shift of about 400 yd remains. Moreover, at Glenhafod Colliery, a northward throwing west-east fault, regarded as antithetic to the Moel Gilau terminates abruptly against one of the cross-faults, implying that the cross-fault was the earlier. In the present state of knowledge these views cannot be reconciled, though a possible solution is that the Moel Gilau system was initiated before the folding and cross-faulting, and was reactivated at a later date.

THE NEATH DISTURBANCE

The Neath Disturbance is a complex zone of fracture which extends in a general south-west to north-east direction along the valley of the River Neath from near its mouth to Pont-nedd-fechan, and thence to beyond Crickhowell in Breconshire. For many miles to the north of the Coalfield, folding associated with, and parallel to, the Disturbance is well authenticated, and has been described in some detail (Robertson 1933, pp. 206–9; Owen 1954, pp. 33–58). Only two miles of the Neath Valley around Clyne lie within the present district. Here, glacial valley-drift and the alluvium of the Neath conceal the solid rocks, and colliery workings have not proved the structure at depth. There is, however, some evidence of anomalous dips on the south-eastern slopes of the valley. West of Clyne Cottages dips of between 10° and 20° towards the valley (i.e. opposed to the regional dip) are visible in the Twrch Brook [80520035], in an old tramway [80280025] and near Ynys-dyfnant Farm [81150132]. This reversal of dip may indicate the presence of an anticline parallel to, and south-east of, the main fracture.

The clearest evidence of the Disturbance is a sinistral tear of about 1450 yd, which shifts a cross-fault with a westerly downthrow of some 200 ft. On the north-west side of the valley the fault lies in the vicinity of Hendre-gledren Farm [79960181] and traverses Craig Ynys-bwllog [801012] before disappearing beneath the alluvium. A fault of similar throw emerges from beneath the valley

---

[1] The displacement of the Ty'n-y-nant Fault shown north of Beddau on six-inch Glamorgan Sheet 36 N.W. is purely hypothetical. The area is concealed beneath thick drift and nothing is known of the relationship between the two sets of faulting.

drift south-west of Melin Court Halt [81370137] and extends southwards towards Blaen-Twrch Farm. Evidence from outside the district shows that between Neath and Glynneath, all the major cross-faults affecting the Coal Measures have a disposition on either side of the valley consonant with a sinistral shift of about the same size. North of the Coalfield, Owen (1954) has described similar lateral movement concentrated along an individual fracture, the Dinas Fault, lying near the south-eastern margin of the zone of disturbance. He also noted a further fault, the Coed-Hir Fault, which has an essentially vertical downthrow to the south-east, and which follows the north-western margin of the compressional belt. Around Clyne a vertical displacement of 350 to 400 ft to the south-east remains after removal of the tear component, and this may point to the existence of a fault analogous in effect to the Coed-Hir Fault.

The Neath Disturbance and the closely similar structure in the Tawe Valley to the west are so opposed in direction and nature to the main Armorican structures that they have been generally considered to post-date that orogeny (see, for example, Strahan 1902, p. 221; Trotter 1947, p. 123), though Owen maintained that they were an integral part of those movements. The present resurvey has thrown little light on this problem. While there is no evidence that the Disturbance had any effect on the sedimentation of the Coal Measures, there are many signs that most of the movement post-dated the cross-faults, which are believed to be of relatively late Armorican age. Its form and orientation suggest that it reflects movements over a long-standing plane of weakness in the basement rocks; such movements may have recurred at intervals over a long period of time.

MINOR THRUSTS IN THE LOWER PENNANT OF THE RHONDDA FAWR, AND ASSOCIATED SLIDES

A series of north to south thrusts, all throwing down to the east, underlies the moorland between the Rhondda Fawr and Avan valleys in the north central part of the district. Exposures on the peat-covered hilltops are poor, and evidence for these faults derives mainly from the workings in the No. 2 Rhondda Seam.

The northernmost thrust has a downthrow to the east of between 100 and 120 ft where proved [90590155] beneath the north-eastern shoulder of Cefntyle-brych. Farther south a second and similar fault separates the No. 2 Rhondda workings of Blaencorrwg and Glyncorrwg Collieries on the west from those of Fernhill Colliery on the east. At one point [90350070] beneath Penwaun Pen-y-coetgae the Fernhill workings passed 25 ft vertically beneath those from Glyncorrwg, proving that the fault is reversed. Reversal was also proved at Graig Level, Blaen-y-cwm, some three-quarters of a mile farther south-east, where workings took place in the overlapping limbs of a parallel thrust. Here the fault-plane dipped at about 30° to 35° to the west, and the throw, where proved [91369975], was 80 ft down east. It is presumably this thrust which repeats the crop of the No. 2 Rhondda on the crags north of Nant y Gwair, and produces nearly vertical dips in the stream bed [91159872]. A north-south fault which limited the No. 2 Rhondda workings at Corrwg Rhondda Colliery, Blaen-gwynfi, is probably of a similar nature; a trial pit [90959767], west of Cwm Selsig, proved its throw to be about 100 ft down east.

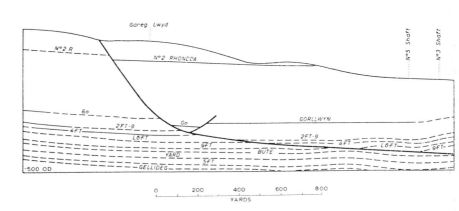

Fig. 50.   *Slide-fault at Fernhill Colliery. The detail in the immediate vicinity of the slide has been considerably simplified*

None of these thrusts has been proved in the deep mine workings, although the Nine-Feet workings from Glyncorrwg and those of the Lower Six-Feet from Tydraw and Glenrhondda should have intersected them had they continued into the main coals.

Closely associated with these thrusts are several west-north-west to east-south-east faults which possess a normal dip, and against which a number of the thrusts end abruptly. The most northerly of these faults dips southwards at 50° to 60° and has been proved in the No. 2 Rhondda near the head of Nant Melyn [90800145], where its downthrow is some 35 ft: to the west of its intersection with the northernmost of the thrusts described above, the throw increases abruptly to 120–150 ft. Another apparently related fracture has a northerly downthrow of 50 to 75 ft on Pen-pych [91879975] and a northerly dip of about 70°. A third such fault forms the northern boundary of the workings from Cwm-cas Level [89009957] and yet another affects the outcrop of the No. 1 Rhondda on Mynydd Blaenafan [90769679], and has been proved to have a downthrow to the north of 21 ft in the No. 2 Rhondda.

One of these normal faults, which outcrops in the district immediately to the north, has been proved in the main coals at Fernhill Colliery, where its behaviour at increasing depth appears to be closely comparable with that of the Jubilee Slide. From the surface to the Gorllwyn workings its dip is maintained at 50° to 60°, but downwards this decreases rapidly until the fault-plane flattens out in the general horizon of the Nine-Feet Seam (see Fig. 50), affecting the measures at least as far south as the colliery shaft. It seems probable that the other fractures in this group are also slides, or, in the case of those which are parallel but of opposing throw, terminate against slides. The comparison with the Jubilee Slide is so close that it seems probable that these faults similarly formed before the onset of the Armorican compression. The associated thrusts may also be of an early age, though it is possible they were initiated much later, but were partly controlled by the pre-existing planes of fracture.

POST-LIASSIC STRUCTURES

Mesozoic rocks, mainly of Triassic age, are limited to small areas along the southern margins of the district, and evidence of later tectonic movements derives mainly from the Bridgend district (Sheet 262) to the south. Folding in post-Liassic (? Miocene) times was responsible for the gentle flexures now evident in the Mesozoic rocks of the Vale of Glamorgan, but the earth movements were mild in comparison with those of the earlier orogeny.

To the south of the Coalfield the post-Liassic folds are orientated more or less east-west, in the same general direction of those of Armorican age. Strahan (1904, pp. 90–3) showed, however, that there is little correspondence in detail between the axes of the two sets of folds and there appears to be little real posthumous rejuvenation along Armorican lines. As far as the present district is concerned dips of 10° to 20° towards the south are registered in the Triassic of the area between Pen-yr-heol and Pyle in juxtaposition with dips of a similar order northwards in the Carboniferous, though over much of this tract major faults separate the two formations. In the Llanharan–Lanelay area, Triassic strata occupying marked hollows in the surface of the Coal Measures again show general southward dips of about 10° to 15°, whereas the underlying rocks dip steeply to the north.

The main faults affecting the Mesozoic strata vary in general direction between west by north and west-north-west. Two such faults have been recognized within the present district. These faults, which have *en echelon* overlap in the area north-east of Bridgend, form the local boundary between the Carboniferous and Mesozoic outcrops and have southward downthrows of at least several hundreds of feet. The westerly fault was proved by boreholes at the British Industrial Solvents' plant about a quarter of a mile west of Water Street. It was also proved at Cribbwr Fawr Colliery where the No. 3 Slant passed from Keuper Marl into Coal Measures at a depth of 121½ ft, the fault-plane being observed dipping southwards at 70°.

While north-north-west faults are common in the Coalfield, none has been mapped in the Mesozoic strata of the district. Although not conclusive this supports the view that these faults are pre-Triassic in age.

A.W.W., W.B.E.

REFERENCES

ANDERSON, E. M. 1951. *The Dynamics of Faulting.* Edinburgh.

DE SITTER, L. U. 1959. *Structural Geology.* London.

DIXEY, F. and SIBLY, T. F. 1918. The Carboniferous Limestone Series on the south-eastern margin of the South Wales Coalfield. *Quart. J. Geol. Soc.,* **73** for 1917, 111–64.

DIXON, E. E. L. *in* CANTRILL, T. C., DIXON, E. E. L., THOMAS H. H. and JONES, O. T. 1916. The Geology of the South Wales Coalfield. Part XII. The Country around Milford. *Mem. Geol. Surv.*
———1921. The Geology of the South Wales Coalfield. Part XIII. The Country around Pembroke and Tenby. *Mem. Geol. Surv.*

EVANS, D. G. and JONES, R. O. 1929. Notes on the Millstone Grit of the North Crop of the South Wales Coalfield. *Geol. Mag.* **66,** 164–76.

EYLES, V. A. 1956. In *Sum. Prog. Geol. Surv.* for 1955, p, 35.

GEORGE, T. N. 1933. The Carboniferous Limestone Series in the West of the Vale of Glamorgan. *Quart. J. Geol. Soc.*, **89**, 221–72.
———1940. The Structure of Gower. *Quart. J. Geol. Soc.*, **96**, 131–98.

KUKUK, P. 1938. *Geologie der Niederrheinisch-Westfälischen Steinkohlengebiete* Berlin.

MOORE, L. R. 1945. The geological sequence of the South Wales Coalfield the 'South Crop' and Caerphilly Basin and its correlation with the Taff Valley sequence. *Proc. S. Wales Inst. Eng.*, **60**, 141–227.

OWEN, T. R. 1954. The structure of the Neath disturbance between Bryniau Gleision and Glynneath, South Wales. *Quart. J. Geol. Soc.*, **109**, 333–65.

ROBERTSON, T. 1927. The Geology of the South Wales Coalfield. Part II Abergavenny. 2nd edit. *Mem. Geol. Surv.*
———1933. The Geology of the South Wales Coalfield. Part V. The Country around Merthyr Tydfil. 2nd edit. *Mem. Geol. Surv.*
——— and GEORGE, T. N. 1929. The Carboniferous Limestone of the North Crop of the South Wales Coalfield. *Proc. Geol. Assoc.*, **40**, 18–40.

STRAHAN, A. 1902. On the Origin of the River-System of South Wales, and its connection with that of the Severn and the Thames. *Quart. J. Geol. Soc.* **58**, 207–25.
——— in STRAHAN, A. and CANTRILL, T. C. 1904. The Geology of the South Wales Coalfield. Part VI. The Country around Bridgend. *Mem. Geol. Surv.*

TROTTER, F. M. 1947. The Structure of the Coal Measures in the Pontardawe-Ammanford area. *Quart. J. Geol. Soc.*, **103**, 89–133.

TRUEMAN, A. E. 1947. Stratigraphical problems in the coalfields of Great Britain. *Quart. J. Geol. Soc.*, **103**, lxv–civ.

WELCH, F. B. A. and TROTTER, F. M. 1961. Geology of the Country around Monmouth and Chepstow. *Mem. Geol. Surv.*

WOODLAND, A. W., EVANS, W. B. and STEPHENS, J. V. 1957. Classification of the Coal Measures of South Wales with special reference to the Upper Coal Measures. *Bull. Geol. Surv. Gt. Brit.*, No. 13, 6–13.

CHAPTER IX

PLEISTOCENE AND RECENT

GLACIAL

DURING THE Pleistocene ice-caps formed over much of mountainous Britain, and descended into and filled many of the surrounding plains. Several such periods of glaciation are generally held to have occurred, separated by inter-glacial periods when the climate was much more temperate, and during which the ice largely if not completely disappeared. Within the present district, however, there remains conclusive evidence for only one of these glaciations, and the freshness of the resultant land-forms makes it likely that this was the most recent of those that affected Britain, reaching its culmination during later Newer Drift, possibly Weichselian, times. While it is possible that during earlier glaciations ice spread into the district, any resulting deposits, if still remaining, can nowhere be separated with certainty from those of the later advance.

The ice that affected this part of South Wales was generated in three distinct areas (see Fig. 51). The most active ice-sheet was that which moved southwards from the Brecknockshire Beacons, and which brought with it an abundance of Old Red Sandstone and Millstone Grit debris. It was however never thick enough to surmount the northwards-facing scarp of the Pennant Measures between the Cynon and Neath valleys, which reaches a height of 1969 ft O.D. near Craig-y-Llyn. Round this scarp, which stood out as a monadnock, the ice divided. One stream following the Cynon Valley to the south-east was separated from a parallel glacier in the Taff Valley by Mynydd Merthyr. At its thickest this ice spread westwards into Cwm Clydach and reached almost to the crest of Cefn Gwyngul. South of Pontypridd it spilled over into the low ground around Llantwit Fardre and Beddau, accumulating here until it surmounted the Pennant scarp to reach the South Crop of the Coalfield east of Llantrisant. A second stream moved south-westwards from Craig-y-Llyn approximately along the line of the Vale of Neath. At one stage it was thick enough to submerge the watershed between the Neath and the Pelena, and to reach the lip of the scarp overlooking the Avan Valley. Farther south-west it lapped around the Pennant hills overlooking Briton Ferry and Port Talbot and spread out into Swansea Bay as a great piedmont-glacier, the eastern extremity of which re-entered the district near Margam and extended as far east as Kenfig Hill.

The coalfield plateau south of the Craig-y-Llyn scarp shows little sign of glaciation, though probably large parts of it were snow-covered, and ice may have accumulated in depressions on its surface. Local glaciers and snowfields nestled in the deep valleys which dissect it, and their deposits are free from Brecknockshire erratics. For the most part the ice appears to have been largely static and not to have given rise to any substantial active valley glaciers. In the Rhondda valleys, and in particular the Rhondda Fawr, however, the ice was thicker; it moved down-valley and joined the Brecknockshire ice near Pontypridd. Along its margins it overtopped the cols into the Ely Valley at Williamstown and

275

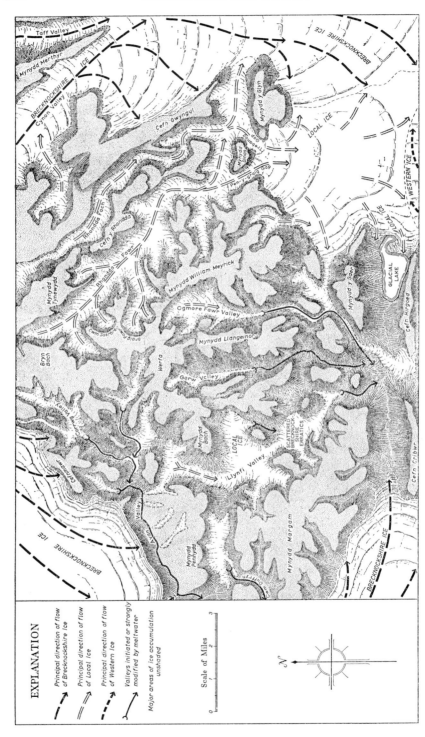

Fig. 51. Distribution and direction of the movement of ice during the last glaciation

Trebanog and debouched into the low ground around Tonyrefail. From here one tongue pushed down the Ogwr Fach and joined feeble glaciers emerging from the Gilfach area, from the Dimbath, the upper Ogwr Fawr and the Garw and Llynfi valleys. The major outflow, however, was southwards, over and through gaps in the Pennant scarp to the South Crop. On the east it was confluent with the Brecknockshire ice overspilling from the Taff Valley and on the west it extended as far as Wern-tarw.

Apart from the Ogmore Valley glacier which at its maximum reached the southern margins of the district, the South Crop between Wern-tarw and Kenfig Hill appears to have been free from ice. No glacial deposits occur within this area, nor is there any reason to suppose that they have been removed by post-glacial erosion, since only minor streams drain the area.

Much of the low ground south of the coalfield was occupied by the eastern extension of the great ice-field which occupied the Irish Sea basin, and whose ultimate provenance is shown by the erratics of Scottish, North Welsh and Pembrokeshire rocks which its deposits contain. This western ice has been held to belong to an earlier glaciation than the one which was responsible for the coalfield glaciation described above; but the topographic forms of its deposits are as fresh as any in the district, and the nature of its contacts with the coalfield ice supports the belief that both were broadly coeval. In the Llanharan area, in particular, the deposits of the two ice-masses rest one against the other and there is no evidence that either one overlies the other at any point.

Although the glacial deposits of the district are considered to be monoglacial in origin, they did not everywhere form at exactly the same time. As the glaciation waned so the general level of the ice in the valleys fell; the till on the higher slopes is usually more eroded and more patchily distributed than that in the valley bottoms where the ice persisted longer. The emergent cols also progressively confined the Brecknockshire ice to the valleys of the Neath, Cynon and Taff, and in places local ice appears to have accumulated after the recession of the main flow from the north.

Only in a few small areas are glacial deposits found on the Pennant plateau; for the most part they are confined to the lower slopes and bottoms of the valleys where they form a single co-extensive deposit ranging from heavy stony clay-till to sandy boulder-gravels. A complete transition exists between these extreme types, and the separation shown on the map between 'Boulder Clay' and 'Glacial Sand and Gravel' is necessarily arbitrary. In general the deposits become increasingly gravelly when traced down-valley. The heads of the enclosed valleys and the tributary side valleys contain boulder clay exclusively. The gravels, when they first appear, are restricted to the valley bottoms, but thereafter they occupy more and more of the valley slopes confining the true boulder clays to ever higher levels. The change from till to gravel is believed to have resulted from the washing effect of melt-waters beneath the ice, which removed much of the clayey matrix, while with the down-valley movement of the ice the contained stones and boulders became increasingly rounded. The washing-out of the 'fines' was increasingly marked towards the valley bottom where the main volume of sub-ice water presumably flowed. At the heads of the valleys, near glacier source, the boulders are completely angular; on the other hand, in the Llantwit basin, far-transported boulder clay contains only rounded and sub-rounded stones. Within the valleys travel of 3 to 4 miles was sufficient to produce well-rounded boulders.

In form, the valley deposits generally consist of a virtually continuous valley-fill, the upper surface of which has a gently concave profile. Much of the urban and industrial development has taken place on the sloping 'terraces' of this fill, and shaft sections show a general thickness of drift beneath these surfaces of 25 to 50 ft. The deposits rarely extend much below the present river beds. Their total volume is sufficiently small to suggest that for the most part the valley glaciers had little erosional effect except around their margins in the upper reaches, where cirques or corries are not uncommon and where lines of landslips appear to have flanked the ice.

Recessional halt moraines, gravelly in nature, associated with the valley glaciers are known only near Llantrisant and at Clyne, in the Ely and Neath valleys respectively, and both these were associated with particularly active ice-flows. The Taff Valley glacier had a similar moraine in the Whitchurch area to the east of the present district, while in the Neath Valley a larger and more distinct bar than that at Clyne lies at Aberdulais only a mile outside the district.

Along the South Crop of the coalfield hummocky morainic forms are more general. The eastern margins of the Neath piedmont glacier around Margam, Pyle, and Kenfig Hill is characterized by an extensive development of kame-hummocks and ridges of sand and gravel. Farther east between Wern-tarw and Llanharan hummocky boulder clay marks the dissolution of lobes of coalfield ice which spilled over the Pennant scarp. Similar chaotic mounds of gravelly drift mark the northern margins of western ice south of Llanharan. These South Crop drifts are appreciably thicker than those in the valleys, 100 ft or more being general.

A temporary glacial lake appears to have been impounded in the Hirwaun Common area by the ice which overspilled the Pennant scarp east of Wern-tarw, and thin banded clays are present over an area of at least a square mile. Similar ponded clays and silts occur sporadically among the hummocky mounds on the South Crop.

GRAVITY ACCUMULATIONS

Gravity accumulations are most apparent in the coalfield valleys, where deep dissection through the Pennant cover into the underlying mudstones has created conditions favourable to mass downhill transport of rock waste; conditions that were aggravated by the oversteepening that resulted from glaciation. Consequently many of the valley sides are draped with the deposits of mass superficial movement, and show every transition between the extreme types represented by landslips and hill-creep. Smaller and less conspicuous accumulations also occur in minor depressions on the plateau. Head deposits are also extensive along the foot of the Pennant scarp between Kenfig Hill and Wern-tarw, in an area unaffected by the last ice, and where they are, in consequence, thicker than on similar slopes in the glaciated areas.

Landslips: These are common in the valleys and particularly affect the lowest major sandstone above the level of the valley-fill. Several distinct types can be recognized, though some individual slips combine elements characteristic of different varieties.

(a) Rotational slips: These are the classic landslips of literature, originating by rotational movement about a curved shear surface. They result in a terraced hillslope, upon which the discrete masses normally show dips inwards to the back scar. They are not common in the district, though several of the larger slips show a certain amount of rotation.

(b) Rock-tumbles: This name has been coined to describe the type most commonly present. Rock-tumbles appear to result from the outward fall of well-jointed masses of sandstone already eroded to near-vertical scarps on the steep valley sides. The joint-fissures behind the rock face were enlarged by a freeze-thaw mechanism, and ultimately slices of sandstone were freed to tumble down slope. Rarely do they present the terraced aspect of rotational slips, and more usually they consist of a jumbled mass of dissociated blocks. When dips are observable they normally are inclined down-slope. The frontal edge of the slips may pass imperceptibly into solifluction flows, which carry detached blocks sometimes as far as the valley floors. Open fissures on the plateau above their scars mark the slices next destined to be detached when the periglacial climate gave way to milder conditions.

(c) Bedding-plane slips: In these movement has taken place roughly along bedding planes; they are therefore confined to dip slopes. Some are broken up, but in other examples the only certain proof of movement is provided by the scars and hollows left behind by the moving blocks.

(d) Founder-slips: In two cases movement appears to have involved large scale foundering en-bloc of an entire hillside, along planes of weakness provided by minor faults, and although modified by later movements, both appear to have been initiated earlier than the other types. The foot of the slip passes beneath the glacial fill and it is not impossible that the slipping was aided by plucking beneath the glacier. Dips within the foundered blocks are essentially the same as those of the surrounding solid, and the various geological horizons can be traced within them without much difficulty.

The perilous appearance of many of the slipped masses is belied by their general stability during historic times, and this suggests that movement for the most part took place during and immediately following glaciation.

Solifluction flows, Screes, Downwash, Hill-creep: These may be conveniently grouped as Head; they generally pass insensibly into one another and even into the landslips described above. In general they overlie boulder clay, but in certain circumstances may be roughly contemporaneous with it. Although hill-creep, downwash, and a few screes are still active, the bulk of the deposits, like the landslips, appear to be the products of a periglacial climate, for even the screes are now largely stabilized and grassed over. Under severe winter conditions some movement may still take place, and it is interesting to note that in a recently active example near Blaina in the Abergavenny district, the solifluction flow took place beneath, rather than over, the vegetative cover.

Movement is less intense on the plateau but deep rotting of the Pennant, of a kind not apparently taking place today, is general, and the resultant material has tended to move downslope into hollows and in places even flowed over the scarp into the valleys below.

The extent of these head deposits is such that no attempt has been made to depict them on the map. If shown they would conceal most of the glacial deposits, and the greater part of the solid of the valley slopes, together with much of the plateau. Indeed it is scarcely an exaggeration to say that apart from the sand-stone crags and the higher rises of the plateau, the entire surface of the district has a cover of several feet of Head, which, in hollows and on the lower slopes, may reach as much as 20 or 30 ft.

PEAT

Hill peat is widespread on the acid soils of the Pennant plateau, especially to the north of the Avan Valley and around the upper reaches of both Rhondda valleys. Here great spreads cover much of the more gently sloping areas over about 1300 ft above sea level. Apart from the altitude the main controlling factor is presumably the high rainfall, these areas remaining saturated throughout most of the year. Two topographic types can be recognized: *blanket peat* drapes much of the long gentle slopes and ridges; and *basin peat* covers the hollows where drainage is impeded. The two types occur in close association. They differ in one important respect: the basin peats are still growing, while the blanket peat in many areas is wasting, possibly because of its better drainage.

The peat varies considerably in thickness. On the more extensive bogs, such as those of Panwen Garreg-wen, north of Maerdy, on the watershed between the upper Rhondda Fawr and the Corrwg valleys, and west of Cefnmawr, it is commonly 6 to 12 ft, and has been credibly reported as reaching as much as 25 ft in some of the hollows.

In the central and southern Pennant areas peat is more sporadic, but numerous small areas have been mapped. Many of these consist of basin peat and are located along mudstone slacks or in hollows associated with faults. In the south of the district basin peat is also present locally, especially associated with the alluvium of the Ewenny Fach Valley, west of Llanharan, and in the hollows between the mounds of the morainic drift in the same general area.

FLUVIAL DEPOSITS

With a few possible exceptions, such as the lower Avan and Duffryn in the extreme west of the district, which may have originated during the Pleistocene as melt-water channels, the main valleys are pre-glacial. Following the recession of the ice most of the rivers excavated comparatively narrow channels in the glacial drift. The width of these new channels reflects the post-glacial meandering of the rivers and their bottoms are filled with alluvium. This is everywhere coarsely gravelly and rudely bedded. Where accumulation has been active in recent years, in the wider stretches of the flood plains and in the lower reaches of the larger streams, the character of the alluvium has changed by the addition of much carbonaceous and shale waste from the numerous colliery washeries. In many valleys where mining has been particularly active long stretches of the alluvium are now buried by the tipping of colliery waste; in these areas the boundaries depicted on the present map have been taken from the original six-inch survey.

In the lower reaches of a few rivers, notably the Taff, the Rhonddas, the Ely and the Ogwr, gravel terraces result from meander-erosion of outwash fluvio-glacial deposits laid down by the melt-waters from ice farther up-valley.

In a few instances, where tributary streams enter main valleys they have built up alluvial fans of some size. The principal of these are located at Aberaman, where the Aman joins the Cynon; at Mountain Ash, where the Pennar brook joins the same river; and at Treorky where Nant Cwm-parc flows into the Rhondda Fawr.

BLOWN SAND

The belt of Blown Sand which extends along the eastern shores of Swansea Bay spreads into the extreme south-west of the district, near Morfa Bach, its landward margin being marked locally by the course of the Afon Kenfig. The sand-dunes attain a height of about 30 ft, and are largely stabilized.

A.W.W., W.B.E.

DETAILS

NORTH-EASTERN VALLEYS

These comprise the Taff Valley between Aber-fan and Pontypridd, the Cynon Valley below Aberaman, and the Clydach Valley. All were filled with Brecknockshire ice, and Old Red Sandstone, Carboniferous Limestone and Millstone Grit erratics are abundant in both till and gravels.

Taff Valley: The ice filling this valley had already travelled a considerable distance before it entered the present district and the resultant deposits are predominantly gravelly in character. Recognizable boulder clay is largely confined to hollows on the higher slopes west of Aber-fan and east of Merthyr Vale and is scarcely known to the south. The general upper limit of the gravel is at about 500 ft O.D. on the east side of the valley, and rather higher, about 600 ft, on the western side; local patches of boulder clay extend upwards to 800 or 900 ft. The Merthyr Vale Colliery shafts were sunk through 68 ft of drift, mostly sands and gravels with boulders at the base, the solid surface thus lying well below the present valley bottom. Between Abercynon and the mouth of the Clydach the gravel in the valley bottom is especially coarse, numerous boulders up to several feet across being observed, for example, near Stormstown Junction [07909350].

A large post-glacial, partly rotational landslip occurs on the west side of the valley just south of Merthyr Vale. The back-scar now forms the massive crags of Taren-y-gigfran [070986] and the slip appears to have been mainly associated with the shales of the Cefn Glas horizon and controlled to some extent at least by the Kilkenny Fault. Tumbled masses of sandstone extend from about the 1000-ft contour to the river 500 ft below, where they rest on valley-fill gravels.

At, and to the north of Merthyr Vale, the river alluvium is up to 400 yd wide, but downstream it narrows considerably. South of Mount Pleasant three meander-terraces are distinguishable. Between Abercynon and Pontypridd a relatively narrow strip of alluvium is flanked by up to four terraces. In general they cannot be matched from one side of the valley to the other and they were probably formed by the erosion of a single gravel deposit as the river meandered. The terrace intervals range from 2 to 10 ft and the highest surface is some 25 to 30 ft above the present river.

Cynon Valley: North of Abercwmboi the valley-fill consists predominantly of boulder clay, but to the south the lower parts become increasingly gravelly. At Mountain Ash the gravels extend from the bottom of the valley up to about 400 to 500 ft O.D., above which they become increasingly clayey in character. Around Tyntetown, and especially on the east side of the valley, the gravels are moundy and may mark a slow down or pause in the retreat of the valley glacier. At Ynys-boeth, where the valley narrows sharply, there has been much scouring of the drift gravels which do not now rise much more than 50 ft above the alluvium. Throughout this stretch the solid floor lies well below the present valley bottom, the various shafts from Lletty Shenkin to Penrikyber recording between 40 and 90 ft of sand and gravel.

Tongues of boulder clay fill most of the tributary valleys. The Aman Valley in particular is filled with till up to about 1100 ft O.D. at its head, and was evidently the site of a strong local tributary to the main through glacier. One of the shafts at Fforchaman Colliery penetrated 69 ft 7 in of drift, including much sand with stones. North-east of Mountain Ash a long tongue of boulder clay lines a pre-existing hollow on the south side of Nant-y-Ffrwd and reaches 1100 ft O.D., well on to the upper slopes of Mynydd Merthyr. Extensive patches of boulder clay are also preserved on shelves well above the general level of the continuous valley-fill. One, half a mile long, smears the broad shelf between 800 and 1000 ft O.D. on the west side of the valley at Blaencwmboi. Another at Daren-las, also on the west side of the valley, at Mountain Ash, reaches to about 800 ft O.D. On the east side of the valley at Penrhiwceiber the shelf beneath the Cefn Glas Seam is covered with boulder clay up to nearly 1000 ft O.D. north of Fforest Isaf [06569786]; and on the opposite side of the valley near Tyntetown the feature below the same seam forms a broad hollow in which boulder clay is preserved up to about 900 ft O.D. This last named tongue appears to mark the track of one of the main spill-ways which carried the Cynon ice with its Brecknockshire erratics into the adjoining Clydach Valley via the col between Perthgelyn [05259718] and Pen-twyn-uchaf [06169617].

On the west side of the valley between Penrhiwceiber and Abercynon there are extensive rock-tumbles. Extending over at least a mile and a half they were formed by the collapse of the oversteepened scarp of the sandstone above the Daren-ddu Seam. Other major slips from the sandstone above the No. 2 Rhondda occur at the head of the Aman Valley.

The alluvium of the Cynon reaches half a mile in width at Aberaman, and large areas are now permanently flooded as a result of mining subsidence. Where the Aman and Pennar streams debouch in the main valley, broad alluvial fans have been built up. Downstream from Mountain Ash, where the main mass of the Pennant dips beneath valley bottom the valley narrows and with it the alluvium, which rarely exceeds about 200 yd in width. Between Ynys-boeth and Abercynon at least two low meander terraces are recognizable.

Cwm Clydach: Although this tributary valley of the Taff rises within the Pennant plateau of the coalfield its glacial deposits nearly everywhere contain Old Red Sandstone and other erratics which show that it was submerged beneath Brecknockshire ice. This presumably spilled over from the Cynon glacier via the cols near Penrhiw-Caradoc [04609798], Perthgelyn and to the north-west of Gilfach-y-rhyd [07169531], all at about 900 to 1000 ft O.D. Tongues of boulder clay preserved in hollows in these areas almost to the top of the ridge suggest that the entire interfluve west of Penrhiwceiber and Abercynon was at one time submerged by ice. At the head of the valley, north of Llanwonno, the boulder clay reaches a height of 1100 ft O.D. and on the west side of the valley generally it is preserved in hollows up to 1000 ft, the main ice stream was probably reinforced by local ice accumulating in the upper reaches of the Clydach itself and its tributaries, the Ysfa, Y Ffrwd and Llys Nant High as the boulder clay reaches on these western slopes, the ice itself clearly did not surmount the ridge separating them from the Rhondda Fach, for no extra-coalfield erratics have been found in that valley.

In the lower parts of the valley below Ynys-y-bwl clayey gravels are present and these extend downstream to the confluence with the Taff. In the Lady Windsor shafts 51½ ft of gravel with large boulders at the base were proved.

Great hummocks of slipped Pennant cover the area around Cyrnau, north of Pen-y-wal [05869227], between Llys Nant and Nant Blaenhenwysyg. The entire hill appears to have slipped down-dip parallel to the bedding and an area of about 800 yd by 600 yd is strewn haphazardly with huge blocks of sandstone; the terminal-scar of the slip is now located on the back slope of the hill.   A.W.W

NORTH-WESTERN VALLEYS

These comprise that part of the north-west of the district which came under the influence of Brecknockshire ice.

Pelena Valley: The south-eastward extent of the Brecknockshire ice is marked by the limit of strewn erratic boulders which follows the lip of the plateau overlooking the Avan Valley. This limiting line falls progressively south-westwards from about 1400 ft O.D. in the extreme north near the headwaters of Nant-y-Felin, to some 1250 ft to the north of Cefnmawr, 1150 ft on Mynydd Nant-y-bar, and about 900 ft on Mynydd Pen-Rhys; a descent of rather less than 100 ft per mile. On the southern side of the Pelena Valley the erratics pass down slope into sandy boulder clay with abundant erratics. In places this deposit is washed almost clean of clay, but these patches are so localized and pass so quickly into clay that it is impractical to separate them out. The erratic-bearing drift is found above about 500 ft O.D.; below this, both in the Pelena Valley and in the tributary Cwmgwenffrwyd, thick boulder clay containing only local pennant is preserved, being particularly well seen in the northern entrance to the Gyfylchi railway tunnel [80979600]. This local valley-fill has presumably accumulated after the withdrawal of the Brecknockshire ice. It falls from about 700 ft O.D. near Fforchdwmuchaf Row [816974] and at the head of Cwmgwenffrwyd [800985], to about 400 ft in the extreme west. Around Ton-mawr it is washed to a boulder gravel, and 49 ft of 'gravel and clay' was recorded during the sinking of the nearby Whitworth shafts. Immediately west of the margin of the district the Pelena enters a steep gorge, which continues from Efail Fach to Pont Rhyd-y-fen. The distribution of Brecknockshire drift strongly suggests that this breach in the Pennant hills originated as a meltwater channel, and possibly the pre-glacial course of the Pelena led westwards into the Crythan and so to the Neath.

Considering the thickness of the Brecknockshire ice it is surprising how little boulder clay has been deposited on the high ground separating the Pelena and Avan valleys from that of the Neath; the upper levels of the ice, at least, must have been relatively clean. On the plateau west of Cefnmawr, however, collapses into old workings have proved the presence of boulder clay, though its exact limits cannot be determined nor depicted on the maps since it is everywhere overlain by thick peat.

Peat is particularly extensive around the headwaters of the Pelena. Some, as on Mynydd Nant-y-bar, is blanket peat, and reaches from the highest ground at about 1600 ft down to about 1200 ft. The spreads around Banwen Tor y Betel [823997] and Twyn Pumerw [827992] also include basin peat, some of which is still forming. On the higher ground north-west of Glyncorrwg blanket peat is again extensive forming the continuation of the spreads on Cefn Ffordd to the north.

Neath Valley: The oversteepened sides of the Neath Valley have provided little lodgement for drift, which is restricted to the upper slopes of the side valleys and to the main valley floor. In the north, around Nant-y-Felin at least 40 ft of boulder clay is preserved, and here extends to the plateau; it is best exposed in the various left-bank streams entering the brook from the south. Along Melin Court Brook, especially on its northern side, stony clay extends almost to the headwaters, and can be seen in the stream banks up to 30 ft thick. Near Cwm-syfirig [83060159] it is at least 40 ft thick, and appears to fill a former stream course running eastwards. In the valley bottom a little gravel lies to the west of the river near Ynysgollan [79570051], and a strip 100 to 200 yd wide follows the south-east of the valley from Melin Court to Ynys-dyfnant [81040139], where its upper limit is at about 200 ft O.D. Around Clyne and near the mouth of the Twrch brook this gravel spreads out until it is 700 yd wide and forms a bar across the valley. Up to 30 ft of dirty gravel full of erratics can be seen in diggings [80070040] east of Syd Cottages. Its irregular moundy form and its erratic content suggest that it is almost certainly a recessional halt-moraine, though a rather ill-defined one.

A large rock-tumble on the west side of Moel yr Hyrddod is located in the sandstone above the Wenallt. Landslips are also extensive west and north-west of Cefnmawr, where the tumble extends so far down the gentle slopes that it may well once have rested on ice.

The alluvium of the Neath is about 400 yd wide, except where it is constricted to the stream bed by the Clyne moraine. Nothing is known of its thickness beneath this part of the valley, but it is probable that a buried gorge of considerable depth underlies part of its spread. Small alluvial fans have been built out by several of the tributaries draining the hills on the west of the river.   W.B.E.

RHONDDA VALLEYS

Rhondda Fach: The river rises to the north of the district on the dip slope south of the Craig-y-Llyn escarpment. Throughout the valley the drift is derived wholly from Coal Measures (mainly Pennant) debris. There is a broad spread of boulder clay around the headwaters, but downstream the river descends from the sandstone plateau in a constricted gorge cut in solid rocks to the north of Castell Nos [96500017]. South of this point the valley opens out again and the glacial drift is virtually continuous to its mouth at Porth. There are no corries at the head of the valley, but several incipient ones lie on the north-eastwards facing slopes between Maerdy and Ferndale.

Above Pont-y-gwaith the valley is almost wholly floored by boulder clay; around Maerdy and Ferndale numerous storm channels cut through about 20 ft of this, the contained pennant debris showing but little rounding in the former area. The thickest known drift in this part of the valley is in the Mardy No. 1 and Ferndale No. 1 pits, in each of which rather more than 50 ft were penetrated. At Ferndale the deposit was described as sand and gravel, and the solid valley bottom lies at a considerably greater depth than in the nearby river which runs over mudstones and shales. Immediately around Tylorstown the valley, here constricted, is virtually drift-free, but downstream it is again choked by drift, with gravels becoming increasingly important and continuous along the lower slopes to Porth, particularly on the east side of the valley. At National Colliery 51 ft of sand and gravel are recorded, and similar, though less thick, deposits were penetrated in all the shafts in this part of the valley.

In the extreme north the drift reaches up to about 1300 ft O.D. and to even greater heights beyond the margin of the district. Down valley the level falls steadily: at Ferndale the main valley-fill reaches up to 950 ft, at Pont-y-gwaith 850 ft, at Ynyshir 700 ft, and at Porth it falls to less than 500 ft. This presumably reflects the general southward fall of the ice surface. Small patches of clayey drift also occur well above this general level, particularly in the tributary of the Nant Llechau at Wattstown, where they reach nearly 900 ft O.D.; and on certain of the higher shelves. These may have formed from small tributary glaciers and local snow- and ice-fields; or they may be the residual remnants of high-level till deposited at an earlier maximum of the glacier. At Ynyshir the top of the main fill lies between 550 and 700 ft O.D. and this level contrasts sharply with that on the western slopes of the Clydach, beyond the watershed to the east. Clearly the ice-level in the Rhondda Fach must have been considerably below that of the Brecknockshire ice.

The present valley is markedly asymmetrical, the river now tending to follow its eastern side. Nearly everywhere the drift rises almost 200 ft higher on the eastward and north-eastward facing slopes. Gravels are best developed on the eastern side of the valley, as might be expected if the melt-waters were concentrated here. This is particularly noticeable below Wattstown where gravels line the east side to the exclusion of recognizable boulder clay, while on the west side till is preserved in strength above the gravels. A further consequence of the asymmetry of the valley is that the surface of the valley-fill forms a wide shelf on the west side of the stream and, as at Maerdy and Ferndale, this has controlled the development of the village settlements.

Landslips are extensively developed throughout the valley and especially between Maerdy and Ferndale; here rock-tumbles from the over-steepened scarp of the No. 2 Rhondda sandstone are virtually continuous on the eastern side. Although presumably active during glaciation on scarps rising above the ice, most of the tumble now visible is post-glacial. No large-scale rotational movement is evident, and there appears to have been constant dilapidation from the steep crags of well-jointed sandstone under intensive frost action in a wet climate. South of Ferndale several similar, but more weathered, and presumably older, slips underlie the scarps of higher Pennant sandstones.

Broad spreads of hill peat lie on Panwen Garreg-wen, north-east of Castell Nos as well as more isolated patches on the mountain-top between Maerdy and Treherbert.

The flood-plain alluvium varies in width up to about 150 yd but is generally absent above Ferndale. Between Wattstown and Porth a low terrace, about 5 ft above the alluvium level, is developed, particularly on the west bank of the stream. It is more than 200 yd wide at Ynyshir where most of the village is built on it.

Rhondda Fawr: Thin patches of clayey drift occur on the surface of the plateau overlooking the head of the Rhondda Fawr. Some may be the products of nivation, though the largest, exposed in the banks near the source of Nant Selsig [91579724], consists of up to 15 ft of stony till. This plateau-drift is everywhere separated from that in the valley by continuous steep sandstone crags, over which the various tributary streams fall in spectacular waterfalls or through which they have cut deep gorges. It seems unlikely that active ice ever crossed this part of the plateau; certainly Breck-nockshire ice never penetrated, for its characteristic erratics are nowhere recorded in the valley.

The ice in the main valley was fed by various tributaries which emerged from the side valleys. West of Cwm-parc and south-west of Treherbert these originated in a group of impressive corries, including Graig-fawr [923961] and Cwm Saerbren [927976], the back-walls of which are several hundred feet high. Within the corries the drift is more hummocky than in the valleys, and in places, as in the exit from Cwm Saerbren, forms small corrie-moraines. Close beneath the back-walls, peat-filled hollows mark the sites of former shallow lakes. Other incipient corries lie at the heads of Cwm Orky [963986] and Nant y Blaidd [943965], and on the north slopes of Mynydd Bwllfa [966939]. The fill within the side valleys extends to 1200 ft O.D. in the major corries and in Cwm Orky, to 1100 ft in Cwm Bodringallt [985964] and Cwm Clydach [958930], to over 1000 ft in Nant-y-Gelli [961941], and to almost 1000 ft in Cwm Hafod [040923]. Wherever seen the fill is boulder clay which is generally appreciably thicker than in the main valley; 95 ft of stony clay were recorded in one of the shafts of Cambrian Colliery in Cwm Clydach, and 103 ft at Glenrhondda, Blaen-y-cwm.

The main valley is everywhere floored by glacial drift. At the head of the valley it extends upwards to about 950 ft O.D. but traced downstream the level declines steadily; between Treorky and Llwyn-y-pia it lies at about 700 to 800 ft O.D. and from Porth to Pontypridd it fluctuates between 400 and 600 ft. Gravels first appear in the valley bottom at Blaenrhondda and are continuous to Porth extending farther and farther up the valley sides. Many small exposures can be seen in stream gullies and stream banks, the best showing 45 ft (of poorly bedded gravel with bands of sand) in the river bank [98519484], opposite Ystrad gasworks, and 60 to 70 ft (of clayey boulder-gravel) in the cut-away banks on the northern side of the river between Dinas and Porth. At Llwynypia Colliery 56 ft of sand and gravel are recorded and most of the pits in the valley floor pass through similar, though thinner, deposits. Where one side of the valley has been steepened more than the other, there is a tendency for it to be relatively free from drift, and for the drift that remains to be more gravelly than on the opposite slope; this is the case at Dinas and again near Llwyn-y-pia. Where tributary glaciers entered the main valley they brought clayey deposits to the valley floor on the downstream side, even though the main glacier was depositing gravels extensively on the upstream side. Thus there is no gravel on the west side of the valley

floor at Pentre just downstream from the Cwm-parc tributary, and the same is true a
Tonypandy immediately downstream from the mouth of Cwm Clydach.

South of Tonypandy the main glacier divided: a tongue of ice surmounted the c
to the west of Mynydd Dinas at Williamstown and descended into the Ely Valley
the main glacier passing down the valley joined the Rhondda Fach ice at Portl
Farther down-valley another tongue left the main stream and passed through th
Trebanog gap into the Ely and Glyn valleys; the main ice continued eastwards to joi
the Taff and Cynon ice at Pontypridd.

Landslips are common throughout the valley. In general they are initiated in th
first major sandstone above the level of the fill, suggesting that for the most pa
this was the approximate upper limit of the ice. Many are from the No. 2 Rhondd
sandstone, as is the case west of Blaenrhondda, around Blaen-y-cwm, in Cwm Orky
and west of Gelli. Around Cymmer another group are derived from the sandston
above the No. 1 Rhondda Rider. Most are rock-tumbles, though a rotational eleme
is often present near the back scar. Much of the movement is post-glacial, for tumble
blocks commonly rest on the drift, and often the boulder clay is involved in the slippin
A major landslip [938976] south-west of Treherbert is different. Here the entire valle
side has foundered as a single block, from a back wall controlled by a branch of th
Dinas Fault. The movement has been hinge-like, for while the north-western part c
the slip is well-defined, the south-eastern part merges indefinitely into the solid abov
Gelli-goch, where it appears to be covered in part by later tumble slips.

Solifluction deposits everywhere drape the valley slopes. They have clearly bee
involved in the general slipping processes, and no definite line can be drawn betwee
landslip and head, or even, in places, between head and boulder clay. This is particularl
the case in the corries, where landslip, scree, head and boulder clay appear to hav
been deposited in one continuous operation. The flow lines of the head are best see
in Cwm Saerbren and on the slope south of Blaen-y-cwm. On gentle slopes they tend t
run parallel to the contours; on the steeper hills they form ridges at right angles to th
contours. At Pentre there is a good example of a solifluction flow [97259672], whicl
from a gathering ground on the plateau of Cefn y Rhondda, poured down into th
valley below, and now is preserved as a strip of head about 100 yd wide and 20 ft higl
extending down on to the glacial fill. All these deposits, except for a few screes, hav
long been inactive.

The river alluvium is about 200 yd wide below Tynewydd, and maintains a widt
of 50 to 350 yd throughout the length of the valley. At Treorky an alluvial fan ha
been built up where Nant Cwm-parc enters the main valley and there is another sma
fan at Gwaun-nyth-bran [03459120] at Hafod. No terraces have been recognize
except between Hopkinstown and Pontypridd.   A.W.W., W.B.

THE ELY VALLEY, LLANTWIT BASIN AND EASTERN SOUTH CROP

Ice from the Rhondda Fawr spilled southwards into the Ely Valley through the co
at Williamstown and Trebanog. In the area south of Pontypridd the large Taff Valle
glacier, swollen by ice from the Rhonddas, spilled over the western sides of the valle
probably over much of the ground between Mynydd y Glyn and Garth Hill, nea
Taff's Well. On the west side of the Ely Valley the ice filled the hollow of Nar
Cae'r-gwerlas, south of Penrhiw-fer, and flowed southwards into the Hendre-Forga
Valley, to the west of Tonyrefail. All these ice streams coalesced and spread ove
the whole of the relatively low ground from the Ogwr Fach to Llantrisant Commo
and Llantwit Fardre. The dividing line between the local and Brecknockshire ic
within this basin is indefinite, but lies roughly along the present course of the Mychyd

The whole area contains the most extensive boulder clay deposits of the distric
only the higher areas of the Pennant being drift-free. For the most part, although th
enclosed stones and boulders have been well rounded, an abundant clayey matri

emains, pointing to the general inactivity of the melt-waters. In form the drift has undergone little or no post-Pleistocene modification and nearly everywhere it exhibits an irregular hummocky topography. The thickest known drift has been proved by the colliery shafts: at Cwm Colliery up to $78\frac{1}{2}$ ft of gravel and clay are recorded, and 53 ft at Coedely.

At its maximum a feeble lobe extended south-westwards along Cwm Ogwr, but the main overspill was due southwards. It must have overtopped the Pennant scarp along a broad front between Mynydd Maendy and Mynydd Garth Maelwg and filled the Ewenny Fach Valley from Penprysg to Llanharan, where it appears to have come into contact with the northern flank of the Irish Sea ice, which filled much of the Vale of Glamorgan. It deposited a trail of morainic boulder clay which fills the hollows of Nant Ciwc, Cwm Llanbad and the Meiros Valley on the higher slopes north-west and north of Llanharan, and is extensively preserved over the lower-lying ground between Hendre-wen and Nant Ton-y-groes, where its hummocky form is particularly fresh. Its western limit is well-defined, the coalescing mounds remaining unmodified. The absence of any comparable glacial drift between here and the Ogmore Valley points strongly to the conclusion that no ice covered that part of the South Crop. Over this area a temporary glacial lake, dammed in the Penprysg area, left behind a thin layer of greenish clays, and solifluction head, which reaches 30 ft in thickness locally, is everywhere considerably thicker than in the glaciated areas.

As the general level of the ice fell, its outflow became restricted to a few gaps, notably those of the Ely, north of Talbot Green, of the Nantcymdda-bach, north of Cross Inn, and of Nant Myddlyn, north-east of Rhiwsaeson. Along the courses of the Ely and its tributary, the Mychydd, the concentration of sub-ice melt-waters produced the only extensive gravel deposits of the area. At Talbot Green a well-marked halt-stage recessional moraine bars the valley. Above it a large spread of alluvium and a flanking terrace, 12 to 15 ft higher, may mark the site of a temporary lake. South of the moraine outwash aprons of terrace-like gravel are developed.

Drift associated with the Irish Sea ice penetrates the district only in the area south of the Ewenny Fach near Llanharan. The gravelly boulder clay contains much Mesozoic material from the Vale of Glamorgan, and the reddish colour contrasts sharply with that of the blue-grey drifts of the coalfield ice. The hummocky morainic form of this western drift is extremely fresh and is best seen in the Llanilid area immediately south of the present district.   A.W.W.

THE CENTRAL VALLEYS

These valleys all have their origins within the district, and apart from a limited area in the lower Llynfi Valley extending to its confluence with the Garw, appear to have been affected only by local ice.

Ogwr Fach: The broad valley around Gilfach Goch formed an area of ice accumulation fed by tongues extending on to the upper slopes of Mynydd Pwll-yr-ebog and Mynydd Maesteg. Beneath the ice thick boulder clay was deposited, probably reaching 100 ft or more, $83\frac{1}{2}$ ft being penetrated in the sinking of the Trane Pit at Britannic Colliery. A narrow tongue, augmented by ice from the area west of Tonyrefail, followed the valley south-westwards. This was in turn joined by small glaciers from the valleys in the northwards facing slopes of Mynydd y Gaer.

Dimbath Valley: Immediately west of Llandyfodwg a small glacier entered the main valley from Cwm Dimbath. The upper reaches of this steep narrow valley are free from drift. Downstream, however, in the Pant-y-fid area [95008810] and the hollow extending eastwards to Nant Llwyn-cae [960879], drift lodgements on the more gentle slopes extend upwards to 700 ft O.D.; presumably in the higher valley the drift has been removed by post-glacial erosion.

Ogwr Fawr: North of Ogmore Vale the valley, floored by largely argillaceou Middle Coal Measures and Llynfi Beds, opens into a large amphitheatre walled by th massive pennant sandstones of the Rhondda Beds. A local glacier nestling in thi hollow produced a marked corrie-like over-steepening of these walls particularly a the head of the valley. In the broad valley boulder clay is widespread up to abou 1300 ft O.D. though at these higher levels it is preserved mainly in the tributary valley: The shafts at Western and Wyndham collieries record 26 and 38 ft respectively of ston drift.

South of Ogmore Vale the valley narrows sharply. Little drift is preserved in thi part of the valley but a narrow strip of boulder clay lines the right bank of the alluviun for about a mile. In the Frithwaun area [929874] this gives way to gravelly deposit which continue to Blackmill where an apparently weak stream of ice coalesced wit' that descending the Ogwr Fach. The resultant glacier continues south-westwards t Brynmenin, gravels again making up the bulk of the deposits.

Landslips are widespread around the amphitheatre at the head of the valley. Mos originate in the sandstones above the No. 2 Rhondda. On the eastern slopes rock tumbles extend almost continuously for 2 miles northwards from Cwm-y-ffosp [946922 and boulder clay is involved in the slipping at the head of Nant Blaenogwr [943931 Similar slips occur above Ogmore Vale. A large slip on the eastern slopes of Mynyd Llangeinor between Nantymoel and Wyndham, although involving much tumb around its edges includes large masses which have moved relatively intact. The bac scar forms three-quarters of a mile of precipitous crags. On the west side of the valle near its head, a large mass of ground has moved relatively intact away from a fau plane on Braich-yr-Hydd. It shows little evidence of any rotational movement, an since it appears to pass beneath boulder clay is probably earlier than the other slips

Garw Valley: Like the Ogwr Fawr, the Garw was the site of a largely enclose glacier, fed from large corries at its head. Drift deposits are remarkably limitec boulder clay being confined to a narrow strip in the bottom of the valley extendin from Pontycymmer to the floor of the Blaengarw corrie at the head of the valley a distance of about 2 miles. From Pontycymmer southwards to its confluence with th Ogwr Fawr the valley is now devoid of glacial deposits; throughout this stretch th valley is narrow and steep-sided, and any lodgements of till or gravel may have bee subsequently scoured. Nevertheless it is apparent that glacial activity was weak an largely confined to the upper basin. The thickest recorded drift is 58 ft at Internationa Colliery, and 53 ft of 'sand and gravel' were also penetrated at Ffaldau. The tributar Garwfechan, also free from glacial drift, may have been similarly unaffected b active ice.

Rock-tumbles from the over-steepened Pennant scarps above the drift-line ar common in the open part of the valley, from Pontycymmer northwards. They ar especially notable in the central and upper parts of the Garwfechan where slips fror the overlying Llynfi Rock cover much of the lower slopes on both sides of the valley

Llynfi Valley: In its upper reaches the Llynfi Valley drift is similar in type an distribution to that in the Garw and Ogmore valleys. There was a gathering groun east of Caerau in which boulder clay now extends to about 900 ft O.D. and 1000 f on the northern flank of Mynydd Bach. West of Caerau drift fills the col to the Avar but it is unlikely that much ice crossed over. Down-valley the level of the fill drop quickly, though extensive patch-workings to the east of Nantyffyllon make i difficult to plot its limits. It reaches up to 600 to 700 ft O.D. west of Nantyffyllor and on the southern side of Mynydd Bach; and rises to almost 800 ft west of Maeste where, at Mount Pleasant [840915], it spreads over the col west of the town. The broa hollow of the Middle Coal Measures outcrop east of Maesteg served as anothe gathering ground, with boulder clay up to 900 ft O.D. east of Bryndefaid Farr [88249160], falling to 500 ft north of Garth Hill and to about 400 ft on its south side In Cwm Cerdin and its tributary, Cwmsychbant and Cwm y Goblyn, boulder cla covers the valley floors up to about 550 to 700 ft O.D. This drift appears to contai

nly local boulders of Pennant and Cockshot. It is everywhere clayey, and reaches
s greatest recorded thickness of 56 ft in the Coegnant shafts. That part of Nant
wmdu lying between Garth Hill and Moelcynhordy is steep-sided and drift-free
nd appears to have formed as a melt-water channel of the ice in the gathering ground
ound Cwmdu-canol [87809130] to the north.

Below Cwm-felin the nature of the drift changes. Where the Llynfi enters the
ennant outcrops, it cuts a deep trench through the valley-fill. This fill forms a
rrace-like shelf which falls steadily downstream, until it is at river level near Cefn
dfa Lodge [88678667]. It is best seen in the tributaries, Nant y Gadlys and Nant
ryn-Cynon, within which it rises to 400 to 500 ft above O.D. Everywhere it contains
recknockshire erratics, which are also scattered above the fill-level, and on Moel
roed-y-rhiw, west of Pont Rhyd-y-cyff, reach to above 700 ft. Exposures [87048760]
f till around Gadlys Factory were noted in the previous editions of this memoir;
can also be well seen in the river bank [87198871] north of Gelli Siriol cottages.
milar drift appears in Cwmcedfyw, north-west of Bettws. South of Shwt the Llynfi,
ke the lower Garw, is almost free from glacial drift until its confluence with the Ogwr,
here boulder clay reappears in the valley bottom and spreads eastwards over the
itcrop of the Lower Coal Measures and Millstone Grit in the Sarn and Bryncoch
eas up to rather more than 300 ft O.D. This drift again contains Brecknockshire
rratics, and extends downstream almost to the southern margin of the district and
pstream to Abergarw. Its relationship to the local gravel brought down the Ogmore
obscure.

How these northern erratics entered the area is far from clear. The general fall in
vel of the deposits southwards appears to preclude the possibility of a southerly
rivation from the Vale of Glamorgan, yet to the north there is a belt of country,
: least 4 miles wide, in which such erratics are unknown. Moreover, the southerly
argin of the Brecknockshire ice north of the Avan is so definite that it is difficult
• believe that it ever extended farther south. It may be that a lobe of ice for a time
oread east from Cwmavon, through the Bryn area and the Sychbant, extending
rther south than did the local Llynfi Valley ice, though for this explanation the same
bjections hold. The possibility that this drift belongs to an earlier glaciation cannot
e discounted. If so, the erratics have presumably been removed from the intervening
ound by the local ice of the later glaciation, and only south of the influence of this
cal ice can 'foreign' erratics be detected. Nevertheless, since the drift below Cwmfelin
opears to be a natural continuation of that above, the immediate source of this
e-flow is still uncertain.

Landslips are almost confined to the slopes of the Llynfi Rock. Rock-tumbles are
mmon around Mynydd Bach, and east of Caerau a major rotational slip extends
om Daren y Bannau [87109390] to the river, several hundred feet below. Nant
wmdu is constricted by another large rotational slip on the eastern side of Garth
ill [875900].

The alluvium of the Llynfi is continuous downstream from Maesteg, attaining
width of about 200 yd in the lower reaches. Where the Llynfi and Garw meet the Ogwr
terrace-like gravel spread has been built up about 10 ft above the alluvium in the
ea between Abergarw and Aberkenfig. Nant-y-cerdin has built up a gravelly delta
here it enters the Llynfi near Cwm-felin, and the height of this above the alluvium
iggests that the delta formed when the course of the Llynfi was some 20 ft above its
resent level.

Avan Valley: At the head of the main valley and in Nant Gwynfi boulder clay
xtends up to about 1300 ft O.D. The fill-level falls to about 1000 ft at Aber-gwynfi;
00 ft north of Craig-y-gelli [881967]; and to some 700 ft beneath Tarren Rhiw-llech
:65961]. Near Gelly Houses [972962] the main valley is joined by Cwm Nant-y-fedw
ithin the upper reaches of which the drift-line again reaches up to some 1300 ft O.D.
a a similar fashion the broad col north of Blaencaerau is filled to almost 900 ft by
oulder clay, which extends to over 1000 ft on the western slopes of Mynydd Caerau.

Everywhere the drift consists of pennant debris set in a clayey matrix, though a North Pit, Avon Colliery, 71 ft of drift are recorded as 'gravel with rock boulders' Particularly good sections can be seen in the roadside from Cymmer eastwards to Gelly Houses and again in the banks of Cwm Nant-y-fedw.

The distribution of drift in the Corrwg Valley is basically similar. In the extreme north, although the western slopes of Cwm Corrwg are in solid, local boulder clay reaches to about 1500 ft O.D. north-west of Cefntyle-brych [895014]. It is well exposed in the tributary brooks of Nant Moel-Yorath, Nant Cwm-cas and Nant-yr-allor and from the last-named extends on to the plateau at about 1500 ft O.D.; it maintains a height of about 1350 ft beneath Maen yr Allor, east of Glyncorrwg. The drift is gravelly around Cwm-cas [885996], but quickly reverts to a more clayey type, though on the west side of the valley it is almost absent.

In the Corrwg-fechan a thin tongue of local boulder clay follows the eastern bank of the stream from about 1500 ft O.D. in the extreme north, to below 1000 ft near Hendre-garreg Farm [87649985]. High on the slopes above this farm, Brecknockshire erratics point to a temporary extension of the Vale of Neath ice over the watershed and the patchy nature of this deposit suggests that it predates the valley drift. At Glyncorrwg village the drift from Cwm Corrwg and the Corrwg-fechan joins another spread generated in a small incipient corrie now drained by Nant-du, in which the till reaches to about 1100 ft O.D. Throughout this upper part of the valley exposures of stony till are common along all the streams, and the shafts at North Pit, Glyncorrwg record 26 ft of 'gravel with large boulders'. South of the village the boulder clay level falls rapidly to the valley floor and only an intermittent narrow tongue follows the valley southwards. At Ynyscorrwg Colliery 53 ft of gravel, sand and boulders are recorded extending to a level well below the alluvium. Farther downstream the Nantewlaeth shafts passed through 60 ft of drift, here called boulder clay, extending to about river level. From here to its confluence with the Avan, the Corrwg flows through an increasingly deep gorge which is everywhere cut down into solid rock. A bank of boulder clay is preserved up to about 650 ft O.D. on the east side of the gorge, but otherwise no glacial deposits remain.

Below Cymmer, where the Avan and Corrwg unite, the gorge continues downstream Boulder clay is preserved on the south of the river west of Cymmer, but near Abercregan the deposit has become sufficiently sandy to be mapped as gravel. At Abercregan the Nant Cregan enters the main river, and the gravel can be traced up this valley to about 600 ft O.D. before passing into boulder clay which extends to about 1100 ft in the upper reaches. The drift is local, but remanie Old Red Sandstone and Millstone Grit erratics mark the limit of the Brecknockshire ice above the western slopes of the side-valleys.

Below Abercregan, gravels continue in the valley bottom to Pont Rhyd-y-fen becoming increasingly fluvial downstream. They reach up to about 500 ft O.D. near Duffryn Rhondda Colliery; to about 440 ft at Cynonafan [825952]; to 350 to 400 ft at Ynys Fawr [809947]; and to 275 ft at Pont Rhyd-y-fen. The river, for the most part, follows their northern edge, and everywhere runs on or near solid. On the north bank of the river opposite Duffryn Rhondda Colliery up to 50 ft of dirty gravel can be seen in the river bank, and north of the river opposite Duffryn village a gravel mound [836960] rising almost to 600 ft O.D. appears to be of morainic origin. Further gravel exposures occur on either bank between Duffryn and Cynonafan. Downstream near the mouth of Cwm Ifan-bach [80049404], the gravel is well bedded with seams of clean sand and appears to have been built up to a nearly flat top at about 350 to 370 ft above O.D. Although this part of the main valley shows little sign of having been occupied by ice, the various small south bank tributary valleys are all floored by boulder clay, presumably derived from small glaciers which never emerged from their gullies. From Nant Trefael [843953] a tongue of boulder clay extends southward and bifurcates; one branch occupying a col at about 820 ft O.D. and being continuous with the drift in the Llynfi Valley, the other extending to about 1100 ft O.D. on the plateau. In Nant-yr-hwyaid [830948] and Blaen-nant [838950] similar boulder clay

ongues reach to about 1000 ft O.D. In Cwm-yr-Argoed south of the stream, a broader
rea of ice-accumulation existed and near the valley mouth [81699494] at least 20 ft
of stony drift can be seen in the roadside. Head and downwash appear to have con-
ributed significantly to the deposits in these side-valleys, as doubtless too did nivation.
At the last mentioned locality up to 10 ft of head overlie the till.

Below Pont Rhyd-y-fen, where the gravel is at about 275 ft O.D. the river breaks
hrough the Pennant sandstones and opens out into a broad valley cut across the
rop of the Lower and Middle Coal Measures. On the east side of this valley drift is
almost absent, though its western side, in the adjacent Swansea district, was occupied
by a tongue of Brecknockshire ice extending from near Baglan.

Landslips are common around the heads of the valleys. On either side of Mynydd
Blaenafan, east of Blaen-gwynfi, slips have originated in the sandstone above the No. 2
Rhondda. That on the southwards facing slope [902963] is largely tumble; that on the
opposite flank of the spur [900966] has a large rotational element. In Nant Gwynfi
here is a large rotational slip below the back-scar of Twyn Pigws [89599740]. In the
pper Corrwg the valley slopes are similarly lined with slips above the fill. East of
North Rhondda Colliery minor-slipping and flow has affected much of the boulder
lay slope and the solid above it. At Taren Pannau [882985], south-east of Glyncorrwg,
rotational slip, including the castellated sandstone pillars of Maen yr Allor, passes
own-hill into rock-tumble and merges imperceptibly into solifluction flows, which
ave involved the boulder clay. On the hillside south of Cymmer much of the dip-slope
of the sandstone above the No. 2 Rhondda shows slight movement on bedding planes,
hough it is rarely disrupted sufficiently to warrant mapping as landslip. Open fissures
long the contours indicate the down-hill gravity movements that have taken place.
A similar bedding-plane slip lies south-west of Cynonafan, though here a distinct
ack-scar [82629457] has resulted. Tarren Forgan [814946] on the other hand is a
ypical rock-tumble with a rotational element along its scar, as are the slips in Cwm
Argoed. In Cwm Nant-y-fedw the boulder clay rests on disrupted sandstone which
ppears to be a pre-glacial landslip [88609570].

As has been mentioned above, head and downwash are widespread, and are
articularly noticeable on the dip-slopes south of the Avan. Stabilized screes line the
teep scarp slopes along much of the valley. Fluvial deposits are almost absent, being
estricted to a 100 to 200-yd wide spread of gravelly alluvium extending a mile
ownstream from Glyncorrwg, and to a few patches in the Avan.

Hill-peat is widespread on Mynydd Ynyscorrwg and along the Avan–Rhondda
'awr watershed, for the most part being restricted to ground above about 1300 ft O.D.
Although small patches occur to the south in favourable situations, the Avan marks
he approximate southward limit of the extensive spreads of hill-peat.

Cwm Farteg: Thin local boulder clay has accumulated over the outcrop of the Middle
Coal Measures from Bryn towards Cwmavon. The deep valley of Cwm Duffryn,
raining southwards, is drift-free and may well have formed as a melt-water channel.

A.W.W., W.B.E.

MARGAM–KENFIG HILL AREA

The eastern margins of the terminal lobe of the Vale of Neath piedmont glacier
xtended into the extreme south-west of the district between Margam and Kenfig Hill.
Glacial deposits reach about 350 ft O.D. beneath the Pennant scarp of Craig Goch
821851] and Coed Tonmawr [826857] and extend as far as Pontrhyd-y-maen [85408404]
a the valley north of Cefn Cribwr, which presumably marks the eastward limit
eached by this ice. To the north-east of Coal Brook and in the area between Kenfig
Hill and Pontrhyd-y-maen the deposits consist largely of boulder clay ground-moraine
ith only sporadic gravelly mounds and pond-silts, but to the west kame-mounds
arking the marginal dissolution of the ice are more general. Throughout the area

west of the main road from Pyle to Margam barely modified hummocks of coarse gravels abound and the drift reaches a general thickness of nearly 100 ft. The erratics everywhere are typical of the Brecknockshire ice. Westwards and south-westwards the gravels pass beneath the alluvium and blown sand marginal to the coast.

Immediately to the north, the Pennant hills south of the Moel Gilau Fault as well as the tract occupied by the outcrop of the Lower and Middle Coal Measures west of the Ogmore are entirely free from glacial deposits and there is no reason to believe that they were ever covered by ice. The slopes of the hills and the low ground generally are draped with thick solifluction deposits.

The broad alluvial area of Margam Moors is covered by grey clay, which is seen to 5 ft or more in many of the ditches. Subsidence from workings at Newlands Colliery has produced large areas of permanently standing water west of Kenfig House [80308420].   A.W.W.

CHAPTER X

ECONOMIC GEOLOGY

COAL MINING

THE FOLLOWING section on the coals of the district has been written by Mr. H. F. Adams, Scientific Department, South-West Division, National Coal Board.

Variation of rank and related properties: Coal has for long been the principal economic product of the district. The seams exhibit a wider regional variation in coal rank and associated properties than is encountered within any other district of comparable extent in Great Britain. In the main lower group of seams the range extends from anthracite, with volatile matter of less than 9 per cent, through low-volatile dry steam and coking steam coal, and medium-volatile prime coking coal, to high-volatile strongly or medium caking coal with volatile matter of 35 per cent. In some of the upper seams the range extends from dry steam coal, through the intermediate classes, to high-volatile medium or weakly caking coal with volatile matter exceeding 37 per cent. This is almost the full range to be found in South Wales, except that the anthracite is not of so high rank as in the north-west of the coalfield. Each of the principal seams exhibits the greater part of this range of property, with a pronounced regional variation. The general direction of increase in rank is from south to north, swinging towards the north-west in the north-western corner of the district. Coal rank and volatile matter are closely related in the coals of South Wales. The regional variation in these properties is illustrated in Fig. 52: Fig. 52A is based upon the Nine-Feet Seam, and is broadly representative of the main productive group of seams; Fig. 52B is representative of seams in the Lower Pennant Measures and is based mainly upon the No. 2 Rhondda and Rock Fawr (No. 3 Rhondda) seams. Lines of equal volatile matter (isovols) are shown, together with the Coal Rank Codes indicating the rank of coal. These codes have the significance indicated in the table given below. Coals of rank higher than code 102 and lower than code 702 are not represented in the district.

Coal Rank Code	Description	Volatile Matter (d.m.m.f.)
102	Anthracite (non-caking)	6·1– 9·0
201 a & b ..	Dry steam coal (usually non-caking)	9·1–13·5
202 & 203 ..	Coking steam coal (usually weakly to medium caking	13·6–17·0
204	Coking steam coal (medium to strongly caking) ..	17·1–19·5
301 a & b ..	Prime coking coal	19·6–32·0
401	High-volatile coal (very strongly caking) ..	32·1–36·0
501/601/701 ..	High-volatile coal (strongly/medium/weakly caking)	32·1–36·0
502/602/702 ..	High-volatile coal (strongly/medium/weakly caking)	Over 36·0

293

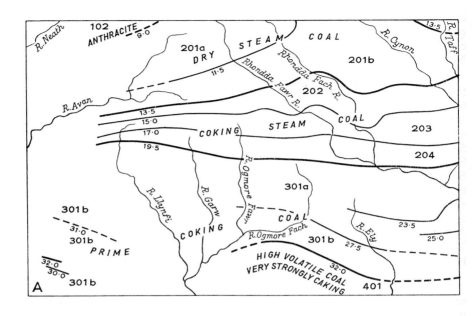

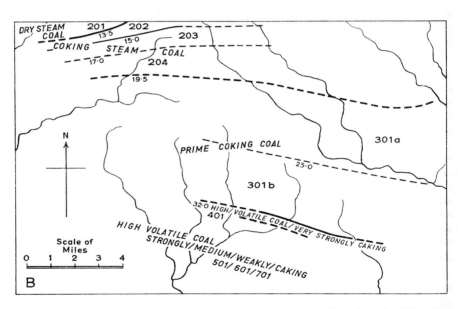

FIG. 52.   A.   Coal Rank and Volatile Matter in lower seams (based on the Nine-Feet)
B.   Coal Rank and Volatile Matter in upper seams (based on the No. 2 Rhondda)
Isovols which separate the main coal types are shown thickened

Apart from the regional variation the seams at any one locality show a general increase in rank with increase of depth, in conformity with 'Hilt's law'. (This generalization does not apply rigidly to seams that are near together.) For example, in the vicinity of Nantymoel the volatile matter of the No. 2 Rhondda, Nine-Feet and Gellideg seams is 26·0, 19·5 and 17·0 per cent respectively.

A very high proportion (over 95 per cent) of the present output of collieries in this district is derived from the main productive group of seams, Two-Feet-Nine to Gellideg inclusive. All of the seams in this group are in production except for the Lower Seven-Feet and Lower Five-Feet. In general the seams in production are clean seams in many of which the ash of the coal, over extensive areas, is 5 per cent or less. In some localities the Five-Feet Seam and the Lower Nine-Feet Seam have up to 8 per cent of ash in the coal. At several collieries the Gellideg Seam has the very low ash of 2·5 per cent or less. The sulphur content of the main coals is generally low, usually in the range 0·5 to 1·5 per cent. Some of the upper seams are of similar quality; others have higher ash or higher sulphur content. In parts of the Rhondda Valley the Pentre Seam of the Middle Coal Measures has a low ash and low sulphur content. The No. 2 Rhondda and No. 3 Rhondda seams in the southern part of the district are generally characterized by high and low sulphur content respectively.

Utilization of coal worked in the district. It is not usual to market the individual seams separately. In many instances coal from several seams, of similar rank and derived from more than one colliery, is mixed and prepared for the market at a central coal preparation plant. Two exceptions are that seams from the upper measures are usually marketed separately from those in the main productive group and that, at some collieries, there is seam selection for the supply of household coal, notably from the Two-Feet-Nine Seam.

A consequence of the wide range of coal rank is that the coals have a wide variety of use. The high-volatile coals of the southern parts of the district, both from the lower and upper seams, including the No. 3 Rhondda (Rock Fawr), were, in former years, used primarily for gas manufacture. Coke-ovens have now become the main source of gas-supply in South Wales, and it is as constituents of coking blends that the high-volatile coals now have their main use. The supply of gas from the carbonization of coal is supplemented by methane drained from the workings of collieries, notably in the Avan valley. The prime coking coals of class 301 (especially 301a) are among the best metallurgical coking coals in Great Britain, and the coals of classes 204, 401 and 501 are valuable as constituents of coking blends. There are modern coking plants at Cwm and Coedely collieries producing foundry-coke, and the district is an important source of supply of coal to coking plants making blast furnace coke, especially those at nearby steel works. Use of these classes of coal is not confined to coke-making; they are used also for steam-raising, including locomotive use, and as house coals.

The principal use of the low-volatile steam coals of classes 202 and 203 (as well as 204) is for steam-raising, both at power stations and in a variety of industrial boiler plants, and for marine bunkers. Other uses are for central heating and other domestic purposes. Every class of coal in the district is, or has been, to a greater or lesser extent, used for steam-raising. This includes, in suitable furnaces, the smaller sizes of dry steam coal (class 201) and anthracite. The coals of this district share the high reputation for steam-raising held by South Wales coals on account of their inherently high calorific value.

The sized grades of dry steam coal (class 201) and anthracite are in demand for domestic use in slow-combustion appliances, and are used also in bakeries and lime-kilns. The supply of fuel for domestic slow-combustion appliances is supplemented by carbonized ovoids. These are manufactured at Aberaman from dry-steam coal duff, much of which is derived from collieries within the boundaries of the present district.   H.F.A.

QUARRYING

The sandstones of the Upper Carboniferous have been extensively quarried in the past. Because of their widespread outcrops, those of the Pennant Measures are by far the most important; but some of the Middle Coal Measures sandstones, especially the Cockshot Rocks, were of economic importance in the Maesteg area, while those of the Millstone Grit Series and Lower Coal Measures were quarried here and there along the South Crop.

Before the first World War most of the buildings on the coalfield were constructed of Pennant sandstones. Practically all divisions of the formation yield suitable stone and most of it was hewn locally. Every settlement had its own quarry, and the larger villages and towns had several major ones nearby. Many individual buildings, such as farms scattered along the hillsides, were built from stone dug out on the spot, as were the dry-stone walls so characteristic of the coalfield scenery.

The thicker beds provided most of the building stone, but the more flaggy beds were worked for paving. The latter were used not only locally but were exported on a large scale and many of the streets of English cities are paved with Pennant flagstones from the South Wales Coalfield. Flaggy sandstones have been quarried from close below the Cefn Glas Seam and its equivalents at widely scattered localities, e.g. at Pen-y-coedcae, south of Pontypridd, on Bryn-y-Wrach, south of Llangeinor, to the south-east of Bettws village, and in the Avan and Pelena valleys. At the first of these localities especially high quality stone was followed underground. Thinly bedded sandstones of 'tilestone' type are developed widely immediately above the Brithdir Rider, especially in the north-eastern parts of the district; these have been quarried at numerous localities.

Rising costs of production during the present century have made the working of the Coal Measures sandstones uncompetitive in relation to other building materials, and nowadays quarrying has almost ceased. Only one major quarry is at present worked within the district—Craig-yr-Hesg Quarry—situated on the west side of the Taff Valley just north of Pontypridd. A 130-ft face exposes beds lying between the Brithdir Rider and the Cefn Glas (Geol. Surv. Phot. A1651–2). The bottom 40 ft is in very massive sandstone, almost devoid of bedding and this is extracted in large blocks for sawing. The principal products include road metal, rough block masonry, architectural and monumental masonry, kerbs and paving; there are also various concrete mouldings (e.g. kerbs and paving) made from crushed stone, cement and sand.

BRICK-MAKING

There are active brickworks at Tondu, Bryncethin, and Coedely Colliery, all using argillaceous strata from the Coal Measures. At Tondu and Bryncethin the raw material is dug from open pits adjoining the kilns; at the former from beds between the Amman Rider and Upper Six-Feet (see pp. 138–40), and at the latter from between the Pentre Coal and the base of the Llynfi Rock (see pp. 167–8). Little selection of the material is practised apart from the rejection of the coals and thicker sandstones. Mudstones, silty mudstones, thin sandstones and seatearths are crushed together. Both works produce bricks for the building trade. In former years Tondu Brickworks also produced a red flooring tile, the raw material being obtained by selective quarrying of shales from below the Gellideg. Coedely Colliery Brickworks produces a building brick from blocky mudstones obtained from above the Two-Feet-Nine during the mining of that seam.

Brick-making was formerly much more widespread. In the Rhondda Fawr good seatearths underlie the Nos. 1, 2 and 3 Rhondda seams, and all were extensively worked during the early decades of the Nineteenth Century. With the sinking of pits to the deeper coals in the middle and later decades of the century, there was a large demand for bricks for shaft-walling, while local coke and gas-works required fire-bricks for furnace-lining. Colliery brickworks were established at Llwyn-y-pia, Bodringallt and Treherbert as well as at Gelliwion, near Pontypridd farther down the valley. So great became the demand for bricks by the coal industry that the products of these Rhondda works were sent all over South Wales. When sinkings became more infrequent towards the close of the century, the industry died out, and has never been revived. Brickworks also existed at Bryn and near Cwmavon, the latter operating until recently, using mudstones from between the Two-Feet-Nine and Upper Nine-Feet.

Refractory bricks are reputed to have been made on the South Crop at Tycribwr just south of Park Slip Colliery, the seatearth of the Gellideg forming the raw material of the industry. Bricks were also made at Cefn Cribbwr Brickworks, the material apparently being obtained from a nearby open pit in strata just below and above the Garw.

At Aberaman in the Cynon Valley a flourishing works has existed for many years making salted glazed pipes and fittings, chimney pots, ducts, etc. The industry made use of the clay of the Clay (Hafod) Seam and the nearby Clay Level was worked until recently with the main object of obtaining the seatearth, which was left in heaps to weather for several years before being used. This mining has now ceased and the raw material is imported as waste from other parts of the coalfield. The same horizon is believed to have been mined in the last century under Garn-wen.

IRON ORE

The early iron industry of South Wales was based on the local iron ores of the Coal Measures. Two varieties of these exist—clay ironstones or 'mine', disposed as a great number of thin bands and nodules associated with the argillaceous rocks between the Garw and the Cockshot Rocks, and Blackband ironstones lying mainly at a few horizons between the Gorllwyn and the Llynfi Rock. The iron-works of the district were sited near the outcrops of these measures.

The measures with clay-ironstones crop only along the South Crop and in the area from Maesteg westwards to Cwmavon. They consist of ferrous carbonate admixed with carbonates of lime and magnesium, clay minerals and small amounts of organic matter. Iron varies from 25 to 35 per cent and sulphur and phosphorus are present in small quantities. The ores were dug by hand in 'patch-works' and the nodules of ironstone hand-picked after weathering. Close to and around Maesteg the principal horizons worked were the Yellow Pin, the Ballsag and the Black Pin Balls, all lying close above the Two-Feet-Nine; the Double Pin, some 6 ft above the Upper Four-Feet; and the Pin Haulkin, rather more than 40 ft above the Six-Feet.

On the South Crop extensive 'patchworks' occur between the crops of the Gellideg and the Two-Feet-Nine between Evanstown (Aberkenfig) and Pont-rhyd-y-maen, south of Aberbaiden. In the Cwm-ffos–Cefn area the ironstones between the Garw and Gellideg, so important along the North Crop, were patched and also mined. These beds were also worked at Bryndu Colliery, north-west of Kenfig Hill (for section, see pp. 97–8). The principal horizons were known as the Little Pin-Garw, the Pin-Garw, the Double Pin-ground, the Red Mine-ground, the Yellow Vein, and the Black Pin Ironstone and Mine-ground. Farther along the South Crop ironstones are also reputed to have been worked at Brynau-gwynion Colliery, a mile west of Llanharan. At Cwmavon the Jack Mine, between the Lower Five-Feet and the Gellideg, and the Spotted Pins, probably the most valuable of all, lying between the Gellideg and Garw, were mined. The Ramshead and Cefn-y-glo, about 30 ft above the Six-Feet and the Coal and Mine, on one of the splits of the Red Vein group were also worked.

The blackband ores contain 30 to 36 per cent of iron and up to 7 per cent of carbonaceous matter. Three distinct horizons were exploited around Maesteg, Bryn, and Cwmavon. These were initially called the Lower Blackband, lying close below the Victoria; the Middle Blackband, associated with the coal under-lying the Lower Cwmgorse Marine Band; and the Upper Blackband on the horizon of the White. This last did not retain a distinctive name for long, the middle of these horizons being more generally referred to as the Upper Black-band, a usage followed in this memoir. This horizon was worked at an early date from Cwmavon eastwards towards Bryn, and has been trenched at outcrop along most of this tract. Its discovery in the Llynfi Valley is said to have been made by De la Beche in a brook at Garn-wen, and if so its exploitation in this valley must post-date the establishment of the ironworks and the patchworks on the clay ironstones. It was subsequently mined extensively beneath Mynydd Bach and Garn-wen and worked opencast from deep trenches on the hillsides around Maesteg. The Lower Blackband was also dug at outcrop in the same general area, and mined on a rather smaller scale beneath Mynydd Bach.

The oldest furnace of any size appears to have been at Melin Court in the Neath Valley. It probably owes its position to the nearby waterfall, for certainly the ore had to be carried to it, and when coal replaced charcoal as a fuel it is said to have been obtained from the Graigola seam on Cefnmawr. By the middle of the nineteenth century furnaces at Maesteg, Cwmavon, Cefn and Tondu had attained a considerable production. They were, however, short-lived and by the end of the century production of the ore had virtually ceased and the ironworks had closed.

LEAD

Although galena has been reported as isolated crystals, usually in septarian nodules, from various parts of the coalfield, its concentration into a form rich enough to merit working is known only in the Dolomitic Conglomerate on the south side of the Nant Myddlyn, near Rhiwsaeson. It occurs associated with iron-pyrites and a little copper-ore in pockets, discontinuous veins, and stringers, and has been worked from numerous shallow pits near Llwynmilwas and from similar pits and quarries near Gwern-Efa, a little to the west. The deposit is undoubtedly of low grade and no working has taken place for many years.

Underground Water Supply

Public supplies are in general obtained from overground sources. Small amounts of ground-water, usually up to a few thousand gallons per hour, have been obtained from a few wells in the Coal Measures and near Margam Moor a little has been pumped from the Drift for industrial purposes.

A considerable quantity of mine-water is pumped from the collieries of the district. In 1961 this amounted to 15·16 m.g.d., made up as follows: Taff and Cynon valleys 5·40 m.g.d.; Rhondda Valley 5·43 m.g.d.; Neath and Avan valleys 1·28 m.g.d.; others 3·05 m.g.d.[1] Little of this water is used for public supply purposes; the only major case being Pwll-y-gwlaw Level near Pont Rhyd-y-fen, which supplies the town of Port Talbot. With the increasing demand for water several authorities are considering the utilization of these sources for industrial purposes.

A.W.W., W.B.E.

---

[1] These figures have been supplied by Dr. J. Ineson of the Geological Survey's Water Department.

APPENDIX I

RECORDS OF PRINCIPAL COLLIERY AND BOREHOLE SECTIONS

Graphic sections of certain colliery shafts in the district were published in Sheets 83, 84 and 85 of the Geological Survey's Vertical Sections, 1900–01. A few were also given in the text or in the Plate of the 1st Edition of this Memoir, 1903, with an addition of about a dozen in the Appendix to the Second Edition, 1917.

All the principal colliery shafts for which sections exist are listed below; those previously published are given in abstract only, a reference to the previous appearance being also stated. For the collieries where records have not previously been published, an abridged section of one of the shafts is usually given, any additional records being in abstract. In a few cases deepenings are listed in extended form, while a few cross-measures or underground pit sections are given where these provide the only record for the measures concerned. Few deep boreholes from surface have been drilled within the district, and lack of space precludes the publication of the many hundreds of underground records.

For some of the shafts several versions showing slight variations of the section exist; in these cases, that which is considered the most reliable is given.

For an explanation of some of the descriptive rock-terms used in the sinkers' logs, see p. 75.

Aberaman Colliery: Downcast Shaft

Ht. above O.D. 437 ft.   6-in SO 00 S.W.; Glam. 18 N.E.

Site 600 yd S. 9° W. of Aberaman church. National Grid Ref. 01590038. Sunk about 1845. Published in 1st edition, pp. 12–14.

Gorllwyn at 136 ft 7 in, Two-Feet-Nine at 347 ft 11 in, Four-Feet at 387 ft 8 in, Upper Six-Feet at 403 ft 8 in, Lower Six-Feet at 440 ft 9 in, Red Vein at 484 ft 10 in, Nine-Feet at 538 ft 9 in, Bute at 574 ft 10 in, Yard at 634 ft 11 in, Seven-Feet at 665 ft 9 in, Five-Feet at 705 ft 2 in, Gellideg at 710 ft 7 in. Sunk to 751 ft 4 in.

Abercwmboi Pit

Ht. above O.D. 436 ft.   6-in ST 09 N.W.; Glam. 18 N.E.

Site 2000 yd E. 43° S. of Aberaman church. National Grid Ref. 03049962.

	Thickness ft in		Depth ft in	
DRIFT				
Surface stuff 118 1			118	1
COAL MEASURES				
Fireclay, black shale and clift with balls of mine .. 40 5			158	6
Coal .. 10 in **10**			**159**	**4**
Fireclay, rock, clift and mine ground 58 6			217	10
CLAY				
Coal .. 29 in **2 5**			**220**	**3**

	Thickness ft in		Depth ft in	
Fireclay, rock, clift and mine ground 80 4			300	7
Coal .. 1 in **1**			**300**	**8**
Fireclay 4 3			304	11
GRAIG				
Coal .. 30 in				
Clod .. 2 in				
Coal .. 5 in				
Clod .. 8 in				
Coal .. 4 in **4 1**			**309**	**0**
Fireclay, clift and mine ground 51 6			360	6

	Thickness ft in	Depth ft in
PENTRE RIDER		
Coal .. 15 in	1 3	361 9
Fireclay, rock, strong clift and mine ground .. 31 4		393 1
PENTRE		
Coal .. 7 in		
Clod .. 3 in		
Coal .. 7 in	1 5	394 6
Fireclay, bastard rock and strong clift with balls of mine 17 10		412 4
LOWER PENTRE		
Coal .. 14 in		
Black stone 3 in		
Coal .. 6 in	1 11	414 3
Fireclay, bastard rock and clift with mine 32 7		446 10
EIGHTEEN-INCH		
Coal .. 13 in	1 1	447 11
Black clay, rock and strong clift with beds of rock and mine 38 2		486 1
Coal .. 7 in	7	486 8
Strong clift with rock .. 8 11		495 7
GORLLWYN		
Coal and rashings 30 in	2 6	498 1
Black fireclay, rock and mine ground 67 7		565 8
Coal .. 2 in	2	565 10
Bastard clay, white rock, mine ground 55 11		621 9
Coal .. 2 in	2	621 11
Fireclay, rock and clift with balls of mine .. 32 2		654 1
THREE COALS		
Coal .. 20 in		
Strong fireclay 21 in		
Coal .. 3 in		
Rock .. 36 in		
Coal .. 7 in	7 3	661 4
Strong clay and clift; beds of rock 26 3		687 7
Black rashings and flap of coal 6 8		694 3
Strong clift and beds of rock 8 8		702 11
Black rashings 3 0		705 11
TWO-FEET-NINE		
Coal .. 34 in		
Clod .. 8 in		
Coal .. 5 in		
Clod .. 18 in		
Coal .. 8 in	6 1	712 0
Fireclay and mine ground with mine 19 0		731 0
FOUR-FEET		
Coal .. 66 in		
Holing .. 6 in		
Coal .. 12 in	7 0	738 0
Strong fireclay and mine ground 18 0		756 0
UPPER SIX-FEET		
Coal .. 24 in	2 0	758 0
Strong fireclay and strong clift 2 6		760 6
LOWER SIX-FEET		
Coal .. 6 in		
Bast .. 8 in		
Coal .. 4 in		
Parting .. 2 in		
Coal .. 36 in	4 8	765 2
Fireclay 12 8		777 10
Bast and Coal .. 14 in	1 2	779 0
Rough fireclay 3 0		782 0
Coal .. 4 in	4	782 4
Coarse fireclay, clod and clift 13 7		795 11
RED VEIN		
Coal .. 10 in		
Fireclay .. 10 in		
Coal .. 2 in		
Fireclay .. 24 in		
Coal .. 9 in	4 7	800 6
Bastard fireclay, rock, clift and mine ground .. 73 2		873 8
NINE-FEET		
Coal 48 to 180 in		
Rashings 24 in	11 6	885 2
Rock and clift 16 1		901 3
BUTE		
Coal .. 26 in		
Fireclay .. 28 in		
Coal .. 26 in	6 8	907 11
Fireclay, strong clift and rock 43 6		951 5
AMMAN RIDER		
Coal .. 15 in		
Clod and Rock .. 36 in		
Coal .. 21 in	6 0	957 5
Fireclay, clift and clod .. 12 10		970 3
YARD		
Coal .. 50 in	4 2	974 5
Fireclay and clift 17 8		992 1
SEVEN-FEET		
Coal .. 66 in	5 6	997 7
Fireclay, strong clift and mine balls 15 0		1012 7
Coal .. 12 in	1 0	1013 7
Fireclay and clift with mine balls 14 0		1027 7

Abergorki Pit

Ht. above O.D. 350 ft.   6-in ST 09 N.E.; Glam. 19 N.W.

Site 950 yd S. 35° E. of Mountain Ash church. National Grid Ref. 05279867.

	Thickness ft in	Depth ft in
DRIFT		
Made ground and drift ..	45 2	45 2
COAL MEASURES		
Fireclay, rock and clift ..	76 6	121 8
Coal .. 3 in	3	121 11
Fireclay and clift	23 6	145 5
Mainly rock	91 1	236 6
Strong clift and bands of rock	37 6	274 0
Rock	18 3	292 3
Rashings 6 in		
Coal .. 24 in	2 6	294 9
Fireclay and clift	25 2	319 11
Coal .. 9½ in		
Clod .. 4½ in		
Coal .. 11 in		
Black shale 29 in		
Coal .. 4 in		
Rashings 3 in	5 1	325 0
Bastard fireclay ..	8 1	333 1
White rock ..	20 8	353 9
Clift	13 4	367 1
Coal .. 15 in	1 3	368 4

	Thickness ft in	Depth ft in
Fireclay and clift	18 6	386 10
Coal .. 15 in		
Stone .. 2 in		
Coal .. 4 in		
Rashings 6 in	2 3	389 1
Fireclay, clift, etc. ..	50 11	440 0
Rock	46 11	486 11
Shale, fireclay and clod ..	2 5	489 4
Coal .. 5 in	5	489 9
Fireclay, clift, etc. ..	10 2	499 11
White rock ..	11 0	510 11
Coal and rashings 5 in		
Coal .. 3 in		
Coal and rashings 11 in	1 7	512 6
Fireclay, rock and clod ..	22 4	534 10
Coal .. 13 in	1 1	535 11
Fireclay and clift	52 7	588 6
CLAY		
Coal .. 24 in	2 0	590 6
Fireclay and rock	7 8	598 2

Abergorky Colliery: probably No. 1 Pit

Ht. above O.D. 691 ft.   6-in SS 99 N.E.; Glam. 18 S.W.

Site 1560 yd N. 6° W. of Treorchy railway station. National Grid Ref. 95819790. Published in Vertical Sections, Sheet 84, 1901.

Abergorky at 42 ft 6 in, Gorllwyn at 276 ft 10 in, Two-Feet-Nine at 459 ft 1 in, Four-Feet (Bottom Coal) at 533 ft, Upper Six-Feet at 558 ft 1 in, Lower Six-Feet at 600 ft 8 in, Upper Nine-Feet at 700 ft 3 in, Lower Nine-Feet at 747 ft 7 in, Bute at 796 ft 3 in, Amman Rider at 816 ft 6 in, Yard at 855 ft 10 in, Middle Seven-Feet at 878 ft 5 in, Lower Seven-Feet at 914 ft 4 in, Five-Feet at 924 ft 11 in, Gellideg at 973 ft 3 in. Sunk to 1011 ft 2 in.

Abergorky Colliery: probably No. 3 Pit

Ht. above O.D. 696 ft.   6-in SS 99 N.E.; Glam. 18 S.W.

Site 1620 yd N. 7° W. of Treorchy railway station. National Grid Ref. 95769795.

Walling and drift to 47 ft, Pentre Rider at 90 ft 9 in, Gorllwyn at 232 ft 8 in, Four-Feet at 511 ft 11 in, Upper Six-Feet at 534 ft 10 in, Lower Six-Feet at 589 ft 11 in, Upper Nine-Feet at 688 ft 8 in, Lower Nine-Feet at 729 ft 10 in, Bute at 778 ft 8 in, Amman Rider at 811 ft 3 in. Yard at 851 ft 1 in, Middle Seven-Feet at 876 ft 4 in, Lower Seven-Feet at 902 ft 6 in, Upper Five-Feet at 907 ft 10 in, Lower Five-Feet at 913 ft 2 in, Gellideg at 962 ft 1 in. Sunk to 965 ft 3 in.

APPENDIX I

Anglo-American Oil Co., Ltd., Pontypridd No. 1 Borehole

Ht. above O.D. 769 ft. 6-in ST 08 N.W.; Glam. 27 S.E.

te 170 yd S. 18° E. of Cottage Bach. National Grid Ref. 01578915. Drilled 1941–42.

	Thickness ft in	Depth ft in		Thickness ft in	Depth ft in
oal Measures to ? Garw Seam	1370 0	1370 0	Dark shale with thin quartzite beds	30 0	2330 0
ale and mudstone with thin sandstones	136 0	1506 0	Dark shale	50 0	2380 0
ne- to medium-grained sandstone	46 0	1552 0	Dark pyritous mudstone (R_2 fossils)	10 0	2390 0
rey shale with bands of sandstone	143 0	1695 0	Hard fine-grained quartzitic sandstone	43 0	2433 0
ne-grained sandstone	26 0	1721 0	Dark shale with bands of sandstone	48 0	2481 0
ark shale with sandstone band	32 0	1753 0	Hard fine- to coarse-grained sandstone	39 0	2520 0
ne-grained sandstone	20 0	1773 0	Hard dark shale with sandstone layers	55 0	2575 0
ark shale	490 0	2263 0	Hard fine-grained quartzitic sandstone with bands of hard dark shale	174 0	2749 0
ard coarse-grained quartzite	37 0	2300 0			

Avon Colliery: South Pit

Ht. above O.D. 953 ft. 6-in SS 89 N.W.; Glam. 17 S.E.

Site 970 yd E. 42° S. Blaengwynfi railway station. National Grid Ref. 89489617. Published in Vertical Sections, Sheet 85, 1901.

Hafod (Bottom Coal) at 510 ft, Abergorky (Top Coal) at 662 ft, Abergorky (Bottom Coal) at 688 ft 6 in, Pentre Rider at 762 ft, Pentre at 826 ft, Gorllwyn at 1021 ft, Two-Feet-Nine at 1299 ft, Four-Feet at 1383 ft 3 in, Upper Six-Feet at 1406 ft 3 in, Lower Six-Feet at 1452 ft 11 in, Caerau at 1496 ft 3 in. Sunk to 1504 ft 4 in.

Avon Colliery: North Pit

Ht. above O.D. 954 ft. 6-in SS 89 N.E.; Glam. 17 S.E.

Site 945 yd E. 34° S. Blaengwynfi railway station. National Grid Ref. 89539627.

Hafod (Bottom Coal) at 538 ft 6 in, Abergorky (Bottom Coal) at 707 ft 6 in, Pentre Rider at 786 ft, Pentre at 853 ft 6 in, Gorllwyn at 1023 ft 6 in. Sunk through the western branch of the Glyncorrwg Fault to ?Lower Nine-Feet at 1269 ft 11 in, ?Bute at 1295 ft 2 in, ?Amman Rider at 1342 ft 8 in, ?Yard at 1413 ft 10 in, ?Middle Seven-Feet at 1521 ft 5 in. Sunk to 1528 ft 2 in.

Blackmill Borehole

Ht. above O.D. 320 ft. 6-in SS 98 N.W.; Glam. 34 N.E.

Site 1350 yd S. 43° W. of Blackmill railway station. National Grid Ref. 92428557. Deviation up to 43° at bottom of hole. In following abstract depths are corrected.

No. 2 Rhondda at 480 ft, ?No. 3 Rhondda at 824 ft, Upper Cwmgorse Marine Band at 994 ft, Lower Cwmgorse Marine Band at 1107 ft, Pentre Rider at 1228 ft, Pentre at 1250 ft, Lower Pentre at 1270 ft.

Bodringallt Colliery: Downcast Shaft

Ht. above O.D. 569 ft. 6-in SS 99 N.E.; Glam. 18 S.W.
Site 955 yd E. 2° S. of Ystrad railway station. National Grid Ref. 98159532.

		Thickness ft in		Depth ft in	
Walling		5	0	5	0
DRIFT					
Clay, gravel and sand ..		45	0	50	0
COAL MEASURES					
Clift and fireclay		36	0	86	0
ABERGORKY					
Coal ..	26 in				
Clod ..	8 in				
Coal ..	14 in	4	0	90	0
Fireclay, rock and clift ..		70	10	160	10
PENTRE RIDER					
Coal ..	14 in	1	2	162	0
Fireclay, rock and clift ..		26	4	188	4
PENTRE					
Coal ..	19 in				
Stone ..	1 in				
Coal ..	12 in	2	8	191	0
Fireclay, clift and shale ..		38	0	229	0
LOWER PENTRE					
Coal ..	14 in				
Stone ..	1½ in				
Coal ..	12 in	2	3½	231	3½
Fireclay		26	6	257	9½
Coal ..	10 in		10	258	7½
Fireclay, clift and rashes..		37	6	296	1½
GORLLWYN RIDER					
Coal ..	11 in		11	297	0½
Fireclay and clift		41	11½	339	0
GORLLWYN					
Coal ..	27 in	2	3	341	3
Fireclay, rock and clift ..		118	9	460	0
Coal ..	2 in		2	460	2
Fireclay		10	0	470	2
Coal ..	20 in	1	8	471	10
Bastard fireclay, clift and mine		43	2	515	0
Coal ..	7 in		7	515	7
Fireclay, rock and clift ..		26	0	541	7
Coal ..	6 in		6	542	1
Fireclay and clift.. ..		4	4	546	5
Coal ..	17 in	1	5	547	10
Fireclay, clift and rock ..		33	0	580	10
Coal ..	24 in	2	0	582	10
Fireclay		9	0	591	10
Coal ..	20 in				
Fireclay ..	24 in				
Coal ..	18 in	5	2	597	0
Fireclay, clift and beds of rock		19	10	616	10
Coal ..	8 in		8	617	6
Fireclay, clift and bands of rock		19	0	636	6
Coal ..	18 in	1	6	638	0
Clift, rock and fireclay ..		14	3	652	3
Rashes and Coal ..	48 in	4	0	656	3

		Thickness ft in		Depth ft in	
Rock and clift		6	0	662	3
Coal ..	9 in		9	663	0
Fireclay and shale ..		3	6	666	6
TWO-FEET-NINE					
Coal ..	60 in	5	0	671	6
Fireclay and rock.. ..		4	0	675	6
Coal ..	12 in				
Rashes ..	3 in				
Coal ..	9 in	2	0	677	6
Fireclay and clift		39	0	716	6
?FOUR-FEET					
Coal ..	36 in	3	0	719	6
Disturbed ground and broken coal		68	0	787	6
Clift and ironstone ..		30	0	817	6
UPPER NINE-FEET					
Coal ..	48 in				
Rashings	6 in				
Coal ..	30 in				
Bast and shale ..	36 in				
Bast ..	6 in	10	6	828	0
Fireclay and clift with ironstone		31	0	859	0
LOWER NINE-FEET					
Coal ..	36 in	3	0	862	0
Fireclay. clift and bastard rock		32	0	894	0
BUTE					
Bast ..	6 in				
Coal ..	66 in	6	0	900	0
Fireclay, shale, clift and rashings		23	3	923	3
AMMAN RIDER					
Coal ..	15 in	1	3	924	6
Fireclay, clift and ironstone		29	0	953	6
Coal ..	18 in	1	6	955	0
Fireclay, clift and ironstone		14	0	969	0
YARD					
Coal ..	40 in				
Rashes ..	9 in				
Coal ..	6 in	4	7	973	7
Fireclay		6	6	980	1
UPPER SEVEN-FEET					
Coal ..	18 in	1	6	981	7
Fireclay, clift and ironstone		26	7	1008	2
MIDDLE SEVEN-FEET					
Coal ..	40 in	3	4	1011	6
Shale, fireclay and bastard rock		18	0	1029	6
LOWER SEVEN-FEET					
Coal ..	20 in	1	8	1031	2
Fireclay, rashes, clift and ironstone ..		23	6	1054	8
Coal ..	4 in				
Shale ..	24 in				
Coal ..	8 in	3	0	1057	8

Bodringallt Colliery: Staple Pit to Five-Feet

Said to be sunk from Nine-Feet, but apparently from Bute.

		Thickness ft in		Depth below Bute ft in	
Clift and rock ..	..	29	0	29	0
AMMAN RIDER					
Coal .. 16 in	1	4		30	4
Measures ..	.. ..	6	0	36	4
Coal .. 12 in	1	0		37	4
Measures ..	..	17	0	54	4
YARD					
Coal .. 39 in					
Parting .. 8 in					
Coal .. 6 in	4	5		58	9
Parting ..	..	2	2	60	11
UPPER SEVEN-FEET					
Coal .. 9 in		9		61	8
Clift and rock ..	..	20	4	82	0

		Thickness ft in		Depth below Bute ft in	
MIDDLE SEVEN-FEET					
Coal .. 28 in	2	4		84	4
Fireclay, clift and rock ..	13	7		97	11
LOWER SEVEN-FEET					
Coal .. 16 in	1	4		99	3
Fireclay and clift ..	..	8	6	107	9
Coal .. 3 in					
Clift .. 54 in					
Coal .. 3 in	5	0		112	9
Fireclay and clift ..	..	33	10	146	7
FIVE-FEET					
Coal .. 41 in					
Shale .. 3 in					
Coal .. 30 in	6	2		152	9

Britannic Colliery: Trane Pit

Ht. above O.D. 782 ft. 6-in SS 98 N.E.; Glam. 27 S.W.

Site 400 yd N. of Gilfach Goch church. National Grid Ref. 98278987.

		Thickness ft in		Depth ft in	
Walling ..	.. ..	23	0	23	0
DRIFT					
Soil, blue clay and gravel with a few boulders ..	83	6		106	6
COAL MEASURES					
Rock and clift ..	..	26	8	133	2
NO. 3 RHONDDA					
Coal .. 34 in	2	10		136	0
Fireclay and clift ..	..	19	11	155	11
Coal .. 2 in		2		156	1
Strong fireclay ..	..	3	0	159	1
Coal .. 1 in		1		159	2
Fireclay, rock and clift ..	38	3		197	5
Coal .. 7 in		7		198	0
Fireclay ..	.. ..	2	9	200	9
Coal .. 10 in		10		201	7
Fireclay, rock and clift ..	23	9		225	4
Coal .. 3 in					
Clift .. 28 in					
Coal .. 6 in	3	1		228	5
Fireclay, rock and clift ..	23	4		251	9
Coal and black rashes 10 in		10		252	7
Fireclay and black rashes	8	2		260	9
Strong clift and mine ..	67	0		327	9
Coal .. 11 in					
Clod .. 8 in					
Coal .. 5 in	2	0		329	9
Clift and rock ..	..	15	1	344	10

		Thickness ft in		Depth ft in	
HAFOD (BOTTOM COAL)					
Coal .. 17 in	1	5		346	3
Fireclay, clift and rock ..	87	0		433	3
Coal .. 15 in	1	3		434	6
Fireclay, clift and rock ..	41	0		475	6
ABERGORKY (TOP COAL)					
Coal .. 17 in	1	5		476	11
Fireclay and clift with rock bands	..	18	7	495	6
ABERGORKY (BOTTOM COAL)					
Coal .. 17 in	1	5		496	11
Fireclay and strong clift with rock bands ..	62	11		559	10
PENTRE RIDER					
Coal .. 17 in	1	5		561	3
Fireclay and clift with rock band	..	24	9	586	0
Black rashes .. 6 in					
Coal .. 2 in					
Clod and rashes .. 10 in					
Coal .. 7 in	2	1		588	1
Clod and clift ..	..	8	6	596	7
PENTRE					
Coal .. 26 in					
Stone .. 1½ in					
Coal .. 10½ in	3	2		599	9
Fireclay and clift with rock	19	5		619	2

	Thickness ft in	Depth ft in
LOWER PENTRE		
Coal .. 18 in		
Clod .. 2 in		
Coal .. 10 in	2 6	621 8
Fireclay and clift with rock bands and mine .. 34 1		655 9
EIGHTEEN-INCH		
Coal .. 16 in	1 4	657 1
Fireclay and rock.. ..	4 5	661 6
Coal .. 8 in	8	662 2
Fireclay and clift with rock bands 41 2		703 4
GORLLWYN		
Black rashes.. 6 in		
Coal .. 33 in	3 3	706 7
Fireclay and clift 111 10		818 5
Coal .. 5 in	5	818 10
Black rashes, fireclay and clift 11 3		830 1
Coal .. 7 in	7	830 8
Fireclay, rock and clift .. 48 8		879 4
Coal .. 15 in	1 3	880 7
Fireclay and clift 12 0		892 7
Coal .. 4 in	4	892 11
Fireclay and clift with mine 8 0		900 11
Coal .. 10 in		901 9
Fireclay and clift with layers of rock 34 11		936 8
TWO-FEET-NINE		
Coal .. 22 in	1 10	938 6

	Thickness ft in	Depth ft in
Fireclay and clift with mine	6 6	945 6
Coal .. 12 in		
Black rashes.. 2 in		
Coal with stone .. 6 in	1 8	946 8
Fireclay and clift with mine	9 10	956 6
Coal .. 4 in	4	956 10
Fireclay and clift ..	4 9	961
Coal .. 20 in	1 8	963 5
Fireclay, clift and rock ..102 7		1065 10
Coal .. 4 in	4	1066 2
Fireclay, clift and stone .. 17 9		1083 11
Coal .. 3 in	3	1084 2
Clift 21 8		1105 10
Coal .. 18 in	1 6	1107 4
Clift and rock 81 9		1189 1
?UPPER SIX-FEET		
Coal .. 20 in	1 8	1190 9
Rashings 15 4		1206 1
?LOWER SIX-FEET AND CAERAU		
Coal .. 120 in		
Rashings 18 in		
Coal .. 26 in	13 8	1219 9
Fireclay and clift 28 4		1248 1
?NINE-FEET		
Coal .. 110 in	9 2	1257 3
Measures 124 7		1381 10
?BUTE		
Coal .. 44 in		
Clod .. 8 in		
Coal .. 10 in	5 2	1387 6
	Sunk to 1398 6	

Britannic Colliery: Llewellyn Pit

Ht. above O.D. 782 ft.   6-in SS 98 N.E.; Glam. 27 S.W.

Site 430 yd N. of Gilfach Goch church. National Grid Ref. 98268987.

No. 3 Rhondda at 131 ft 8 in, Pentre at 588 ft 6 in, Lower Pentre at 610 ft 4 in, Lower Six-Feet disturbed to 1240 ft 7 in; rest of section disturbed. Sunk to 1535 ft.

Bryncethin (Barrow) Colliery: No. 3 Pit

Ht. above O.D. 261 ft.   6-in SS 98 S.W.; Glam. 34 S.E.

Site 450 yd E. 6° N. of St. Theodore's Church, Bryncethin. National Grid Ref. 91808429. Published in 2nd Edition, p. 48. Strata dipping north at 35°. Correlation uncertain.

	Thickness ft in	Depth ft in
DRIFT		
Yellow and blue clay and gravel 18 6		18 6
Cliff with beds of rock .. 91 3		109 9
Coal .. 18 in	1 6	111 3
Fireclay and cliff 29 9		141 0
Coal .. 9 in	9	141 9
Fireclay and cliff 11 3		153 0
Coal .. 9 in	9	153 9

	Thickness ft in	Depth ft in
Fireclay 3 6		157 3
Coal .. 15 in	1 3	158 6
Fireclay and cliff with mine 27 8		186 2
Coal .. 60 in	5 0	191 2
Fireclay 23 0		214 2
Bastard rock 55 5		269 7
Shale, fireclay and cliff .. 44 5		314 0
Coal .. 14 in	1 2	315 2

	Thick-ness ft in	Depth ft in		Thick-ness ft in	Depth ft in
Cliff with bands of rock and mine 63 10		379 0	Coal .. 59 in		
Coal 18 ft 2 in **18 2**		**397 2**	Shale .. 16 in		
Shale and cliff with mine ..127 10		525 0	Coal .. 24 in		
Coal 17 ft 9 in **17 9**		**542 9**	Clay .. 2 in		
Cliff, shale and bastard rock 40 9		583 6	Coal .. 72 in **18 1**		**778 11**
Cannal coal 12 in **1 0**		**584 6**	Shale 57 9		836 8
Fireclay and shale .. 17 8		602 2	Coal .. 30 in		
Coal .. 12 in **1 0**		**603 2**	Fireclay .. 12 in		
Shale 7 7		610 9	**Coal** .. 6 in **4 0**		**840 8**
Coal .. 12 in **1 0**		**611 9**	Fireclay, rock and shale .. 44 2		884 10
Rock and shale 35 5		647 2	Bastard rock 30 8		915 6
Coal 8 ft 10 in **8 10**		**656 0**	Shale and bastard rock with mine 134 11		1050 5
Fireclay and shale .. 33 7		689 7	**Coal** .. 36 in **3 0**		**1053 5**
Coal .. 19 in **1 7**		**691 2**	Fireclay and shale ..111 4		1164 9
Fireclay, shale and bastard rock 69 8		760 10	Rock and bastard rock .. 15 6		1180 3
Coal .. 16 in			Shale and ironstone .. 76 0		1256 3
Fireclay .. 28 in			**Coal** .. 5 in **5**		**1256 8**
			Rock and bastard rock with shale layers .. 43 0		1299 8

Bryn Colliery

Ht. above O.D. 468 ft.   6-in SS 89 S.W.; Glam. 25 N.E.

Site 180 yd N. 44° E. of Bryn church. Published in 2nd Edition, p. 134.

Two-Feet-Nine at 113 ft, Four-Feet at 149 ft 11 in, Six-Feet at 315 ft 1 in. Sunk through disturbed ground including Jubilee Slide to Lower Seven-Feet at 1001 ft 10 in, Upper Five-Feet at 1026 ft 11 in, ?Lower Five-Feet at 1141 ft 10 in. Sunk to 1144 ft 11 in.

Bryndu Colliery

6-in SS 88 S.W.; Glam. 33 S.E.   1 mile N.N.E. of Pyle church.

Section from Garw Seam to neighbourhood of Six-Feet in North and South Drifts published in 2nd Edition, pp. 54–5. Sequence broadly similar to that of Newlands Colliery.

Bute Colliery: Cwmsaerbren Pit

Ht. above O.D. 660 ft.   6-in SS 99 N.W.; Glam. 17 N.E.

Site 490 yd W. 22° N. of Treherbert railway station. National Grid Ref. 93509842.

	Thick-ness ft in	Depth ft in		Thick-ness ft in	Depth ft in
DRIFT			Fireclay, shale and iron-stone 73 6		143 10
Clay and gravel 31 0		31 0	**Coal** .. 5 in **5**		**144 3**
			Fireclay, shale and sand-stone 48 5		192 8
COAL MEASURES			**Coal** .. 12 in **1 0**		**193 8**
Sandstone and shale .. 24 8		55 8	Fireclay, shale and sand-stone 30 0		223 8
GORLLWYN RIDER			**Coal** .. 10 in **10**		**224 6**
Coal .. 19 in **1 7**		**57 3**	Fireclay, shale and iron-stone 30 0		254 6
Fireclay, shale and sand-stone 11 3		68 6	**Coal** .. 10 in **10**		**255 4**
GORLLWYN			Fireclay and shale .. 11 2		266 6
Coal .. 22 in **1 10**		**70 4**			

	Thickness ft in	Depth ft in
THREE COALS		
Coal .. 4 in		
Shale .. 3 in		
Coal .. 3 in		
Shale .. 30 in		
Coal .. 12 in		
Shale .. 2 in		
Coal .. 5 in		
Fireclay and shale .. 20 in		
Coal .. 4 in	6 11	273 5
Fireclay, shale and ironstone	22 8	296 1
Clod and Coal .. 13 in	1 1	297 2
TWO-FEET-NINE		
Coal .. 44 in	3 8	300 10
Clod, fireclay, shale and ironstone, bands of sandstone	23 1	323 11
Coal .. 23 in	1 11	325 10
Fireclay, sandstone and shale	8 7	334 5
Coal .. 12 in		
Clod .. 13 in		
Coal .. 12 in	3 1	337 6
Shale and ironstone ..	18 11	356 5

	Thickness ft in	Depth ft in
FOUR-FEET (TOP COAL)		
Coal .. 13 in		
Clod .. 5 in		
Coal .. 12 in	2 6	358 11
Fireclay and shale ..	4 11	363 10
FOUR-FEET (BOTTOM COAL)		
Coal .. 64 in	5 4	369 2
Fireclay and shale with sandstone bands ..	25 10	395 0
UPPER SIX-FEET		
Coal .. 30 in	2 6	397 6
Fireclay, sandstone and shale	11 5	408 11
Coal .. 1 in	1	409 0
Fireclay and shale ..	18 2	427 2
LOWER SIX-FEET		
Coal .. 48 in		
Clod .. 17 in		
Coal .. 8 in	6 1	433 3
Clod, clift and stone, disturbed	58 4	491 7
UPPER NINE-FEET		
Coal .. 50 in		
Bast .. 13 in		
Coal .. 42 in	8 9	500 4

Bute Colliery: Lady Margaret Pit

Ht. above O.D. 628 ft.   6-in SS 99 N.W.; Glam. 18 S.W.

Site 285 yd S. 33° E. Treherbert railway station. National Grid Ref. 94009794.

	Thickness ft in	Depth ft in
Walling	5 5	5 5
DRIFT		
Clay and stones	21 9	27 2
COAL MEASURES		
Clift and fireclay	12 5	39 7
Coal and fireclay 15 in	1 3	40 10
Fireclay, clift and rock ..	45 3	86 1
GORLLWYN RIDER		
Coal .. 22 in	1 10	87 11
Rock and clift	17 4	105 3
GORLLWYN		
Coal .. 9 in		
Shale .. 1 in		
Coal .. 9 in		
Shale .. 1 in		
Coal .. 10 in	2 6	107 9
Fireclay, clift and mine ..	72 1	179 10
Coal .. 5 in	5	180 3
Clift and fireclay	22 3	202 6
Blue and white rock ..	27 0	229 6
Coal .. 9 in	9	230 3

	Thickness ft in	Depth ft in
Fireclay, rock and clift ..	28 4	258 7
Coal .. 13 in	1 1	259 8
Fireclay, rock and clift ..	34 2	293 10
Coal .. 12 in	1 0	294 10
Fireclay with ironstone ..	9 8	304 6
Coal .. 9 in		
Shale .. 15 in		
Coal .. 9 in	2 9	307 3
Fireclay, rock and clift ..	16 11	324 2
TWO-FEET-NINE		
Coal .. 38 in	3 2	327 4
Fireclay, bastard rock and clift	23 5	350 9
Coal .. 22 in		
Clod .. 12 in		
Coal .. 9 in	3 7	354 4
Fireclay and clift	11 3	365 7
Coal .. 11 in	11	366 6
Fireclay, clift and bastard rock	35 7	402 1
FOUR-FEET		
Rashes .. 29 in		
Coal .. 58 in	7 3	409 4
Fireclay, clift and bastard rock	36 0	445 4

	Thick-ness ft in	Depth ft in
UPPER SIX-FEET		
Coal .. 59 in	4 11	450 3
Clift 17 4		467 7
Coal .. 3 in	3	467 10
Clift with mine 20 3		488 1
LOWER SIX-FEET		
Coal .. 54 in		
Shale .. 41 in		
Coal .. 11 in	8 10	496 11
Fireclay and clift 16 6		513 5
RED VEIN		
Coal .. 3 in		
Fireclay .. 33 in		
Coal .. 36 in	6 0	519 5
Fireclay, clift and rock sands 22 1		541 6
Coal .. 9 in	9	542 3
Fireclay, clift and rock bands 20 8		562 11
Coal .. 3 in	3	563 2

	Thick-ness ft in	Depth ft in
Fireclay, clift and rock bands 18 11		582 1
LOWER NINE-FEET		
Coal .. 2 in		
Fireclay .. 17 in		
Coal .. 27 in		
Clod .. 3 in		
Coal .. 5 in	4 6	586 7
Fireclay 5 11		592 6
Coal .. 3 in		
Fireclay .. 7 in		
Coal .. 6 in	1 4	593 10
Fireclay, rock and clift .. 28 7		622 5
?BUTE		
Coal .. 22 in		
Thin parting		
Coal .. 22 in	3 8	626 1
Fireclay 9 0		635 1

Caerau Colliery: No. 1 (North) Pit

Ht. above O.D. 769 ft.   6-in SS 89 S.E.; Glam. 26 N.W.

Site 1220 yd E. 13° N. Caerau railway station. National Grid Ref. 86599455. Published in Vertical Sections, Sheet 85, 1901.

Upper Blackband Ironstone at 148 ft 9 in, Victoria at 306 ft 10 in, Upper Yard at 387 ft 10 in, Two-and-a-Half at 423 ft 10 in, Caedavid at 481 ft, Two-Feet-Nine at 813 ft 7 in, Four-Feet at 898 ft 7 in. Sunk through small fault to Upper Six-Feet at 971 ft 1 in, Lower Six-Feet at 1020 ft 6 in, Caerau Vein at 1051 ft 6 in. Sunk to 1066 ft 6 in.

Caerau Colliery: South Pit

Ht. above O.D. 769 ft.   6-in SS 89 S.E.; Glam. 26 N.W.

Site 1220 yd E. 10° N. Caerau railway station. National Grid Ref. 86609448.

Victoria at 235 ft 2 in, Upper Yard at 328 ft 8 in, Two-and-a-Half at 360 ft 4 in, Caedavid (bottom coal) at 419 ft 9 in. Sunk through small fault to Two-Feet-Nine at 753 ft, Four-Feet at 823 ft 9 in, Upper Six-Feet at 923 ft 9 in, Lower Six-Feet at 983 ft 11 in, Caerau Vein at 1015 ft 2 in. Sunk to 1052 ft 7 in.

Caerau Colliery: No. 3 Pit

Ht. above O.D. 769 ft.   6-in SS 89 S.E.; Glam. 26 N.W.

Site 1300 yd E. 13° N. Caerau railway station. National Grid Ref. 86669456. Sunk 1889.

Clay at 56 ft 7 in, Victoria at 322 ft 10 in, Two-and-a-Half at 436 ft 11 in, Caedavid (bottom coal) at 499 ft 9 in. Sunk to 517 ft.

Cambrian Colliery: No. 2 Pit

Ht. above O.D. c. 890 ft.   6-in SS 99 S.E.; Glam. 27 N.W.

Site 1350 yd W. 13° S. of Blaen Clydach church. National Grid Ref. 97049272.

No. 3 Rhondda at 167 ft 1 in, Pentre at 621 ft 9 in, Lower Six-Feet at 1159 ft 1 in, Upper Nine-Feet at 1247 ft 3 in, Lower Nine-Feet and Bute at 1314 ft 9 in.

Cambrian Colliery: No. 3 Pit

Ht. above O.D. 887 ft.   6-in SS 99 S.E.; Glam. 27 N.W.

Site 1450 yd W. 15° S. of Blaen Clydach church. National Grid Ref. 96939265. Published in Vertical Sections, Sheet 84, 1901.

Pentre at 650 ft 10 in, Two-Feet-Nine at 1021 ft 5 in. Lower Six-Feet at 1198 ft 2 in, Caerau at 1205 ft 8 in, Upper Nine-Feet at 1288 ft 10 in. Lower Nine-Feet and Bute at 1358 ft 4 in, Yard at 1431 ft 4 in, Middle Seven-Feet at 1465 ft 5 in, Five-Feet at 1542 ft 10 in, Gellideg at 1577 ft 1 in. Sunk to 1665 ft 7 in.

Cambrian Colliery: No. 4 Pit

Ht. above O.D. c. 890 ft.   6-in SS 99 S.E.; Glam. 27 N.W.

Site 1370 yd W. 14° S. of Blaen Clydach church. National Grid Ref. 97039270.

Pentre at 614 ft 4 in, Lower Pentre at 651 ft 5 in, Two-Feet-Nine at 992 ft 1 in, Lower Six-Feet and Caerau at 1178 ft 9 in, Upper Nine-Feet at 1228 ft 8 in, Lower Nine-Feet and Bute at 1292 ft 5 in, Yard at 1369 ft 7 in.

Cardiff Navigation Colliery: Downcast Shaft

Ht. above O.D. 159 ft.   6-in ST 08 S.W.; Glam. 36 S.W.

Site 400 yd W. 30° S. of Mwyndy Junction. National Grid Ref. 03128208.

		Thickness ft	in	Depth ft	in
DRIFT					
Gravel and sand		64	8	64	8
COAL MEASURES					
Black shale and grey metal		17	0	81	8
White and grey sandstone		51	7	133	3
Coal ..	20 in	1	8	134	11
Blue metal..		16	7	151	6
Coal ..	13 in	1	1	152	7
Fireclay sandstone and grey and blue metal ..		102	1	254	8
THICK SEAM[1]					
Coal with three bands..	96 in				
Fireclay and blackband	42 in				
Coal ..	34 in	14	4	269	0
Sandstone and blue metal		21	10	290	10
GORLLWYN					
Coal ..	26 in				
Fireclay..	36 in				
Coal ..	17 in	6	7	297	5
Sandstone and blue metal		59	6	356	11
Grey sandstone and black and white rock with metal bands		77	2	434	1
Blue metal..		11	0	445	1
TWO-FEET-NINE					
Coal ..	66 in				
Fireclay ..	42 in				
Coal ..	34 in	11	10	456	11
Hard fireclay		15	0	471	11
Coal ..	23 in	1	11	473	10
Clift and rock		34	2	508	0
Faulty ground, fireclay and clift		53	0	561	0
Coal ..	9 in		9	561	9
Fireclay, clift and mine ..		48	10	610	7
Coal ..	62 in	5	2	615	9
Fireclay, bastard rock and clift		14	11	630	8
Coal ..	19 in	1	7	632	3
Shale, fireclay and clift ..		6	0	638	3
Coal ..	36 in	3	0	641	3
Fireclay and clift		42	0	683	3
Roll of blackband, coal and mixed fireclay ..		13	6	696	9
Coal ..	37 in	3	1	699	10
Clift, rock		194	8	894	6

Cilely Colliery: No. 1 Pit

Ht. above O.D. 561 ft.   6-in ST 08 N.W.; Glam. 27 S.E.

Site 2000 yd W. 40° S. of Cymmer church. National Grid Ref. 01158960. Section unreliable.

Drift to 24 ft, Hafod at 105 ft, Lower Four-Feet at 714 ft, Six-Feet and Caerau at 794 ft 10 in, Upper Nine-Feet at 879 ft, Yard at 1206 ft 11 in. Sunk to 1231 ft 6 in.

---

[1] See Llanháran Colliery, pp. 334–5.

Clyne Pit

Ht. above O.D. c. 300 ft.   6-in SN 80 S.W.; Glam. 16 N.E.

Site 820 yd E. 43° S. of Hermon Chapel, Clyne. National Grid Ref. 80720025.

	Thickness ft in	Depth ft in		Thickness ft in	Depth ft in
Walling 15 0		15 0	Clift 2 1		177 1
			Coal .. 9 in **9**		**177 10**
COAL MEASURES			Clift and fireclay 6 1		183 11
Rock with thin beds of			Coal .. 9 in **9**		**184 8**
clift 113 9		128 9	Shale, fireclay and clift		
			mixed with rock .. 22 1		206 9
?BRITHDIR			Rock, with thin beds of		
Coal .. 10 in **10**		**129 7**	clift and discontinuous		
Fireclay, rock and clift .. 9 11		138 6	streaks of coal .. 418 5		625 2
Coal .. 1 in **1**		**138 7**	NO. 1 RHONDDA		
Fireclay, clift and rock .. 12 0		150 7	Coal .. 18 in		
White rock 24 5		175 0	Clod .. 15 in		
			Coal .. 9 in **3 6**		**628 8**

Coedely Colliery: No. 1 Pit

Ht. above O.D. 345 ft.   6-in ST 08 N.W.; Glam. 35 N.E.

Site 2330 yd S. 8° E. of St. David's Church, Tonyrefail. National Grid Ref. 01568613. Published in 2nd Edition, pp. 134–35. Sunk 1913.

Drift to 45 ft 4 in, No. 1 Rhondda at 561 ft 7 in, No. 2 Rhondda at 879 ft, No. 3 Rhondda at 1133 ft 9 in, Hafod at 1282 ft, Abergorky at 1363 ft 3 in, Pentre Rider and Pentre at 1449 ft 5 in, Two-Feet-Nine at 1598 ft 2 in, Upper Four-Feet at 1641 ft 4 in, Lower Four-Feet at 1668 ft, Six-Feet at 1736 ft 9 in, Caerau at 1745 ft 11 in, Upper Nine-Feet at 1763 ft 4 in, Lower Nine-Feet and Bute at 1884 ft, Yard at 1952 ft 7 in, Five-Feet at 2029 ft 8 in. Sunk to 2101 ft 10 in.

Coedely Colliery: No. 2 Pit

Ht. above O.D. 345 ft.   6-in ST 08 N.W.; Glam. 35 N.E.

Site 2370 yd S. 8° E. of St. David's Church, Tonyrefail. National Grid Ref. 01568609. Published in 2nd Edition, p. 136.

Drift to 53 ft, No. 2 Rhondda at 859 ft 2 in, No. 3 Rhondda at 1113 ft 5 in, Hafod at 1262 ft 10 in, Pentre Rider, Pentre and Lower Pentre at 1440 ft 4 in, Two-Feet-Nine at 1591 ft 7 in, Six-Feet and Caerau at 1728 ft 9 in, Lower Nine-Feet and Bute at 1879 ft 10 in, Yard at 1916 ft 6 in, Five-Feet at 2000 ft 3 in, Garw at 2128 ft 3 in. Sunk to 2288 ft 8 in.

Coegnant Colliery: North Pit

Ht. above O.D. 607 ft.   6-in SS 89 S.E.; Glam. 26 N.W.

Site 1126 yd S. 4° E. Caerau railway station. National Grid Ref. 85549329. Published in Vertical Sections, Sheet 85.

Two-Feet-Nine at 255 ft 4 in, Four-Feet at 324 ft 4 in, Six-Feet at 429 ft 8 in, Caerau at 451 ft 8 in. Initially sunk to 534 ft 5 in.

Coegnant Colliery: South Pit

Ht. above O.D. 607 ft.   6-in SS 89 S.E.; Glam. 26 N.W.

Site 1155 yd S. 4° E. Caerau railway station. National Grid Ref. 85559326.

Lower Six-Feet at 429 ft 8 in, Caerau at 456 ft 9 in, Upper Nine-Feet at 601 ft 2 in, Lower Nine-Feet at 688 ft 5 in, Bute at 733 ft 10 in, Amman Rider at 778 ft 10 in, disturbed ground to Yard at 919 ft 10 in, Upper Seven-Feet at 934 ft 10 in, Middle Seven-Feet at 1000 ft 10 in, Lower Seven-Feet at 1045 ft 10 in, Upper Five-Feet at 1069 ft 10 in. Sunk to 1121 ft 8 in.

Cwm Colliery: Margaret Shaft

Ht. above O.D. 436 ft.   6-in ST 06 N.E.; Glam. 36 N.W.

Site 1470 yd N. 34° E. of Beddau cross-roads. National Grid Ref. 06618639. Published in 2nd Edition, pp. 138–140. Sunk 1911–14.

Daren-ddu at 84 ft 5 in, Cefn Glas at 309 ft 7 in, No. 2 Rhondda at 1508 ft 1 in, No. 3 Rhondda at 1699 ft 4 in, Hafod at 1841 ft 10 in, Abergorky at 1915 ft 10 in, Pentre at 1975 ft 8 in, Upper Four-Feet at 2207 ft 2 in, Lower Four-Feet at 2230 ft 5 in, Six-Feet at 2286 ft 3 in, Upper Nine-Feet at 2356 ft 11 in, ?Lower Nine-Feet at 2362 ft 11 in, ?Bute at 2371 ft 5 in. Sunk to 2384 ft 9 in.

Cwm Colliery: Mildred Shaft

Ht. above O.D. 436 ft.   6-in ST 06 N.E.; Glam. 36 N.W.

Site 1450 yd N. 32° E. of Beddau cross-roads. National Grid Ref. 06578639. Sunk 1911–14.

No. 2 Rhondda at 1501 ft 8 in, No. 3 Rhondda at 1691 ft 7 in, Six-Feet at 2297 ft 9 in, Caerau at 2307 ft 2 in. Sunk to 2315 ft 9 in.

Cwmaman Colliery: Sheppard's Pit

Ht. above O.D. 676 ft.   6-in SS 99 N.W.; Glam. 18 N.E.

Site 950 yd W. 9° N. of Cwmaman church. National Grid Ref. 99379948.

	Thickness ft in	Depth ft in		Thickness ft in	Depth ft in
DRIFT			PENTRE		
Walling 75 0		75 0	Coal .. 34 in 2 10		362 8
			Fireclay and mine ground 10 0		372 8
COAL MEASURES			LOWER PENTRE		
Mine ground 17 0		92 0	Coal .. 23 in 1 11		374 7
			Fireclay, bastard rock and		
CLAY			mine ground 41 5		416 0
Coal .. 24 in 2 0		94 0	Coal .. 12 in 1 0		417 0
Fireclay, rock and clift .. 57 8		151 8	Rock, clift and mine		
Big rock 49 0		200 8	ground 45 5		462 5
Mine ground 27 0		227 8	GORLLWYN RIDER		
Coal .. 3 in 3		227 11	Coal .. 18 in 1 6		463 11
Strong fireclay and clift .. 18 4		246 3	Fireclay and mine ground 15 3		479 2
GRAIG			GORLLWYN		
Coal .. 24 in 2 0		248 3	Coal .. 24 in 2 0		481 2
Fireclay and mine ground 15 0		263 3	Fireclay and mine ground 41 6		522 8
Coal .. 16 in 1 4		264 7	Coal .. 3 in 3		522 11
Fireclay, mine ground and			Strong fireclay and clift .. 36 0		558 11
clift 48 0		312 7	Coal .. 5 in 5		559 4
PENTRE RIDER			Clift, fireclay, bastard rock		
Coal .. 9 in 9		313 4	and mine ground .. 38 0		597 4
Fireclay and clift 13 6		326 10	Coal .. 14 in 1 2		598 6
Rock 33 0		359 10			

	Thickness ft in		Depth ft in	
Fireclay and clift with beds of rock	15	8	614	2
Coal .. 6 in		6	614	8
Fireclay	6	6	621	2
THREE COALS				
Coal .. 9 in				
Clod .. 5 in				
Coal .. 9 in				
Rock .. 6 in				
Coal .. 10 in	3	3	624	5
Fireclay, clift and rock ..	36	0	660	5
TWO-FEET-NINE				
Coal .. 33 in	2	9	663	2
Fireclay and mine ground	25	8	688	10
FOUR-FEET				
Coal .. 70 in				
Rashings 23 in	7	9	696	7
Strong shales and rock ..	18	2	714	9
UPPER SIX-FEET				
Coal .. 24 in	2	0	716	9
Fireclay, shales and rock..	60	2	776	11
LOWER SIX-FEET				
Coal .. 40 in	3	4	780	3
Clift, fireclay and rock ..	25	6	805	9
RED VEIN				
Coal .. 17 in				
Fireclay .. 24 in				
Coal .. 18 in				
Clod .. 6 in				
Coal .. 18 in	6	11	812	8
Clift and rock	47	0	859	8
NINE-FEET				
Coal .. 27 in	2	3	861	11
Fireclay, clift, rock, black shale and mine ground..	48	9	910	8

	Thickness ft in		Depth ft in	
BUTE				
Coal .. 15 in				
Clod .. 11 in				
Coal .. 24 in	4	2	914	10
Fireclay and rashes ..	6	8	921	6
Coal .. 1½ in		1½	921	7½
Bastard clift with balls of mine, hard rock and clift with mine	25	8¼	947	4
AMMAN RIDER				
Coal .. 24 in	2	0	949	4
Rashes and hard clay ..	4	0	953	4
Coal .. 6 in		6	953	10
Hard clay and clift ..	17	9	971	7
YARD				
Coal .. 33 in				
Fireclay .. 69 in				
Coal .. 20 in	10	2	981	9
Clift, shale and clod ..	11	5	993	2
SEVEN-FEET				
Coal .. 33 in				
Soft fireclay 24 in				
Coal .. 13 in	5	10	999	0
Soft and hard clift ..	23	8	1022	8
Coal .. 8 in		8	1023	4
Clift and bastard rock ..	15	0	1038	4
FIVE-FEET AND GELLIDEG				
Coal .. 30 in				
Clift .. 54 in				
Coal .. 22 in				
Clod .. 1 in				
Coal .. 7 in				
Clod .. 4 in				
Coal .. 24 in	11	10	1050	2
Fireclay and clift	20	6	1070	8
Coal .. 4 in		4	1071	0
Rock and clift	43	2	1114	2
Disturbed ground ..	17	0	1131	2

Cwm Cynon Colliery: No. 1 Pit

Ht. above O.D. 355 ft.   6-in ST 09 N.E.; Glam. 19 S.W.

Site 230 yd N. 30° E. of Penrhiwceiber church. National Grid Ref. 05949792.

	Thickness ft in		Depth ft in	
DRIFT				
Gravel, sand and clay ..	87	0	87	0
COAL MEASURES				
Coal .. 6 in		6	87	6
Fireclay, clift and rock ..	43	3	130	9
Coal .. 19 in				
Rashings 16 in				
Coal .. 18 in	4	5	135	2
Fireclay and clift	10	0	145	2

	Thickness ft in		Depth ft in	
White rock	13	0	158	2
Clift	34	8	192	10
Rock	35	0	227	10
NO. 1 RHONDDA RIDER (NO. 3 RHONDDA)				
Coal .. 26 in				
Black rashings 12 in	3	2	231	0
Fireclay, rock and clift ..	76	0	307	0
Coal .. 2 in		2	307	2
Clift	4	4	311	6

	Thickness ft in	Depth ft in
NO. 1 RHONDDA		
Coal .. 18 in	1 6	313 0
Bastard fireclay	2 3	315 3
Rock	10 7	325 10
Clift, fireclay and rock beds 18	1	343 11
Rock with clift beds .. 77	3	421 2
Clift and bastard rock .. 28	7	449 9
Rock with coal scar ..110	4	560 1
Coal .. 14 in	1 2	561 3
Fireclay, clift and rock .. 33	5	594 8
?NO. 2 RHONDDA		
Coal .. 32 in		
Black rock		
and		
rashings 38 in		
Coal .. 5 in	6 3	600 11
Bastard fireclay and clift.. 15	6	616 5
White and black rock .. 20	3	636 8
Clift	4 0	640 8
Coal .. 11 in	11	641 7
Fireclay, clift and rock .. 26	4	667 11
Coal .. 22 in		
Rashings 3 in	2 1	670 0
Clift, rock and mine ground 91	10	761 10
Coal .. 2 in	2	762 0
Clift	2 1	764 1
Coal .. 8 in	8	764 9
Fireclay, rock and clift .. 10	5	775 2
Coal .. 2 in	2	775 4
Fireclay and clift 21	7	796 11
Coal .. 12 in	1 0	797 11
Fireclay, clift and bastard		
rock 57	9	855 8
CLAY		
Coal .. 27 in	2 3	857 11
Fireclay, rock and clift ..	7 1	865 0
Rock	10 6	875 6
Black shale, clift and mine 54	11	930 5
Coal .. 3 in	3	930 8
Rock and clift	14 6	945 2
GRAIG		
Coal .. 21 in	1 9	946 11
Fireclay, rock and clift .. 57	9	1004 8
PENTRE RIDER		
Coal .. 19 in		
Rashings 7 in	2 2	1006 10
Fireclay, rock and clift .. 18	4	1025 2
PENTRE		
Coal .. 11 in	11	1026 1
Fireclay, clift and mine .. 27	8	1053 9

	Thickness ft in	Depth ft in
LOWER PENTRE		
Coal .. 11 in		
Stone .. 2 in		
Coal .. 6 in	1 7	1055 4
Fireclay, rock and clift .. 25	5	1080 9
EIGHTEEN-INCH		
Coal .. 12 in	1 0	1081 9
Fireclay and clift 33	7	1115 4
Coal .. 15 in		
Fireclay .. 45 in		
Coal .. 7 in		
Rashings 6 in	6 1	1121 5
Fireclay, clift and mine .. 77	2	1198 7
Coal .. 3 in	3	1198 10
Fireclay, clift, mine ground 40	8	1239 6
Coal .. 1 in	1	1239 7
Clift, fireclay and beds of		
rock 34	6	1274 1
THREE COALS		
Coal .. 20 in		
Clod .. 18 in		
Coal .. 4 in		
Rashings 10 in		
Coal .. 3 in		
Bastard		
fireclay 30 in		
Coal .. 12 in	8 1	1282 2
Clod	3 5	1285 7
TWO-FEET-NINE		
Coal .. 32 in	2 8	1288 3
Fireclay, clift and rock .. 38	5	1326 8
FOUR-FEET		
Coal .. 66 in		
Rashings 4 in		
Coal .. 20 in	7 6	1334 2
Clift with rock and mine .. 30	0	1364 2
SIX-FEET		
Coal .. 124 in	10 4	1374 6
Clod, fireclay and clift .. 63	1	1437 7
Coal .. 6 in	6	1438 1
Fireclay, rock and clift .. 45	0	1483 1
Coal .. 2 in	2	1483 3
Bastard fireclay and clift .. 12	4	1495 7
NINE-FEET		
Coal .. 50 in		
Coal .. 56 in		
Shale and		
bast 3 in		
Coal .. 58 in	13 11	1509 6
Fireclay and bastard rock 13	6	1523 0

Cwmgwyneu Colliery: No. 1 Pit

Ht. above O.D. 264 ft.   6-in SS 89 S.W.; Glam. 25 S.E.

Site 2415 yd S. 43° W. of Bryn church. National Grid Ref. 80279049.

Field Vein at 480 ft, Wernddu at 604 ft.

Cwmneol Colliery: South Pit

Ht. above O.D. 513 ft.   6-in ST 09 N.W.; Glam. 18 N.E.

Site 700 yd N.E. of Cwmaman church. National Grid Ref. 00729978.

Clay at 177 ft 2 in, Graig at 269 ft 3 in, Two-Feet-Nine at 675 ft 7 in, Four-Feet at 714 ft 4 in, Upper Six-Feet at 733 ft 10 in, Lower Six-Feet at 778 ft 7 in, Red Vein at 804 ft 7 in, Nine-Feet at 867 ft, Bute at 903 ft 4 in, Yard at 974 ft 1 in, Seven-Feet at 993 ft 3 in. Sunk to 1041 ft 3 in.

Cwmneol Colliery: North Pit

Ht. above O.D. about 513 ft.   6-in ST 09 N.W.; Glam. 18 N.E.

Site 700 yd N. 43° E. of Cwmaman church. National Grid Ref. 00699981.

	Thickness ft	in	Depth ft	in
DRIFT				
Walling	47	0	47	0
COAL MEASURES				
Fine ground	10	4	57	4
Coal .. 11 in		11	58	3
Fireclay, rock and mine ground	52	5	110	8
Coal .. 7 in		7	111	3
Fireclay and mine ground	52	1	163	4
CLAY				
Coal .. 24 in	2	0	165	4
Fireclay, clift, rock, etc. ..	100	8	266	0
CRAIG				
Coal .. 20 in	1	8	267	8
Fireclay, rock, clift and mine ground	75	4	343	0
Coal .. 11 in		11	343	11
Fireclay and hard clift ..	14	8	358	7
Coal and clod .. 11 in		11	359	6
Rock	5	8	365	2
Coal .. 3 in		3	365	5
Fireclay, clift and rock ..	22	0	387	5
Coal and clod .. 17 in	1	5	388	10
Fireclay and mine ground	13	8	402	6
Coal and clod .. 19 in	1	7	404	1
Fireclay and clift	7	10	411	11
Coal and clod .. 6 in		6	412	5
Fireclay, clift, rock and mine ground	16	3	428	8
Coal .. 13 in	1	1	429	9
Fireclay, clift, rock and mine ground	32	4	462	1
Coal .. 12 in	1	0	463	1
Fireclay, clift and mine ground	60	0	523	1
Coal .. 4 in		4	523	5
Fireclay, clift, rock and mine ground	46	2	569	7
Coal .. 4 in		4	569	11
Fireclay and clift with rock bands	15	1	585	0
Coal .. 6 in		6	585	6
Fireclay, clift, rock and mine ground	15	3	600	9
Coal with balls of mine 41 in	3	5	604	2
Clift	4	0	608	2
Coal .. 12 in	1	0	609	2
Fireclay, rock and clift ..	9	11	619	1
THREE COALS				
Coal .. 10 in				
Clod .. 7 in				
Coal .. 8 in				
Clod .. 7 in				
Coal .. 5 in				
Rock .. 9 in				
Coal .. 3 in	4	1	623	2
Fireclay, rock and clift with balls of mine ..	35	5	658	7
TWO-FEET-NINE				
Coal .. 29 in	2	5	661	0
Fireclay, rock and clift ..	12	6	673	6
Coal .. 14 in				
Clod .. 3 in				
Coal .. 4 in	1	9	675	3
Rock and mine ground ..	19	10	695	1
FOUR-FEET				
Coal .. 63 in				
Clod .. 4 in				
Coal .. 11 in	6	6	701	7
Clift, rock and mine ground	17	11	719	6
UPPER SIX-FEET				
Coal .. 19 in	1	7	721	1
Fireclay and clift with mine balls	43	4	764	5
LOWER SIX-FEET				
Coal .. 2½ in				
Clod .. ½ in				
Coal .. 39 in	3	6	767	11
Fireclay and strong clift ..	21	0	788	11
RED VEIN				
Coal .. 16 in				
Rashings 2½ in				
Coal .. 12½ in				
Rashings 5 in				
Coal .. 24 in	5	0	793	11
Fireclay and clift with rock beds and mine	53	0	846	11
NINE-FEET				
Coal .. 66 in	5	6	852	5
Clift and rock	89	6	941	11
YARD AND ?AMMAN RIDER				
Coal and rashings 30 in				
Coal .. 24 in	4	6	946	5
Strong fireclay, clift and mine	37	6	983	11
SEVEN-FEET				
Coal .. 14 in				
Clod .. 24 in				
Coal .. 30 in				
Clod .. 3 in				
Coal .. 14 in	7	1	991	0
Fireclay and clift	9	0	1000	0
Coal .. 7 in		7	1000	7
Fireclay, rock and clift ..	78	0	1078	7

Cymmer Colliery

Ht. above O.D. 321 ft.   6-in SS 09 S.W.; Glam. 27 S.E.

Site 250 yd N. of Cymmer church. National Grid Ref. 02659107. Published in Vertical Sections, Sheet 84, 1901.

No. 2 Rhondda at 37 ft 6 in, No. 3 Rhondda at 245 ft 8 in, Hafod at 413 ft 3 in, Abergorky at 502 ft 3 in, Pentre at 586 ft 7 in, Two-Feet-Nine at 811 ft 11 in, Four-Feet at 898 ft 3 in, Six-Feet and Caerau at 962 ft 6 in, Upper Nine-Feet at 1030 ft 9 in, Lower Nine-Feet at 1101 ft 1 in, Bute at 1109 ft 9 in, Yard at 1208 ft 7 in, Upper and Middle Seven-Feet at 1215 ft 4 in, Five-Feet at 1299 ft 5 in, Gellideg at 1309 ft 9 in. Sunk to 1366 ft 11 in.

Dare Colliery: Ystradfechan Shaft

Ht. above O.D. 619 ft.   6-in SS 99 N.E.; Glam. 18 S.W.

Site 200 yd S. 24° E. of Cwm-parc church. National Grid Ref. 95109590.

	Thickness ft in		Depth ft in	
Walling	16	8	16	8
DRIFT				
Gravel	6	0	22	8
COAL MEASURES				
Shale with ironstone ..	22	6	45	2
Coal .. 5 in		5	45	7
Fireclay, shale and ironstone	13	6	59	1
Coal .. 8 in		8	59	9
Fireclay, shale, ironstone and rock	53	7	113	4
Coal .. 10 in		10	114	2
Fireclay, shale, ironstone and rock	22	10	137	0
Coal .. 6 in		6	137	6
Fireclay	4	10	142	4
Coal .. 18 in	1	6	143	10
Fireclay, shale and ironstone	34	11	178	9
Coal .. 15 in	1	3	180	0
Fireclay, shale and sandstone	34	2	214	2
THREE COALS				
Coal .. 13 in				
Shale .. 45 in				
Coal .. 12 in				
Clod .. 17 in				
Coal .. 13 in				
Shale and sandstone 56 in				

	Thickness ft in		Depth ft in	
Coal .. 16 in	14	4	228	6
Clod and shale	13	0	241	6
TWO-FEET-NINE				
Coal .. 50 in	4	2	245	8
Fireclay and sandstone ..	3	6	249	2
Coal .. 17 in	1	5	250	7
Clay, shale and ironstone	14	3	264	10
FOUR-FEET (TOP COAL)				
Coal .. 8 in				
Clod .. 7 in				
Coal .. 13 in	2	4	267	2
Fireclay, shale and mine ..	30	6	297	8
FOUR-FEET				
Coal .. 17 in				
Fireclay .. 16 in				
Coal .. 26 in				
Clod .. 34 in				
Coal .. 12 in	8	9	306	5
Clod, fireclay, shale and ironstone	26	2	332	7
UPPER SIX-FEET				
Coal .. 28 in	2	4	334	11
Fireclay, shale, ironstone and sandstone	56	4	391	3
LOWER SIX-FEET				
Coal .. 63 in	5	3	396	6
Fireclay	10	6	407	0
Coal .. 24 in	2	0	409	0
Shale	7	0	416	0
Sandstone	15	0	431	0

Deepening of **Downcast Pit** from Lower Six-Feet given as at 401 ft

	Thickness ft in		Depth ft in	
Fireclay	8	0	409	0
Coal .. 11 in		11	409	11
Fireclay and shale ..	5	9	415	8
Rock with strong shale ..	49	9	465	5
Shale, rock and fireclay ..	25	7	491	0
UPPER NINE-FEET				
Coal .. 12 in				

	Thickness ft in		Depth ft in	
Clod .. 24 in				
Coal .. 36 in				
Bast .. 12 in				
Coal .. 24 in	9	0	500	0
Fireclay, rock and shale ..	20	8	520	8
Coal .. 2 in		2	520	10
Fireclay, ironstone and shale	12	8	533	6

	Thickness ft in	Depth ft in		Thickness ft in	Depth ft in
OWER NINE-FEET			Fireclay, shale, rock and		
Coal .. 27 in			mine 32 9		657 5
Clod .. 2 in			YARD AND UPPER SEVEN-FEET		
Coal .. 3 in	2 8	536 2	Coal .. 36 in		
Fireclay and shale.. .. 14 6		550 8	Clod .. 3 in		
Coaly rashes 1 0		551 8	Coal .. 10 in		
Shale, bastard rock and			Clod .. 2 in		
bast 19 10		571 6	Coal .. 10 in	5 1	662 6
			Fireclay and shale .. 24 7		687 1
UTE			Coal .. 6 in	6	687 7
Coal .. 47 in	3 11	575 5	Fireclay and clod .. 4 1		691 8
Fireclay, shale, mine and			[MIDDLE SEVEN-FEET]		
rock 30 6		605 11	Coal .. 24 in	2 0	693 8
			Fireclay 10 0		703 8
MMAN RIDER					
Coal .. 12 in	1 0	606 11	[LOWER SEVEN-FEET]		
Fireclay, shale, mine and			Coal .. 22 in	1 10	705 6
rashes 16 7		623 6	Fireclay, shale and rock .. 11 0		716 6
Coal .. 14 in	1 2	624 8			

Deep Duffryn Colliery

Ht. above O.D. 380 ft.   6-in ST 09 N.W.; Glam. 19 N.W.

Site 270 yd W. 36° N. of Mountain Ash church. National Grid Ref. 04559954. Sunk to 845 ft in 1850; deepened to 1167 ft in 1924.

	Thickness ft in	Depth ft in		Thickness ft in	Depth ft in
RIFT			THREE COALS		
Clay, gravel and sand .. 63 0		63 0	Coal and		
			dirt .. 48 in	4 0	812 2
OAL MEASURES			Fireclay 6 0		818 2
Rock 26 0		89 0			
Coal .. 30 in	2 6	91 6	TWO-FEET-NINE		
Clod, clift and rock .. 77 4		168 10	Coal .. 34 in	2 10	821 0
Coal .. 17 in	1 5	170 3	Mine ground 21 0		842 0
Fireclay 6 0		176 3			
Coal .. 4 in			FOUR-FEET		
Soft clod 6 in			Coal .. 72 in	6 0	848 0
Coal .. 4 in	1 2	177 5	Bastard fireclay, clift with		
Fireclay, clift, mine ground			ironstone, rock with beds		
and rock 105 1		282 6	of clift 61 8		909 8
Coal .. 17 in	1 5	283 11			
Clod, rock and mine			SIX-FEET		
ground 29 0		312 11	Rashings 2 in		
Coal .. 18 in	1 6	314 5	Coal .. 37 in		
Rock and clift 45 0		359 5	Bast .. 14 in		
			Coal .. 10 in		
LAY			Bast .. 10 in		
Coal .. 15 in	1 3	360 8	Coal .. 36 in	9 1	918 9
Clod, mine ground and			Black band, fireclay, rock		
rock 125 0		485 8	and clift with balls of		
			mine 38 6		957 3
RAIG			Coal .. 4 in	4	957 7
Coal .. 18 in	1 6	487 2	Clift and faulty ground .. 42 11		1000 6
Fireclay and mine ground 66 0		553 2			
Coal .. 12 in	1 0	554 2	NINE-FEET		
Fireclay, mine ground and			Coal .. 66 in		
rock 74 0		628 2	Dirt .. 84 in		
Coal .. 18 in	1 6	629 8	Coal .. 68 in		
Fireclay and mine ground 63 0		692 8	Rashings 16 in	19 6	1020 0
Coal .. 10 in	10	693 6	Clift 15 10		1035 10
Fireclay, mine ground and					
rock 114 8		808 2			

		Thickness ft in	Depth ft in			Thickness ft in	Depth ft i
BUTE				**YARD**			
	Rashings 16 in				Coal .. 34 in	2 10	1105
	Coal .. 23 in				Fireclay, clift and rock .. 37 9		1143
	Clift .. 32 in			**SEVEN-FEET**			
	Coal .. 20 in	7 7	1043 5		Coal .. 14 in		
Fireclay, clift and rock .. 28 6			1071 11		Stone .. 1 in		
					Coal .. 25 in		
AMMAN RIDER					Stone .. 2 in		
	Coal .. 18 in				Coal .. 18 in	5 0	1148
	Dirt .. 24 in			Fireclay and clift 10 0			1158
	Coal .. 4 in	3 10	1075 9		Coal .. 9 in	9	1159
Fireclay, clift and rock .. 27 2			1102 11	Fireclay 7 9			1167

Dinas Colliery: Lower Pit

Ht. above O.D. 366 ft.   6-in ST 09 S.W.; Glam. 27 N.E.

Site 800 yd E. 17° S. of Dinas railway station. National Grid Ref. 01289172.

No. 3 Rhondda at 123 ft, Six-Feet and Caerau at 1075 ft. Sunk to 1110 ft.

Dinas Colliery: Middle Pit

Ht. above O.D. 382 ft.   6-in ST 09 S.W.; Glam. 27 N.E.

Site 200 yd E. 35° S. of Dinas railway station. National Grid Ref. 00769183.

No. 3 Rhondda at 249 ft 10 in, Lower Four-Feet at 918 ft 11 in, Six-Feet at 993 ft, Upper Nine-Feet at 1108 ft 4 in. Sunk to 1305 ft.

Dinas Main Colliery: North Pit

Ht. above O.D. 831 ft.   6-in SS 98 N.E.; Glam. 27 S.W.

Site 1050 yd N. 30° W. of Gilfach Goch church. National Grid Ref. 97869037.

Drift to 35 ft 9 in, No. 3 Rhondda at 288 ft, Hafod at 507 ft 5 in, Pentre at 761 ft 10 in, Lower Pentre at 795 ft 11 in, Bute at 1439 ft 2 in, Yard at 1472 ft 3 in, Five-Feet at 1546 ft 8 in. Sunk to 1586 ft 3 in.

Dinas Main Colliery: South Pit

Ht. above O.D. c. 830 ft.   6-in SS 98 N.E.; Glam. 27 S.W.

Site 1000 yd N. 30° W. of Gilfach Goch church. National Grid Ref. 97899033.

No. 3 Rhondda at 279 ft, Pentre at 764 ft 2 in, Bute at 1435 ft, Yard at 1473 ft 5 in, Five-Feet at 1552 ft, ?Garw at 1754 ft 1 in. Sunk to 1796 ft 9 in.

Duffryn Madoc Colliery: Duffryn Pit

Ht. above O.D. c. 580 ft.   6-in SS 89 S.W.; Glam. 26 S.W.

Site 835 yd N. 44° W. of Maesteg railway station. National Grid Ref. 84799204. Published as lower part of Duffryn Pit in Vertical Sections, Sheet 85, 1901.

Two-Feet-Nine at 215 ft, Four-Feet at 278 ft, Six-Feet at 445 ft, Upper Nine-Feet at 554 ft.

Duffryn Madoc Colliery: Gin Pit

Ht. above O.D. c. 520 ft.   6-in SS 89 N.W.; Glam. 26 S.W.

Site 600 yd W. 8° N. of Maesteg railway station. National Grid Ref. 84779158. Section compiled from imperfect graphic log.

Two-Feet-Nine at 100 ft, Four-Feet at 156 ft, Six-Feet at 266 ft, Caerau at 288 ft, Red Vein at 328 ft.

Duffryn Rhondda Colliery: No. 1 Shaft

Ht. above O.D. 508 ft.   6-in SS 89 N.W.; Glam. 17 S.W.

Site 1160 yd S. 37° W. of Abercregan school. National Grid Ref. 84209575. Section to Caedavid published in 2nd Edition, p. 140.

Wernddu at 489 ft, Tormynydd at 919 ft, Clay at 1100 ft, Albert at 1282 ft 6 in, Victoria at 1360 ft 6 in, Two-and-a-Half at 1526 ft 1 in, Caedavid at 1556 ft 9 in.

	Thickness ft in	Depth ft in			Thickness ft in	Depth ft in
Deepening from Caedavid at		1556 9	FOUR-FEET			
Fireclay, bastard rock and clift	111 0	1667 9	Coal ..	10 in		
Coal	3 in 3	1668 0	Clod ..	3 in		
Fireclay and clift ..	28 9	1696 9	Coal ..	21 in		
Coal ..	3 in 3	1697 0	Rashes ..	5 in		
Fireclay and clift ..	55 9	1752 9	Coal ..	24 in		
Coal ..	7 in 7	1753 4	Bast ..	2 in		
Fireclay and clift ..	15 10	1769 2	Coal ..	4 in		
Coal ..	11 in 11	1770 1	Bast ..	4 in		
Fireclay ..	9 7	1779 8	Coal ..	20 in		
Coal ..	14 in 1 2	1780 10	Rashes ..	7 in 8 4	2010 0
Fireclay and clift ..	38 10	1819 8	Fireclay ..	8 10	2018 10
Coal ..	13 in 1 1	1820 9	Coal ..	2 in 2	2019 0
Fireclay and clift ..	67 8	1888 5	Fireclay ..	7 2	2026 2
Rock ..	30 0	1918 5	Coal ..	5 in 5	2026 7
?TWO-FEET-NINE			Fireclay, clift and rock ..	76 11	2103 6
Coal ..	20 in		SIX-FEET			
Rashes ..	3 in 1 11	1920 4	Coal ..	27 in		
Fireclay ..	2 9	1923 1	Parting ..	3 in		
Coal ..	8 in 8	1923 9	Coal ..	72 in 8 6	2112 0
Fireclay, clift and rock ..	77 11	2001 8	Fireclay and clift ..	12 0	2124 0

No. 2 Shaft 45 yd E. of No. 1 Shaft. Deepening from Four-Feet.

	Thickness ft in	Depth below Four-Feet ft in			Thickness ft in	Depth below Four-Feet ft in
FOUR-FEET			SIX-FEET			
Rashings	3 in		Coal ..	21 in		
Coal ..	12 in		Rashings	1 in		
Rashings	3 in		Coal ..	9 in		
Coal ..	12 in		Parting			
Rashings	10 in		Coal ..	42 in		
Coal ..	25 in		Siderite ..	1 in		
Mudstone	1 in		Coal ..	37 in		
Coal ..	24 in		Rashings	7 in		
Fireclay and mudstone ..	12 0	12 0	Coal ..	4 in 10 2	77 6
Coal ..	12 in 1 0	13 0	Seatearth, mudstone and siltstone ..	20 9	98 3
Fireclay, mudstone	54 4	67 4	Sandstone ..	6 1	104 4

	Thickness ft in	Depth below Four-Feet ft in		Thickness ft in	Depth below Four-Feet ft in
Silty mudstone	5 0	109 4	Seatearth, rashings and mudstone	42 9	259
Sandstone	30 3	139 7	?BUTE		
Mudstone and silty mudstone	34 5	174 0	Coal .. 40 in	3 4	262
Coal .. 17 in	1 5	175 5	Rashings, seatearth and mudstone sheared near		
Seatearth and mudstone ..	19 8	195 1	base	50 9	313
Coal and rashings 2 in	2	195 3			
Silty mudstone	16 8	211 11			
?LOWER NINE-FEET			?LOWER NINE-FEET		
Coal .. 44 in			Coal .. 44 in		
Rashes .. 4 in			Rashes .. 2 in		
Coal .. 6 in	4 6	216 5	Coal .. 6 in	4 4	317

Eastern Colliery

Ht. above O.D. 715 ft.   6-in SS 99 S.E.; Glam. 27 N.W.

Site 650 yd S. 7° E. Ton church. National Grid Ref. 97089434.

	Thickness ft in	Depth ft in		Thickness ft in	Depth ft in
COAL MEASURES			GORLLWYN		
NO. 3 RHONDDA			Coal .. 26 in	2 2	606
Coal .. 36 in	3 0	3 0	Fireclay, rashes, clift and		
Fireclay, clift, rock and			rock	..107 6	714
shale	70 3	73 3	Coal .. 6 in	6	714
Coal .. 19 in	1 7	74 10	Fireclay and clift	24 3	738 1
Fireclay, rock and clift ..	25 3	100 1	Coal .. 6 in	6	739
Coal .. 6 in	6	100 7	Fireclay, clift and rock ..	18 10	758
Fireclay, clift and rock ..	79 0	179 7	Coal .. 14 in	1 2	759
Coal .. 9 in	9	180 4	Fireclay, clift and rock ..	47 5	806
Fireclay and clift	32 3	212 7	Coal .. 17 in	1 5	808
			Fireclay and shale ..	10 1	818
HAFOD (BOTTOM COAL)			THREE COALS		
Coal .. 12 in	1 0	213 7	Coal .. 11 in		
Fireclay and clift	96 0	309 7	Fireclay .. 39 in		
Coal .. 9 in	9	310 4	Coal .. 11 in		
Fireclay and clift	48 9	359 1	Fireclay .. 15 in		
			Coal .. 8 in	7 0	825
ABERGORKY			Fireclay, rock and shale ..	37 11	863
Coal .. 18 in			Coal .. 18 in	1 6	864
Stone and rashes .. 14 in			Fireclay and shale ..	4 6	869
Coal .. 17 in	4 1	363 2	TWO-FEET-NINE		
Fireclay, sandstone and			Coal .. 60 in	5 0	874
clift	68 3	431 5	Fireclay	3 4	877
PENTRE RIDER			Coal .. 8 in		
Coal .. 18 in	1 6	432 11	Shale .. 2 in		
Fireclay and clift	19 0	451 11	Coal .. 12 in	1 10	879
Coal .. 29 in	2 5	454 4	Fireclay and shale ..	21 10	901
Fireclay and clift	18 0	472 4	Coal .. 15 in	1 3	902
Coal .. 24 in	2 0	474 4	Shale and fireclay ..	16 10	919
Fireclay, clift and rock ..	53 8	528 0	Coal .. 27 in	2 3	921
Coal .. 29 in	2 5	530 5	Shale, fireclay and rashes..	10 8	932
Fireclay, rock and clift ..	25 6	555 11	FOUR-FEET (BOTTOM COAL)		
Blackband 12 in			Coal .. 42 in		
Coal .. 14 in			Rashes .. 15 in		
Fireclay .. 32 in			Coal .. 18 in	6 3	938
Coal .. 8 in	5 6	561 5	Shale, fireclay and rock ..	13 9	952
Fireclay, rock and clift ..	43 0	604 5			

	Thickness ft in	Depth ft in		Thickness ft in	Depth ft in

FOUR-FEET (BOTTOM COAL)
Repeated:

Coal .. 30 in			
Rashes .. 18 in			
Coal .. 42 in			
Rashes .. 15 in			
Coal .. 27 in	11 0	963 2	
Shale and rock	29 0	992 2	

UPPER SIX-FEET

Coal .. 34 in	2 10	995 0
Fireclay, shale and main ground 46 6	1041 6	

LOWER SIX-FEET

Coal .. 82 in	6 10	1048 4
Shale 2 0	1050 4	
Coal .. 4 in	4	1050 8
Fireclay and mine .. 6 3	1056 11	
Coal .. 17 in	1 5	1058 4
Fireclay and shale .. 43 7	1101 11	
Coal .. 24 in	2 0	1103 11
Clift and shale with mine.. 14 9	1118 8	
Coal .. 5 in	5	1119 1
Clod 2 5	1121 6	

UPPER NINE-FEET

Coal .. 48 in		
Bast .. 17 in		
Coal .. 19 in		
Rashes .. 3 in		
Coal .. 6 in	7 9	1129 3
Fireclay and clift with beds of rock 23 6	1152 9	

LOWER NINE-FEET

Rashes .. 5 in		
Coal .. 5 in	10	1153 7
Fireclay and shale with beds of rock 29 10	1183 5	

BUTE

Rashes and coal .. 36 in		
Coal .. 23 in		
Rashes .. 24 in		
Coal .. 25 in		
Rashes .. 9 in	9 9	1193 2
Fireclay and rock.. .. 11 11	1205 1	

?AMMAN RIDER

Coal .. 23 in	1 11	1207 0
Fireclay, rock and clift .. 27 3	1234 3	
Coal .. 18 in	1 6	1235 9
Fireclay, clift and rock .. 81 3	1317 0	

YARD AND UPPER SEVEN-FEET

Coal .. 42 in		
Clod .. 8 in		
Coal .. 9 in		
Clod .. 5 in		
Coal .. 9 in	6 1	1323 1
Fireclay, rock and clift .. 27 10	1350 11	

MIDDLE SEVEN-FEET

Coal .. 40 in	3 4	1354 3
Fireclay and clift 13 8	1367 11	

LOWER SEVEN-FEET

Coal .. 20 in	1 8	1369 7
Fireclay and clift 19 1	1388 8	
Coal .. 2 in		
Shale .. 11 in		
Coal .. 1 in		
Shale .. 1 in		
Coal .. 5 in	1 8	1390 4
Shale and clift 8 2	1398 6	
Coal .. 1 in		
Shale .. 10 in		
Coal .. 1 in	1 0	1399 6
Fireclay, clift and shale .. 17 5	1416 11	

FIVE-FEET

Coal .. 34 in		
Stone .. 2 in		
Coal .. 6 in		
Stone .. 4 in		
Coal .. 28 in		
Shale .. 2 in		
Coal .. 4 in	6 8	1423 7
Fireclay and rashes .. 6 0	1429 7	

GELLIDEG

Coal .. 6 in		
Rashes .. 7 in		
Coal .. 50 in	5 3	1434 10
Fireclay and clift 10 7	1445 5	
Coal .. 1 in	1	1445 6
Clift and rock 48 11	1494 5	

Ely Pit

Ht. above O.D. 616 ft.   6-in SS 99 S.E.; Glam. 27 S.E.

Site 450 yd S. by W. of Pen-y-graig church. National Grid Ref. 99939100. Published to Upper-Nine-Feet in Vertical Sections, Sheet 84, 1901.

	Thickness ft in	Depth ft in		Thickness ft in	Depth ft in
No. 2 RHONDDA .. at		54 3	LOWER NINE-FEET AND BUTE		
No. 3 RHONDDA .. at		312 4	Coal .. 36 in		
CENTRE .. .. at		689 11	Brass .. 2 in		
LOWER SIX-FEET .. at		1187 9	Coal .. 40 in		
UPPER NINE-FEET .. at		1300 3	Rashes .. 2 in		
Clift and rock 50 0	1350 3	Coal .. 10 in		
			Rashes .. 12 in		

	Thickness ft in	Depth ft in
Coal .. 48 in		
Fireclay .. 24 in		
Coal .. 12 in	15 6	1365 9
Fireclay, clift and rock ..	20 0	1385 9
AMMAN RIDER		
Coal .. 24 in	2 0	1387 9
Fireclay, clift and rock ..	58 7	1446 4
YARD, UPPER AND MIDDLE SEVEN-FEET		
Coal .. 66 in		
Fireclay .. 30 in		

	Thickness ft in	Depth ft in
Coal .. 18 in		
Rashes .. 8 in		
Coal .. 24 in	12 2	1458 6
Rock, fireclay and clift ..	12 8	1471 2
LOWER SEVEN-FEET		
Coal .. 12 in	1 0	1472 2
Fireclay, clift and rock ..	74 6	1546 8
FIVE-FEET		
Coal .. 60 in	5 0	1551 8
Clift and fireclay	12 0	1563 8
GELLIDEG		
Coal .. 66 in	5 6	1569 2

Ferndale Colliery: No. 1 Pit

Ht. above O.D. 698 ft.   6-in ST 09 N.W.; Glam. 18 S.E.

Site 300 yd E. 38° S. of Ferndale church. National Grid Ref. 00239690. Sunk 1861.

Drift to 53 ft 6 in, Two-Feet-Nine at 765 ft 3 in, Four-Feet at 796 ft 5 in.

Ferndale Colliery: No. 2 Pit

Ht. above O.D. 796 ft.   6-in SS 99 N.E.; Glam. 18 S.E.

Site 1200 yd W. 27° N. of Ferndale church. National Grid Ref. 99039763. Sunk 1884.

Four-Feet at 667 ft, Lower Six-Feet at 741 ft, Upper Nine-Feet at 823 ft, Lower Nine-Feet at 850 ft, Bute at 869 ft, Yard at 923 ft, Middle Seven-Feet at 961 ft, Five-Feet at 1008 ft, Gellideg at 1027 ft. Sunk to 1050 ft.

Ferndale Colliery: No. 3 Fan Pit

Ht. above O.D. 755 ft.   6-in SS 99 N.E.; Glam. 18 S.E.

Site 600 yd N. 44° W. of Ferndale church. National Grid Ref. 99679749.

	Thickness ft in	Depth ft in
DRIFT		
Peat and gravel	12 0	12 0
COAL MEASURES		
Blackband and clift with mine	81 8	93 8
Fireclay and rock.. ..	18 8	112 4
TORMYNYDD		
Coal .. 10 in	10	113 2
Fireclay, rock and clift ..	22 2	135 4
Coal .. 3 in	3	135 7
Fireclay and clift	24 0	159 7
Coal .. 2 in	2	159 9
Fireclay and blackband ..	5 6	165 3
Coal .. 1 in	1	165 4
Fireclay and clift with beds of rock	64 6	229 10
HAFOD		
Coal .. 18 in		
Fireclay .. 48 in		
Coal .. 5 in	5 11	235 9
Strong clift	15 0	250 9
Rock and clift	48 7	299 4
Clod and clift with mine ..	51 10	351 2
ABERGORKY		
Coal .. 26 in		
Fireclay .. 24 in		
Coal .. 14 in	5 4	356 6
Fireclay and clift with beds of rock	59 3	415 9
PENTRE RIDER		
Coal .. 5 in	5	416 2
Fireclay and clift mine ..	27 0	443 2

		Thickness ft	Thickness in	Depth ft	Depth in

NTRE

		Thickness	Thickness ft	Thickness in	Depth ft	Depth in
Coal	..	8 in				
Clod	..	1 in				
Coal	..	8 in	1	5	444	7
ireclay and clift with thin beds of rock	..	..	35	0	479	7

WER PENTRE

Coal	..	12 in				
Stone	..	2 in				
Coal	..	10 in	2	0	481	7
ireclay and clift with mine			28	3	509	10

GHTEEN-INCH

Coal	..	10 in		10	510	8
ireclay and clift with beds of rock and mine		..	40	6	551	2

ORLLWYN RIDER

Coal	..	18 in	1	6	552	8
ireclay and clift with beds of rock and mine		..	16	4	569	0

ORLLWYN

Coal	..	10 in				
Clod	..	1 in				
Coal	..	12 in	1	11	570	11
ireclay and clift with beds of rock and mine		..	56	2	627	1
Coal	..	2 in		2	627	3
ireclay with beds of rock and clift	..	..	19	6	646	9
Coal	..	18 in	1	6	648	3
ireclay, rock and clift	..	41	11		690	2
Coal	..	1 in		1	690	3
ireclay and clift	..	..	17	0	707	3
Coal	..	1 in		1	707	4
ireclay and rock..		..	17	0	724	4
Coal	..	12 in	1	0	725	4
ireclay and clift with bedrock	..	..	30	0	755	4

HREE COALS

Coal	..	18 in				
Fireclay	..	36 in				
Coal with brass	..	12 in				
Fireclay	..	36 in				
Coal	..	12 in				
Fireclay	..	28 in				
Coal	..	4 in				
Fireclay	..	20 in				
Coal	..	10 in	14	8	770	0
trong fireclay	..	..	3	0	773	0

WO-FEET-NINE

Coal	..	42 in	3	6	776	6
ireclay with rock		..	4	6	781	0
Coal	..	8 in				
Clod	..	2 in				
Coal	..	6 in	1	4	782	4
ireclay and clift with mine			17	4	799	8

FOUR-FEET

Gob	..	60 in				
Clod	..	7 in				
Coal	..	8 in	6	3	805	11
Fireclay and clift with mine		23	0		828	11

UPPER SIX-FEET

Coal	..	18 in	1	6	830	5
Fireclay, rock and clift	..	44	10		875	3

LOWER SIX-FEET

Gob	..	36 in				
Clod	..	7 in				
Coal	..	8 in	4	3	879	6
Fireclay and clift	..	..	35	2	914	8
Coal	..	12 in				
Fireclay	..	42 in				
Coal	..	3 in	4	9	919	5
Fireclay and clift with mine		22	0		941	5

UPPER NINE-FEET

Bast and coal (disturbed)		21	0		962	5
Fireclay, clift and rock	..	32	6		994	11

LOWER NINE-FEET

Coal	..	30 in	2	6	997	5
Fireclay and clift with beds of rock	..	..	25	6	1022	11

BUTE

Coal	..	16 in				
Clod	..	12 in				
Coal	..	30 in	4	10	1027	9
Fireclay, rock and clift	..	51	8		1079	5

AMMAN RIDER

Coal	..	21 in				
Clod	..	6 in				
Coal	..	9 in	3	0	1082	5
Fireclay	..	..	6	6	1088	11

YARD

Coal	..	36 in				
Clod	..	7 in				
Coal	..	8 in	4	3	1093	2
Fireclay	..	..	2	10	1096	0

UPPER SEVEN-FEET

Coal	..	9 in		9	1096	9
Fireclay and clift	..	..	19	0	1115	9

MIDDLE AND LOWER SEVEN-FEET

Coal	..	32 in				
Fireclay	..	20 in				
Coal	..	10 in	5	2	1120	11
Fireclay and clift with beds of rock and mine		..	69	0	1189	11

FIVE-FEET

Coal	..	60 in	5	0	1194	11
Fireclay	..	..	6	6	1201	5

GELLIDEG

Coal	..	41 in	3	5	1204	10
Fireclay, rock and clift	..	22	6		1227	4

Ferndale Colliery: Pendyrys North
(No. 6) Pit

Ht. above O.D. 625 ft.   6-in ST 09 N.W.; Glam. 18 S.E.

Site 1800 yd S. 43° E. of Ferndale church. National Grid Ref. 01129589. Published in Vertical Sections, Sheet 83, 1900.

Drift to 33 ft 4 in, No. 2 Rhondda at 123 ft, Two-Feet-Nine at 972 ft, Four-Feet at 999 ft, Upper Six-Feet at 1041 ft, Lower Six-Feet at 1086 ft, Upper Nine-Feet at 1163 ft, Lower Nine-Feet at 1209 ft, Bute at 1234 ft 2 in, Yard at 1284 ft, Middle Seven-Feet at 1316 ft 6 in, Five-Feet at 1380 ft, Gellideg at 1389 ft. Sunk to 1392 ft.

Ferndale Colliery: Cynllwyn-du South
(No. 8) Pit

Ht. above O.D. 571 ft.   6-in ST 09 N.W.; Glam. 18 S.E.

Site 2550 yd S. 28° E. of Ferndale church. National Grid Ref. 01069502. Published in Vertical Sections, Sheet 83, 1900.

Drift to 29 ft 8 in, Two-Feet-Nine at 1223 ft 3 in, Four-Feet at 1253 ft 9 in, Upper Six-Feet at 1325 ft 3 in, Lower Six-Feet at 1376 ft 2 in, Upper Nine-Feet (disturbed) at 1525 ft 11 in, Lower Nine-Feet at 1572 ft 3 in, Bute at 1603 ft 5 in, Yard at 1672 ft 7 in, Middle Seven-Feet at 1707 ft 10 in, Five-Feet at 1811 ft 1 in, Gellideg at 1819 ft 2 in.

Ferndale Colliery: Cynllwyn-du North
(No. 9) Pit

Ht. above O.D. 580 ft.   6-in ST 09 N.W.; Glam. 18 S.E.

Site 2500 yd S. 29° E. of Ferndale church. National Grid Ref. 01079507. Sunk 1904–07.

Drift to 39 ft, No. 2 Rhondda at 215 ft 3 in, Two-Feet-Nine at 1232 ft 6 in, Four-Feet at 1262 ft 3 in, Upper Six-Feet at 1301 ft 4 in, Lower Six-Feet at 1347 ft, Upper Nine-Feet (disturbed) at 1441 ft 4in, Lower Nine-Feet at 1486 ft 1 in, Bute at 1525 ft, ?Yard at 1651 ft 1 in.

Fernhill Colliery: No. 2 Shaft

Ht. above O.D. 804 ft.   6-in SN 90 S.W.; Glam. 17 N.E.

Site 500 yd N. 2° E. of Blaenrhondda school. National Grid Ref. 92650045.

	Thickness ft in		Depth ft in			Thickness ft in		Depth ft in	
DRIFT					Sandstone 88	4		340	5
Gravel, sand and clay .. 49	6		49	6	Coal .. 3 in	3		340	8
					Underclay, shale and sandstone bands 30	2		370	10
COAL MEASURES									
Shale and sandstone .. 29	6		79	0					
TORMYNYDD					ABERGORKY				
Coal .. 21 in 1	9		80	9	Coal .. 22 in				
Fireclay, sandstone and					Clod .. 24 in				
shale 38	3		119	0	Coal .. 18 in 5	4		376	2
Coal .. 1 in	1		119	1	Underclay, sandstone, shale				
Underclay and sandstone 32	9		151	10	and ironstone 51	2		427	4
Coal .. 6 in	6		152	4					
Underclay, shale and sandstone 78	4		230	8	PENTRE RIDER				
Coal .. 3 in	3		230	11	Coal .. 18 in 1	6		428	10
Underclay, sandstone and					Underclay, shale and sandstone 18	11		447	9
shale 17	8		248	7	Coal .. 14 in 1	2		448	11
					Underclay, sandstone and				
HAFOD (BOTTOM COAL)					shale 17	6		466	5
Coal .. 12 in 1	0		249	7	Coal .. 2 in	2		466	7
Underclay 2	6		252	1	Fireclay and shale .. 10	1		476	8

	Thickness ft in	Depth ft in
LOWER PENTRE		
Coal .. 2 in		
Shale .. 1 in		
Coal .. 1 in		
Shale .. 1 in		
Coal .. 3 in		
Shale .. 12 in		
Coal .. 6 in	**2 2**	**478 10**
Underclay, sandstone, shale and ironstone	33 0	511 10
Coal .. 2 in	**2**	**512 0**
Underclay	2 0	514 0
Sandstone	37 9	551 9
Shale	10 9	562 6
GORLLWYN RIDER		
Coal .. 21 in	**1 9**	**564 3**
Underclay and shale ..	15 10	580 1
GORLLWYN		
Coal .. 9 in		
Thin parting		
Coal .. 8 in		
Holing .. 2 in		
Coal .. 11 in	**2 6**	**582 7**
Underclay, shale and mine ground	59 10	642 5
Coal .. 6 in	**6**	**642 11**
Underclay	6 3	649 2
Sandstone	27 2	676 4
Shale	3	676 7
Coal .. 8 in	**8**	**677 3**
Underclay, sandstone, shale and mine	25 2	702 5
Coal .. 7 in	**7**	**703 0**
Underclay and shale ..	25 0	728 0
Coal .. 5 in	**5**	**728 5**
Underclay with ironstone	8 6	736 11
Coal .. 5 in	**5**	**737 4**

	Thickness ft in	Depth ft in
Underclay and shale ..	3 4	740 8
Coal .. 6 in	**6**	**741 2**
Underclay and shale ..	12 4	753 6
Sandstone	17 6	771 0
Shale	14 0	785 0
TWO-FEET-NINE		
Coal .. 34 in	**2 10**	**787 10**
Shale	15 1	802 11
Coal .. 2 in		
Underclay 39 in		
Coal .. 7 in	**4 0**	**806 11**
Underclay, shale and sandstone bands	46 9	853 8
Coal .. 8 in	**8**	**854 4**
Shale and ironstone (inferior)	4 8	859 0
FOUR-FEET		
Coal .. 11 in		
Shale .. 1 in		
Coal .. 11 in		
Shale .. 42 in		
Coal (inferior) 9 in		
Coal .. 66 in		
Shale .. 12 in		
Coal (inferior) 22 in	**14 6**	**873 6**
Underclay and shale ..	24 2	897 8
UPPER SIX-FEET		
Coal .. 51 in	**4 3**	**901 11**
Underclay, shale and ironstone	15 1	917 0
LOWER SIX-FEET		
Coal .. 4 in		
Shale .. 15 in		
Coal .. 50 in	**5 9**	**922 9**

Fernhill Colliery: No. 3 Shaft

Ht. above O.D. 844 ft.   6-in SN 90 S.W.; Glam. 17 N.E.

Site 785 yd N. 13° W. of Blaenrhondda school. National Grid Ref. 92500069.

Abergorky at 378 ft 1 in, Gorllwyn at 570 ft 5 in, Two-Feet-Nine at 814 ft 5 in, Four-Feet at 889 ft 8 in, Upper Six-Feet at 938 ft 8 in, Lower Six-Feet at 958 ft 7 in, Red Vein at 1007 ft 1 in, Upper Nine-Feet at 1131 ft 5 in.

Fernhill Colliery: No. 5 Shaft

Ht. above O.D. 873 ft.   6-in SN 90 S.W.; Glam. 17 N.E.

Site 930 yd N. 14° W. of Blaenrhondda school. National Grid Ref. 92460082.

Abergorky at 500 ft 2 in, Gorllwyn at 688 ft 8 in. Sunk to 816 ft 9 in. Deepening: Two-Feet-Nine at 928 ft, Four-Feet at 956 ft; disturbed ground and inclined strata to ?Upper Nine-Feet at 1054 ft, ?Bute at 1158 ft. Sunk to 1239 ft.

Ffaldau Colliery: Downcast Shaft

Ht. above O.D. 519 ft.   6-in SS 99 S.W.; Glam. 26 S.E.

Site 520 yd N. 37° W. of St. Theodore's Church, Pontycymmer. National Grid Ref. 90369157. Published in Vertical Sections, Sheet 85, 1901.

Upper Yard (Pentre) at 52 ft 3 in, Two-and-a-Half at 98 ft 3 in, Eighteen-Inch at 112 ft 10 in, Caedavid at 149 ft, Two-Feet-Nine at 419 ft 2 in, Upper Six-Feet at 570 ft 11 in, Lower Six-Feet at 645 ft 1 in. Sunk to 663 ft 7 in.

Ffaldau Colliery: Upcast Shaft

Ht. above O.D. 518 ft.   6-in SS 99 S.W.; Glam. 26 S.E.

Site 550 yd N. 34° W. of St. Theodore's Church, Pontycymmer. National Grid Ref. 90369160.

Caedavid at 117 ft 7 in, Two-Feet-Nine at 407 ft 11 in, Upper Six-Feet at 549 ft 10 in, Lower Six-Feet at 638 ft 4 in, Caerau at 646 ft, Lower Nine-Feet at 721 ft 3 in, Bute at 758 ft 9 in, Yard at 863 ft 11 in, Middle Seven-Feet at 925 ft 10 in, Lower Seven-Feet at 954 ft 11 in, Upper Five-Feet at 973 ft 7 in. Sunk to 995 ft.

Ffaldau Colliery: Victoria or Braichycymmer Shaft

Ht. above O.D. 528 ft.   6-in SS 99 S.W.; Glam. 26 S.E.

Site 800 yd N. 42° W. of St. Theodore's Church, Pontycymmer. National Grid Ref. 90179172. Published in 2nd Edition, pp. 141–2.

Caedavid at 173 ft 10 in, Two-Feet-Nine at 534 ft 2 in, Upper Six-Feet at 661 ft 2 in, Lower Six-Feet at 752 ft 3 in, Caerau at 759 ft 1 in, Lower Nine-Feet at 820 ft 4 in, Bute at 877 ft 9 in, Yard at 974 ft 5 in, Middle Seven-Feet at 1033 ft 1 in, Lower Seven-Feet at 1068 ft 4 in, Five-Feet at 1091 ft 5 in. Sunk to 1104 ft 11 in.

Fforchaman Colliery: Downcast Shaft

Ht. above O.D. 642 ft.   6-in SS 99 N.E.; Glam. 18 N.E.

Site 300 yd W. 23° N. of Cwmaman church. National Grid Ref. 99989946. Sunk about 1850.

	Thickness ft	in	Depth ft	in
Made Ground	33	1	33	1
DRIFT				
Sand and blue clay ..	69	7	102	8
COAL MEASURES				
Mine ground	6	0	108	8
?GRAIG				
Coal .. 9 in				
Clod .. 6 in				
Coal .. 9 in	2	0	110	8
Fireclay, rock and strong clift	42	2	152	10
?PENTRE RIDER				
Coal .. 7 in		7	153	5
Fireclay, strong clift, clift and rock ..	51	9	205	2
Coal .. 5 in		5	205	7
Fireclay and strong clift ..	17	6	223	1
Coal .. 18 in	1	6	224	7
Fireclay, strong clift and beds of rock	34	0	258	7
Coal .. 11 in		11	259	6
Fireclay, rock and clift ..	41	3	300	9
GORLLWYN RIDER				
Coal .. 16 in	1	4	302	1
Strong clift and beds of rock	12	8	314	9
GORLLWYN				
Coal .. 23 in	1	11	316	8
Strong clift and mine ground	51	1	367	9
Coal .. 3 in		3	368	0
Fireclay, rock, strong clift and mine ground ..	26	3	394	3
Coal .. 3 in		3	394	6
Rock, mine ground and fireclay	19	8	414	2
Coal .. 5 in		5	414	7
Fireclay, rock, strong clift with mine ..	25	6	440	1
Coal .. 12 in	1	0	441	1
Fireclay	8	0	449	1

		Thickness ft in	Depth ft in
THREE COALS			
Coal	.. 9 in		
Clod	.. 24 in		
Coal	.. 9 in		
Rock	.. 14 in		
Coal	.. 3 in	4 11	454 0
Fireclay and rock..	.. 32 2		486 2
TWO-FEET-NINE			
Coal	.. 29 in	2 5	488 7
Fireclay and beds of rock	6 9		495 4
Coal	.. 6 in		
Clod	.. 6 in		
Coal	.. 4 in	1 4	496 8
Dark clod and mine ground	20 6		517 2
FOUR-FEET			
Coal	.. 66 in	5 6	522 8
Clod, bastard rock, fireclay, clift and mine ..	.. 24 0		546 8
UPPER SIX-FEET			
Coal	.. 24 in	2 0	548 8
Clod, clift, rock and mine	47 4		596 0
LOWER SIX-FEET			
Coal	.. 48 in	4 0	600 0
Bastard fireclay ..	.. 10 4		610 4
Coal	.. 11 in	11	611 3
Fireclay and clift with balls of mine	.. 8 7		619 10
RED VEIN			
Coal	.. 15 in		
Clod	.. 6 in		
Coal	.. 16 in	3 1	622 11
Fireclay, clift, rock and rashings.. ..	.. 57 6		680 5

		Thickness ft in	Depth ft in
NINE-FEET			
Coal mixed with rashings	144 in		
Bast	.. 24 in		
Rashings	19 in		
Coal	.. 66 in	21 1	701 6
Fireclay, rock and bastard clift	.. 44 5		745 11
BUTE			
Coal	.. 17 in		
Clod	.. 10 in		
Coal	.. 28 in	4 7	750 6
Fireclay, bastard clift, rock and clift ..	.. 57 5		807 11
Coal	.. 24 in		
Bast	.. 20 in	3 8	811 7
Fireclay	.. 2 9		814 4
YARD			
Coal	.. 38 in	3 2	817 6
Fireclay	.. 7 3		824 9
SEVEN-FEET			
Coal	.. 10 in		
Bast	.. 12 in		
Coal	.. 36 in		
Fireclay..	28 in		
Coal	.. 24 in	9 2	833 11
Fireclay, rock and clift ..	29 2		863 1
FIVE-FEET			
Coal	.. 25 in		
Fireclay..	21 in		
Coal	.. 6 in		
Fireclay..	22 in		
Coal	.. 30 in	8 8	871 9
Fireclay, clift and rock ..	32 8		904 5
GELLIDEG			
Coal	.. 33 in	2 9	907 2
Bastard rock	.. .. 10 9		917 11
Coal	.. 3 in	3	918 2
Rock	.. .. .. 3 7		921 9

Fforchaman Colliery: Upcast Shaft

Ht. above O.D. 617 ft.   6-in SS 99 N.E.; Glam. 18 N.E.

Site 360 yd W. 21° N. of Cwmaman church. National Grid Ref. 99949947.

Drift to 72 ft, Two-Feet-Nine at 475 ft 4 in, Four-Feet at 516 ft 5 in, Upper Six-Feet at 542 ft 5 in, Lower Six-Feet at 582 ft 2 in, Red Vein at 600 ft 10 in, Nine-Feet at 686 ft 2 in, Bute at 736 ft 3 in, Yard at 794 ft 8 in, Seven-Feet at 810 ft 2 in. Sunk to 832 ft 2 in.

Fforchwen Shaft

Ht. above O.D. 771 ft.   6-in SS 99 N.E.; Glam. 18 N.E.

Site 1570 yd W. 5° N. of Cwmaman church. National Grid Ref. 98809949.

Graig at 291 ft, Abergorky at 378 ft, Two-Feet-Nine at 684 ft, Four-Feet at 723 ft, Upper Six-Feet at 750 ft, Lower Six-Feet at 810 ft, Nine-Feet at 900 ft, Bute at 939 ft, Yard at 984 ft, Seven-Feet at 1011 ft, Five-Feet at 1065 ft, Gellideg at 1115 ft.

Garth Merthyr (Celtic) Colliery: South Pit

Ht. above O.D. 402 ft.   6-in SS 88 N.E.; Glam. 26 S.W.

Site 2200 yd S. 43° E. of Maesteg railway station. National Grid Ref. 86668999.

Victoria at 174 ft, Two-and-a-Half at 336 ft, Caedavid at 410 ft, Two-Feet-Nine at 672 ft, Upper Four-Feet at 702 ft, Lower Four-Feet at 731 ft, Lower Six-Feet at 885 ft, Caerau at 942 ft, Upper Nine-Feet at 1031 ft, Lower Nine-Feet at 1106 ft, Bute at 1120 ft.

Garw Colliery: Downcast Shaft

Ht. above O.D. 699 ft.   6-in SS 99 S.W.; Glam. 26 N.E.

Site 670 yd N. 34° E. of Mount Zion Church, Blaengarw. National Grid Ref. 90549311. Published in Vertical Sections, Sheet 85, 1901.

Victoria at 88 ft 7 in, Upper Yard at 152 ft 10 in, Caedavid at 248 ft 5 in, Two-Feet-Nine at 551 ft 4 in, Lower Six-Feet at 714 ft 1 in, Caerau at 728 ft 1 in, Upper Nine-Feet at 813 ft 5 in, Lower Nine-Feet at 898 ft 5 in, Bute 946 ft 11 in, Yard at 1071 ft 11 in, Middle Seven-Feet at 1110 ft 5 in, Lower Seven-Feet at 1137 ft, Upper Five-Feet at 1157 ft 2 in, Lower Five-Feet at 1179 ft 5 in. Sunk to 1201 ft 10 in.

Garw Colliery: Upcast Shaft

Ht. above O.D. c. 700 ft.   6-in SS 99 S.W.; Glam. 26 N.E.

Site 650 yd N. 40° E. of Mount Zion Church, Blaengarw. National Grid Ref. 90609307.

Two-Feet-Nine at 566 ft 1 in, Lower Six-Feet at 699 ft 1 in, Upper Nine-Feet at 802 ft 3 in, Lower Nine-Feet at 875 ft 6 in, Yard at 1098 ft 3 in. Sunk to 1170 ft.

Gelli Colliery

Ht. above O.D. 520 ft.   6-in SS 99 S.E.; Glam. 27 N.W.

Site 1050 yd E. 3° S. of Ton church. National Grid Ref. 98109487. Published to Lower Six-Feet in Vertical Sections, Sheet 84, 1901.

No. 3 Rhondda at 132 ft 6 in, Abergorky at 465 ft, Two-Feet-Nine at 952 ft 2 in, Upper Six-Feet at 1050 ft 3 in, Lower Six-Feet at 1084 ft 10 in. Sunk to 1134 ft 4 in.

	Thickness ft in	Depth ft in		Thickness ft in	Depth ft in
DOWNCAST PIT			LOWER NINE-FEET		
Deepening (1914–15) from		1127 7	Coal .. 33 in		
			Rashes .. 17 in		
RED VEIN			Coal .. 6 in		
Coal .. 20 in			Fireclay .. 24 in		
Rashes .. 24 in			Coal .. 6 in		
Coal .. 15 in			Bast .. 8 in	7 10	1332 11
Rashes .. 75 in			Rock: 0 to 4 ft		
Coal .. 24 in	13 2	1140 9			
Very disturbed rashes, clift and coal.. .. .. 49 0		1189 9	BUTE		
			Coal .. 46 in		
UPPER NINE-FEET			Rashings 10 in		
Disturbed			Coal .. 37 in		
Coal 19 ft 5 in	19 5	1209 2	Fireclay .. 6 in		
Fireclay and rashes .. 6 3		1215 5	Coal .. 2 in	8 5	1341 4
Coal .. 5 in	5	1215 10	Fireclay, rock and clift .. 26 6		1367 10
Fireclay, clift and rock .. 26 7		1242 5			
Coal .. 5 in	5	1242 10	AMMAN RIDER		
Fireclay and rashings .. 6 1		1248 11	Coal .. 10 in	10	1368 8
Coal .. 5 in	5	1249 4	Fireclay, clift and rock .. 84 6		1453 2
Fireclay, rashings and bastard rock 29 6		1278 10	YARD		
Coal .. 5 in	5	1279 3	Coal .. 53 in		
Rock, clift and ironstone.. 15 8		1294 11	Parting .. 6 in		
Coal .. 2 in	2	1295 1	Coal .. 7 in	5 6	1458 8
Clift and rock 30 0		1325 1	Fireclay 2 0		1460 8

	Thickness ft in	Depth ft in
UPPER SEVEN-FEET		
Coal .. 12 in	1 0	**1461 8**
Fireclay, clift and rock ..	31 2	1492 10
MIDDLE SEVEN-FEET		
Coal .. 30 in	2 6	**1495 4**
Fireclay, rock and clift ..	24 0	1519 4
LOWER SEVEN-FEET		
Coal .. 20 in	1 8	**1521 0**

	Thickness ft in	Depth ft in
Fireclay, rashings and clift	21 6	1542 6
Coal .. 5 in	**5**	**1542 11**
Clift and clod	28 10	1571 9
FIVE-FEET AND GELLIDEG		
Coal .. 170 in	**14 2**	**1585 11**
Fireclay, clift with iron-stone	81 8	1667 7

Glenrhondda Colliery: No. 2 Pit

Ht. above O.D. 827 ft.   6-in SS 99 N.W.; Glam. 17 N.E.

Site 450 yd S. 36° W. of Blaen-y-cwm school. National Grid Ref. 91689855.

	Thickness ft in	Depth ft in
Walling	10 0	10 0
DRIFT		
Gravel and clay	103 0	113 0
COAL MEASURES		
Clift	20 0	133 0
Coal .. 6 in	6	133 6
Clift	7 3	140 9
Coal .. 4 in	4	141 1
Fireclay	2 6	143 7
Rock and bastard rock ..	18 3	161 10
Clift	62 9	224 7
Coal .. 8 in	8	225 3
Fireclay, rashings and clift	47 1	272 4
ABERGORKY		
Coal .. 35 in	**2 11**	**275 3**
Fireclay and clift	10 7	285 10
Rock and bastard rock ..	46 3	332 1
Clift, with beds of rock ..	16 2	348 3
PENTRE RIDER		
Coal .. 21 in	**1 9**	**350 0**
Fireclay, clift and rock ..	22 10	372 10
Coal .. 14 in	**1 2**	**374 0**
Fireclay, clift and bastard rock	11 1	385 1
Coal .. 2 in	**2**	**385 3**
Fireclay and clift	21 0	406 3
LOWER PENTRE		
Coal .. 13 in		
Bast .. 2 in		
Coal .. 5 in	**1 8**	**407 11**
Fireclay and clift	28 7	436 6
Coal .. 3 in	**3**	**436 9**
Fireclay and clift	11 0	447 9
Rock	26 9	474 6

	Thickness ft in	Depth ft in
GORLLWYN RIDER		
Coal .. 8 in	**8**	**475 2**
Fireclay and bastard rock	28 3	503 5
GORLLWYN		
Coal .. 18 in	**1 6**	**504 11**
Fireclay, clift and rock ..	99 3	604 2
Rock, with a thin clift band	33 3	637 5
Coal .. 7 in	**7**	**638 0**
Fireclay, rock and clift ..	62 3	700 3
Coal .. 8 in	**8**	**700 11**
Bastard fireclay	10 4	711 3
Coal .. 6 in		
Clod .. 20 in		
Coal .. 7 in	**2 9**	**714 0**
Fireclay and clift	7 7	721 7
TWO-FEET-NINE		
Coal .. 31 in	**2 7**	**724 2**
Fireclay, bastard rock and clift	43 9	767 11
Coal .. 7 in		
Rashings 5 in		
Coal .. 4 in	**1 4**	**769 3**
Fireclay, bastard rock and clift	23 11	793 2
FOUR-FEET [TOP COAL]		
Coal .. 11 in	**11**	**794 1**
Fireclay and clift	47 4	841 5
FOUR-FEET [BOTTOM COAL]		
Coal .. 4 in		
Parting,		
Coal .. 48 in		
Rashings 6 in		
Coal .. 9 in	**5 7**	**847 0**
Fireclay, rock and clift ..	30 8	877 8
UPPER SIX-FEET		
Coal .. 28 in	**2 4**	**880 0**
Fireclay and clift	18 11	898 11

	Thickness ft	in	Depth ft	in
LOWER SIX-FEET				
Coal .. 49 in				
Stone .. 1 in				
Coal .. 6 in	**4**	**8**	**903**	7
Fireclay and clift 14		5	918	0
Coal .. 5 in		**5**	**918**	**5**
Fireclay and clift 17		6	935	11
RED VEIN				
Coal .. 6 in				
Rashings 12 in				
Coal .. 31 in	**4**	**1**	**940**	**0**
Fireclay, clift and rashings 24		4	964	4
Coal .. 2 in		**2**	**964**	**6**
Rashings 8		6	973	0

	Thickness ft	in	Depth ft	in
UPPER NINE-FEET				
Coal .. 54 in	**4**	**6**	**977**	
Fireclay and rashes ..	6	4	983	10
Coal .. 6 in		**6**	**984**	
Clift and stone 27		5	1011	
LOWER NINE-FEET				
Coal .. 28 in	**2**	**4**	**1014**	
Fireclay, rashings and shale	5	7	1019	8
Coal .. 3 in		**3**	**1019**	11
Fireclay, rock and clift .. 24		4	1044	

Glyncorrwg Colliery: North Pit

Ht. above O.D. c. 910 ft.　　　6-in SN 80 S.E.; Glam. 17 N.W.

Site 1730 yd E. 32° N. of Glyncorrwg church. National Grid Ref. 88780010.

?Daren Rhestyn horizon at 40 ft, ?No. 2 Rhondda horizon at 221 ft 5 in; sunk through eastern branch of Glyncorrwg Fault to Abergorky at 467 ft, Pentre Rider at 546 ft 2 in; sunk through reversed fault of c. 133 ft to Pentre Rider at 679 ft; Lower Pentre at 721 ft 9in, Gorllwyn at 821 ft 2 in, ?Two-Feet-Nine at 1080 ft 5 in, Four-Feet at 1125 ft 1 in, Six-Feet at 1263 ft 1 in. Total depth 1283 ft 1 in.

Glyncorrwg Colliery: South Pit

Ht. above O.D. 888 ft.　　　6-in SN 80 S.E.; Glam. 17 N.W.

Site 1625 yd E. 27° N. of Glyncorrwg church. National Grid Ref. 88780010. Published in 2nd Edition, p. 142.

?Daren Rhestyn at 50 ft. No. 2 Rhondda at 279 ft 3 in; sunk through eastern branch of Glyncorrwg Fault to Four-Feet at 1093 ft 10 in, Six-Feet at 1209 ft 7 in. Total depth 1222 ft.

Great Western Colliery: No. 1 Pit

Ht. above O.D. 264 ft.　　　6-in ST 09 S.E.; Glam. 28 S.W.

Site 1650 yd W. 26° N. of Pontypridd church. National Grid Ref. 05819084. Sunk 1923.

	Thickness ft	in	Depth ft	in
Made ground 19		3	19	3
DRIFT				
Soil and pebble clay .. 10		5	29	8
COAL MEASURES				
Rock 30		2	59	10
Coal .. 9 in		**9**	**60**	**7**
Fireclay and shale .. 5		4	65	11
NO. 1 RHONDDA				
Coal .. 28 in	**2**	**4**	**68**	**3**
Fireclay and clift 21		5	89	8
Coal .. 5 in		**5**	**90**	**1**

	Thickness ft	in	Depth ft	in
Fireclay and shale ..	9	5	99	
Coal .. 10 in		**10**	**100**	
Shale and clift 68		10	169	
Pennant rock 50		8	219	1
FFOREST FACH				
Clod .. 4 in				
Coal .. 8 in				
Clod .. 7 in				
Coal .. 5½ in				
Parting .. ½ in				
Coal .. 7 in	**2**	**8**	**222**	
Fireclay and clift with beds				
of rock 23		9	246	

	Thickness ft	in	Depth ft	in
NO. 2 RHONDDA				
Coal .. 24 in	2	0	248	3
Fireclay and clift	7	0	255	3
Rock	30	6	285	9
Clift	7	9	293	6
Coal .. 6 in		6	294	0
Fireclay and clift	21	9	315	9
Coal .. 6 in		6	316	3
Clift with beds of rock ..	16	8	332	11
Dark rashings	3	0	335	11
Fireclay, clift and bastard rock	34	7	370	6
Coal .. 18 in	1	6	372	0
Hard rock	54	9	426	9
Clift with rock bands ..	60	3	487	0
Coal .. 11 in		11	487	11
Fireclay, clift and rock ..	21	2	509	1
Coal .. 8 in		8	509	9
Fireclay, shale and clift ..	8	1	517	10
Coal .. 4 in		4	518	2
Fireclay and shale ..	11	8	529	10
Coal .. 6 in		6	530	4
Fireclay and clift with bands of rock	67	5	597	9
HAFOD				
Coal .. 18 in				
Clod .. 3 in				
Coal .. 6 in	2	3	600	0
Rock, fireclay and clift ..	71	5	671	5
Coal .. 6 in		6	671	11
Fireclay	4	7	676	6
ABERGORKY				
Coal .. 14 in	1	2	677	8
Fireclay and clift with rock bands	48	10	726	6
PENTRE RIDER				
Coal .. 12 in	1	0	727	6
Fireclay, rock and clift ..	23	0	750	6
PENTRE				
Coal .. 17 in	1	5	751	11
Fireclay and clift	17	7	769	6
LOWER PENTRE				
Coal .. 18 in	1	6	771	0
Fireclay, rock and clift ..	25	0	796	0

	Thickness ft	in	Depth ft	in
EIGHTEEN-INCH				
Coal .. 10 in		10	796	10
Fireclay and clift 19		7	816	5
?GORLLWYN RIDER				
Coal .. 13 in	1	1	817	6
Fireclay and hard clift .. 13		7	831	1
Hard rock with bands of clift at top	63	11	895	0
Clift and fireclay with ironstone	41	0	936	0
Coal .. 11 in		11	936	11
Hard rock	16	0	952	11
Coal .. 4 in		4	953	3
Fireclay and rock with soft bands	21	4	974	7
Coal in faulty ground ..		10	975	5
Rock rashings and clift ..	12	5	987	10
TWO-FEET-NINE				
Coal .. 17 in				
Clod .. 2 in				
Coal .. 14 in	2	9	990	7
Fireclay and bastard rock with clift bands ..	12	2	1002	9
Rashings 5 in				
Coal .. 10 in	1	3	1004	0
Fireclay, clift and rock ..	23	10	1027	10
FOUR-FEET				
Gob .. 75 in	6	3	1034	1
Fireclay, rock and clift ..	59	2	1093	3
SIX-FEET				
Gob .. 48 in				
Clod .. 6 in				
Coal .. 42 in	8	0	1101	3
Fireclay and clift	14	0	1115	3
Coal .. 22 in	1	10	1117	1
Clift with layers of rock ..	66	10	1183	11
UPPER NINE-FEET				
Gob .. 60 in	5	0	1188	11
Fireclay and rashings ..	5	6	1194	5
Disturbed ground				
Coal .. 90 in	7	6	1201	11
Fireclay	7	6	1209	5
Coal and fireclay in mixed ground	9	6	1218	11
Fireclay	4	0	1222	11
Coal .. 72 in	6	0	1228	11
Fireclay, rock and clift ..	20	0	1248	11
Coal .. 17 in	1	5	1250	4
Fireclay and clift	13	4	1263	8

Great Western Colliery: Hetty Pit, No. 2 Pit, Trial Pit

Ht. above O.D. 273 ft. 6-in ST 09 S.W.; Glam. 28 S.W.

Site 2000 yd W. 25° N. of Pontypridd church. National Grid Ref. 05489091. Based on combined section.

No. 3 Rhondda at 457 ft 7 in, Hafod at 628 ft 1 in, Four-Feet at 1046 ft 10 in, Six-Feet at 1098 ft 2 in, Upper Six-Feet at 1174 ft 3 in, Lower Nine-Feet at 1228 ft 9 in, Bute at 1265 ft 1 in, Yard at 1322 ft 5 in, Five-Feet at 1420 ft 1 in, Gellideg at 1424 ft 10 in, Garw at 1524 ft.

Great Western Colliery: Ty-mawr Pit

Ht. above O.D. 293 ft.   6-in ST 09 S.E.; Glam. 28 S.W.

Site 1550 yd W. 30° N. of Pontypridd church. National Grid Ref. 05909088.

No. 1 Rhondda at 141 ft 3 in, No. 2 Rhondda at 293 ft, Four-Feet at 1065 ft 8 in, Six-Feet at 1134 ft 6 in, Nine-Feet at 1240 ft 7 in, Bute at 1258 ft, Yard at 1311 ft 11 in, Upper and Middle Seven-Feet at 1323 ft, Five-Feet at 1418 ft 2 in, Gellideg at 1423 ft 3 in, sunk to 1458 ft 8 in.

Gyfeillon Pit

Ht. above O.D. c. 280 ft.   6-in ST 09 S.W.; Glam. 28 S.W.

Site 2150 yd W. 25° N. of Pontypridd church. National Grid Ref. 05369097.

No. 1 Rhondda at 73 ft 4 in, No. 2 Rhondda at 259 ft 7 in, No. 3 Rhondda at 445 ft 3 in.

International Colliery: No. 1 (Downcast) Pit

Ht. above O.D. 737 ft.   6-in SS 89 S.E.; Glam. 26 N.E.

Site 850 yd N. 38° W. of Mount Zion Church, Blaengarw. National Grid Ref. 89749322. Published (to 927 ft 5 in) in Vertical Sections, Sheet 85, 1901; deepening to 1436 ft 9 in. in 2nd Edition, pp. 143–4.

Victoria at 231 ft 5 in, Upper Yard (Pentre) at 294 ft 1 in, Caedavid at 367 ft 7 in, Two-Feet-Nine at 683 ft 3 in, Lower Six-Feet at 873 ft 8 in, Upper Nine-Feet at 1033 ft 11 in, Lower Nine-Feet at 1115 ft 1 in, Bute at 1161 ft 11 in, Yard at 1276 ft 5 in, Middle Seven-Feet at 1309 ft 4 in, Lower Seven-Feet at 1344 ft 10 in, Upper Five-Feet at 1359 ft 2 in, Lower Five-Feet at 1391 ft 10 in, pit bottom at 1436 ft 9 in.

International Colliery: No. 2 (Upcast) Pit

Ht. above O.D. 739 ft.   6-in SS 89 S.E.; Glam. 26 N.E.

Site 850 yd N. 41° W. of Mount Zion Church, Blaengarw. National Grid Ref. 89719319.

Two-Feet-Nine at 681 ft, Lower Six-Feet at 874 ft, Upper Nine-Feet at 1034 ft, Bute at 1162 ft, sunk to 1171 ft.

International Colliery: No. 3 Pit

Ht. above O.D. 737 ft.   6-in SS 89 S.E.; Glam. 26 N.E.

Site 850 yd N. 35° W. of Mount Zion Church, Blaengarw. National Grid Ref. 89769324.

Victoria at 231 ft, Caedavid at 367 ft.

Lady Windsor Colliery: Downcast Shaft

Ht. above O.D. 491 ft.   6-in ST 09 S.W.; Glam. 28 N.W.

Site 480 yd E. 14° S. of Ynys-y-bwl church. National Grid Ref. 06299426. Published in Vertical Sections, Sheet 83, 1900.

Four-Feet at 1632 ft 2 in, Six-Feet at 1684 ft 6 in, Nine-Feet at 1804 ft, Bute at 1830 ft 7 in, sunk to 1849 ft 6 in.

Lady Windsor Colliery: Borehole down from Nine-Feet

6-in ST 09 S.W.; Glam. 28 N.W.   National Grid Ref. 06489386.

		Thickness ft in	Depth below Nine-Feet ft in
Broken ground and mudstone		36 6	36 6
BUTE			
Coal .. 31 in			
Shale .. 8 in			
Coal .. 24 in		5 3	41 9
Seatearth and mudstone ..		16 3	58 0
AMMAN RIDER			
Coal .. 7 in		7	58 7
Seatearth, siltstone and mudstone		8 2	66 9
Coal .. ½ in		½	66 9½
Seatearth, mudstone, siltstone and shale		21 8½	88 6
YARD			
Coal .. 38 in		3 2	91 8

		Thickness ft in	Depth below Nine-Feet ft in
Seatearth, siltstone and mudstone		16 4	108 0
SEVEN-FEET			
Coal .. 38 in			
Seatearth 5 in			
Coal .. 3 in		3 10	111 10
Seatearth, sandstone, mudstone and shale.. ..		92 10	204 8
FIVE-FEET AND GELLIDEG			
Coal .. 40 in			
Fireclay .. 2 in			
Coal .. 6 in			
Coal-bearing rashings 22 in			
Coal .. 24 in		7 10	212 6
Seatearth, mudstone, sandstone		182 6	395 0

Lewis Merthyr Colliery

Ht. above O.D. 303 ft.   6-in ST 09 S.W.; Glam. 28 S.W.

Site 670 yd E. 23° S. of St. Luke's Church, Porth. National Grid Ref. 03969113. Published in Vertical Sections, Sheet 83, 1900.

Drift to 14 ft, No. 2 Rhondda at 124 ft, No. 3 Rhondda at 346 ft 3 in, Hafod at 511 ft 3 in, Four-Feet at 990 ft 6 in, Six-Feet at 1064 ft 1 in, Upper Nine-Feet at 1131 ft 3 in, Lower Nine-Feet at 1173 ft 6 in, Bute at 1196 ft 11 in, Yard at 1263 ft 5 in, Upper and Middle Seven-Feet at 1281 ft 8 in, Five-Feet at 1353 ft 5 in, Gellideg at 1358 ft 7 in, sunk to 1418 ft 11 in.

Lewis Merthyr Colliery: Lady Lewis Pit

Ht. above O.D. 361 ft.   6-in ST 09 S.W.; Glam. 27 N.E.

Site 620 yd S. 1° E. of Ynyshir church. National Grid Ref. 02559233. Published in 2nd Edition, pp. 147–9.

No. 2 Rhondda at 149 ft 5 in, No. 3 Rhondda at 362 ft 6 in, Hafod at 525 ft 3 in, Lower Four-Feet at 1070 ft 7 in, Six-Feet at 1133 ft, Upper Six-Feet at 1214 ft 11 in, Lower Nine-Feet at 1278 ft 2 in, Bute at 1287 ft 1 in, Fault, Yard at 1294 ft 6 in, Upper and Middle Seven-Feet at 1307 ft 10 in, Five-Feet at 1404 ft, Gellideg at 1426 ft 6 in.

Llanharan Colliery: South (Upcast) Pit

Ht. above O.D. 250 ft. 6-in SS 98 S.E.; Glam. 35 S.E.

Site 1000 yd S. 40° W. of Llanharan church. National Grid Ref. 99678254. Published to 432 ft 6 in. in 2nd Edition, p. 40 and Plate.

	Thickness ft	in	Depth ft	in
DRIFT				
Soil, gravel and clay ..	11	6	11	6
COAL MEASURES				
Shale with sandstone bands	59	4	70	10
Coal .. 4 in		4	71	2
Sandstone, bastard rock and shale	22	10	94	0
Coal .. 4 in		4	94	4
Fireclay and shale with ironstone	19	11	114	3
Coal .. 21 in	1	9	116	0
Fireclay and shale ..	25	2	141	2
Sandstone in beds ..	36	6	177	8
Shale with ironstone ..	23	6	201	2
TWO-FEET-NINE				
Cannel .. 4 in				
Coal .. 120 in	10	4	211	6
Shale	8	8	220	2
UPPER FOUR-FEET				
Coal .. 20 in				
Shale .. 6 in				
Coal .. 18 in	3	8	223	10
Fireclay and shale with ironstone	8	0	231	10
Sandstone	8	0	239	10
Shale with ironstone ..	19	7	259	5
Sandstone	15	0	274	5
Carbonaceous shale .. 3 in				
Coal .. 19 in	1	10	276	3
Fireclay and sandstone ..	23	9	300	0
LOWER FOUR-FEET				
Coal .. 8 in				
Shale .. 6 in				
Coal .. 24 in				
Shale .. 4 in				
Coal .. 24 in	5	6	305	6
Fireclay and sandstone ..	25	0	330	6
Shale with ironstone and sandstone bands ..	61	10	392	4
SIX-FEET				
Coal .. 84 in	7	0	399	4
Fireclay and shale ..	15	6	414	10
Coal .. 12 in				
Shale .. 24 in				
Coal .. 5 in	3	5	418	3
Llanharan Thrust hereabouts.				
Fireclay and sandy shale with rock beds	54	6	472	9
Hard rock and shale ..	39	7	512	4
HAFOD				
Coal .. 29 in	2	5	514	9
Shale and sandy shale ..	88	8	603	5
Rashings 2 in				
Coal .. 19 in	1	9	605	2
Fireclay and shale ..	36	7	641	9

	Thickness ft	in	Depth ft	in
ABERGORKY				
Rashings 3 in				
Coal .. 20 in				
Rashings 10 in	2	9	644	(
Sandy shale ..	39	4	683	1(
Hard rock	15	11	699	
Sandy shale and shale with ironstone	53	6	753	
THICK SEAM (= PENTRE RIDER, PENTRE, LOWER PENTRE, TWO-AND-A-HALF AND EIGHTEEN-INCH)				
Coal .. 34 in				
Shale .. 6 in				
Coal .. 77 in				
Fireclay .. 33 in				
Coal .. 9 in				
Rashings 11 in				
Coal .. 6 in				
Rashings 2 in				
Coal .. 29 in	17	3	770	
Fireclay, rock and shale ..	30	3	800	
GORLLWYN				
Coal .. 19 in				
Fireclay .. 54 in				
Coal .. 13 in				
Rashings 4 in	7	6	808	
Fireclay, shale and rock ..	85	9	894	
Coal .. 6 in		6	894	
Sandy fireclay and shale with ironstone and veins of coal	28	9	923	
Coal .. 3 in		3	923	
Fireclay and sandy shale ..	17	3	940	
Coal .. 20 in	1	8	942	
Faulty ground to about 1032 ft.				
Fireclay	1	3	943	
TWO-FEET-NINE (disturbed)				
Coal .. 51 in				
Mixed ground 60 in				
Coal .. 53 in	13	8	957	
Fireclay, shale and rock ..	14	7	971	1
?FOUR-FEET				
Coal .. 8 in				
Rashings 6 in				
Coal .. 1 in				
Fireclay .. 45 in				
Coal .. 6 in				
Fireclay .. 14 in				
Coal .. 10 in				
Rashings 10 in				
Coal .. 24 in	10	4	982	
Fireclay, shale and mixed ground	60	8	1042	11

	Thickness (in)	Thickness ft	in	Depth ft	in
X-FEET					
Coal ..	90 in				
Rashings	2 in				
Coal ..	37 in				
Rashings	3 in	11	0	1053	11
Fireclay		6	3	1060	2
AERAU					
Coal ..	7 in				
Fireclay ..	16 in				
Coal ..	24 in				
Rashings and fireclay	29 in				
Coal ..	6 in	6	10	1067	0
Fireclay, rock and shale ..		14	5	1081	5
ED VEIN					
Coal ..	18 in				
Rashings	3 in				
Coal ..	15 in	3	0	1084	5
Fireclay and rock..		14	8	1099	1
Shale		28	0	1127	1
PPER NINE-FEET					
Coal ..	66 in	5	6	1132	7
Fireclay		4	0	1136	7
OWER NINE-FEET					
Coal ..	7 in				
Rashings	2 in				

	Thickness (in)	Thickness ft	in	Depth ft	in
Coal ..	27 in				
Rashings	1 in				
Coal ..	24 in	5	1	1141	8
Fireclay and shale ..		10	11	1152	7
BUTE					
Coal ..	9 in				
Rashings	12 in				
Coal ..	11 in				
Fireclay ..	12 in				
Coal ..	54 in	8	2	1160	9
Fireclay and sandy shale ..		20	1	1180	10
Fireclay and bastard fire-clay		8	6	1189	4
Shale with ironstone ..		27	6	1216	10
YARD					
Coal ..	42 in	3	6	1220	4
Fireclay and shale ..		31	11	1252	3
Coal ..	27 in	2	3	1254	6
Fireclay and shale with rock beds		87	10	1342	4
FIVE-FEET					
Coal ..	66 in	5	6	1347	10
Rotten fireclay		4	9	1352	7
GELLIDEG					
Coal ..	72 in	6	0	1358	7
Rashings and fireclay ..		4	0	1362	7

Llanharan Colliery: North (Downcast) Pit

Ht. above O.D. 214 ft.   6-in SS 98 S.E.; Glam. 35 S.E.

Site 950 yd W. 43° S. of Llanharan church. National Grid Ref. 99588267.

Two-Feet-Nine at 278 ft 6 in, Upper Four-Feet at 297 ft, Lower Four-Feet at 341 ft 1 in, Llanharan Thrust at about 370 ft, Two-Feet-Nine at 1004 ft 7 in, Upper Four-Feet at 1022 ft 11 in, Lower Four-Feet at 1051 ft 4 in, Six-Feet at 1128 ft 5 in.

Llantrisant Colliery: No. 1 Pit

Ht. above O.D. 253 ft.   6-in ST 08 S.W.; Glam. 36 S.W.

Site 1600 yd W. 27° N. of Llantrisant church. National Grid Ref. 03328410.

	Thickness (in)	Thickness ft	in	Depth ft	in
DRIFT					
Made ground and sandy clay		51	0	51	0
COAL MEASURES					
Pennant rock ..		309	0	360	0
NO. 1 RHONDDA RIDER					
Coal ..	15 in	1	3	361	3
Fireclay and sandy shale ..		33	3	394	6
Hard rock		29	0	423	6

	Thickness (in)	Thickness ft	in	Depth ft	in
NO. 1 RHONDDA					
Coal ..	18 in				
Fireclay ..	6 in				
Coal ..	30 in				
Rashings	24 in	6	6	430	0
Fireclay and sandy shale ..		39	6	469	6
Hard rock and conglomerate		313	3	782	9
Strong shale and clod ..		6	6	789	3

	Thickness ft in	Depth ft in
NO. 2 RHONDDA		
Coal .. 69 in	**5 9**	**795 0**
Sandy fireclay and hard rock 27 9		**822 9**
Coal .. 3 in	**3**	**823 0**
Sandy fireclay, rock and sandy shale 56 8		**879 8**
Coal .. 12 in	**1 0**	**880 8**
Fireclay, hard rock and shale 33 6		**914 2**
Coal .. 22 in		
Fireclay .. 10 in		
Coal .. 2 in	**2 10**	**917 0**
Fireclay, sandy shale, rock and shale 44 8		**961 8**
Coal .. 6 in		
Rashings 6 in		
Coal .. 6 in	**1 6**	**963 2**
Fireclay, rock and shale .. 20 3		**983 5**
Rashings 5 in		
Coal .. 14 in	**1 7**	**985 0**
Fireclay and hard rock .. 49 5		**1034 5**
NO. 3 RHONDDA		
Coal .. 23 in	**1 11**	**1036 4**
Fireclay, rock and shale .. 56 9		**1093 1**
TORMYNYDD		
Coal .. 1 in		
Shale .. 11 in		
Coal .. 22 in	**2 10**	**1095 11**
Sandy fireclay, rock and shale 31 10		**1127 9**
Coal .. 4 in	**4**	**1128 1**
Fireclay and shale .. 12 10		**1140 11**
WHITE		
Coal .. 20 in	**1 8**	**1142 7**
Fireclay and shale with ironstone and rock band 71 10		**1214 5**
HAFOD		
Rashings 2 in		
Coal .. 36 in	**3 2**	**1217 7**
Fireclay, rock and shale .. 63 4		**1280 11**
Coal .. 17 in	**1 5**	**1282 4**
Fireclay, rock and shale .. 21 8		**1304 0**
ABERGORKY		
Coal .. 32 in	**2 8**	**1306 8**
Rashings, fireclay, shale and rock 64 3		**1370 11**
PENTRE RIDER		
Coal .. 24 in	**2 0**	**1372 11**
Clod 1 0		**1373 11**
PENTRE		
Coal .. 27 in	**2 3**	**1376 2**
Strong fireclay 8 6		**1384 8**
Coal .. 9 in	**9**	**1385 5**
Strong fireclay and sandy shale 37 4		**1422 9**
Coal .. 20 in	**1 8**	**1424 5**
Fireclay and sandy shale .. 32 4		**1456 9**

	Thickness ft in	Depth ft in
GORLLWYN		
Coal .. 21 in		
Fireclay .. 6 in		
Coal .. 25 in	**4 4**	**1461 1**
Fireclay, shale and sandy shale 88 10		1549 11
Coal .. 2 in	**2**	**1550 1**
Fireclay and shale .. 18 8		1568
Fireclay and shale with bands of rock .. 15 9		1584
Coal .. 5 in	**5**	**1584 11**
Rock and shale with ironstone 39 7		1624
TWO-FEET-NINE		
Coal .. 86 in	**7 2**	**1631**
Fireclay and rock.. .. 21 1		1652
?UPPER FOUR-FEET		
Coal .. 35 in	**2 11**	**1655**
Fireclay, shale and rock .. 18 1		1673
?LOWER FOUR-FEET		
Coal .. 19 in		
Rashings 9 in		
Coal .. 20 in		
Fireclay .. 4 in		
Coal .. 2 in	**4 6**	**1678**
Fireclay, rock and shale .. 47 3		**1725**
?SIX-FEET		
Rashings 10 in		
Coal .. 11 ft		
Band .. 1 in		
Coal mixture 18 in	**13 5**	**1738 1**
Fireclay and shale .. 5 2		1744
Coal .. 12 in	**1 0**	**1745**
Fireclay, shale and rock .. 85 7		1830
?UPPER NINE-FEET		
Coal .. 72 in		
Rashings 30 in		
Coal .. 6 in	**9 0**	**1839**
Fireclay 3 6		1843
Coal .. 12 in		
Fireclay .. 9 in		
Coal .. 8 in	**2 5**	**1845**
Fireclay with shale .. 7 0		1852
?LOWER NINE-FEET		
Coal .. 20 in	**1 8**	**1854**
Fireclay and shale with rock bands 35 7		1889 10
?BUTE		
Coal .. 36 in		
Fireclay .. 6 in		
Coal .. 37 in	**6 7**	**1896**
Fireclay, rock and shale .. 49 2		1945
Borehole:—		
Strong shale with ironstone and rock bands 154 6		2100

Llantrisant Colliery: No. 2 Pit

Ht. above O.D. c. 250 ft.   6-in ST 08 S.W.; Glam. 36 S.W.

Site 1570 yd W. 25° N. of Llantrisant church. National Grid Ref. 03338402.

No. 1 Rhondda at 354 ft 2 in, No. 2 Rhondda at 727 ft 9 in, No. 3 Rhondda at 975 ft 8 in, Hafod at 1149 ft 2 in, Two-Feet-Nine at 1562 ft. Sunk to 1876 ft 8 in.

Llantrisant Colliery: No. 3 Pit

Ht. above O.D. c. 250 ft.   6-in ST 08 S.W.; Glam. 36 S.W.

Site 1550 yd W. 22° N. of Llantrisant church. National Grid Ref. 03338394.

No. 1 Rhondda at 240 ft 11 in, No. 2 Rhondda at 597 ft 10 in. Sunk to 613 ft 10 in.

Lletty Shenkin Colliery: Upper Pit

Ht. above O.D. 582 ft.   6-in SO 00 S.W.; Glam. 18 N.E.

Site 1450 yd E. 16° N. of Aberaman church. National Grid Ref. 02960127. Sunk to 801 ft 6 in; bored to 884 ft 7 in.

	Thickness ft	in	Depth ft	in
DRIFT				
Clay and gravel	24	0	24	0
COAL MEASURES				
Mine ground, shale and fireclay	53	8	77	8
CRAIG				
Coal .. 20 in	1	8	79	4
Fireclay, shale, sandstone and mine ground ..	67	4	146	8
PENTRE RIDER				
Coal .. 16 in	1	4	148	0
Fireclay, shale and mine ground	21	0	169	0
PENTRE				
Coal .. 12 in	1	0	170	0
Shale, fireclay, strong clift and mine ground ..	23	4	193	4
LOWER PENTRE				
Coal .. 9 in				
Shale .. 4 in				
Coal .. 11 in	2	0	195	4
Strong clift and mine ground	23	11	219	3
EIGHTEEN-INCH				
Coal .. 9 in		9	220	0
Fireclay, rock, strong shale and mine ground ..	33	2	253	2
GORLLWYN				
Coal .. 12 in				
Strong fireclay 108 in				
Coal .. 4 in				
Clod .. 14 in				
Coal .. 18 in	13	0	266	2
Fireclay, mine ground and strong shale	65	4	331	6
Coal .. 2 in		2	331	8
Fireclay, clift and mine ground	30	4	362	0
Coal .. 2 in		2	362	2
Blackband, shale and rock	13	3	375	5
Coal .. 2 in		2	375	7
Mine ground, shale, rock and fireclay	109	5	485	0
TWO-FEET-NINE				
Coal .. 30 in	2	6	487	6
Clod and blue ground with big balls of ironstone ..	20	6	508	0
FOUR-FEET				
Coal .. 78 in	6	6	514	6
Measures	25	6	540	0
UPPER SIX-FEET				
Coal .. 18 in	1	6	541	6
Measures	54	5	595	11
LOWER SIX-FEET				
Coal .. 5 in				
Clod .. 2 in				
Coal .. 48 in	4	7	600	6
Measures with RED VEIN ..	74	10	675	4
NINE-FEET				
Coal .. 48 in				
Bast .. 12 in				
Coal .. 78 in	11	6	686	10
Fireclay, rock and clift ..	20	6	707	4
BUTE				
Coal .. 24 in				
Clod .. 18 in				
Coal .. 22 in	5	4	712	8
Fireclay and clift	23	0	735	8

AMMAN RIDER	Thickness ft in	Depth ft in		Thickness ft in	Depth ft in
AMMAN RIDER			Fireclay 15 0		816 6
Coal .. 18 in			Coal .. 6 in 6		817 0
Fireclay .. 12 in			Bastard rock and clift .. 24 0		841 0
Coal .. 6 in 3 0		738 8	**FIVE-FEET**		
Fireclay, rock and clift .. 32 4		771 0	Coal .. 31 in		
YARD			Clod .. 34 in		
Coal .. 24 in 2 0		773 0	Coal .. 24 in 7 5		848 5
Fireclay, rock and clift .. 24 0		797 0	Shale 30 0		878 5
SEVEN-FEET			**GELLIDEG**		
Coal .. 54 in 4 6		801 6	Coal .. 2 in 2		878 7
			Shale 6 0		884 7

Lletty Shenkin Colliery: Lower Pit

Ht. above O.D. 380 ft.   6-in SN 00 S.W.; Glam. 18 N.E.

Site 900 yd E. 9° N. of Aberaman church. National Grid Ref. 02530105.

Drift to 60 ft 10 in. Two-Feet-Nine at 271 ft 4 in. Sunk to Four-Feet at 295 ft 10 in.

Llwyncelyn Pit

Ht. above O.D. 290 ft.   6-in ST 09 S.W.; Glam. 28 S.W.

Site 370 yd S. 5° E. of St. Luke's Church, Porth. National Grid Ref. 03429100.

No. 2 Rhondda at 81 ft 7 in. No. 3 Rhondda at 305 ft 4 in.

Llwynypia Colliery: No. 1 Pit

Ht. above O.D. 455 ft.   6-in SS 99 S.E.; Glam. 27 N.E.

Site 650 yd N. 14° E. of St. Andrew's Church, Llwyn-y-pia. National Grid Ref. 99479339. Published in Vertical Sections, Sheet 84, 1901.

No. 2 Rhondda at 98 ft, No. 3 Rhondda at 324 ft 6 in, Pentre at 690 ft 3 in, Two-Feet-Nine at 974 ft 2 in, Upper Six-Feet at 1078 ft 7 in, Lower Six-Feet at 1122 ft 1 in, Upper Nine-Feet at 1253 ft 3 in, Lower Nine-Feet and Bute at 1373 ft 7 in, Yard at 1476 ft 1 in, Five-Feet at 1520 ft 7 in, Gellideg at 1551 ft 7 in.

Llwynypia Colliery: No. 6 Pit

Ht. above O.D. 454 ft.   6-in SS 99 S.E.; Glam. 27 N.E.

Site 500 yd N. 31° E. of St. Andrew's Church, Llwyn-y-pia. National Grid Ref. 99569323.

No. 2 Rhondda at 78 ft, No. 3 Rhondda at 321 ft 9 in, Two-Feet-Nine at 1058 ft 6 in, Four-Feet (split) at 1093 ft 10 in, Lower Six-Feet at 1160 ft 3 in. Sunk to ?Lower Nine-Feet and Bute at 1326 ft 7 in.

Lower Duffryn Colliery

Ht. above O.D. 496 ft.   6-in ST 00 S.W.; Glam. 19 N.W.

Site 1070 yd N. 40° W. of Mountain Ash church. National Grid Ref. 04130014.

	Thickness ft in	Depth ft in		Thickness ft in	Depth ft in
COAL MEASURES			**CLAY**		
			Coal .. 27 in 2 3		279 3
CASTELL-Y-WIEVER at pit top level .			Fireclay, clift, rock and mine ground 80 4		359 7
			Coal .. 1 in 1		359 8
Measures 277 0		277 0	Fireclay 4 3		363 11

		Thickness ft in	Depth ft in
CRAIG			
Coal .. 30 in			
Strong clod .. 2 in			
Coal .. 5 in			
Clod .. 8 in			
Coal .. 4 in		**4 1**	**368 0**
Fireclay and mine ground 51 6			419 6
PENTRE RIDER			
Coal .. 15 in		**1 3**	**420 9**
Fireclay and clift with rock bands 31 4			452 1
PENTRE			
Coal .. 7 in			
Clod .. 3 in			
Coal .. 7 in		**1 5**	**453 6**
Clift with rock bands .. 17 10			471 4
LOWER PENTRE			
Coal .. 14 in			
Black stone .. 2 in			
Coal .. 6 in		**1 10**	**473 2**
Fireclay, clift and beds of rock 32 7			505 9
EIGHTEEN-INCH			
Coal .. 13 in		**1 1**	**506 10**
Black clay, strong clift and beds of rock 36 7			543 5
Bast with blackband .. 1 9			545 2
GORLLWYN RIDER			
Coal .. 7 in		**7**	**545 9**
Strong clift 8 11			554 8
GORLLWYN			
Coal and rashings 30 in		**2 6**	**557 2**
Fireclay, rock, mine ground and clift.. .. .. 67 7			624 9
Coal .. 3 in		**3**	**625 0**
Fireclay, bands of rock, clift and mine ground .. 55 10			680 10
Coal .. 2 in		**2**	**681 0**
Fireclay, clift and mine ground with rock beds.. 32 2			713 2
THREE COALS			
Coal .. 20 in			
Fireclay .. 21 in			
Coal .. 3 in			
Hard rock 36 in			
Coal .. 7 in		**7 3**	**720 5**
Fireclay, strong clift, rock bands and rashings .. 44 7			765 0
TWO-FEET-NINE			
Coal .. 34 in			
Clod .. 4 in			
Coal .. 5 in			
Clod .. 10 in			
Coal .. 3 in		**4 8**	**769 8**
Fireclay and mine ground 19 2			788 10

		Thickness ft in	Depth ft in
FOUR-FEET			
Coal .. 60 in		**5 0**	**793 10**
Fireclay and mine ground 16 0			809 10
UPPER SIX-FEET			
Coal .. 24 in		**2 0**	**811 10**
Fireclay and clift with beds of rock 29 6			841 4
LOWER SIX-FEET			
Coal .. 5 in			
Bast .. 10 in			
Coal .. 39 in			
Rashings 24 in		**6 6**	**847 10**
Fireclay 9 0			856 10
Coal .. 8 in			
Rashings 8 in		**1 4**	**858 2**
Strong fireclay .. 4 0			862 2
Coal .. 6 in			
Rashings 17 in		**1 11**	**864 1**
Strong fireclay and clift .. 11 11			876 0
Rashings 16 in			
Coal .. 4 in			
Fireclay and rashings 28 in			
Coal .. 8 in			
Rashings 12 in		**5 8**	**881 8**
Strong rock, clift and mine ground 42 4			924 0
NINE-FEET			
Coal .. 45 in			
Rashings 4 in			
Bast .. 5 in			
Coal .. 62 in			
Rashings 18 in		**11 2**	**935 2**
Bastard fireclay, clift and bed of rock 15 8			950 10
BUTE			
Coal .. 22 in			
Fireclay .. 36 in			
Coal .. 26 in		**7 0**	**957 10**
Fireclay, clift and mine .. 21 6			979 4
AMMAN RIDER			
Coal .. 18 in			
Fireclay .. 8 in			
Coal .. 8 in		**2 10**	**982 2**
Fireclay, clift and mine .. 25 8			1007 10
YARD			
Coal .. 36 in		**3 0**	**1010 10**
Fireclay and clift 22 10			1033 8
SEVEN-FEET			
Coal .. 12 in			
Stone .. 2 in			
Coal .. 30 in			
Stone .. 1 in			
Coal .. 12 in		**4 9**	**1038 5**

APPENDIX I

Maindy Colliery

Ht. above O.D. 622 ft. 6-in SS 99 S.E.; Glam. 27 N.W.

Site 550 yd W. 3° N. of Ton church. National Grid Ref. 96529498.

	Thickness ft in		Depth ft in	
DRIFT				
Clay and gravel	18	0	18	0
COAL MEASURES				
Mine ground	9	0	27	0
Coal .. 9 in		9	27	9
Fireclay, rock and clift ..	38	1	65	10
ABERGORKY (TOP COAL)				
Coal .. 20 in	1	8	67	6
Fireclay and clift	11	6	79	0
ABERGORKY (BOTTOM COAL)				
Coal .. 16 in	1	4	80	4
Fireclay, rock, clift and mine ground	74	7	154	11
PENTRE RIDER				
Coal .. 18 in	1	6	156	5
Fireclay, rock and clift ..	19	10	176	3
PENTRE				
Coal .. 31 in	2	7	178	10
Fireclay, rock and clift ..	56	5	235	3
LOWER PENTRE				
Coal .. 12 in				
Parting .. 2 in				
Coal .. 3 in	1	5	236	8
Fireclay, clift and blackband	38	7	275	3
Coal .. 6 in				
Fireclay .. 26 in				
Coal .. 2 in				
Fireclay .. 48 in				
Coal .. 1 in	6	11	282	2
Fireclay, rock, clift and mine ground	49	11	332	1
GORLLWYN RIDER				
Coal .. 26 in	2	2	334	3
Fireclay and black ground	5	5	339	8
GORLLWYN				
Coal .. 4 in				
Rashes .. 18 in				
Coal .. 6 in	2	4	342	0
Fireclay, clift, rock and mine ground	103	7	445	7
Coal .. 6 in		6	446	1
Fireclay, clift and beds of rock	24	8	470	9
Coal .. 5 in		5	471	2

	Thickness ft in		Depth ft in	
Rock and clift	20	0	491	2
Coal .. 9 in		9	491	11
Rock and clift	34	0	525	11
Coal and bast .. 2 in		2	526	1
Fireclay and mine	16	2	542	3
Coal .. 13 in				
Stone .. 1 in				
Coal .. 10 in	2	0	544	3
Fireclay, rock and clift ..	23	3	567	6
Coal .. 16 in	1	4	568	10
Fireclay and clift	12	11	581	9
Coal and bast .. 9 in				
Fireclay .. 36 in				
Coal .. 9 in	4	6	586	3
Fireclay and clift	9	0	595	3
Coal .. 12 in	1	0	596	3
Fireclay, rock and clift ..	21	1	617	4
Coal .. 27 in	2	3	619	7
Fireclay, shale and mine ..	15	6	635	1
TWO-FEET-NINE				
Coal .. 66 in	5	6	640	7
Fireclay and shale ..	5	0	645	7
Coal .. 16 in				
Shale .. 1 in				
Coal .. 11 in	2	4	647	11
Fireclay, shale and mine ground	24	0	671	11
FOUR-FEET (TOP COAL)				
Coal .. 9 in				
Fireclay .. 7 in				
Coal .. 6 in				
Shale .. 2 in				
Coal .. 6 in				
Fireclay .. 26 in				
Coal .. 3 in				
Shale .. 6 in				
Coal .. 27 in	7	8	679	7
Fireclay and mine ..	6	0	685	7
FOUR-FEET (BOTTOM COAL)				
Coal .. 42 in				
Fireclay .. 15 in				
Coal .. 18 in				
Shale .. 3 in				
Coal .. 3 in	6	9	692	4
Fireclay, clod and shale ..	27	0	719	4
Coal .. 3 in		3	719	7

No record from here to Lower Six-Feet

Downcast Shaft

	Thickness ft in	Depth below Lower Six-Feet ft in
LOWER SIX-FEET		
Coal .. 78 in	6 6	6 6
Measures and drift .. 31 3		37 9
Coal .. 24 in	2 0	39 9
Fireclay, clift and rock .. 72 0		111 9
Coal .. 4 in		
Bast .. 13 in		
Coal .. 12 in	2 5	114 2
Fireclay, clift and rock .. 27 6		141 8
Coal and brass .. 24 in		
Rashes .. 12 in	3 0	144 8
Fireclay, clift and rock .. 48 4		193 0
BRUNTS		
Coal .. 25 in		
Clod .. 2 in		
Coal .. 25 in	4 4	197 4
Fireclay, rock and clift .. 26 4		223 8
AMMAN RIDER		
Coal .. 11 in	11	224 7
Fireclay and clift .. 34 5		259 0
Rock 28 7		287 7
Clift 25 1		312 8
YARD		
Rashes and coal .. 2 in		
Coal .. 40 in		
Clod .. 7 in		
Coal .. 10 in	4 11	317 7
Clod	8	318 3
UPPER SEVEN-FEET		
Coal .. 8 in	8	318 11
Fireclay and clift 34 3		353 2
MIDDLE SEVEN-FEET		
Coal .. 38 in		
Rashes .. 10 in	4 0	357 2
Fireclay and clift 11 9		368 11
LOWER SEVEN-FEET		
Coal .. 22 in	1 10	370 9
Fireclay, clift and clod .. 15 6		386 3
Coal .. 1 in	1	386 4
Shales, clift and rock .. 26 1		412 5
Coal .. 3 in		
Clod .. 3 in		
Coal .. 9 in	1 3	413 8
Fireclay and clift 6 10		420 6
FIVE-FEET		
Coal .. 41 in		
Clod .. 7 in		
Coal .. 28 in		
Rashes .. 1 in		
Coal .. 4 in		
Rashes .. 3 in		
Coal .. 4 in	7 4	427 10
Fireclay 6 6		434 4
Coal .. 4 in		
Rashes .. 6 in		
Coal .. 3 in	1 1	435 5
Clod 1 6		436 11
GELLIDEG		
Coal .. 63 in	5 3	442 2
Fireclay and clod .. 21 0		463 2
Coal .. 1 in	1	463 3
Fireclay, rock and clift .. 31 9		495 0

Mardy Colliery: No. 1 Pit

Ht. above O.D. 949 ft.   6-in SS 99 N.E.; Glam. 18 N.W.

Site 300 yd N. 33° W. of Maerdy railway station. National Grid Ref. 97369885.

	Thickness ft in	Depth ft in
DRIFT		
Subsoil 45 0		45 0
COAL MEASURES		
Fireclay 7 6		52 6
Rock and clift .. 10 6		63 0
TORMYNYDD		
Coal .. 24 in	2 0	65 0
Fireclay and hard clift .. 30 0		95 0
Coal .. 3 in	3	95 3
Rock 15 0		110 3
Fireclay and shale .. 14 9		125 0
?BLACKBAND		
Coal .. 18 in	1 6	126 6
Fireclay, rock and clift .. 25 6		152 0
Stone and mine ground .. 36 0		188 0
Coal .. 9 in	9	188 9
Fireclay and clift 25 3		214 0
HAFOD (BOTTOM COAL)		
Coal .. 15 in	1 3	215 3
Fireclay and rock.. .. 31 9		247 0
Clift and mine ground .. 64 6		311 6
Coal .. 1 in	1	311 7
Fireclay and clift 8 0		319 7
Rock 4 5		324 0
Mine ground and clift .. 41 0		365 0

```
                              Thick-                              Thick-
                               ness    Depth                      ness    Depth
                              ft  in   ft  in                     ft  in   ft  in
ABERGORKY                              TWO-FEET-NINE
      Coal     .. 22 in                      Coal     .. 20 in
      Dirt     .. 13 in                      Soft shale  55 in
      Coal     .. 16 in  4  3   369  3       Coal  with
Fireclay and clift with iron-                  brass ..  22 in
  stone    ..      ..  .. 64 9  434  0        Fireclay ..  30 in
                                             Coal     .. 17 in  12  0  820  9
PENTRE RIDER                           Shale and ironstone  .. 22  9  843  6
      Coal     .. 17 in  1  5   435  5
Fireclay and shale with                FOUR-FEET
  ironstone   ..     .. 17 8   453  1         Coal     .. 18 in
                                             Clod     .. 12 in
PENTRE                                       Coal     .. 66 in
      Coal     .. 13 in                       Rashes   ..  6 in
      Clod     ..  3 in                       Coal     .. 12 in  9  6  853  0
      Coal     .. 11 in  2  3   455  4 Fireclay and  shale  with
Clay, rock and clift   ..  7  4 462  8    rock band    ..     .. 18  9  871  9
      Coal     .. 18 in  1  6   464  2
Fireclay, clift with rock              UPPER SIX-FEET
  bands and ironstone .. 25 7  489  9         Coal     .. 33 in  2  9  874  6
                                       Blackband and black rashes 21  5  895 11
LOWER PENTRE
      Coal     .. 12 in                LOWER SIX-FEET
      Stone    ..  2 in                      Coal     .. 38 in  3  2  899  1
      Coal     ..  6 in                Fireclay  ..      ..   .. 3 10  902 11
      Stone    ..  1 in                Black rashes and coal  ..  2  4  905  3
      Coal     ..  1 in  1 10   491  7 Hard shale with thin rock
Fireclay, clift and shale .. 32 3  523 10  and ironstone  ..   .. 38  4  943  7
                                             Coal     ..  4 in
EIGHTEEN-INCH                                Clod     ..  4 in
      Coal     .. 12 in  1  0   524 10       Coal     .. 26 in  2 10  946  5
Fireclay and clift with rock           Clod, shale and rashes  ..  5  5  951 10
  bands    ..      ..  .. 35 2  560  0
                                       NINE-FEET
GORLLWYN RIDER                               Soft coal   75 in
      Coal     .. 13 in  1  1   561  1       Clod     ..  6 in
Fireclay and clift with rock                 Coal     .. 10 in  7  7  959  5
  bands   ..      ..  .. 21 7   582  8 Black rashes   ..      .. 3  9  963  2
                                       Fireclay, shale and rock .. 25  0  988  2
GORLLWYN
      Coal     ..  1 in                ?AMMAN RIDER
      Soft clod  2 in                        Coal     .. 25 in  2  1  990  3
      Coal     .. 21 in  2  0   584  8 Fireclay  and  shale  with
Fireclay, clift and shale .. 69 7  654  3  mine   ..      ..   .. 29  3  1019  6
      Coal     ..  2 in     2   654  5
Clay and clift with rock               YARD
  band and mine..     .. 50 8   705  1       Coal     .. 35 in
      Coal     ..  7 in     7   705  8       Stone    ..  5 in
Fireclay and  shale  with                    Coal     ..  6 in  3 10  1023  4
  mine   ..      ..   .. 26 0   731  8 Fireclay, shale and rashes 28  4  1051  8
      Coal     .. 13 in  1  1   732  9
Fireclay and clift with rock           SEVEN-FEET
  beds and ironstone  .. 27 5   760  2       Coal     .. 42 in
                                             Fireclay ..  52 in
THREE COALS                                  Coal     .. 20 in  9  6  1061  2
      Coal     .. 22 in                Fireclay,  shale  and  clift
      Fireclay with                      with mine   ..     .. 30  4  1091  6
        mine ..   58 in
      Coal with                        FIVE-FEET
        brass ..  10 in                      Coal     .. 29 in
      Shale and                              Stone    ..  2 in
        fireclay 28 in                       Coal     ..  5 in
      Coal     ..  8 in                      Clod     .. 29 in
      Rashes   ..  6 in                       Coal     .. 27 in  7  8  1099  2
      Coal     .. 10 in  11 10  772  0 Fireclay, clift and shale  ..  8  7  1107  9
Fireclay, clift and rock .. 36 9  808  9 Rock   ..      ..   .. 29  1  1136 10

                                       GELLIDEG
                                             Coal       .. 26 in  2  2  1139  0
                                       Fireclay, rock and clift .. 21  7  1160  7
```

Mardy Colliery: No. 3 Pit

Ht. above O.D. 1064 ft.   6-in SS 99 N.E.; Glam. 18 N.W.

Site 1870 yd N.W. of Maerdy railway station. National Grid Ref. 96379987. Sunk 1892–94. Published in Vertical Sections, Sheet 83, 1900.

Drift to 34 ft 11 in, Gorllwyn at 779 ft 8 in, Two-Feet-Nine at 1022 ft 7 in, Four-Feet at 1070 ft 5 in, Upper Six-Feet at 1087 ft 1 in, Lower Six-Feet at 1142 ft 10 in, Red Vein at 1160 ft, Nine-Feet at 1235 ft 6 in, Bute at 1279 ft 8 in, Yard at 1359 ft 11 in, Upper and Middle Seven-Feet at 1373 ft 1 in, Five-Feet at 1423 ft 11 in, Gellideg at 1455 ft 5 in. Sunk to 1499 ft 4 in.

Margam Park No. 1 Borehole

Ht. above O.D. 286 ft.   6-in SS 88 N.W.; Glam. 33 N.E.

Site 2050 yd E. 14° S. of St. Mary's Church, Margam Park. National Grid Ref. 81948581. Average dip 15°. Drilled 1954.

Cefn Coed Marine Band at 207 ft 9 in, Britannic Marine Band at 296 ft 10 in, Hafod Heulog Marine Band at 332 ft, repeated at 370 ft 10 in, Two-Feet-Nine at 551 ft 7 in, Upper Four-Feet at 595 ft 7 in, Lower Four-Feet at 630 ft 7 in, ?Six-Feet at 723 ft 2 in, ?Red Vein at 798 ft 1 in, ?Upper Nine-Feet at 835 ft, ?Lower Nine-Feet at 873 ft 3 in, Bute at 1013 ft 3 in, Amman Marine Band at 1053 ft 1 in, Yard at 1175 ft 6 in, Lag fault at about 1225 ft, Gellideg at 1342 ft 6 in, Garw at 1503 ft 6 in. For section to base of Coal Measures at 2650 ft, see pp. 77–9. Drilled to 3550 ft.

Margam Park No. 2 Borehole

Ht. above O.D. 320 ft.   6-in SS 88 N.W.; Glam. 33 N.E.

Site 1070 yd E. 30° N. of St. Mary's Church, Margam Park. National Grid Ref. 81118632. Average dip 15°. Drilled 1954–55.

Lower Cwmgorse Marine Band at 105 ft 6 in, Five Roads Marine Band at 223 ft, Foraminifera Marine Band at 295 ft 7 in, ?Pentre at 346 ft 3 in, Britannic Marine Band at 494 ft, Hafod Heulog Marine Band at 528 ft 6 in, Two-Feet-Nine at 736 ft 6 in, Upper Four-Feet at 833 ft 8 in, Lower Four-Feet at 874 ft 10 in, Six-Feet at 994 ft 10 in, Caerau at 1018 ft 10 in, ?Red Vein at 1084 ft 6 in, Upper Nine-Feet at 1152 ft 9 in, Lower Nine-Feet at 1221 ft 7 in, Lag fault at about 1370 ft, Lower Five-Feet at 1430 ft 7 in, Gellideg at 1580 ft, Garw at 1747 ft 7 in. Drilled to 2029 ft 8 in.

Margam Park No. 3 Borehole

Ht. above O.D. 359 ft.   6-in SS 88 N.W.; Glam. 33 N.E.

Site 1950 yd E. 35° N. of St. Mary's Church, Margam Park. National Grid Ref. 81628730. Average dip 10°. Drilled by percussion to 389 ft 3 in, the record to that depth being of an earlier borehole on the same site. Drilled 1954–55.

	Thickness ft in	Depth ft in		Thickness ft in	Depth ft in
DRIFT			Mudstone and silty mudstone	16 7	405 10
Red clay and boulders ..	8 0	8 0	Pennant-type sandstone ..	5 5	411 3
COAL MEASURES			Mudstone, silty mudstone and striped beds with sandstone bands	30 3	441 6
Coal .. 9 in	9	8 9			
Rock 49	3	58 0	Sandstone, mainly pennant-type, with bands of siltstone, striped beds and sporadic mudstone ..177	10	619 4
Fireclay 3	0	61 0			
Rock with coal markings in places 142	0	203 0	Mudstone with two coal streaks 1	3	620 7
Fireclay and clift 31	6	234 6			
Sandstone 40	0	274 6			
Sandy clift with sandstone bands 114	9	389 3			

	Thickness ft in	Depth ft in
Seatearth, passing into silty mudstone and striped beds	23 0	643 7
Coal .. 6 in	6	644 1
Seatearth and mudstone ..	8 5	652 6
Coal .. 2 in		
Mudstone 11 in		
Coal .. 1½ in		
Seatearth and shale .. 3 in		
Coal .. 3 in		
Seatearth 2½ in		
Coal .. 1 in	2 0	654 6
Seatearth and mudstone ..	9 2	663 8
Coal .. 7 in	7	664 3
Shale and mudstone ..	4 0	668 3
Black rashings 4 in		
Coal .. 3 in	7	668 10
Seatearth and mudstone ..	10 5	679 3

WHITE

	Thickness ft in	Depth ft in
Coal .. 24 in	2 0	681 3
Seatearth and silty mudstone with ironstone ..	25 0	706 3

UPPER CWMGORSE MARINE BAND

	Thickness ft in	Depth ft in
Silty and shaly mudstone; *Lingula*; Nuculids; *Curvirimula*; *Anthracoceras?*; fish ..	13 9	720 0
Silty mudstone and siltstone	10 0	730 0
Shale and mudstone with coal streak at top and base	1 9	731 9
Mudstone and shale; *Spirorbis sp.; Naiadites* cf. *hindi*	2 9	734 6
Mudstone and silty mudstone; plants ..	23 2	757 8
Coal .. 2 in	2	757 10
Seatearth and mudstone with silty and sandy bands	51 3	809 1

HAFOD (BOTTOM COAL)

	Thickness ft in	Depth ft in
Rashings 6 in		
Coal .. 24 in	2 6	811 7
Seatearth	3 5	815 0
Sandstone and siltstone with mudstone bands ..	33 9	848 9
Mudstone and silty mudstone with ironstone; *Planolites ophthalmoides* near top; *Euestheria sp.* at 879 ft; *Naiadites sp.* and *Geisina? subarcuata* 887 ft 2 in to 888 ft ..	62 3	911 0
Siltstone with quartzitic bands	4 0	915 0

LOWER CWMGORSE MARINE BAND

	Thickness ft in	Depth ft in
Mudstone with ironstone; foraminifera; *P. ophthalmoides; Lingula; Myalina?; Euestheria; Hollinella*; fish	32 4	947 4
Coal .. 10 in	10	948 2
Seatearth and black shale	6 8	954 10
Coal .. 6 in	6	955 4
Seatearth passing into mudstone with ironstone; *Naiadites sp.* and *G.? subarcuata* at 960 ft to 962 ft 6 in; *N. sp.* and *Euestheria sp.* 990 ft to 1003 ft	60 1	1015 5

ABERGORKY (TOP COAL)

	Thickness ft in	Depth ft in
Coal .. 23 in	1 11	1017 4
Mudstone with ironstone; *Euestheria sp.*; fish ..	9 1	1026 5
Coal .. 8 in		
Rashings 6 in	1 2	1027 7
Mudstone, shale and seatearth with silty bands; *Naiadites sp.* at base ..	26 0	1053 7

ABERGORKY (BOTTOM COAL)

	Thickness ft in	Depth ft in
Coal .. 17 in	1 5	1055 6
Seatearth and mudstone, silty at base	11 10	1066 10

FIVE ROADS MARINE BAND

	Thickness ft in	Depth ft in
Mudstone with siltstone bands; *Myalina*	5 5	1072 3
Mudstone, siltstone and sandstone	9 2	1081 5
Mudstone	59 7	1141 0

FORAMINIFERA MARINE BAND

	Thickness ft in	Depth ft in
Smooth mudstone with small ironstone nodules; *P. ophthalmoides*, foraminifera	10 7	1151 7

PENTRE RIDER

	Thickness ft in	Depth ft in
Coal .. 17 in	1 5	1153 6
Seatearth and mudstone with siltstone and sandstone layers	29 2	1182 2
Mudstone, seatearth and rashings	6 5	1188 7
Mudstone and shale ..	5 9	1194 4
Coal .. 9 in	9	1195 1
Seatearth, rashings and mudstone	2 6	1197 7
Sandstone and siltstone ..	11 2	1208 9
Mudstone; plants.. ..	5 7	1214 4
Coal .. 12 in	1 0	1215 4
Seatearth and mudstone with siltstone band ..	27 7	1242 11
Coal .. 19 in	1 7	1244 6

	Thickness ft	in	Depth ft	in
Seatearth and mudstone ..	8	9	1253	3
Strong siltstone	8	1	1261	4
Coal .. 6 in				
Rashings 8 in	**1**	**2**	**1262**	**6**
Mudstone and striped beds with sandstone layers ..	44	2	1306	8
Seatearth with rashings band	8	9	1315	5
Mudstone, with bands of ganister-like seatearth, sheared in places ..	27	11	1343	4
Coal, dirty 3 in		**3**	**1343**	**7**
Seatearth and mudstone, sandy at top; many ironstone layers; *P. ophthalmoides* at 1384 ft to 1388 ft	99	0	1442	7

CEFN COED MARINE BAND

	Thickness ft	in	Depth ft	in
Dark shaly mudstone; crinoids; Chonetids, *Lingula, Orbiculoidea;* gastropods; lamellibranchs; *Anthracoceras sp.;* trilobites	4	9	1447	4
Coal .. 4 in		**4**	**1447**	**8**
Seatearth, passing to sandstone and siltstone ..	10	3	1457	11
Mudstone with sphaerosiderite	14	3	1472	2
Coal .. 2 in		**2**	**1472**	**4**
Shale and mudstone with silty and sandy layers ..	21	10	1494	2
Coal .. 2 in		**2**	**1494**	**4**
Seatearth and mudstone, sandy near top, much ironstone towards base; *P. ophthalmoides* at 1547 ft to 1549 ft	56	10	1551	2

BRITANNIC MARINE BAND

	Thickness ft	in	Depth ft	in
Dark shaly mudstone; ?foraminifera; fish ..	1	11	1553	1
Cannelly shale with **coal** streaks		6	1553	7
Cannel and cannelly shale .. 2 in		**2**	**1553**	**9**
Shale and mudstone ..	13	1	1566	10
Sandstone	6	9	1573	7
Mudstone with ironstone	32	4	1605	11
Coal .. 2 in		**2**	**1606**	**1**
Seatearth and mudstone; *Anthraconaia rubida* at 1614 ft 8 in	18	0	1624	1
Sandstone	7	2	1631	3
Mudstone with ironstone	45	3	1676	6
Coal .. 2 in		**2**	**1676**	**8**
Seatearth	4	3	1680	11
Coal .. 11 in		**11**	**1681**	**10**
Seatearth and mudstone with roots and sphaerosiderite	31	5	1713	3
Mudstone with silty layers	15	6	1728	9

	Thickness ft	in	Depth ft	in
Sandstone and siltstone with mudstone layers ..	33	3	1762	0
Mudstone with ironstone, sheared at base.. ..	32	4	1794	4

TWO-FEET-NINE

	Thickness ft	in	Depth ft	in
Coal .. 70 in	**5**	**10**	**1800**	**2**
Mudstone and seatearth, sheared in places, with sandy layers	60	2	1860	4
Coal .. 72 in	**6**	**0**	**1866**	**4**
Seatearth and mudstone, highly sheared in places, with a few sandy bands	70	10	1937	2
Coal .. 108 in	**9**	**0**	**1946**	**2**
Seatearth and mudstone ..	22	10	1969	0
Rashings		10	1969	10
Seatearth and mudstone with silty layers; *Anthraconaia pulchella* near base	27	4	1997	2

?RED VEIN

	Thickness ft	in	Depth ft	in
Coal .. 22 in				
Seatearth 26 in				
Coal .. 48 in	**8**	**0**	**2005**	**2**
Seatearth and mudstone; *Anthracosia* cf. *phrygiana*	20	0	2025	2
Coal .. 27 in	**2**	**3**	**2027**	**5**
Seatearth and mudstone with a little sandstone ..	56	11	2084	4

?UPPER NINE-FEET

	Thickness ft	in	Depth ft	in
Coal .. 53 in	**4**	**5**	**2088**	**9**
Seatearth and mudstone with ironstone	48	2	2136	11
Sandstone and siltstone ..	10	10	2147	9
Mudstone with a little sandstone and siltstone	64	9	2212	6

?LOWER NINE-FEET

	Thickness ft	in	Depth ft	in
Coal .. 51 in				
Seatearth 34 in				
Coal .. 15 in	**8**	**4**	**2220**	**10**
Mudstone and soft clay (? fault)	13	3	2234	1
Coal .. 24 in				
(? repeat of last coal) ..	**2**	**0**	**2236**	**1**
Mudstone with ironstone; cannelly towards base; crushed shells including *Anthracosphaerium turgidum* at 2253 ft to 2257 ft 3 in	24	0	2260	1

BUTE

	Thickness ft	in	Depth ft	in
Coal .. 71 in	**5**	**11**	**2266**	**0**
Seatearth and mudstone with sandy and silty layers; *Anthracosia* cf. *ovum* and *A.* cf. *aquilinoides* at 2300 ft to 2300 ft 8 in	34	8	2300	8

AMMAN MARINE BAND

	Thickness ft	in	Depth ft	in
Smooth mudstone; *Lingula mytilloides* ..	3	2	2303	10

AMMAN RIDER	Thickness ft in	Depth ft in
Coal .. 8 in		
Seatearth 8 in		
Coal .. 10 in	2 2	2306 0
Seatearth and mudstone with silty and sandy layers; *Spirorbis sp.; Naiadites sp.* near base..	49 8	2355 8
Coal .. 15 in	1 3	2356 11
Mudstone with silty and sandy layers; ironstone *Anthracosia regularis* at 2388 ft 7 in to 2389 ft ..	45 1	2402 0
?YARD		
Coal .. 15 in	1 3	2403 3
Seatearth and mudstone ..	5 10	2409 1
?MIDDLE SEVEN-FEET		
Coal .. 66 in		
Seatearth 43 in		
Coal .. 11 in		
Seatearth 3 in		
Coal .. 1 in	10 4	2419 5
Seatearth and mudstone with striped layers ..	16 9	2436 2
?LOWER SEVEN-FEET		
Coal .. 42 in	3 6	2439 8
Seatearth and mudstone, cannelly near base ..	21 11	2461 7
Coal .. 6 in	6	2462 1
Seatearth and mudstone with ironstone; sandy and silty layers.. ..	49 11	2512 0
UPPER FIVE-FEET		
Coal .. 81 in	6 9	2518 9
Seatearth and mudstone ..	9 0	2527 9
Coal .. 2 in	2	2527 11
Seatearth and mudstone ..	27 3	2555 2
Siltstone and sandstone with mudstone bands ..	10 11	2566 1
?LOWER FIVE-FEET		
Coal .. 12 in		
Seatearth 3 in		
Coal .. 1 in	1 4	2567 5
Seatearth and mudstone with sporadic ironstone and sandy layers ..	83 9	2651 2

	Thickness ft in	Depth ft in
Striped beds, siltstone and sandstone	24 6	2675 8
Dark mudstone; *Anthraconaia sp. nov.* cf. *wardi, Carbonicola* cf. *pseudorobusta, Curvirimula* cf. *subovata*	4 8	2680 4
GELLIDEG		
Rashings 2 in		
Coal .. 101 in	8 7	2688 11
Mudstone	10 4	2699 3
Siltstone and sandstone ..	5 0	2704 3
Mudstone and shale; *P. ophthalmoides* and *Carbonicola sp.* at 2709 ft; *C.* aff. *communis, C. sp.* of *communis/pseudorobusta* group, *Curvirimula subovata* and *Geisina arcuata* at 2725 ft 9 in to 2731 ft 7 in	32 4	2736 7
Black shale with fish ..	5	2737 0
Cannel 2 in	2	2737 2
Black shale with fish ..	1 6	2738 8
Mudstone	6 10	2745 6
Sandstone and siltstone ..	21 8	2767 2
Mudstone with much ironstone; *Spirorbis sp.; C.* cf. *pseudorobusta, Curvirimula sp.; G. arcuata;* fish scales	32 5	2799 7
Siltstone with mudstone bands	18 5	2818 0
Mudstone, with much ironstone, black near base; *C.* cf. *communis* and *G. arcuata* at 2830 ft 8 in to 2838 ft; *Spirorbis sp., Curvirimula sp., G. arcuata* and fish remains at 2844 ft 9 in to 2853 ft 4 in	44 0	2862 0
GARW		
Coal .. 19 in	1 7	2863
Seatearth and mudstone ..	19 5	2883

Margam Park No. 4 Borehole

Ht. above O.D. 327 ft.   6-in SS 88 N.W.; Glam. 34 S.W.

Site 2620 yd N. 2° E. of St. Theodore's Church, Kenfig Hill. National Grid Ref. 84148530. Average dip 15°–20°. Drilled 1955.

Hafod (bottom coal) at 163 ft 4 in, Lower Cwmgorse Marine Band at 286 ft 11 in, Abergorky (top coal) at 366 ft 10 in, Abergorky (bottom coal) at 401 ft 5 in, Foraminifera Marine Band at 488 ft 1 in, Cefn Coed Marine Band at 855 ft 7 in, Hafod Heulog Marine Band at 1050 ft, Two-Feet-Nine at 1254 ft 10 in, Upper Four-Feet at 1306 ft 6 in, Lower Four-Feet at 1348 ft 9 in, Six-Feet at 1453 ft 4 in, ?Caerau at 1524 ft 9 in, ?Red Vein at 1555 ft 5 in, Upper Nine-Feet at 1602 ft 1 in, Lower Nine-Feet at 1638 ft 9 in, Bute at 1670 ft, Amman Marine Band at 1702 ft 4 in, Yard at 1768 ft 1 in, ?Middle Seven-Feet at 1788 ft 4 in, Lower Seven-Feet at 1856 ft 6 in, Upper Five-Feet at 1974 ft 8 in, Lower Five-Feet at 2136 ft 6 in, Gellideg at 2286 ft 3 in, Garw at 2495 ft 4 in. Drilled to 2530 ft 2 in.

Maritime Colliery: Downcast Shaft

Ht. above O.D. 264 ft.   6-in ST 08 N.E.; Glam. 28 S.W.

Site 600 yd S. 35° W. of Pontypridd church. National Grid Ref. 06828968. Published in Vertical Sections, Sheet 84, 1901. Sunk to Six-Feet about 1872; deepened 1904.

No. 3 Rhondda at 171 ft 9 in, Hafod at 310 ft 8 in, Six-Feet and Caerau at 842 ft 7 in, Upper Nine-Feet at 890 ft 5 in. Section below disturbed. ?Lower Nine-Feet and Bute at 1168 ft 7 in. Sunk to 1170 ft 5 in.

Meiros Colliery

Ht. above O.D. 397 ft.   6-in ST 08 S.W.; Glam. 35 S.E.

Site 850 yd N. 10° E. of Llanharan church. National Grid Ref. 00408398. Composite section of shaft and cross-measures drift.

	Thick-ness ft in	Depth ft in		Thick-ness ft in	Depth ft in
COAL MEASURES			Clift 23 9		603 5
Measures with two thin			Coal .. 12 in 1 0		604 5
coals 225 2		225 2	Fireclay, clift and rock ..161 9		766 2
			Coal .. 10 in 10		767 0
No. 3 RHONDDA			Fireclay 17 4		784 4
Coal .. 34 in 2 10		228 0	Coal .. 9 in 9		785 1
Measures 48 0		276 0	Clift 27 4		812 5
TORMYNYDD OR WHITE			Coal .. 30 in 2 6		814 11
Coal .. 8 in			Shale and layers of iron-		
Holing .. 2 in			stone 37 6		852 5
Coal .. 14 in 2 0		278 0	'FOUR-FEET'		
Rock 96 8		374 8	Coal .. 63 in 5 3		857 8
HAFOD			Clift with bands of rock		
Coal .. 28 in 2 4		377 0	and black shale .. 12 0		869 8
Clift 10 0		387 0	Coal .. 48 in 4 0		873 8
Rock 53 0		440 0	Clift 8 0		881 8
Coal .. 27 in 2 3		442 3	Coal .. 18 in 1 6		883 2
Fireclay and clift 24 0		466 3	Clift 55 0		938 2
BERGORKY			Coal .. 24 in 2 0		940 2
Coal .. 21 in 1 9		468 0	Shale 16 6		956 8
Fireclay, rock and clift .. 53 10		521 10	Coal .. 12 in 1 0		957 8
ENTRE RIDER			Rock, clift and rashings .. 11 0		968 8
Coal .. 21 in 1 9		523 7	Coal .. 21 in 1 9		970 5
Fireclay and clift with			Hard clift and bands of		
rashings band 54 11		578 6	rock 20 0		990 5
Coal .. 14 in 1 2		579 8	'SIX-FEET'		
			Coal .. 72 in 6 0		996 5

Merthyr Vale Colliery: No. 1 Pit

Ht. above O.D. 442 ft.   6-in SO 00 S.E.; Glam. 19 N.W.

Site 700 yd N. 28° W. of Merthyr Vale railway station. National Grid Ref. 07370009. Sunk to 1484 ft 10 in. in 1873; deepened to 1626 ft 2 in. in 1930. Published in Vertical Sections, Sheet 83, 1900.

Drift to 68 ft, Four-Feet at 1275 ft 10 in, Six-Feet at 1335 ft 10½ in, Nine-Feet at 1442 ft 9½ in, Yard at 1534 ft 8 in, Seven-Feet at 1565 ft 3 in, Gellideg at 1614 ft 4 in.

Middle Duffryn Pit

Ht. above O.D. 390 ft.   6-in SO 00 S.W.; Glam. 18 N.E.

Site 1550 yd E. 24° S. of Aberaman church. National Grid Ref. 03010035.

Two-Feet-Nine at 519 ft, Four-Feet at 540 ft, Lower Six-Feet at 594 ft, Red Vein at 620 ft 9 in, Nine-Feet at 699 ft, Bute at 735 ft 8 in, Yard at 782 ft, Seven-Feet at 815 ft 10 in, Five-Feet and Gellideg at 878 ft 8 in.

APPENDIX I

Mynachdy Borehole

Ht. above O.D. c. 650 ft.   6-in ST 09 N.E.; Glam. 19 S.W.

Site 1000 yd N. 41° W. of Ynys-y-bwl church. National Grid Ref. 05089508.

	Thickness ft in	Depth ft in
DRIFT		
Soil and sand	10 0	10 0
COAL MEASURES		
Pennant rock with thin shale bands ..	254 8	264 8
BRITHDIR RIDER		
Coal .. 9 in	9	265 5
Light grey seggar	21 0	286 5
Pennant rock with thin shale bands ..	175 10	462 3
BRITHDIR		
Coal .. 3 in		
Black stone .. 1 in		
Seggar .. 3 in		
Coal .. 5 in		
Seggar and black stone .. 11 in		
Coal .. 34 in	4 9	467 0
Seggar and shale with bands of rock	45 0	512 0
Pennant rock with shale bands	18 10	530 10
Coal .. 11 in	11	531 9
Seggar and shale with bands of rock	47 2	578 11
Pennant rock with shale bands	84 2	663 1
Coal .. 5 in		
Seggar and black stone .. 8½ in		
Coal .. 4 in		
Seggar with coal streaks 3½ in		
Coal .. 3 in	2 0	665 1
Seggar and shale	19 8	684 9
Pennant rock	12 0	696 9
Dark shale and seggar ..	43 6	740 3
Pennant rock	6 1	746 4
Dark shale and seggar ..	9 11	756 3
Coal .. 21 in	1 9	758 0
Dark seggar ..	7 10	765 10
Pennant rock and shale ..	13 2	779 0
Dark grey shale	15 4	794 4
Coal .. 2 in	2	794 6
Seggar and shale	19 11	814 5
Grey rock with shale bands	238 11	1053 4

Nantewlaeth Colliery: No. 2 Pit

Ht. above O.D. 680 ft.   6-in SS 89 N.E.; Glam. 17 S.W.

Site 2615 yd N. 11° E. of Cymmer Avan railway station. National Grid Ref. 86349776.

	Thickness ft in	Depth ft in
Walling	9 11	9 11
DRIFT		
Clay and boulders ..	60 0	69 11
COAL MEASURES		
Clift and rock	12 1	82 0
Rock	173 6	255 6
?NO. 1 RHONDDA RIDER		
Coal .. 14 in	1 2	256 8
Fireclay and clift	10 4	267 0
Coal .. 5 in	5	267 5
Fireclay and clift	15 7	283 0
Rock	33 0	316 0
?NO. 1 RHONDDA		
Coal .. 21 in	1 9	317 9
Fireclay and clift	47 3	365 0
Coal .. 12 in	1 0	366 0
Fireclay and rashings ..	12 0	378 0
Coal .. 10 in	10	378 10
Rashings and clift ..	25 2	404 0
?DAREN RHESTYN		
Coal .. 16 in		
Stone .. 2 in		
Coal .. 10 in	2 4	406 4
Clift with thin rock bands 47 0		453 4
Rock with thin bands of clift	194 8	648 0
Clift	9 0	657 0
NO. 2 RHONDDA		
Coal .. 12 in		
Holing .. 11 in		
Coal .. 8 in		
Fireclay .. 30 in		
Coal .. 24 in	7 1	664 1
Fireclay, bastard rock and clift	43 5	707 6
WERNPISTYLL RIDER		
Coal .. 18 in	1 6	709 0
Fireclay, clift and rock ..	44 6	753 0

	Thickness ft in	Depth ft in
WERNPISTYLL		
Coal .. 18 in	**1 6**	**755 0**
Fireclay	4 10	759 10
Bastard rock and rock ..	79 11	839 9
Clift and clod	8 8	848 5
Coal .. 6 in		
Clod .. 18 in		
Coal .. 7 in	**2 7**	**851 0**
Fireclay and clift with bastard rock	45 6	896 6
Coal .. 12 in	**1 0**	**897 6**
Fireclay and shale ..	7 4	904 10
?NO. 3 RHONDDA		
Coal .. 11 in	**11**	**905 9**
Fireclay	4 3	910 0
Rock with thin band of clift	190 7	1100 7
CORMYNYDD/WHITE		
Coal .. 8 in		
Rock .. 74 in		
Coal .. 5 in		
Rock .. 24 in		
Coal .. 18 in		
Rashings 12 in	**11 9**	**1112 4**
Fireclay and clift with bands of rock	77 5	1189 9
CLAY RIDER		
Coal .. 6 in		
Clift and fireclay 48 in		
Coal .. 7 in	**5 1**	**1194 10**
Fireclay, clift and rock ..	32 6	1227 4
CLAY		
Coal .. 12 in	**1 0**	**1228 4**
Fireclay, clift and rock ..	99 9	1328 1
UPPER BLACKBAND		
Coal .. 10 in	**10**	**1328 11**
Fireclay, clift and rock ..	48 9	1377 8
BERGORKY (TOP COAL)		
Coal .. 16 in	**1 4**	**1379 0**
Fireclay and clift	14 3	1393 3
ALBERT		
Coal .. 9 in	**9**	**1394 0**
Fireclay, rock and bastard rock with bands of clift	62 11	1456 11
Clift	23 3	1480 2
VICTORIA		
Coal .. 24 in	**2 0**	**1482 2**
Fireclay, clift and mine ..	14 10	1497 0
Coal and rashes.. 11 in	**11**	**1497 11**
Fireclay	3 6	1501 5
Rock	26 7	1528 0
Clift and fireclay ..	15 8	1543 8
UPPER YARD		
Coal .. 12 in		
Shale .. 3 in		
Coal .. 6 in	**1 9**	**1545 5**

	Thickness ft in	Depth ft in
Fireclay, rock and clift ..	29 8	1575 1
Coal .. 5 in	**5**	**1575 6**
Fireclay and clift	9 4	1584 10
TWO-AND-A-HALF		
Coal .. 12 in	**1 0**	**1585 10**
Fireclay, bastard rock and clift	36 5	1622 3
Coal .. 11 in	**11**	**1623 2**
Fireclay, rock and clift ..	25 0	1648 2
CAEDAVID		
Coal .. 17 in	**1 5**	**1649 7**
Fireclay, rashings, rock and clift	91 9	1741 4
Rashes	6	1741 10
Rock and clift ..	19 10	1761 8
Coal and rashes.. 12 in	**1 0**	**1762 8**
Rock and clift ..	21 6	1784 2
Coal and rashes.. 11 in	**11**	**1785 1**
Fireclay	5 3	1790 4
Cockshot	47 2	1837 6
Clift	3 1	1840 7
Coal .. 13 in	**1 1**	**1841 8**
Fireclay and clift with beds of rock	36 6	1878 2
Coal .. 5 in	**5**	**1878 7**
Fireclay	2 5	1881 0
Cockshot and black rock..	20 10	1901 10
Coal .. 16 in	**1 4**	**1903 2**
Fireclay and rashes ..	8 0	1911 2
Coal .. 3 in	**3**	**1911 5**
Fireclay, bastard rock and clift	48 9	1960 2
Rashes	9	1960 11
Clift	23 1	1984 0
Rashes and clift ..	1 0	1985 0
Clift and bastard rock ..	31 6	2016 6
FOUR-FEET		
Coal .. 8 in		
Holing .. 3 in		
Coal .. 56 in	**5 7**	**2022 1**
Fireclay, clift and beds of rock	112 3	2134 4
SIX-FEET		
Coal .. 15 in		
Holing .. 3 in		
Coal .. 69 in	**7 3**	**2141 7**
Fireclay	12 2	2153 9
Rashing and clift ..	9 2	2162 11
Fireclay and clift, beds of rock	64 7	2227 6
Rashes and clift ..	3 9	2231 3
Rock and clift ..	8 0	2239 3
Coal .. 13 in	**1 1**	**2240 4**
Fireclay, clift and rashes ..	50 11	2291 3
UPPER NINE-FEET		
Coal .. 56 in	**4 8**	**2295 11**

National Colliery

Ht. above O.D. 453 ft.   6-in ST 09 S.W.; Glam. 27 N.E.
Site 1330 yd E. 14° S. of Pontygwaith church. National Grid Ref.
02129378. Published in Vertical Sections, Sheet 83, 1900.

Drift to 17 ft 5 in, No. 2 Rhondda at 298 ft 6 in, Two-Feet-Nine at
1151 ft 3 in, Four-Feet at 1223 ft 10 in, Upper Six-Feet at 1254 ft 6 in,
Lower Six-Feet at 1284 ft 7 in, Upper Nine-Feet at 1366 ft 2 in, Lower
Nine-Feet and Bute at 1434 ft 11 in, Yard at 1506 ft 1 in, Upper and
Middle Seven-Feet at 1533 ft 6 in, Five-Feet at 1653 ft 11 in, Gellideg
at 1663 ft 9 in. Sunk to 1684 ft 3 in.

Naval Colliery: Nantgwyn Pit

Ht. above O.D. 693 ft.   6-in SS 99 S.E.; Glam. 27 N.E.
Site 1080 yd S. 12° W. of St. Andrew's Church, Llwyn-y-pia. National
Grid Ref. 99129187.

No. 3 Rhondda at 203 ft 1 in, Pentre at 596 ft 11 in, Six-Feet at
1079 ft 9 in, Upper Nine-Feet at 1161 ft 6 in, Lower Nine-Feet at
1221 ft 8 in, Bute at 1235 ft 7 in, Yard at 1335 ft 8 in, Five-Feet at
1454 ft 8 in, Gellideg at 1481 ft 4 in.

Naval Colliery: Pandy Pit

Ht. above O.D. 425 ft.   6-in SS 99 S.E.; Glam. 27 N.E.
Site 1100 yd S. 35° E. of St. Andrew's Church, Llwyn-y-pia. National
Grid Ref. 99889199. Published in Vertical Sections, Sheet 84, 1901.

No. 3 Rhondda at 261 ft 7 in, Pentre at 670 ft 2 in, Two-Feet-Nine
at 1009 ft 10 in, Upper Six-Feet at 1108 ft, Lower Six-Feet at 1121 ft 6 in,
Upper Nine-Feet at 1241 ft 10 in. Deepened to Gellideg.

	Thickness ft in		Depth ft in			Thickness ft in		Depth ft in	
					UPPER AND MIDDLE SEVEN-FEET				
UPPER NINE-FEET given as at	–	–	1246	5	Coal .. 18 in				
Fireclay and clift with beds					Rashes .. 12 in				
of rock	76	3	1322	8	Coal .. 24 in	4	6	1437	8
					Bastard quar, fireclay and				
LOWER NINE-FEET AND BUTE					clift	12	9	1450	5
Rashes .. 4 in					LOWER SEVEN-FEET				
Coal with					Coal .. 13 in	1	1	1451	6
partings 15 ft 9 in	16	1	1338	9	Fireclay and clift with bed				
Fireclay, rock and clift ..	36	4	1375	1	of rock	75	3	1526	9
AMMAN RIDER					FIVE-FEET				
Coal .. 21 in	1	9	1376	10	Coal .. 60 in				
Fireclay, clift and bastard					Rashes .. 15 in	6	3	1533	0
quar	48	6	1425	4	Fireclay with mine ..	10	9	1543	9
YARD					GELLIDEG				
Coal .. 64 in	5	4	1430	8	Coal .. 67 in	5	7	1549	4
Fireclay	2	6	1433	2					

Naval Colliery: Anthony Pit

Ht. above O.D. 425 ft.   6-in SS 99 S.E.; Glam. 27 N.E.
Site 1100 yd S. 35° E. of St. Andrew's Church, Llwyn-y-pia. National
Grid Ref. 99909198.

No 3 Rhondda at 256 ft 2 in, Pentre at 672 ft 7 in, Lower Six-Feet at
1115 ft 7 in, Upper Nine-Feet at 1234 ft 9 in, Lower Nine-Feet and
Bute at 1324 ft 4 in, Yard at 1412 ft 11 in, Five-Feet at 1522 ft 9 in,
Gellideg at 1534 ft 3 in. Sunk to 1544 ft 9 in.

Navigation Colliery

Ht. above O.D. 368 ft.   6-in ST 09 N.E.; Glam. 19 N.W.

Site 630 yd S. 32° E. of Mountain Ash church. National Grid Ref. 05069890.

		Thickness ft in	Depth ft in
Walling		16 9	16 9
DRIFT			
Gravel and sand ..	..	38 4	55 1
COAL MEASURES			
Soft blue clay		1 3	56 4
Measures, mainly pennant rock	..215 7		271 11
NO. 2 RHONDDA			
Coal ..	10 in		
Clod ..	4 in		
Coal ..	12 in		
Clod ..	18 in		
Coal ..	10 in	4 6	276 5
Measures with several thin coals	..279 8		556 1
CLAY			
Coal ..	23 in	1 11	558 0
Fireclay, clift and rock ..	95 9		653 9
Coal ..	2 in	2	653 11
Clod	..	4 6	658 5
GRAIG			
Coal ..	30 in	2 6	660 11
Fireclay, rock and clift ..	50 4		711 3
PENTRE RIDER			
Coal ..	17 in	1 5	712 8
Fireclay and clift ..	.. 21 0		733 8
PENTRE			
Coal ..	14 in	1 2	734 10
Fireclay and mine	.. 18 5		753 3
LOWER PENTRE			
Coal ..	8 in		
Blackband	3 in		
Coal ..	6 in	1 5	754 8
Fireclay and clift ..	.. 29 3		783 11
EIGHTEEN-INCH			
Coal ..	11 in	11	784 10
Fireclay, rock and clift ..	39 3		824 1
GORLLWYN RIDER			
Bast ..	11 in		
Coal ..	7 in	1 6	825 7
Bastard fireclay ..	.. 7 4		832 11
GORLLWYN			
Coal ..	12 in		
Clod ..	4 in		
Coal ..	8 in	2 0	834 11
Fireclay and mine ground	74 0		908 11
Coal ..	3 in	3	909 2

		Thickness ft in	Depth ft in
Fireclay, mine ground and rock	40 2		949 4
Coal ..	2 in	2	949 6
Fireclay	.. 17 3		966 9
Black shale with coal ..	5 8		972 5
Clift, rock and mine ground	16 3		988 8
Coal ..	8 in		
Clod ..	10 in		
Coal ..	8 in	2 2	990 10
Fireclay, mine ground and rock	14 0		1004 10
Coal ..	8 in	8	1005 6
Fireclay and rock..	.. 7 6		1013 0
TWO-FEET-NINE			
Coal with parting	10 in		
Coal ..	24 in	2 10	1015 10
Black shale, bastard rock, clift and mine ground ..	39 5		1055 3
FOUR-FEET			
Coal ..	70 in		
Holing ..	1 in		
Coal ..	16 in	7 3	1062 6
Fireclay, clift and mine ground	.. 20 4		1082 10
SIX-FEET			
Coal ..	24 in		
Clod ..	1 in		
Coal ..	8 in		
Clod ..	10 in		
Coal ..	12 in		
Bast ..	7 in		
Coal ..	36 in	8 2	1091 0
Fireclay, clift, rock	..102 1		1193 1
'RED VEIN'			
Coal ..	17 in	1 5	1194 6
Fireclay, clift ..	.. 35 9		1230 3
NINE-FEET			
Coal ..	..111 in	9 3	1239 6
Fireclay and disturbed ground	.. 69 6		1309 0
YARD			
Coal ..	34 in	2 10	1311 10
Fireclay and clift ..	.. 29 2		1341 0
SEVEN-FEET			
Coal ..	53 in	4 5	1345 5
Fireclay	.. 5 0		1350 5

Navigation Colliery: North Pit

Ht. above O.D. 370 ft. 　　　　 6-in ST 09 N.E.; Glam. 19 N.W.

Site 500 yd S. 35° E. of Mountain Ash church. National Grid Ref. 05029902.

Four-Feet at 1008 ft, Six-Feet at 1037 ft 9 in, Nine-Feet at 1155 ft 7 in, Bute at 1213 ft, Yard at 1296 ft 7 in, Seven-Feet at 1326 ft 9 in. Sunk to 1333 ft 2 in.

Newlands Colliery

6-in SS 88 S.W.; Glam. 33 S.E. 　　　　 Site 1½ miles W.N.W. of Pyle.

The measures from the Gellideg (Cribbwr Fawr) Seam to the Caerau (Caegarw) are described in the following section which is a composite one made up of the Main Drift, No. 1 Scheme, Upper Nine-Feet (South Fawr) to Caerau and the No. 3 Drift, Gellideg to Upper Nine-Feet.

	Thickness ft in	Depth below Caerau ft in
CAERAU (CAEGARW)		
Coal .. 62 in		
Clay .. 4 in		
Coal .. 9 in		
Rashings and clay 7 in		
Coal .. 12 in		
Rashings and clay 16 in		
Coal .. 16 in		
Clay and rashings 5 in	10 11	10 11
Fireclay and clift with bands of rock and ironstone 28 8		39 7
RED VEIN (NORTH FAWR)		
Coal .. 18 in		
Rashings 2 in		
Coal .. 3 in		
Rashings and clay 5 in		
Coal with brass .. 48 in		
Rashings 16 in		
Coal .. 2 in		
Rashings 5 in	8 3	47 10
Fireclay and clift with mine 60 3		108 1
UPPER NINE-FEET (SOUTH FAWR)		
Coal .. 30 in		
Rashings 1 in		
Coal .. 60 in		
Rashings 1 in		
Coal .. 39 in	10 11	119 0
Fireclay and clift 13 2		132 2
Coal .. 1 in	1	132 3
Fireclay, shale and clift with mine 60 3		192 6
LOWER NINE-FEET (AIL)		
Coal .. 16 in		
Rashings 12 in		
Coal .. 30 in		
Rashings 2 in		
Coal .. 2 in	5 2	197 8
Fireclay, rock and shale .. 16 6		214 2
Coal .. 8 in		
Rashings 2 in	10	215 0
Fireclay and clift with bands of rock 73 6		288 6
BUTE (DRYDYDD)		
Coal .. 56 in		
Clay and rashes.. 27 in		
Coal .. 11 in	7 10	296 4
Fireclay and clift with ironstone 37 6		333 10
AMMAN RIDER (WYTHIEN FACH)		
Coal .. 11 in		
Fireclay .. 12 in		
Coal .. 13 in		
Rashings 1 in		
Coal .. 4 in	3 5	337 3
Fireclay and clift 14 9		352 0
Coal .. 18 in	1 6	353 6
Fireclay, shale and clift with bands of rock .. 75 6		429 0
YARD (SIX-FEET)		
Coal .. 51 in	4 3	433 3
Fireclay 5 11		439 2
Coal .. 3 in		
Shale .. 8 in		
Coal .. 2 in	1 1	440 3
Fireclay and clift with ironstone 35 7		475 10
MIDDLE SEVEN-FEET (SLATOG FACH)		
Coal .. 36 in		
Fireclay .. 12 in		
Coal .. 12 in		
Fireclay .. 6 in		
Coal .. 6 in	6 0	481 10
Fireclay and clift with mine 10 8		492 6
Coal .. 8 in	8	493 2
Fireclay, rock and clift with mine 41 11		535 1

	Thickness ft in	Depth below Caerau ft in
LOWER SEVEN-FEET (SLATOG FAWR)		
Coal .. 28 in		
Rashings 1 in		
Coal .. 18 in		
Rashings 3 in	4 2	539 3
Fireclay and shale with coal streak 22 4		561 7
Coal .. 9 in		
Fireclay and rashings 13 in		
Coal .. 6 in	2 4	563 11
Fireclay, clift and rock .. 19 5		583 4
Cannel .. 30 in	2 6	585 10
Strong clift and shale .. 52 8		638 6
UPPER FIVE-FEET (NINE-FEET)		
Coal .. 40 in		
Brass .. 5 in		
Coal .. 33 in		
Rashings 1 in		
Coal .. 26 in		
Clay .. 1 in		
Coal .. 34 in	11 8	650 2
Fireclay, clift, shale and rock 9 2		659 4
?FIERY		
Coal .. 14½ in		
Rashings ½ in		
Coal .. 16 in	2 7	661 11
Fireclay and shale .. 19 0		680 11

	Thickness ft in	Depth below Caerau ft in
LOWER FIVE-FEET (FIVE-QUARTER)		
Disturbed:		
Coal (dirty).. 18 in		
Fireclay and shale with streaks of coal .. 54 in		
Coal .. 16 in		
Brass .. 4 in		
Coal .. 17 in		
Rashings 1 in		
Coal .. 14 in		
Fireclay and shale 17 ft 2 in		
Coal .. 44 in		
Shale and rashings 36 in		
Coal .. 9 in	34 11	715 10
Fireclay, sandy shale, shale and rock bands .. 34 1		749 11
GELLIDEG (CRIBBWR FAWR)		
Coal .. 110 in		
Rashings 3 in		
Coal .. 3 in	9 8	759 7

Below the Upper Five-Feet the measures are disturbed.

Oakwood Colliery: Downcast Shaft

Ht. above O.D. 402 ft. 6-in SS 89 S.E.; Glam. 26 S.W.

Site 1200 yd S. 44° E. of Maesteg railway station. National Grid Ref. 86069068. Published in Vertical Sections, Sheet 85, 1901.

Several versions exist, the Two-Feet-Nine being given at various depths from 454 ft to 568 ft 6 in, and the depths to the other seams being equally variable.

Two-Feet-Nine at 484 ft, Upper Four-Feet at 514 ft, Lower Four-Feet at 544 ft, Six-Feet at 644 ft, Caerau at 649 ft, Upper Nine-Feet at 800 ft, Lower Nine-Feet at 855 ft. Sunk to Bute at 900 ft.

Old Duffryn Pit

Ht. above O.D. c. 380 ft. 6-in SO 00 S.W.; Glam. 18 N.E.

Site 920 yd E. 18° N. of Aberaman church. National Grid Ref. 02500116.

Two-Feet-Nine at 222 ft 9 in, Four-Feet at 247 ft 4 in.

Park Colliery: No. 1 Pit

Ht. above O.D. 680 ft. 6-in SS 99 N.W.; Glam. 18 S.W.

Site 825 yd W. 38° S. of Cwm-parc church. National Grid Ref. 94429562.

Lower Pentre at 16 ft 6 in, Gorllwyn at 119 ft 11 in, Two-Feet-Nine at 356 ft 7 in, Four-Feet at 424 ft 6 in, Upper Six-Feet at 466 ft 6 in, Lower Six-Feet at 529 ft 6 in, Red Vein at 566 ft 8 in.

Park Colliery: No. 2 Pit

Ht. above O.D. 693 ft.   6-in SS 99 N.W.; Glam. 18 S.W.

Site 800 yd W. 35° S. of Cwm-parc church. National Grid Ref. 99419565. Record of deepening from Six-Feet.

Red Vein at 47 ft 10 in, Middle Seven-Feet at 265 ft 10 in, Lower Seven-Feet at 277 ft 5 in, Five-Feet at 330 ft 1 in, Gellideg at 407 ft 1 in. Sunk to 485 ft 2 in below Six-Feet.

Park Colliery: New Upcast Shaft

Ht. above O.D. 1082 ft.   6-in SS 99 S.W.; Glam. 17 S.E.

Site 1948 yd W. 15° S. of Cwm-parc church. National Grid Ref. 93309463.

	Thickness ft	in	Depth ft	in
DRIFT				
Sand and clay	63	0	63	0
COAL MEASURES				
Rock	18	0	81	0
Coal .. 2 in		2	81	2
Bastard fireclay, rock and clift	12	5	93	7
Coal .. 3 in				
Stone .. 2 in				
Coal .. 3 in				
Coal and rashes.. 15 in	1	11	95	6
Fireclay and clift	11	7	107	1
Coal .. 2 in		2	107	3
Fireclay and clift	9	6	116	9
Coal .. 4 in				
Coal and rashes.. 20 in	2	0	118	9
Fireclay, rock and clift ..	82	4	201	1
Coal .. 2 in		2	201	3
Fireclay and clift	22	0	223	3
HAFOD (BOTTOM COAL)				
Coal .. 7 in		7	223	10
Fireclay and clift	95	7	319	5
Coal .. 5 in		5	319	10
Clift and rock	62	8	382	6
ABERGORKY				
Coal .. 19 in				
Clod .. 11 in				
Coal .. 17 in	3	11	386	5
Fireclay, rock and clift ..	59	3	445	8
PENTRE RIDER				
Coal .. 14 in	1	2	446	10
Fireclay, clift and rock ..	34	0	480	10
PENTRE				
Coal .. 15 in	1	3	482	1
Fireclay, clift, and bands of rock	41	4	523	5
LOWER PENTRE				
Coal .. 11 in				
Stone .. 2 in				
Coal .. 6 in	1	7	525	0
Fireclay, clift, blackband and bands of rock ..	35	11	560	11
Coal .. 6 in		6	561	5
Fireclay and clift	31	4	592	9
Rock	28	0	620	9
Clift		6	621	3

	Thickness ft	in	Depth ft	in
GORLLWYN RIDER				
Coal .. 21 in	1	9	623	0
Fireclay and clift	14	10	637	10
GORLLWYN				
Coal .. 18 in	1	6	639	4
Fireclay, clift and rock bands	72	4	711	8
Coal .. 1 in		1	711	9
Fireclay and clift	27	9	739	6
Coal .. 6 in		6	740	0
Fireclay, rock and clift ..	50	2	790	2
Coal .. 1 in		1	790	3
Fireclay, rock and clift ..	36	8	826	11
Coal .. 4 in				
Bast .. 3 in		7	827	6
Bastard fireclay and clod ..	4	0	831	6
Rock	19	0	850	6
Clift		10	851	4
Coal .. 11 in		11	852	3
Clift	12	10	865	1
Coal .. 11 in				
Fireclay .. 22 in				
Coal .. 12 in				
Fireclay and rashes.. 23 in				
Coal .. 8 in	6	4	871	5
Fireclay, clift, mine and rock bands	55	1	926	6
TWO-FEET-NINE				
Coal .. 38 in				
Coal .. 27 in	5	5	931	11
Fireclay, clift and rock bands	7	10	939	9
Coal .. 12 in	1	0	940	9
Fireclay, clift and mine ..	14	6	955	
Coal .. 10 in		10	956	
Fireclay and clift	6	8	962	
Coal .. 1 in		1	962	10
Fireclay, clift and mine ..	44	9	1007	
FOUR-FEET				
Coal .. 18 in				
Rashings 4 in				
Coal .. 34 in				
Rashings 6 in				
Coal .. 9 in				
Rashings 5 in	6	4	1013	1
Fireclay, clift and rock bands	33	9	1047	

	Thickness ft in	Depth ft in
UPPER SIX-FEET		
Coal .. 27 in	2 3	1049 11
Fireclay, clift and rock bands	19 5	1069 4
Coal .. 4 in	4	1069 8
Clift, mine and rock bands	28 5	1098 1
LOWER SIX-FEET		
Gob .. 38 in	3 2	1101 3
Fireclay, mine and clift ..	26 4	1127 7
RED VEIN		
Coal .. 11 in		
Fireclay.. 24 in		
Coal .. 34 in	5 9	1133 4
Fireclay, clift, rock bands and mine ground ..	74 0	1207 4
Rashings with pockets of coal	9	1208 1
Clift and mine ground ..	19 5	1227 6
Shale and rashings ..	2 3	1229 9
Fireclay, clift, bastard rock, rock and mine ground—disturbed	83 2	1312 11
BUTE		
Coal .. 38 in	3 2	1316 1
Fireclay, clift and rock ..	31 5	1347 6
AMMAN RIDER		
Coal .. 12 in	1 0	1348 6
Clift	11 6	1360 0
Rock and bastard rock ..	46 2	1406 2
Clift with rock bands ..	20 1	1426 3

	Thickness ft in	Depth ft in
YARD AND UPPER SEVEN-FEET		
Coal .. 36 in		
Stone .. 2 in		
Coal .. 11 in	4 1	1430 4
Fireclay, shale and clift ..	63 8	1494 0
MIDDLE SEVEN-FEET		
Coal .. 6 in		
Rashings 14 in		
Coal .. 27 in		
Rashings 6 in		
Coal .. 2 in	4 7	1498 7
Fireclay	10 0	1508 7
Rock	16 6	1525 1
LOWER SEVEN-FEET		
Coal .. 27 in	2 3	1527 4
Fireclay, clift and rock bands	17 3	1544 7
FIVE-FEET		
Coal .. 21 in		
Stone and rashings 7 in		
Coal .. 32 in	5 0	1549 7
Fireclay, clift and rock bands	24 0	1573 7
Rock	11 0	1584 7

Penrhiw-fer Colliery: No. 1 Pit

Ht. above O.D. 812 ft.   6-in SS 99 N.E.; Glam. 27 S.E.

Site 1850 yd S. 10° W. of Pen-y-graig church. National Grid Ref. 99698974. Published in 1st Edition, pp. 26–28.

No. 3 Rhondda at 246 ft 11 in, Hafod at 449 ft 11 in, Pentre and Lower Pentre at 731 ft 5 in. Sunk to 1299 ft.

Penrikyber Colliery: No. 2 Pit

Ht. above O.D. 360 ft.   6-in ST 09 N.E.; Glam. 19 S.W.

Site 700 yd S. 27° E. of Penrhiwceiber church. National Grid Ref. 06129707. Published graphically in Vertical Sections, Sheet 83, 1900.

No. 1 Rhondda Rider at 503 ft, Four-Feet at 1581 ft 11 in, Six-Feet at 1638 ft 3 in, Nine-Feet at 1750 ft 9 in, Gellideg at 1966 ft 7 in.

Penrikyber Colliery: No. 3 Pit

Ht. above O.D. 366 ft.   6-in ST 09 N.E.; Glam. 19 S.W.

Site 820 yd S. 28° E. of Penrhiwceiber church. National Grid Ref. 06179697.

Drift to 35 ft 6 in, No. 1 Rhondda at 543 ft 10 in. Sunk to 553 ft 8 in.

Pentre Colliery: Downcast Pit

Ht. above O.D. 622 ft.   6-in SS 99 N.W.; Glam. 18 S.W.

Site 410 yd S. 37° E. of Pentre church. National Grid Ref. 97179588. Published in Vertical Sections, Sheet 84, 1901.

Pentre at 190 ft 7 in, Two-Feet-Nine at 594 ft 2 in, Four-Feet at 640 ft, Upper Six-Feet at 663 ft 8 in, Lower Six-Feet at 711 ft 8 in, Upper Nine-Feet at 808 ft, Lower Nine-Feet at 848 ft 10 in, Bute at 893 ft 6 in.

Continuation of section in Staple Pit, 1890

Amman Rider at 945 ft 10 in, Yard at 985 ft 5 in, Middle Seven-Feet at 1015 ft 2 in, Lower Seven-Feet at 1027 ft 3 in, Five-Feet at 1102 ft 11 in.

Pen-y-rhiw Colliery

Ht. above O.D. 393 ft.   6-in ST 08 N.E.; Glam. 28 S.W.

Site 1550 yd S. 32° W. of Pontypridd church. National Grid Ref. 06408894.

No. 1 Rhondda at 203 ft 11 in, No. 2 Rhondda at 377 ft 3 in, No. 3 Rhondda at 665 ft 1 in. Sunk to 756 ft 11 in.

Rhondda Main Colliery: Catherine Pit

Ht. above O.D. 385 ft.   6-in SS 98 N.W.; Glam. 27 S.W.

Site 1280 yd S. 20° E. of Ogmore Vale church. National Grid Ref. 93608906. Published in 2nd Edition, pp. 144–45.

No. 2 Rhondda at 749 ft. Sunk to 817 ft 7 in.

Rhondda Main Colliery: Anne Pit

Ht. above O.D. 382 ft.   6-in SS 98 N.W.; Glam. 27 S.W.

Site 1400 yd S. 20° E. of Ogmore Vale church. National Grid Ref. 93648895.

No. 2 Rhondda at 777 ft 6 in. Sunk to 1134 ft 10 in.

Rhondda Merthyr (Tynewydd) Colliery: Downcast Shaft

Ht. above O.D. 837 ft.   6-in SS 99 N.W.; Glam. 17 N.E.

Site 1070 yd N. 20° W. of Treherbert railway station. National Grid Ref. 93569909. No record below ?RED VEIN.

	Thickness ft in		Depth ft in			Thickness ft in		Depth ft in	
DRIFT					**ABERGORKY**				
Gravel 18	0		18	0	Coal .. 24 in				
					Coal .. 7 in				
					Parting .. 3 in				
COAL MEASURES					Coal .. 14 in	4	0	184	0
Coal .. 3 in	3		18	3	Fireclay, mine ground,				
Shale and rock 30	0		48	3	shale and rock 53	0		237	0
HAFOD (BOTTOM COAL)					**PENTRE RIDER**				
Coal .. 9 in	9		49	0	Coal .. 6 in				
Fireclay, rock, shale and					Clod .. 3 in				
mine ground 84	5		133	5	Coal .. 9 in	1	6	238	6
Coal .. 1 in	1		133	6	Fireclay, rock and shale .. 21	0		259	6
					Coal .. 5 in	5		259	11
Fireclay, rock and clod .. 5	6		139	0	Fireclay, rock and shale .. 20	0		279	11
Coal .. 8 in	8		139	8	Coal .. 14 in	1	2	281	1
Fireclay, shale and rock .. 40	4		180	0	Fireclay, mine and shale .. 21	0		302	1

LOWER PENTRE		Thickness ft in		Depth ft in	
Coal	.. 3 in				
Shale	.. 36 in				
Coal	.. 7 in	3	10	305	11
Fireclay, rock, shale and mine	.. ..	33	0	338	11
Coal	.. 1 in		1	339	0
Mine ground	.. ..	7	0	346	0
Coal	.. 3 in		3	346	3
Fireclay, rock and shale	..	20	0	366	3
GORLLWYN RIDER					
Coal	.. 12 in	1	0	367	3
Fireclay, rock and shale	..	20	0	387	3
GORLLWYN					
Coal	.. 22 in	1	10	389	1
Fireclay and mine ground		28	0	417	1
Coal	.. 1 in		1	417	2
Fireclay, mine and rock	..	72	0	489	2
Rashes	.. ..		6	489	8
Fireclay, shale and rock	..	50	0	539	8
Coal	.. 9 in		9	540	5
Clod and shale	.. ..	15	2	555	7
THREE COALS					
Coal	.. 6 in				
Fireclay	.. 84 in				
Coal	.. 4 in				
Clod	.. 24 in				
Coal	.. 7 in	10	5	566	0
Rock	.. ..	24	0	590	0

TWO-FEET-NINE		Thickness ft in		Depth ft in	
Coal	.. 33 in	2	9	592	9
Shale and rock	..	..102	0	694	9
Coal	.. 7 in		7	695	4
Shale and mine	..	13	10	709	2
Coal	.. 20 in	1	8	710	10
Clod	.. ..	4	4	715	2
FOUR-FEET					
Coal	.. 44 in				
Parting	.. 6 in				
Coal	.. 12 in	5	2	720	4
Fireclay, shale and mine	..	21	6	741	10
UPPER SIX-FEET					
Coal	.. 31 in	2	7	744	5
Fireclay, rock, shale and mine	.. ..	35	5	779	10
LOWER SIX-FEET					
Coal	.. 48 in				
Fireclay	.. 30 in				
Coal	.. 5 in	6	11	786	9
Shale and mine	..	10	0	796	9
?RED VEIN					
Coal	.. 64 in	5	4	802	1

Rhondda Merthyr Colliery: Upcast Shaft

Ht. above O.D. c. 837 ft. 6-in SS 99 N.W.; Glam. 17 N.E.

Site 1040 yd N. 19° W. of Treherbert railway station. National Grid Ref. 93589907. Continuation of section below 802 ft 1 in.

Strata		Thickness ft in		Depth ft in	
Strata	.. ..	15	0	817	1
RED VEIN					
Coal	.. 28 in				
Fireclay	.. 36 in				
Coal	.. 32 in	8	0	825	1

Sandstone, shale and mine		Thickness ft in		Depth ft in	
ground	.. ..	64	6	889	7
UPPER NINE-FEET					
Coal	.. 30 in				
Clod	.. 2 in				
Coal	.. 22 in	4	6	894	1

Rhondda Merthyr Colliery: Staple Pit

Sunk from Bute.

Fireclay and clift		Thickness ft in		Depth below Bute ft in	
Fireclay and clift	.. ..	35	0	35	0
AMMAN RIDER					
Coal	.. 9 in		9	35	9
Clift	.. ..	67	0	102	9
YARD					
Coal	.. 33 in	2	9	105	6
Clod	.. ..		9	106	3

SEVEN-FEET		Thickness ft in		Depth below Bute ft in	
Coal	.. 15 in				
Fireclay	.. 18 in				
Coal	.. 24 in				
Clod	.. 8 in				
Coal	.. 12 in	6	5	112	8
Fireclay, clift and rock	..	45	6	158	2
UPPER FIVE-FEET					
Coal	.. 36 in	3	0	161	2

St. John's Colliery: South Pit

Ht. above O.D. 638 ft.   6-in SS 89 S.E.; Glam. 26 S.W.

Site 2445 yd E. 6° N. of Maesteg railway station. National Grid Ref. 87569166. Published in 2nd Edition, p. 145.

Two-Feet-Nine at 457 ft 9 in, Upper Four-Feet at 495 ft 9 in, Lower Four-Feet at 561 ft, Upper Six-Feet at 656 ft 2 in, Lower Six-Feet at 716 ft 1 in, Caerau at 742 ft 7 in. Sunk through disturbed ground including Jubilee Slide to Lower Seven-Feet at 1088 ft 3 in, Upper Five-Feet at 1117 ft 9 in. Sunk to 1135 ft 9 in.

St. John's Colliery: North Pit

Ht. above O.D. 638 ft.   6-in SS 89 S.E.; Glam. 26 S.W.

Site 2460 yd E. 7° N. of Maesteg railway station. National Grid Ref. 87589171.

Two-Feet-Nine at 495 ft 11 in, Upper Four-Feet at 532 ft 5 in, Lower Four-Feet at 623 ft 7 in, Upper Six-Feet at 701 ft 9 in, Six-Feet at 759 ft 6 in. Sunk through disturbed ground including Jubilee Slide to Upper Five-Feet at 1159 ft 5 in. Sunk to 1165 ft.

South Rhondda Colliery: Downcast Shaft

Ht. above O.D. 495 ft.   6-in SS 98 S.E.; Glam. 35 S.E.

Site 2300 yd N. 33° W. of Llanharan church. National Grid Ref. 99208488.

	Thickness ft in		Depth ft in			Thickness ft in		Depth ft in
COAL MEASURES					Clift with **coal** and rashes			
Rock and clift ..	.. 27	0	27	0	at base	.. 12	3	182 4
					Clift	.. 22	5	204 9
NO. 2 RHONDDA					**Coal** .. 6 in		6	205 3
Coal .. 48 in	4	0	31	0	Clift with rock bands .. 86		1	291 4
Rock and clift ..	..138	9	169	9	NO. 3 RHONDDA			
Coal .. 4 in		4	170	1	**Coal** .. 32 in	2	8	294

Standard Colliery: No. 2 Pit

Ht. above O.D. 429 ft.   6-in ST 09 S.W.; Glam. 27 N.E.

Site 400 yd N. 14° W. of Ynyshir church. National Grid Ref. 02449326. Published in Vertical Sections, Sheet 83, 1900.

Drift to 51 ft, No. 2 Rhondda at 229 ft 11 in, Two-Feet-Nine at 1040 ft 11 in, Four-Feet at 1096 ft 8 in, Upper Six-Feet at 1163 ft 2 in, Lower Six-Feet at 1187 ft 6 in, Upper Nine-Feet at 1270 ft 2 in, Bute at 1350 ft 9 in, Yard at 1399 ft 1 in, Upper and Middle Seven-Feet at 1413 ft 7 in, Five-Feet at 1482 ft 8 in, Gellideg at 1490 ft, Garw at 1570 ft 11 in. Sunk to 1577 ft 7 in.

Trewen Shaft

Ht. above O.D. 786 ft.   6-in SS 99 N.E.; Glam. 18 N.E.

Site 1450 yd W. 13° N. of Cwmaman church. National Grid Ref. 98949958. Sunk 1910.

Made ground to 26 ft 6 in, Drift to 74 ft 6 in, Two-Feet-Nine at 696 ft 3 in, Four-Feet at 733 ft 8 in, Upper Six-Feet at 761 ft 9 in, Lower Six-Feet at 818 ft 7 in, Red Vein at 852 ft 7 in, Nine-Feet at 910 ft 3 in, Bute at 969 ft 6 in, Yard at 1017 ft 6 in, Lower Seven-Feet at 1046 ft 7 in. Sunk to 1064 ft 2 in.

Tydraw (Dunraven) Colliery: No. 1 Pit

Ht. above O.D. 694 ft.   6-in SS 99 N.W.; Glam. 17 N.E.

Site 1550 yd N. 32° W. of Treherbert railway station. National Grid Ref. 92689894.

		Thickness ft	in	Depth ft	in
DRIFT					
Gravel		19	2	19	2
COAL MEASURES					
Mine ground		30	2	49	4
PENTRE RIDER					
Coal .. 18 in		1	6	50	10
Fireclay and mine ground		21	9	72	7
Coal .. 15 in		1	3	73	10
Fireclay, clift, mine ground and rock bands ..		24	5	98	3
Coal .. 2 in			2	98	5
Fireclay and clod		10	0	108	5
LOWER PENTRE					
Coal .. 14 in					
Clod .. 2 in					
Coal .. 5 in		1	9	110	2
Fireclay, mine ground and rock bands		26	9	136	11
Bast .. 8 in			8	137	7
Mine ground		4	9	142	4
Bast .. 7 in			7	142	11
Mine ground, fireclay, rock and clift		25	1	168	0
Rock		21	0	189	0
Mine ground		10	7	199	7
GORLLWYN RIDER					
Coal .. 24 in		2	0	201	7
Fireclay and clift		13	10	215	5
GORLLWYN					
Coal .. 7 in					
Clod parting					
Coal .. 12 in					
Clod .. 1 in					
Coal .. 11 in		2	7	218	0
Fireclay, clift, mine ground and rock		76	9	294	9
Coal .. 6 in					
Clod .. 1 in					
Coal . 7 in		1	2	295	11
Fireclay, rock and clift ..		49	7	345	6
Coal .. 9 in			9	346	3
Fireclay, rock and clift ..		30	3	376	6
Coal .. 8 in			8	377	2
Fireclay, clift and mine ground		28	8	405	10
Coal .. 8 in			8	406	6
Fireclay, rashes and mine ground		11	11	418	5
THREE COALS					
Coal .. 4 in					
Fireclay and clod .. 18 in					
Coal .. 8 in					
Fireclay .. 8 in					
Coal .. 1 in		3	3	421	8
Fireclay, clift, rock and bast		16	4	438	0
TWO-FEET-NINE					
Coal .. 37 in		3	1	441	1
Fireclay, clift and mine ..		26	10	467	11
Rock		9	10	477	9
Coal .. 12 in					
Clod .. 3 in					
Coal .. 2 in		1	5	479	2
Clod and fireclay		7	11	487	1
Coal .. 11 in					
Clod .. 2 in					
Coal .. 2 in		1	3	488	4
Clod and mine ground ..		28	10	517	2
FOUR-FEET (TOP COAL)					
Coal .. 12 in					
Clod .. 5 in					
Coal .. 5 in					
Clod .. 1 in					
Coal .. 7 in		2	6	519	8
Fireclay and clod ..		5	0	524	8
FOUR-FEET (BOTTOM COAL)					
Coal .. 3 in					
Clod .. 9 in					
Coal .. 5 in					
Clod .. 1 in					
Coal .. 44 in					
Holing .. 4 in		5	6	530	2

Tydraw: Downcast Shaft (probably No. 1 Pit). Alternative Version

Gorllwyn at 217 ft 5 in, Two-Feet-Nine at 436 ft 11 in, Four-Feet at 527 ft 2 in, Upper Six-Feet at 551 ft 11 in, Lower Six-Feet at 615 ft, Red Vein at 624 ft, Nine-Feet at 729 ft.

Tydraw: Downcast Shaft

	Thickness ft in		Depth ft in	
Section below Upper Nine-Feet—said to be at ..			677	1
Fireclay and clift	5	4	682	5
Rock	12	0	694	5
LOWER NINE-FEET				
Coal .. 35 in	2	11	697	4
Clift, fireclay and rock ..	82	0	779	4
BUTE				
Coal .. 37 in	3	1	782	5
Fireclay, clift, shale and mine	25	3	807	8
AMMAN RIDER				
Coal .. 6 in		6	808	2
Shale and clift	30	3	838	5
Rock	15	8	854	1
Clift, thin rock bands ..	54	3	908	4
YARD				
Coal .. 31 in				
Stone .. 5 in				
Coal .. 8 in				
Stone .. 1 in				
Coal .. 6 in	4	3	912	7
Rashes	3	0	915	7
Coal .. 14 in	1	2	916	9
Rashes and fireclay ..	7	6	924	3

	Thickness ft in		Depth ft in	
MIDDLE SEVEN-FEET				
Coal .. 24 in				
Stone .. 1 in				
Coal .. 3 in	2	4	926	7
Fireclay and clift ..	16	0	942	7
LOWER SEVEN-FEET				
Coal .. 25 in	2	1	944	8
Fireclay and clift ..	11	5	956	1
UPPER FIVE-FEET				
Coal .. 8 in				
Thin parting,				
Coal .. 37 in	3	9	959	10
Fireclay	9	0	968	10
LOWER FIVE-FEET				
Coal .. 21 in	1	9	970	7
Fireclay and clift ..	48	10	1019	5
GELLIDEG				
Coal .. 29 in	2	5	1021	10
Fireclay and clift ..	16	3	1038	1
Coal .. 3 in		3	1038	4
Fireclay, clift and rock ..	25	6	1063	10
Clift and mine ground ..	50	0	1113	10
Black shale full of shells ..	1	2	1115	0
Black shale, mine ground, fireclay and thin rock bands	15	11	1130	11

Tylecoch Colliery

Ht. above O.D. 576 ft.   6-in SS 99 N.E.; Glam. 18 S.W.

Site 635 yd W. 30° N. of Treorchy railway station. National Grid Ref. 95449677. Section affected by structural disturbances for about 200 ft below Lower Six-Feet and again near the Bute.

	Thickness ft in		Depth ft in	
DRIFT				
Soil, clay and gravel ..	31	0	31	0
COAL MEASURES				
Clift	25	7	56	7
LOWER PENTRE				
Coal .. 15 in				
Stone .. 1 in				
Coal .. 11 in	2	3	58	10
Fireclay, rock, clift and mine	30	8	89	6
Coal .. 5 in				
Fireclay .. 24 in				
Coal .. 13 in	3	6	93	0
Clift, rock, shale and mine ground	34	1	127	1
GORLLWYN RIDER				
Coal .. 18 in	1	6	128	7
Fireclay, rock and clift ..	16	8	145	3

	Thickness ft in		Depth ft in	
GORLLWYN				
Coal .. 2i in	1	9	147	0
Fireclay, clift, mine ground and rock	80	2	227	2
Coal .. 8 in		8	227	10
Fireclay	6	2	234	0
Coal .. 6 in		6	234	6
Rock	17	0	251	6
Rock, clift and mine ground	31	0	282	6
Coal .. 10 in		10	283	4
Clift, mine ground and rock	18	0	301	4
Fireclay	5	4	306	8
Mine ground	9	0	315	8
Coal .. 20 in	1	8	317	4
Fireclay, clift and rock ..	32	7	349	11

	Thickness ft in		Depth ft in	
THREE COALS				
Coal .. 23 in				
Fireclay and clod .. 104 in				
Coal .. 11 in				
Clod .. 36 in				
Coal .. 12 in	15	6	365	5
Clift, rock and ironstone ..	26	5	391	10
TWO-FEET-NINE				
Coal .. 32 in				
Parting .. 1 in				
Coal .. 18 in	4	3	396	1
Fireclay, rock and mine ..	4	3	400	4
Coal .. 14 in	1	2	401	6
Clift and mine	13	2	414	8
FOUR-FEET (TOP COAL)				
Coal .. 12 in				
Clod .. 2 in				
Coal .. 13 in	2	3	416	11
Fireclay, clift and shale ..	34	4	451	3
FOUR-FEET (BOTTOM COAL)				
Coal and clod .. 15 in				
Coal .. 15 in				
Clod .. 43 in				
Soft coal.. 2 in				
Coal .. 12 in	7	3	458	6
Fireclay, clift and mine, bands of rock	29	3	487	9
Coal .. 24 in	2	0	489	9
Fireclay, clift and mine ..	41	4	531	1
UPPER SIX-FEET				
Coal .. 32 in	2	8	533	9
Fireclay and clift	14	8	548	5
Coal .. 4 in		4	548	9
Shale and mine ground ..	63	11	612	8
LOWER SIX-FEET				
Coal .. 57 in				
Holing .. 3 in				
Coal .. 15 in	6	3	618	11

	Thickness ft in		Depth ft in	
Fireclay and rashes ..	10	0	628	11
Coal .. 26 in				
Rashes .. 9 in				
Coal .. 38 in				
Fireclay .. 32 in				
Coal .. 48 in	12	9	641	8
Clift, mine ground and rock	15	0	656	8
Coal .. 36 in	3	0	659	8
Fireclay, rock and mine ..	2	8	662	4
Deepening in 1922 from kerb at 662 ft 5 in.				
Clift and shale	15	2	677	6
Bast .. 8 in				
Coal .. 8 in	1	4	678	10
Fireclay		9	0	687	10
Coal .. 42 in	3	6	691	4
Clift, rock and mine ground, highly disturbed	18	9	710	1
Coal .. 21 in	1	9	711	10
Clift and mine ground ..	55	11	767	9
Bast .. 3 in				
Coal .. 42 in				
Stone .. 7 in				
Coal .. 39 in	7	7	775	4
Rashes, fireclay	6	11	782	3
Coal .. 3 in		3	782	6
Fireclay, clift and rock ..	13	0	795	6
Coal .. 4 in		4	795	10
Fireclay, clift and rock ..	15	10	811	8
Coal .. 36 in	3	0	814	8
Fireclay, clift and rock ..	45	2	859	10
BUTE				
Coal .. 23 in				
Parting,				
Coal .. 28 in	4	3	864	1
Clift and rock, highly disturbed	62	6	926	7
BUTE				
Coal .. 23 in				
Parting .. 1 in				
Coal .. 31 in	4	7	931	2
Fireclay, clift and rock ..	21	11	953	1

Tynybedw Colliery: Downcast Pit

Ht. above O.D. 575 ft.   6-in SS 99 N.E.; Glam. 18 S.W.

Site 460 yd W. 34° N. of Pentre church. National Grid Ref. 96609642.
Section below Lower Six-Feet published in Vertical Sections, Sheet 84, 1901.

	Thickness ft in		Depth ft in	
DRIFT				
Walling and gravel ..	32	6	32	6
COAL MEASURES				
Clift	17	0	49	6
Coal .. 14 in	1	2	50	8
Fireclay, rock and clift ..	27	7	78	3

	Thickness ft in		Depth ft in	
LOWER PENTRE				
Coal .. 30 in	2	6	80	9
Fireclay, rock and clift ..	26	9	107	6
Coal .. 8 in		8	108	2
Fireclay, clift and rock ..	48	10	157	0
GORLLWYN RIDER				
Coal .. 20 in	1	8	158	8
Fireclay and clift	18	4	177	0

		Thickness ft	in	Depth ft	in
GORLLWYN					
Coal .. 23 in		**1**	**11**	**178**	**11**
Clift, rock and mine ground 100	5			279	4
Coal .. 8 in			**8**	**280**	**0**
Fireclay, rock and clift .. 42	4			322	4
Coal .. 8 in			**8**	**323**	**0**
Fireclay, clift and rock .. 42	10			365	10
Coal .. 24 in		**2**	**0**	**367**	**10**
Fireclay, rock and clift .. 27	4			395	2
Coal .. 18 in		**1**	**6**	**396**	**8**
Fireclay, rock and clift .. 14	2			410	10
THREE COALS					
Coal .. 10 in					
Fireclay and clift .. 54 in					
Coal .. 11 in					
Fireclay .. 33 in					
Coal .. 13 in					
Fireclay and clift .. 66 in					
Coal .. 15 in		**16**	**10**	**427**	**8**
Fireclay, clift and mine .. 24	0			451	8
TWO-FEET-NINE					
Coal .. 50 in		**4**	**2**	**455**	**10**
Fireclay, rock and clift ..		4	6	460	4
Coal .. 18 in		**1**	**6**	**461**	**10**
Fireclay, clift and mine .. 22	6			484	4

		Thickness ft	in	Depth ft	in
FOUR-FEET					
Probably disturbed:					
Coal .. 18 in					
Rashes .. 6 in					
Coal .. 8 in					
Rashes .. 2 in					
Coal .. 12 in					
Fireclay .. 42 in					
Coal .. 24 in					
Rashes .. 6 in					
Coal .. 24 in					
Rashes .. 6 in					
Coal .. 30 in		**14**	**10**	**499**	**2**
Clift and clay		7	0	506	2
FOUR-FEET					
Repeated:					
Coal .. 24 in					
Rashes .. 3 in					
Coal .. 25 in					
Clod .. 12 in					
Coal .. 8 in		**6**	**0**	**512**	**2**
Fireclay, rock and clift .. 19	2			531	4
UPPER SIX-FEET					
Coal .. 24 in		**2**	**0**	**533**	**4**
Fireclay, clift and mine .. 51	11			585	3
LOWER SIX-FEET					
Coal .. 65 in		**5**	**5**	**590**	**8**

Shaft deepening from Lower Six-Feet, said to be at 579 ft 6 in. Red Vein at 639 ft 6 in, Upper Nine-Feet at 735 ft 7 in, Lower Nine-Feet at 782 ft, Bute at 856 ft 4 in, Amman Rider at 893 ft 6 in, Yard at 933 ft 10 in, Middle Seven-Feet at 988 ft 2 in, Lower Seven-Feet at 1008 ft 6 in. Sunk to 1011 ft 6 in.

Tynewydd Pit

Ht. above O.D. c. 310 ft.   6-in SS 09 S.W.; Glam. 27 S.E.

Site 900 yd N. 20° W. of Cymmer church. National Grid Ref. 02359162.

		Thickness ft	in	Depth ft	in
DRIFT					
Gravel 28	0			28	0
COAL MEASURES					
Rock 25	0			53	0
NO. 2 RHONDDA					
Coal .. 34 in		**2**	**10**	**55**	**10**
Fireclay, rock and clift .. 21	0			76	10
Coal (in three layers) 30 in		**2**	**6**	**79**	**4**
Fireclay, rock and clift ..130	6			209	10
Coal .. 18 in					
Clod .. 36 in					
Coal .. 8 in		**5**	**2**	**215**	**0**
Fireclay and clift 52	0			267	0
NO. 3 RHONDDA					
Coal .. 36 in		**3**	**0**	**270**	**0**
Sump 6	0			276	0

Tyn-y-waun Pit

Ht. above O.D. c. 380 ft.   6-in SS 98 S.W.; Glam. 34 S.E.

Site 2000 yd E. 6° N. of St. Theodore's Church, Bryncethin. National Grid Ref. 93188445. Said to be 900 ft deep; following section begins at horizon of Hafod Top Coal.

	Thickness ft in	Depth below Hafod Top Coal ft in
HAFOD (TOP COAL)		
Coal .. 12 in	1 0	1 0
Fireclay, shale and rock .. 33	0	34 0
HAFOD (BOTTOM COAL)		
Coal .. 18 in	1 6	35 6
Fireclay and shale .. 81	6	117 0
Coal .. 24 in	2 0	119 0
Fireclay, rock and shale .. 42	4	161 4
ABERGORKY (TOP COAL)		
Coal .. 8 in	8	162 0
Fireclay, shale and rock .. 20	7	182 7
ABERGORKY (BOTTOM COAL)		
Coal .. 17 in	1 5	184 0
Fireclay, rock and shale .. 79	10	263 10
PENTRE RIDER		
Coal .. 29 in		
Clod .. 16 in		
Coal .. 3 in		
Rashes .. 8 in	4 8	268 6
Fireclay and shale ..	6 0	274 6
PENTRE		
Coal .. 54 in	4 6	279 0
Fireclay and clift 16	6	295 6
LOWER PENTRE		
Rashes .. 6 in		
Coal .. 24 in	2 6	298 0
Fireclay, rock and clift .. 25	0	323 0
Coal .. 18 in	1 6	324 6
Fireclay and clift 13	6	338 0
Coal .. 6 in	6	338 6
Shale 4	6	343 0
Coal .. 12 in	1 0	344 0
Fireclay and clift 12	8	356 8
Coal .. 4 in	4	357 0
Fireclay and rock.. .. 49	2	406 2
?GORLLWYN RIDER		
Coal .. 10 in	10	407 0
Fireclay and clift 21	2	428 2
?GORLLWYN		
Clod .. 6 in		
Coal .. 12 in		
Clod .. 4 in		
Coal .. 12 in	2 10	431 0
Fireclay, rock and shale with rashes band in middle 106	4	537 4
Coal .. 8in	8	538 0
Rock and shale 11	4	549 4
Coal .. 8 in	8	550 0
Fireclay and shale .. 28	8	578 8
Coal .. 16 in	1 4	580 0
Fireclay and shale .. 15	3	595 3
Coal .. 15 in	1 3	596 6
Fireclay, rock and shale .. 46	6	643 0
Coal .. 24 in	2 0	645 0
Shale and ironstone .. 21	5	666 5
Coal .. 7 in		
Rashes .. 24 in	2 7	669 0
Fireclay, shale and rock .. 51	8	720 8
Coal .. 4 in	4	721 0
Shale 6	6	727 6
?TWO-FEET-NINE		
Coal .. 54 in	4 6	732 0
Shale and rock 26	6	758 6
?UPPER FOUR-FEET		
Coal .. 30 in	2 6	761 0
Shale and rock with two layers of rashes.. .. 42	0	803 0
?LOWER FOUR-FEET		
Coal .. 36 in	3 0	806 0
Fireclay and clift 19	0	825 0
	(about)	(about)

Werntarw Colliery: No. 1 Pit

Ht. above O.D. 387 ft.   6-in SS 98 S.E.; Glam. 35 S.W.

Site 2300 yd E. 8° N. of Heol-y-Cyw school. National Grid Ref. 96638469.

	Thickness ft in	Depth ft in
DRIFT		
Surface clay 18	7	18 7
COAL MEASURES		
Soft dark shale 29	5	48 0
HAFOD		
Coal .. 40 in	3 4	51 4
Clift and 'pennant' rock .. 14	0	65 4
Clift and clay joints .. 35	5	100 9
Coal .. 12 in	1 0	101 9
Bastard clay and clift with bands of rock and clay joints 41	2	142 11
Dark shale 1	0	143 11

	Thick-ness ft in	Depth ft in
ABERGORKY		
Coal .. 28 in	**2 4**	**146 3**
Bastard fireclay and strong 'pennant' rock	66 10	213 1
Clift	9	213 10
PENTRE RIDER		
Coal .. 19 in	**1 7**	**215 5**
Strong clift	7 0	222 5
Hard 'pennant' rock with dark bands	31 9	254 2
PENTRE		
Coal .. 1 in		
Rashings 3 in		
Coal .. 44 in	**4 0**	**258 2**
Fireclay, clift and rock ..	18 1	276 3
LOWER PENTRE		
Coal .. 12 in	**1 0**	**277 3**
Fireclay and clift with beds of rock	14 11	292 2
Rashings 6 in		
Coal .. 11 in	**1 5**	**293 7**
Fireclay and strong clift ..	29 10	323 5
Coal .. 14 in		
Rashings 6 in	**1 8**	**325 1**
Fireclay and clift with bands of rock	87 7	412 8
Rashings	4 0	416 8
Strong fireclay, clift and rock	17 0	433 8
Rashings	3	433 11
Clift and shaly clift ..	3 9	437 8
Strong fireclay and clift with bands of rock ..	39 0	476 8
Coal .. 12 in	**1 0**	**477 8**

	Thick-ness ft in	Depth ft in
Strong fireclay and mixed ground with mine ..	24 0	501 8
Coal .. 12 in	**1 0**	**502 8**
Strong bastard fireclay and strong clift with iron-stone	18 0	520 8
Coal .. 12 in	**1 0**	**521 8**
Strong fireclay and clift ..	7 0	528 8
Coal .. 7 in		
Rashings 8 in		
Coal .. 12 in	**2 3**	**530 11**
Mixed clift, fireclay and mine	13 0	543 11
Black pan 8 in		
Coal .. 24 in	**2 8**	**546 7**
Fireclay, clift with mine and rashings	46 1	592 8
Coal .. 10 in		
Rashings 5 in	**1 3**	**593 11**
Fireclay and strong clift with ironstone	12 6	606 5
Coal .. 21 in		
Rashings 12 in	**2 9**	**609 2**
Fireclay, clift and mixed ground	54 6	663 8
Rashings 12 in		
Soft fireclay 5 in		
Coal disturbed 12 ft to 19 ft say	20 0	683 8
Fireclay and mixed ground	20 0	703 8
Coal .. 96 in		
Rashings 24 in		
Coal 44 to 78 in say	**16 0**	**719 8**
Fireclay, rashings and strong clift	10 7	730 3

Section of No. 1 West Hard Heading: Yard to Six-Feet

	Thick-ness ft in	Depth below Six-Feet ft in
SIX-FEET		
Coal .. 2 in		
Rashings 1 in		
Coal .. 28 in		
Brass .. 6 in		
Coal .. 36 in		
Clod .. 4 in		
Coal .. 14 in		
Rashings 2 in	7 9	7 9
Fireclay	4 6	12 3

	Thick-ness ft in	Depth below Six-Feet ft in
CAERAU		
Rashings 2 in		
Coal .. 28 in		
Rashings 4 in		
Coal .. 24 in		
Rashings 3 in	5 1	17 4
Fireclay and dark shale ..	10 0	27 4
Coal .. 8 in		
Rashings 2 in	10	28 2
Clift with ironstone ..	14 6	42 8
Coal .. 17 in	1 5	44 1
Fireclay and clift	41 6	85 7

	Thickness ft in	Depth below Six-Feet ft in
PPER NINE-FEET		
Stone .. 2 in		
Coal .. 3 in		
Stone .. 2 in		
Coal .. 46 in		
Rashings 4 in		
Clod .. 2 in	4 11	90 6
ireclay and strong clift ..	19 0	109 6
OWER NINE-FEET		
Clod .. 4 in		
Coal .. 24 in	2 4	111 10
ireclay and clift with rock	58 0	169 10
UTE		
Rashings 6 in		
Coal .. 20 in		
Rashings 1 in		
Coal .. 12 in		
Rashings 2 in		
Coal .. 24 in		
Parting .. 1 in		
Coal .. 62 in	10 8	180 6
Fireclay and clift	20 0	200 6
AMMAN RIDER		
Coal .. 23 in	1 11	202 5
Clift with bands of rock and ironstone	43 6	245 11
YARD		
Coal .. 62 in	5 2	251 1

Western Colliery: Downcast Shaft

Ht. above O.D. 728 ft.   6-in SS 99 S.W.; Glam. 27 N.W.

Site 400 yd S.E. of Nant-y-moel church. National Grid Ref. 93809260. Published in Vertical Sections, Sheet 85, 1901.

Hafod (bottom coal) at 69 ft, Abergorky (bottom coal) at 238 ft 2 in, Pentre Rider at 298 ft 1 in, Pentre at 339 ft 9 in, Lower Pentre at 363 ft 9 in, Caedavid at 456 ft 4 in, Two-Feet-Nine at 776 ft 6 in, Upper Six-Feet at 896 ft 6 in, Lower Six-Feet at 958 ft 10 in, Caerau at 965 ft 4 in, Upper Nine-Feet at 1047 ft.

Whitworth Colliery: No. 2 Pit

Ht. above O.D. c. 472 ft.   6-in SS 79 N.E.; Glam. 16 S.E.

Site 635 yd N. 11° W. of Ton-mawr school. National Grid Ref. 79929684.

	Thickness ft in	Depth ft in
lade ground	21 0	21 0
RIFT		
ravel and clay	49 0	70 0
OAL MEASURES		
andstone with thin beds of clift	202 7	272 7
hale	4 0	276 7
BRITHDIR RIDER		
Coal .. 3 in		
Rashes .. 31 in		
Coal .. 2 in	3 0	279 7
ireclay and shale ..	8	280 3
andstone	62 8	342 11
hale	6 0	348 11
RAIG		
Coal .. 6 in		
Rashes .. 7 in		
Coal .. 6 in	1 7	350 6
ireclay and shale ..	29 8	380 2
andstone	143 1	523 3
BRITHDIR		
Coal and rashes.. 24 in	2 0	525 3
ireclay and shale ..	7 0	532 3
andstone	26 0	558 3
hale	28 3	586 6
Coal .. 8 in		
Bast .. 1 in		
Coal .. 12 in	1 9	588 3
Fireclay	4 0	592 3
Sandstone and bastard rock	253 0	845 3
Shale	10 0	855 3
NO. 1 RHONDDA RIDER		
Coal .. 18 in		
Fireclay .. 12 in		
Coal .. 6 in	3 0	858 3
Fireclay	5 0	863 3
Sandstone	32 0	895 3
Clift	31 0	926 3
Sandstone and bastard rock	131 2	1057 5
Coal .. 4 in	4	1057 9
Fireclay	4 0	1061 9
Sandstone	25 0	1086 9
Disturbed ground, with coal 3 in. in disturbance at 1163 ft 2 in	126 5	1213 2
Sandstone	228 9	1441 11
Sandstone with streaks of coal	29 8	1471 7
FIELD VEIN		
Coal .. 15 in		
Rashes .. 5 in		
Coal .. 8 in	2 4	1473 11
Fireclay and clift	33 2	1507 1
WERNDDU		
Coal .. 34 in	2 10	1509 11
Fireclay and clift	10 2	1520 1
Coal .. 7 in	7	1520 8

Wyndham Colliery: No. 2 Pit

Ht. above O.D. 618 ft.   6-in SS 99 S.W.; Glam. 26 N.E.

Site 1020 yd S. 2° E. of Nant-y-moel railway station. National Grid Ref. 93369207. Published in Vertical Sections, Sheet 85, 1901.

Two-Feet-Nine at 481 ft 11 in, Six-Feet (in overlap) at 692 ft 5 in, Caerau at 703 ft 9 in, Upper Nine-Feet at 831 ft 10 in, Lower Nine-Feet at 891 ft 1 in, Bute at 938 ft. Sunk to 1076 ft 11 in.

Ynysawdre Pit

Ht. above O.D. c. 150 ft.   6-in SS 98 S.W.; Glam. 34 S.E.

Site 830 yd E. 3° N. of Tondu church. National Grid Ref. 90028418. Published in 2nd Edition, Plate.

Drift to 29 ft 6 in, 'Bodwr Fawr' (?Caerau) at 34 ft 6 in, 'North Fawr' (?Upper Nine-Feet) at 189 ft 5 in, 'South Vein' at 228 ft 8 in, '2nd South Vein' at 253 ft 11 in, '3rd South Vein' at 295 ft 5 in, 'Six-Feet' at 445 ft 5 in, 'South Nine-Feet' at 479 ft 5 in, 'Fiery' at 555 ft 8 in, 'Five-Quarters' at 589 ft 1 in, 'Cribbwr Fawr' at 623 ft 4 in.

Ynyscorrwg-Blaengwynfi Colliery: Ynyscorrwg Pit

Ht. above O.D. 700 ft.   6-in SS 89 N.E.; Glam. 17 N.W.

Site 630 yd S. 5° E. of Glyncorrwg church. National Grid Ref. 87489871. Published in 2nd Edition, p. 146.

?No. 1 Rhondda Rider at 312 ft 6 in, ?No. 1 Rhondda at 379 ft 9 in, ?Daren Rhestyn at 433 ft 7 in, No. 2 Rhondda at 680 ft 4 in. Sunk to 712 ft 3 in.

*Ynysfeio Colliery: No. 3 Pit

Ht. above O.D. 698 ft.   6-in SS 99 N.W.; Glam. 18 S.W.

Site 1000 yd E. 18° S. of Treherbert railway station. National Grid Ref. 94749787.

	Thickness ft in		Depth ft in	
Walling	23	3	23	3
DRIFT				
Gravel and clay	31	6	54	9
COAL MEASURES				
Clift	56	3	111	0
PENTRE RIDER				
Coal .. 9 in		9	111	9
Fireclay and clift	15	10	127	7
Coal .. 12 in	1	0	128	7
Fireclay and clift	17	0	145	7
Coal .. 17 in	1	5	147	0
Fireclay and clift	17	4	164	4
LOWER PENTRE				
Coal .. 20 in	1	8	166	0
Fireclay, clift and rock ..	30	11	196	11
Coal .. 6 in		6	197	5
Fireclay and clift	6	10	204	3
Rock	19	0	223	3
Clift	14	7	237	10
GORLLWYN RIDER				
Coal .. 20 in	1	8	239	6
Fireclay, clift and rock ..	17	9	257	3

	Thickness ft in		Depth ft in	
GORLLWYN				
Coal .. 15 in	1	3	258	6
Fireclay and clift ..	37	8	296	2
Coal .. 5 in		5	296	7
Fireclay and clift ..	74	9	371	4
Coal .. 8 in		8	372	0
Fireclay, rock and clift ..	28	9	400	9
Coal .. 13 in	1	1	401	10
Fireclay and clift ..	34	4	436	2
Coal .. 12 in	1	0	437	2
Fireclay, rock and clift ..	16	3	453	5
Coal .. 11 in				
Fireclay .. 10 in				
Coal .. 5 in	2	2	455	7
Fireclay and clift ..	21	10	477	5
Coal and rashes .. 16 in	1	4	478	9
Fireclay and clift ..	19	6	498	3
Coal .. 18 in	1	6	499	9
Fireclay, rock and clift ..	12	7	512	4
FOUR-FEET				
Coal .. 28 in				
Clift .. 48 in				
Coal .. 27 in	8	7	520	11
Fireclay, clift and rashes ..	10	5	531	4
Coal .. 18 in	1	6	532	10
Fireclay, clift, mine and rock	30	9	563	7

* Minor discrepancies throughout in record.

	Thickness ft	in	Depth ft	in
UPPER SIX-FEET				
Coal .. 27 in	2	3	565	10
Rock and fireclay..	16	5	582	3
Coal .. 3 in		3	582	6
Clift and rock ..	24	0	606	6
LOWER SIX-FEET				
Coal .. 48 in	4	0	610	6
Fireclay, clift and rashes ..	24	2	634	8
RED VEIN				
Coal .. 5 in				
Ironstone 9 in				
Coal .. 12 in				
Rashes .. 12 in				
Coal .. 40 in	6	6	641	2
Fireclay, clift and mine ..	47	2	688	4
Coal and rashes.. 24 in	2	0	690	4
Clift and fireclay ..	7	0	697	4
UPPER NINE-FEET				
Coal .. 57 in	4	9	702	1
Rashes, fireclay and clift ..	27	8	729	9
LOWER NINE-FEET				
Coal .. 22 in	1	10	731	7
Fireclay, clift and rock ..	37	8	769	3
BUTE				
Coal .. 42 in	3	6	772	9
Fireclay, rock and clift ..	20	0	792	9

	Thickness ft	in	Depth ft	in
AMMAN RIDER				
Coal .. 21 in	1	9	794	6
Fireclay and mine ..	16	9	811	3
YARD				
Coal .. 27 in				
Rashes .. 2 in				
Coal .. 12 in	3	5	814	8
Fireclay, clift and rock ..	88	1	902	9
LOWER SEVEN-FEET				
Coal .. 25 in	2	1	904	10
Fireclay and clift ..	19	1	923	11
FIVE-FEET				
Coal .. 5 in				
Clod .. 2 in				
Coal .. 35 in				
Fireclay .. 45 in				
Coal .. 3 in				
Clod .. 3 in				
Coal .. 24 in	9	9	933	8
Fireclay and clift ..	16	0	949	8
Rock	21	6	971	2
GELLIDEG				
Coal .. 33 in				
Clod .. 4 in				
Coal .. 5 in	3	6	974	8
Clift and mine ..	16	4	991	0

Ynysfeio Colliery: No. 4 Pit

Ht. above O.D. 698 ft.   6-in SS 99 N.W.; Glam. 18 S.W.

Site 1030 yd E. 18° S. of Treherbert railway station. National Grid Ref. 94769785.

Red Vein at 650 ft 8 in, Upper Nine-Feet at 709 ft 8 in, Lower Nine-Feet at 742 ft 10 in, Bute at 787 ft 4 in, Amman Rider at 809 ft 8 in, Yard at 848 ft 1 in, Lower Seven-Feet at 915 ft 10 in, Five-Feet at 937 ft 5 in, Gellideg at 980 ft 4 in.

Ynyshir Pit

Ht. above O.D. 388 ft.   6-in ST 09 S.E.; Glam. 27 N.E.

Site 600 yd S. 2° E. of Ynyshir church. National Grid Ref. 02559262.

	Thickness ft	in	Depth ft	in
DRIFT				
Gravel	9	0	9	0
COAL MEASURES				
Rock	12	0	21	0
Clift with rock band ..	19	0	40	0
Coal .. 1 in		1	40	1
Fireclay and clift ..	10	0	50	1

	Thickness ft	in	Depth ft	in
Rock	99	0	149	1
FOREST FACH				
Coal .. 10 in		10	149	11
Fireclay and clift ..	21	3	171	2
NO. 2 RHONDDA				
Coal .. 34 in	2	10	174	0

APPENDIX II

LIST OF GEOLOGICAL SURVEY PHOTOGRAPHS
(One-Inch Sheet 248)

Copies of these photographs are deposited for reference in the library of the Geological Survey and Museum, South Kensington, London, S.W.7.

Prints and lantern slides may be supplied at a fixed tariff.

All numbers belong to Series A.

1645	View of Rhondda Fawr Valley, from Treorchy Station.
1646–7	Panoramic view of the same.
1648	Gelli-Goch, west of Treherbert Station.
1649	Craig-yr-Hesg, from the Common, Pontypridd.
1650	The same: from the Taff River, Pontypridd.
1651–2	Blue Pennant Stone quarries, Craig-yr-Hesg.
1653	Mynydd-y-Glyn, from the Common, Pontypridd.
1654	Potholes in Pennant Sandstone, River Taff below Craig-yr-Hesg.
1655	Cynon Valley at Penrhiwceiber.
1657–60	Sand-burrows, Kenfig.
1663	Bryncethin Brickworks: Coal Measures shales.
9048–9	Melin Court, Neath Valley.
9050–1	Bryn Brick-pit.
9052	Melin Court Waterfall: waterfall over sandstone above Brithdir Seam.
9053–4	Panoramic view from Port Talbot Road, Maesteg.
9055	View west from Bryn Road, Maesteg, over Bryn towards Swansea Bay.
9056	View east-north-east from Nantyffyllon, showing distant scarp of Llynfi Rock.
9057	Quarry in Pennant Sandstone near Maesteg Cottage Hospital.
9058–9	River gravels of River Neath, at Clyne.
9060–1	Eagle Brick-pit, Cwmavon.
9062	Quarry in Llynfi Rock, near Blaen Cwmdu Farm, Maesteg.
9063	Typical scenery at head of Nant Cwmdu, near Blaen Cwmdu Farm.
9064	Fault-line scarps: Pant-y-ffald, near Torcerryg Farm, Maesteg.
9065–6	Overthrusting in Middle Coal Measures, Eagle Brick-pit, Cwmavon (9065 is Plate VIB).
9067	East side of Avan Valley, south of Pont Rhyd-y-fen, looking south-south east.
9068	Fluvioglacial gravels in south bank of River Avan, Pont Rhyd-y-fen.
9742	Pen Puch, from Blaenrhondda Station.
9743	View of Rhondda Valley, with Pen Puch in distance.
9766	View north-east from Bwlch-y-Clawdd, along Cwm Parc (Plate IIA).
9767	General view of Trealaw, Tonypandy and Llwyn-y-pia, Rhondda Fawr Valley.
9768	General view of Tylorstown and Pontygwaith, Rhondda Fach Valley, looking south-east from Blaen-llechau (Frontispiece, Plate I).
9769	The lower Rhondda Fawr Valley at Pontypridd, looking west from Coed-pen-maen Common.
9770	View looking west from Bryn-y-wrach along the Moel Gilau–Felin Arw Fault trough.
9771	General view of Cwm Parc.

772   General view of headwaters of River Avan.

773   View across the Moel Gilau–Felin Arw Fault trough and the Garw Valley between Llangeinor and Pont-y-rhyl.

774   The Ely Valley at Tonyrefail.

775   The Rhondda Fawr Valley at Dinas (between Porth and Trealaw).

776   The Rhondda Fach Valley south of Maerdy.

777   The Rhondda Fach Valley between Maerdy and Ferndale.

778–9   Views of landslip on west side of Taff Valley at Mount Pleasant.

780   The Rhondda Fach Valley at Ferndale.

781–2   The head of the Ogmore Valley, north of Nantymoel.

783   The South Crop of the South Wales Coalfield at Margam, looking north.

784   The South Crop of the South Wales Coalfield at Aberbaiden, view of Pennant escarpment from Cefn Cribwr.

785   International Colliery and Nant Gwineu, Blaengarw, Garw Valley.

786   The Pennant escarpment at Cwm Llanbad, north-west of Llanharan.

787   Hillside south-east of Tonyrefail: view across Ely Valley towards Tylcha Fach.

788   View of upper reaches of Cwm Clydach (Rhondda Fawr), with Cambrian Colliery.

789   North side of Cwm Clydach (Rhondda Fawr) at Clydach Vale.

790   Rhondda Fawr Valley looking north-north-west from hillside south of Dinas.

791   Taff Valley above Pontypridd, looking north.

792   Nant Cwm-gwyn, at head of Garw Valley.

793   Craig Wen crags on southern slopes of Mynydd y Cymmer, Rhondda Fawr.

794   Steeply dipping Pennant Sandstone on roadside between Tonyrefail and Trebanog.

795   General view of Pennant Measures on Pen-y-foel, east of Ynsy-y-bwl.

796   Section in Boulder Clay in cut-away bank on River Llynfi, near Pont Rhyd-y-cyff, south-east of Maesteg.

797   Forested Pennant on the south-east slopes of Mynydd Garth Maelwg, east of Llantrisant.

798   View along Nant Gwynfi with part of Blaengwynfi in foreground.

812   Rolling plateau of the watershed between the Rhondda Fach and Cwmaman valleys (Plate IIB).

813   The mountains enclosing the head of the Garw Valley (Plate VIA).

INDEX

370

AGAR, R. M. C., 37
agle Brick-pit, 132
ast Lodge Opencast Site, 137, 245
dmondia, 62, 77, 156, 166; *E. lowickensis*,
 15; cf. *E. sulcata*, 167; *E. transversa*, 167
dmondstown, 204
lonichthys, 77
DWARDS, W. N., 26
fail Fach, 283
fail Moelgilau, 230
glwys-Nynnid, 135
ly, River, 146, 220, 241
—— Valley, 65–6, 70, 170, 177, 194, 202,
 205, 220, 226, 242, 275; drift deposits
 in, 286–7
mmonsia parasitica, 115
ngine coal, 75
otrapezium depressum, 236
phippioceras, 13
remopteris artemisaefolia, 215
rw Hir, 214, 221
Estheria ', 18, 26, 110, 127, 130, 148–9,
 151, 156, 164, 166, 168–9, 204
uestheria, 22, 88, 110, 181; *E. simoni*, 70,
 196
upecopteris volkmanni, 110
uphemites, 79–80, 166, 170; *E.
 anthracinus*, 78, 121, 154, 167
VANS, D. G., 7, 9, 238
VANS, W. B., v, xiii
VANS, W. H., 24, 62, 92, 161, 181
vanstown, 98, 298
wenny Fach Valley, 280, 287
YLES, V. A., v, 63, 238

airy Glen, Nant-y-moel, 184
arewell Rock, 6, 33, 79
arteg-fawr, 269
elin Arw Fault, 268
—— —— Trough, 72
erndale, 48, 58, 65–6, 70, 82–92, 111,
 173–4, 198–9, 216, 284–5
forchdwmuchaf Row, 283
forch-wen (Garw Valley), 180
fordd-y-gyfraith, 136–7
forest Isaf, 282
forest-uchaf, 215, 217
fos-felin, 79
fwyl Wood, 79
fynnon Daren Goch, 184–5
fynnon-dwym, 222
fyrnant, 195
ive-Roads Marine Band, 24, 63, 165–6
LETT, J. S., 23
oel Fawr, 206, 208

Foel-goch, 200
Foel Gwilym Hywel, 186, 265
Foel Trawsnau, 211
Foel y Duffryn, 185–6, 206–7
Folding, 241–3,263
Foraminifera, 18, 36–7, 59, 62, 128, 149,
 151, 153–4, 156, 160, 162, 165–6, 168,
 170
—— Marine Band, 24, 62, 148, 150, 152,
 162, 165–6, 168–70
Fountain, 81
FRANCIS, E. H., 236
Frithwaun, 227–8, 288

Gadlys Factory, 289
Gadlys Farm (Dimbath Valley), 227
Gadlys Farm (Llangynwyd), 229
Gadlys Valley, 220
Galena, 233, 235
Gallt-y-cwm Farm, 219
Garn Fawr, 79
Garn Wen, 158, 160–2, 181, 185–7, 297–8
Garth, 62, 66, 158, 181
Garth Hall, 220
Garth Hill (Maesteg), 160, 181, 183, 288–9
Garth Hill (Taff's Well), 286
Garth Uchaf, 235
Garth Villas, borehole near, **235**
Garwfechan, 181–3, 206–7, 288
Garw, River, 2
—— Valley, 48, 51, 54–5, 58, 60, 62–3,
 65–6, 227, 258, 263, 277; Brithdir Beds
 in, 217–8; drift deposits in, 288;
 Llynfi Beds in, 180–6; Lower Coal
 Measures in, 91–4; Middle Coal
 Measures in, 122–9, 157–61; Rhondda
 Beds in, 206–8
Gastrioceras, 15, 78, 129, 161; *G. corona-
 tum*, 81; *G. crencellatum*, 14; *G.
 crenulatum*, 15; *G. cumbriense*, 7, 11,
 15; *G. lineatum*, 14; *G. listeri*, 35, 77,
 81; *G. subcrenatum*, 6, 33, 79; *G. sp.
 nov.* (aff. *cumbriense*), 79; *G. sp.* (with
 faint spiral ornament), 35, 77
Gastrioceras cancellatum Marine Band,
 11, 14–5
Gastrioceras cumbriense Marine Band, 7,
 11, 14–5
Gastrioceras subcrenatum Marine Band,
 4, 6, 15, 33, 79
Geisina arcuata, 24, 35, 37, 78, 86, 96, 99,
 101
Geisina? subarcuata, 63, 154, 156, 160,
 163–4, 167, 181
Gelli, 48, 62–3, 153, 177–8, 200, 202–3,
 205, 286

(87540)   2B 2

Moel Gallt-y-cwm Farm, 219
Moel Gilau Fault, 241, 268–70
Moelgilau-fawr, 217
Moelsychbant, 219
Moel Ton-mawr, 212, 220
Moel Troed-y-rhiw, 220, 289
Moel-y-Fen, 187
Moel Yorath, 210
Moel yr Hyrddod, 1, 219, 284
MOORE, L. R., 24, 26, 45, 88, 108, 199, 238
Moraines, at Clyne, 283; at Talbot Green, 287
Morfa Bach, 281
Mountain Ash, 39, 51, 54, 63–5, 67, 106, 108, 146, 149, 172, 195–6, 221, 223, 281–2
Mount Pleasant, Maesteg, 269, 288
Mount Pleasant, Merthyr Vale, 221, 281
Mussel' faunas: ' Amman fauna ', 25, 45, 104, 109, 112, 115, 122, 130, 140, 143; ' Bute fauna ', 25, 48, 112, 123, 130, 134; ' Four-Feet fauna ', 25, 55, 58, 107–8, 111, 114, 120, 126–7, 138, 145; ' Nine-Feet fauna ', 25, 51, 106, 109, 112, 117, 124, 130; ' pseudorobusta fauna ', 24, 37, 86, 96, 101; ' Red Vein fauna ', 25, 51, 54, 117, 124, 130–1, 134, 144; ' Six-Feet fauna ', 25, 55, 107, 119, 125–6, 132, 136, 144; ' Two-Feet-Nine fauna ', 25, 58, 114, 127, 137, 145; ' Yard fauna ', 24, 43, 84–5, 88, 94–5, 100
Mussel ' zones, see zones
Mwyndy Junction, 146, 245
Myalina, 36, 63, 64, 156, 164–5, 169; M. compressa, 166, 169–70; M. verneuili, 77
Mychydd Valley, 225
Mynachdy, 222
—— Borehole, 197, 215, **348**
Mynydd Bach (east of Nantyfyllon), 62, 159–60, 181, 183, 255, 288–9, 298
Mynydd Bach (south of Bryn), 212, 219, 270
Mynydd Baiden, 212–3, 220
Mynydd Blaenafan, 69, 272, 291
Mynydd Blaenafon, 224
Mynydd Blaengwynfi, 210
Mynydd Blaenrhondda, 201
Mynydd Brith-weunydd, 203, 206
Mynydd Bwllfa, 202, 205, 285
Mynydd Bychan, 132
Mynydd Caerau, 2, 186, 206–8, 289
Mynydd Coed Bychan, 213
Mynydd Corrwg-Fechan, 218
Mynydd Dinas, 206, 286
Mynydd Garth Maelwg, 247, 287

Mynydd Gelliwion, 216–7
Mynydd Hywel–Dio, 170
Mynydd Llangeinor, 206–7, 265, 288
Mynydd Maendy (Ogwr Fach), 204, 242, 253
Mynydd Maendy (Rhondda), 176, 178
Mynydd Maendy (South Crop), 228, 253, 286
Mynydd Maesteg, 204, 287
Mynydd Margam, 213, 220, 242
Mynydd Merthyr, 215, 275, 282
Mynydd Moelgilau, 206–8, 217
Mynydd Nant-y-bar, 219, 224, 283
Mynydd Pen-hydd, 67, 187, 209, 211
Mynydd Pen-Rhys, 218, 283
Mynydd Pen-y-graig, 205–6, 250
Mynydd Pwll-yr-hebog, 287
Mynydd Ton, 202–3, 255
Mynydd Ton-mawr, 213
Mynydd Troed-y-rhiw, 178, 202, 206, 216
Mynydd Ty Isaf, 201
Mynydd Tyle-coch, 176
Mynydd Tynewydd, 1, 201
Mynydd Ty'n-tyle, 199
Mynydd Ty-talwyn, 212–3, 220
Mynydd William Meyrick, 2, 207–8
Mynydd y Castell, 189
Mynydd y Cymmer, 70, 204–5
Mynydd y Gaer, 214, 220, 287
Mynydd y Gelli, 200, 202
Mynydd y Gilfach, 177
Mynydd y Glyn, 70, 206, 216-7, 242, 250, 286
Mynydd y Gwair, 208, 217–8
Mynydd Ynyscorrwg, 291
Mynydd Ynysfeio, 176
Mynydd yr Aber, 242
Mynydd yr Eglwys, 200–1
Mynydd Ystradffernol, 201
Myriophyllites gracilis, 110, 116

Naiadites, 26, 45, 55, 58, 60, 64, 84, 86, 90, 92, 99, 110, 112–5, 117, 119–20, 124–6, 130–2, 136–7, 144–5, 151–2, 164–8, 174–5, 181; N. alatus, 107-8, 120, 131–2; N. angustus, 136–7, 145; N. daviesi, 63, 66, 160, 181, 187; N. flexuosus, 100–1; N. hindi, 62, 152, 154, 159, 164; N. melvillei, 162, 167; N. sp. nov. cf. modiolaris, 48, 123, 130; N. obliquus, 107–8, 113–4, 120, 125–8, 132, 136–7; N. productus, 62, 106, 110, 114, 117, 124–5, 130, 134, 138–9, 144; N. sp. cf. productus, 155, 159; N. quadratus, 48,

104, 109, 123; *N. subtruncatus*, 84, 104, 117, 124, 144; *N. triangularis*, 84, 88; *N. sp.* of *productus/quadratus* group, 94; *N. sp.* of *productus/subtruncatus* group, 54; *N. sp.* intermediate between *productus* and *quadratus*, 100, 117, 124; *N. sp.* intermediate between *productus* and *subtruncatus*, 110
Nant Abercerdin, 177
Nant Blaenhenwyg, 217, 282
Nant Blaenogwr, 288
Nant Boeth, 218
Nant Bryn-Cynon, 229, 289
Nant Cae'r-gwerlas, 157, 177–8, 286
Nant Cedfyw, 228–9
Nant Ciwc, 287
Nant Coly, 154
Nant Craig yr Aber, 212
Nant Crymlyn, 142, 213
Nant Cwm-cas, 290
Nant Cwmdu (Llynfi), 161, 289
Nantcwmdu (Garwfechan), 181
Nant Cwm-farteg, 132
Nant Cwm-llwyd, 213
Nant Cwm-parc, 286
Nant Cwm Phillip 191
Nantcymdda-bach, 287
Nantcynon, 210
Nant-du, 290
Nant Farteg-fach, 211
Nant Ffrwd-amws, 204
Nant Gelli, 177
Nant Gelli-wion, 204, 242
Nant Graig-ddu, 204
Nant Gwineu, 185
Nant Gwyn, 188, 209–10
Nant Gwyn Bach, 159, 161
Nant Gwynfi, 210, 218–9, 289, 291
Nantgwyn Hard Heading, 117, 121, 155
Nant Ian, 205
Nantiorweth-goch, 97
Nant Llan, 226
Nant Llanilid, 227
Nant Llechau, 284
Nant Llech Plant Beds, 33
Nant Llwyn-cae, 287
Nant Melyn (Cynon Valley), 33
Nant Melyn (Rhondda Fawr), 272
Nant Moel-Yorath, 290
Nantmychydd, 226, 241–2, 286–7
Nant Myddlyn, 171, 194, 231, 287, 298
Nant Orky, 153–4, 200
Nant Saerbren, 176
Nant Selsig, 285
Nant Sychbant, 212, 219
Nant Ton-y-groes, 287

Nant Trefall, 290
Nant-ty'r-arlwydd, 225
Nantybar Fault, 255
Nant y Blaidd, 285
Nant y Castell, 229
Nant-y-cerdin, 289
Nantyfedw (Avan Valley), 211
Nant-y-fedw (Cynon Valley), 221
Nant-y-Felin, 283
Nant-y-fforest, 183
Nant-y-Ffrwd, 282
Nantyffyllon, 159, 181, 183, 185–6, 288
Nant y Gadlys, 242, 289
Nant-y-Gelli, 285
Nant y Gwair, 176, 271
Nant-y-maen Farm, 196
Nant-y-moel, 161, 182–3, 208, 253
—— Farm, 161
Nant y Pentre, 153–4
Nant-yr-allor, 290
Nant-yr-hwyaid, 290
Nant Ystrad-ffernol, 154, 175
' *Natica* ' *oppelii*, 236
Natica pylensis, 236
Naticopsis, 78
Neath Disturbance, 239, 270–1
Neath, River, 2
Neath Valley (Vale of Neath), 69–71, 209–11, 218–9, 223–5, 275; drift deposits in, 283–4
Neuropteris flexuosa, 204, 209, 215; *N. gigantea*, 78, 88, 91, 109–10, 116, 120, 135, 167; *N. heterophylla*, 85, 88, 90–1, 106–7, 109, 116–7, 120, 144, 167, 191, 202, 207; *N. hollandica*, 99; *N. microphylla*, 87; *N. obliqua*, 85, 87, 99–100, 116, 120, 135, 155; *N. ovata*, 202, 207, 209; *N. ovata* var. *flexuosa*, 207, 214; *N. pseudogigantea*, 86, 99, 135, 137, 144; *N. rarinervis*, 167, 191, 204, 207, 209, 214–5, 231; *N. scheuchzeri*, 167, 202, 204, 207, 214, 222, 224, 231; *N. schlehani*, 33, 78, 87, 99; *N. tenuifolia*, 87, 99, 137, 144, 155, 167, 191, 202
Newer Drift, 275
Newlands Thrust, 245–7
Newpark, 213, 235
Newtown (Llantwit Fardre), 230, 232
Newtown (Mountain Ash), 215
No. 2 Rhondda, conglomerate above, 195–6, 198, 200, 207, 209, 213
—— Sandstone, 196, 201, 207, 285
Northern Rhondda Fawr Area, Brithdir Beds in, 216; Llynfi Beds in, 175–7; Lower Coal Measures in, 86–8; Middle

Pin, 75
—— Garw Ironstone, 98, 298
—— Haulkin Ironstone, 298
Pinnularia capillacea, 106, 116
Pistyllarian, 79
Pistyll-goleu, 222
Planolites, 11, 24; *P. montanus*, 25, 86–7, 114, 125, 127, 155; *P. ophthalmoides*, 16, 18, 21, 33, 35–7, 55, 59–60, 62, 77–80, 86, 96, 110–1, 113–4, 121–2, 128, 131–2, 134, 140, 148, 151, 153–4, 156, 160, 163–6, 168
Platyconcha, 167, 169; *P. hindi*, 154
Pleistocene, 5
Pleurophorella, 133; *P. sesquiplicata*, 121
Pleuroplax, 143, 169; *P. attheyi*, 115
PLUMMER, E., 223–4
POCOCK, R. W., v
Polygnathodella, 13
Polygnathus communis, 11
Pont-nedd-fechan, 270
Pont Rhyd-y-cyff, 72, 183, 186, 220, 268–9, 289
Pont Rhyd-y-fen, 69–71, 186, 188, 209, 211, 218–9, 243, 263, 267, 283, 290–1, 299
Pontrhyd-y-maen, 291, 298
Pontycymmer, 48, 159–60, 182–5, 207, 242–3, 288; Bridgend Road, section on, 180
Ponty-y-parc, 146, 235
Pontypridd, 28, 48, 54, 63, 67, 69, 72, 120, 177, 197, 202, 206, 216–7, 222–3, 241, 275, 281, 285–6
—— Anticline, 202, 216, 242
Porth, 51, 54, 59–60, 63–4, 67, 70, 120, 177, 179, 202, 284–6
Port Talbot, 299
Posidonia, 9, 10, 77, 79, 81, 114, 167; *P. corrugata*, 10, 11; *P. gibsoni*, 14, 81; *P. sulcata*, 128–9
Posidoniella, 11, 13–5; *P. minor*, 13
Pricetown, 182
PRINGLE, J., 9
Procardia rhaetica, 236
Promytilus, 79
Prothyris, 154
Pruvostina? (*Carbonita* pars), 214
' *Pseudamussium* ', 15
Pterinopecten papyraceus, 161
Ptychopteris unitus, 215
Pullastra arenicola, 236
Pwllandras, 163
Pwll-y-gwlaw, 187, 299
Pygmaeoceras sigma, 10, 14
Pyle, 98, 133–4, 233, 236–7, 273, 278, 292

Quaker's Yard, 221
Quarella Sandstone, 11, 236

Rackett Wood, 225
Raglan, 192
RAMSBOTTOM, W. H. C., v, 9, 60
Ram's Head Mine, 94, 298
RANDALL, H. J., v.
Rashings, rashes, 75
Rectocornuspira?, 77, 134
Red Mine Ground, 97, 298
Reticuloceras, 13; *R. adpressum*, 13; *R. moorei*, 11; *R. nodosum*, 13; *R. paucicrenulatum*, 13; *R. reticulatum* s.s. 13; *R. stubblefieldi*, 11
Reticycloceras, 11, 13
Rhabdoderma, 15, 77, 121, 155, 161, 163, 170, 214
Rhadinichthys, 81, 96, 167–8, 170
Rhaetavicula (*Pteria*) *contorta*, 236
Rhaetic, 4–5, 234; facies of, 236
Rhipidomella carbonaria, 114
Rhiw-glyn, 218
Rhiw-perra Cottage, 213
Rhiwsaeson, 2, 69, 133, 171, 194, 213–4, 221, 233, 287, 298
Rhizodopsis, 114, 156
RHODES, F. H. T., 10
Rhondda Beds, 4, 28; general stratigraphy of, 67–70; thickness of, 67
Rhondda Fach River, 2
—— —— Valley, 55, 59, 65–7, 69–71, 258, 261; Brithdir Beds in, 216; drift deposits in, 284–5; Llynfi Beds in, 173–5; Lower Coal Measures in, 84–6; Middle Coal Measures in, 108–11, 149–51; Rhondda Beds in, 197–9
Rhondda Fawr, River, 2, 179,
—— —— Valley, 48, 55, 62–7, 69–70, 210, 243, 258, 263, 297; Brithdir Beds in, 216–7; drift deposits in, 285–6; Llynfi Beds in, 175–80; Lower Coal Measures in, 86–91; Middle Coal Measures in, 111–22, 151–7; Rhondda Beds in, 200–6
—— Valleys, 3, 28, 37, 39, 45, 51, 54, 58, 60, 62, 65, 67, 70–1, 216, 275
Rhos-gwawr, 195
Rhyd-y-melinydd, 214, 235
Rider, 75
River Terrace Deposits, 3, 280–2, 285–6
ROBERTSON, T., 7, 9–11, 28, 33, 37, 97, 209, 238, 270
Rudry, 33, 37, 238
—— Borehole, 33

(87540)   Wt. 4354/0878   K 10   2/64   Hw.

S.O. Code No. 62–290–4–63*